AF882194

Boundary Element Methods for Heat Transfer with Phase Change Problems: Theory and Application

Authored by

S. G. Ahmed

Faculty of Engineering
Zagazig University
Egypt

Boundary Element Methods for Heat Transfer with Phase Change Problems: Theory and Application

Author: S. G. Ahmed

ISBN (Online): 978-981-5040-88-3

ISBN (Print): 978-981-5040-89-0

ISBN (Paperback): 978-981-5040-90-6

First published in 2022

Bentham Science Publishers Pte. Ltd.
80 Robinson Road #02-00
Singapore 068898
Singapore
Email: subscriptions@benthamscience.net

BENTHAM SCIENCE

CONTENTS

PREFACE

Free and moving boundaries are among the problems of the world of engineering, industry, technology, and even theoretical sciences. These applications are widespread, and the phase change problems are among the most important types of these problems. In modern science, discussing phase transition problems receives a lot of attention, and as technology advances, we discover that interest in the topic is growing, with a wide range of applications. If we want to know the initial emergence of this science, we find it emerged from a very simple thing, which is how to look at an ice cube and how it takes the form of melting until the ice cube reaches a liquid state, and this is what the scientist Stefan started with, through which this science developed and took multiple applied fields among them, for example, are not limited to, and in engineering fields, we mention only some of them:

Industrial and engineering applications:

- Diffusion problems
- Solidification problems
- Freezing Problems
- Groundwater proliferation problems
- Road paving problems
- Problems of the food industry

In the medical fields:

- The diffusion of oxygen in the cell
- The spread of cancer in the body
- In the field of epidemics

In the fields of the environment:

- In the field of wastewater treatment
- In the desalination of sea water

- In the field of groundwater dispersal, tracking, and utilization

In the nuclear field

- Laser industry
- In nuclear reactors in many stages of the nuclear bomb
- In the stages of generating the hydrogen bomb

The current book, in fact, consists of two parts, and each part is comprises more than 500 pages. It is a theoretical and practical study of one type of application of free and moving boundary problems, namely Phase change problems with different methods of dealing with these applications through an advanced numerical method, namely, the Boundary Element Methods. The first part is divided into eight chapters and four appendices and can be classified as follows:

- Chapter One: it is a theoretical study of phase change issues in terms of the theoretical framework, historical overview, and various mathematical models.
- Chapter Two: An analytical study of the limit element methods and their application to positional issues
- Chapter Three: Application of boundary element methods to phase change issues in general
- Chapter Four : Application of boundary element methods to fixed-band propagation problems
- Chapter Five: Application of boundary element methods to mobile-band propagation problems
- Chapter Six: A new proposal and a vision for methods of borderline elements
- Chapter Seven: Numerical studies and analysis of results
- Chapter Eight: An extensive study of one of the applications of free phase change problems.

The appendix consists of four sub-appendices.

The current work is the result of more than 35 years of academic research in this field, and we pray for God's blessing.

CONSENT FOR PUBLICATION

None.

CONFLICT OF INTEREST

The authors declare no conflict of interest.

S. G. Ahmed
Faculty of Engineering
Zagazig University
Egypt

ACKNOWLEGEMENT

I dedicate this work to the immortal soul that I am yet to meet, to my professors who taught me how to conduct scientific research, and to those who taught me how to distinguish between a job and a mission. I dedicate this humble work to my children, and may God make it useful for researchers worldwide.

Dedicated to the soul of my Professor Dr. Mohamed Afout Abdel Majid, who taught me when I was an undergraduate student, taught me until I became Full Professor, 2006 and taught me after that, Is my textbook enough for my teacher?

S. G. Ahmed
Faculty of Engineering
Zagazig University
Egypt

CHAPTER 1

Heat Transfer Problems with Phase Change and Their Applications

Abstract: The present chapter mainly focuses on the definition of phase change problems and is the entrance for both free and moving boundary problems and their practical applications. In this chapter, we will start from the fundamentals, definitions, and formulations; an overview of the phase change problem in both one, two, and three dimensions is developed with physical and mathematical formulations.

Keywords: Free boundary problem, Heat transfer, Moving boundary problem, Mushy zones.

PARTIAL DIFFERENTIAL EQUATIONS

In mathematics, differential equations consist of two major branches, ordinary differential equations (ODE) and partial differential equations (PDE). The partial differential equation involves an unknown function (or functions) of several independent variables and their partial derivatives with respect to those variables [1-4]. Partial differential equations are used to formulate, and thus aid the solution of, problems involving functions of several variables, such as the propagation of sound or heat, electrostatics, electrodynamics, fluid flow, and elasticity. Seemingly distinct physical phenomena may have identical mathematical formulations and thus be governed by the same underlying dynamic [5-10]. The solution of the differential equation in a certain domain satisfying some given conditions is referred to as the ***boundary-value problem***. If one or more of the boundaries is not known and moving with time and space, the problem is then referred to as the ***moving boundary problem*** [11-15]. When both the governing equation and the boundary condition are **time-independent**, the problem is **referred** to as a ***free boundary problem*** [16-20].The phase change phenomena play important roles in natural processes. Solidification and melting processes are present in our day-to-day lives and affect us in several ways, directly or indirectly, whether or not we are aware of them. To mention some examples of the importance of these processes in nature, one can start recalling that undoubtedly they have played a decisive part in the formation of the Earth. Since near-homogeneous accretion was almost certain how the Earth was formed: melting

S. G. Ahmed

and solidification were key processes at work in the initial period of planetary growth and differentiation of the interior [21-25]. Melting and solidification processes have also provided the necessary continuous segregation mechanism for the tectonic activity underlying magma migration and volcanism. Melting and solidification of the ice caps near the poles are critical in stabilizing the world's climate. Snow, which accumulates on mountains in the winter, is a means provided by nature to store the water [26-30]. The melting of this snow is important to the distribution of water for human needs. In the present-day engineering, important applications such as the solidification of castings, crystal growth, scrap melting, and ablation of metals using laser beams in metallurgy; the cooling, freezing, and cold storage of foodstuffs; ablation of thermal shields of hypersonic missiles, rockets, and spaceships caused by aerodynamic heating during the so-called "re-entry" problem in the space sciences; refrigeration; solar heat storage in Latent Heat of Fusion Thermal Energy Storage equipment; cryopreservation of biological cells; are but a few examples of how important are the solidification and melting processes [31-35]. *All of the above problems have in common the characteristic of an interface or a moving boundary, which moves into the solid (melting) or into the liquid region (solidification) depending on the relative magnitudes of the temperature gradients on either side of it.* At any point of this interface, there is a temperature gradient (flux) discontinuity which is related to its normal velocity by an equation balancing the rate of heat flow with the energy rate required to create a fresh amount of solid (or liquid) per unit time. It is assumed that the thermal behavior of the media on either side of the interface is governed by the heat conduction equation, also known as the diffusion equation. Despite this partial differential equation being linear, the above problems are non-linear due to the presence of the moving boundary or interface [36-40]. Partial differential equations have gained the great interest of many researchers due to their efficiency in describing the mathematical model of several physical phenomena and engineering applications. Partial differential equations are used to formulate and thus aid the solution of problems involving functions of several variables. The differential equation in a certain domain satisfying some given conditions is referred to as a boundary-value problem. If one or more of the boundaries are not known and moving with time, the problem is referred to as the moving boundary problem. Furthermore, if the governing equation is time-independent as well as the boundary condition, then the problem is referred to as free boundary problem. Partial differential equations are found and used in different fields of science and technology; however, the great spread of partial differential equations and analytical solutions are restricted to very few simple case problems. The majority of partial differential equations have no analytical solutions due to the nature of these equations as well as the nonlinear behavior encountered. In the next sub-sections, some of the recent numerical methods will be presented. The present chapter is an introduction to partial differential equations, including theirmathematical behavior, order, degree, linear, non-linear, and dimensions, etc. To understand the concept of the above items, let us take a very simple partial differential equation as an example of illustration.

Classification of the Linear Second Order P.D.E.

Linear Partial differential equation of the second order has the following general form:

$$A\frac{\partial^2 u}{\partial x^2} + 2B\frac{\partial^2 u}{\partial x \partial y} + C\frac{\partial^2 u}{\partial y^2} + D\frac{\partial u}{\partial x} + E\frac{\partial u}{\partial y} + Fu = G \tag{1}$$

Equation (1) in some other references can be re-written as follows:

$$A\frac{\partial^2 u}{\partial x^2} + B\frac{\partial^2 u}{\partial x \partial y} + C\frac{\partial^2 u}{\partial y^2} + D\frac{\partial u}{\partial x} + E\frac{\partial u}{\partial y} + Fu = G \tag{2}$$

According to equation (1), one can classify the partial differential equations in a general manner into the following three categories:

Case (1) ***Hyperbolic***

If $B^2 - 4AC > 0$ **(3)**

Case (2) ***Parabolic***

If $B^2 - 4AC = 0$

(4)

Case (3) ***Elliptic***

If $B^2 - 4AC < 0$ **(5)**

Some examples for the aforementioned categories are as follow:

$$\frac{\partial^2 u}{\partial t^2} = c^2 \frac{\partial^2 u}{\partial x^2}$$ 1-D wave equation **(6)**

$$\frac{\partial^2 u}{\partial t^2} = c^2 \left(\frac{\partial^2 u}{\partial x^2} + \frac{\partial^2 u}{\partial y^2} \right)$$ 2-D wave equation **(7)**

$$\frac{\partial u}{\partial t} = c^2 \frac{\partial^2 u}{\partial x^2}$$ 1-D heat equation **(8)**

$$\frac{\partial u}{\partial t} = c^2 \left(\frac{\partial^2 u}{\partial x^2} + \frac{\partial^2 u}{\partial y^2} \right)$$ 2-D heat equation **(9)**

$$\frac{\partial^2 u}{\partial x^2} + \frac{\partial^2 u}{\partial y^2} = 0$$ 2-D Laplace's equation **(10)**

$$\frac{\partial^2 u}{\partial x^2} + \frac{\partial^2 u}{\partial y^2} = f(x, y)$$ 2-D Poisson's equation **(11)**

Example

Obtain all cases for the coefficients in the following partial differential equation $au_{xx} + bu_{yy} + cu_x + du_y = 0$ to be elliptic.

Solution

First of all, let us make the following general classification:

1- The given P.D.E. is homogeneous due to the absence of the function, G.

2- The given P.D.E. is linear because all coefficients are constants.

Now, let us write the general form of the second order linear partial differential equation:

$$A \frac{\partial^2 u}{\partial x^2} + 2B \frac{\partial^2 u}{\partial x \partial y} + C \frac{\partial^2 u}{\partial y^2} + D \frac{\partial u}{\partial x} + E \frac{\partial u}{\partial y} + Fu = G \quad \textbf{(a)}$$

And the given equation is:

$$au_{xx} + bu_{yy} + cu_x + du_y = 0 \quad \textbf{(b)}$$

By making a comparison between equations (a) and (b), leads to:

$$A = a$$
$$B = 0 \quad \textbf{(c)}$$
$$C = b$$

The general condition is:

$$\text{General formula} = B^2 - 4AC \quad \textbf{(d)}$$

By making use of equation (c) into (d), one can get:

$$\text{General formula} = B^2 - 4AC = -4ac \quad \textbf{(e)}$$

Now then, we have the following different cases:

1- If $a, b > 0$, this will correspond to $B^2 - 4AC < 0$ and so, the equation will be elliptic type.

2- If a, b will have the opposite sign, this will correspond to $B^2 - 4AC > 0$, and so the equation will be hyperbolic type.

Example

Show that $au_{xx} + bu_{xy} + cu_y = 0$ is hyperbolic for all possible cases of its constants.

Solution

First of all, let us make the following general classification:

1- The given P.D.E. is homogeneous due to the absence of the function, G.

2- The given P.D.E. is linear because all coefficients are constants.

Now, let us write the general form of the second order linear partial differential equation:

$$A\frac{\partial^2 u}{\partial x^2}+2B\frac{\partial^2 u}{\partial x\partial y}+C\frac{\partial^2 u}{\partial y^2}+D\frac{\partial u}{\partial x}+E\frac{\partial u}{\partial y}+Fu=G \quad \textbf{(a)}$$

And the given equation is:

$$au_{xx}+bu_{xy}+cu_{y}=0 \quad \textbf{(b)}$$

By making a comparison between equations (a) and (b), leads to:

$$\begin{aligned} A&=a\\ B&=b\\ C&=b \end{aligned} \quad \textbf{(c)}$$

The general condition is:

$$\text{General formula}=B^2-4AC \quad \textbf{(d)}$$

By making use of equation (c) into (d), one can get:

$$\text{General formula}=B^2-4AC=b^2>0 \quad \textbf{(e)}$$

From equation (e), we see that B^2-4AC is always positive for any value positive or negative for b; therefore, the equation is of hyperbolic type.

More Definitions

In what follows, we will do some definitions, and to do so, let us consider the following partial differential equation:

$$A(x,y,t)\frac{\partial u}{\partial t}+B(x,y,t)\frac{\partial u}{\partial x}+C(x,y,t)\frac{\partial^2 u}{\partial y^2}=f(x,y,t) \quad \textbf{(12)}$$

In equation (12):

(x,y,t): Spatial and time-independent variables, respectively,

A, B, C: Usually known functions,

$u(x, y, t)$: Dependent unknown variable usually called potential.

Order Definition

The order is defined as the highest derivative that appears in the given partial differential equation.

Degree Definition

The degree is defined as the highest power of the highest derivative that appears in the given partial differential equation.

Dimension Definition

The dimension of the given partial differential equation is defined as the number of spatial variables it contains.

Linearity Definition

The given partial differential equation is called linear if the potential and its derivatives are of the first order and there are no products involving more than these terms.

P.D.E. and Applications

No one can make a separation between life and partial differential equations. Who can forget the ceramic industry? Who can forget the food industry? Who can forget the steel industry? Who can forget the aircraft industry? One can say that partial differential equations describe many natural laws in our daily life, such as the conservation of heat and mass. One can remember that both Poisson and Laplace equations are used to model the steady-state temperature distribution in a plane. The one-dimensional heat equation is used to model time-dependent temperature distribution along the heated bar. The wave equation is used to model the vibration of the beam or slab according to the dimension of the problem. Finally, but not last, the advection-diffusion equation is used to model the transport of a pollutant in a river.

Boundary Conditions

There are three different types of boundary conditions that any practical problem will be subjected to, according to the case underhand.

Dirichlet Boundary Condition

$$u(x, y, t) = \bar{u}(x, y, t) \tag{13}$$

Neumann Boundary Condition

$$\frac{\partial u(x, y, t)}{\partial n} = \bar{q}_N(x, y, t) \tag{14}$$

Robin Boundary Condition

$$\frac{\partial u(x, y, t)}{\partial n} = q_R(x, y, t) \tag{15}$$

But no one can say no more than such conditions, may or may not, there will be different boundary conditions encountered according to the practical application formulation.

Well-posed Problem

Any partial differential equation needs boundary and initial condition – if it is time-dependent problem – in the case when a partial differential equation is given associated with its boundary and initial condition, then it is called well-posed problem.

Ill-posed Problem

If it is given a partial differential equation with too many conditions, then there will be no solution, and if too few conditions are given, then the solution will be unique. Also, the ill-posed problem refers to the wrong formulation in boundary or initial conditions.

Solution Domain

May be someone, ask, what does it mean by domain solution? The answer is so simple; the solution domain is the region in which the solution is exist, converges and is stable. Many authors define the solution domain from different points of view. In what fellow, we will define the solution domain briefly from two different points of view, the first one is from a fluid flow point of view, and the second is a more general form mathematical point of view.

Fluid Flow Point of View

Hyperbolic Problems

In the case of hyperbolic problems, the point $P(x_0, t_0)$ can be influenced by the points in the region bounded by the two characteristic lines $x + ct = \text{const}$ and $x - ct = \text{const}$ and for all times $t < t_0$, see Fig. (**1**).

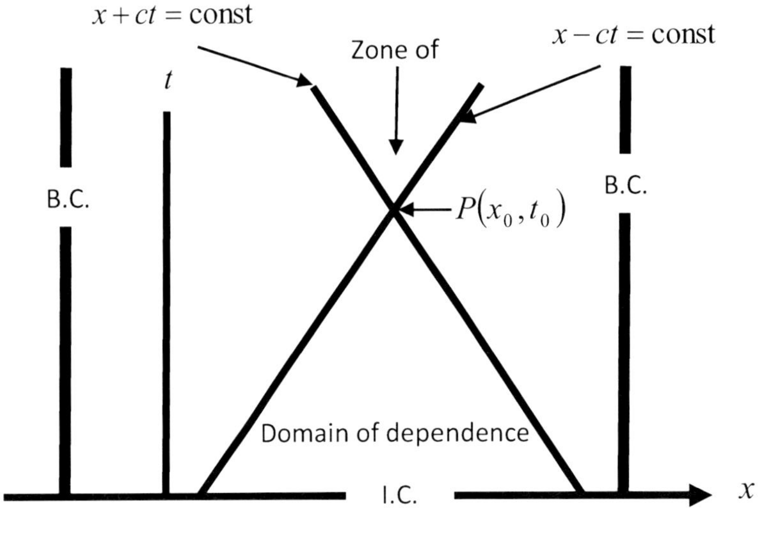

Fig. (1). Domain of Hyperbolic problems.

Parabolic Problems

In the case of parabolic problems, the domain of dependence of the point is the region $t < t_0$ and the zone of influence is $t > t_0$ see Fig. **(2)**.

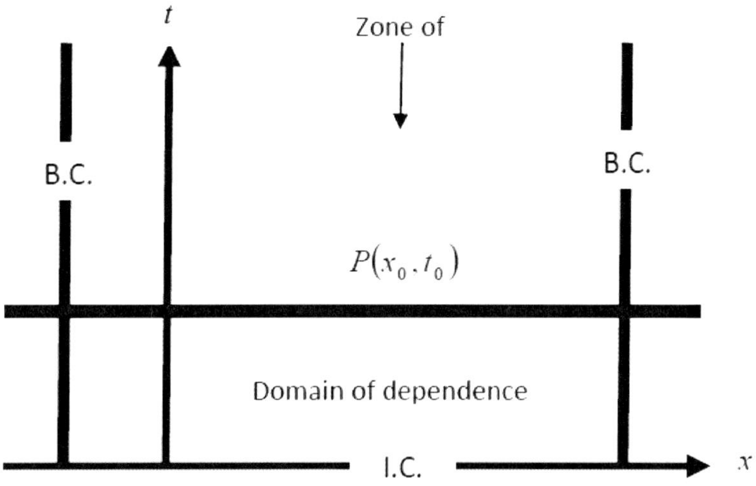

Fig. (2). Domain of parabolic problems.

Elliptic Problems

In the case of elliptic problems, the solution at the point influences all points within the domain and *vice versa*, see Fig. **(3)**.

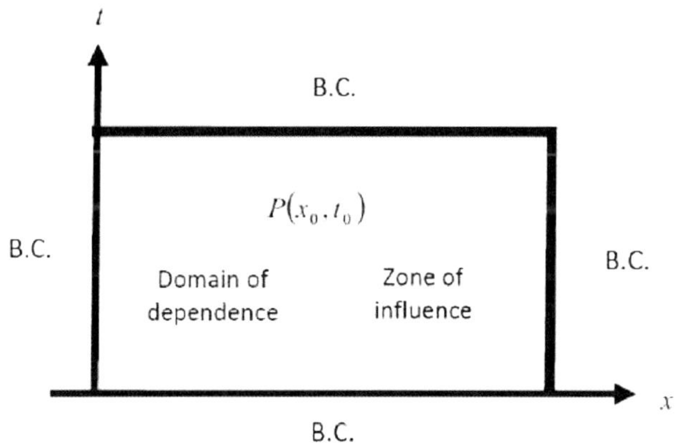

Fig. (3). Domain of Elliptic problems.

SERIES FUNDAMENTALS FOR PDE

Partial differential equations have gained the great interest of many researchers due to their efficiency in describing the mathematical model of several physical phenomena and engineering applications. Partial differential equations are used to formulate and thus aid the solution of problems involving functions of several variables. The differential equation in a certain domain satisfying some given conditions is a boundary-value problem. If one or more of the boundaries are not known and moving with time, the problem is referred to as the moving boundary problem. Furthermore, if the governing equation is time independent as well as the boundary condition, then the problem referred to as free boundary problem. Partial differential equations are found and used in different fields of science and technology; however, the great spread of partial differential equations, and analytical solutions are restricted to very few simple case problems. The majority of partial differential equations have no analytical solutions due to the nature of these equations as well as the nonlinear behavior encountered. In the next sub-sections, some of the recent numerical methods will be presented. We will start the numerical treatment of the partial differential equation by approximating the derivatives by the finite differences approximations, and then we will follow up the other classical methods of solutions [41-45].

Forward and Backward Finite Difference for 1st Derivative

We know that the differential equation in general, ordinary or partial, consists of dependent and independent variables in addition to derivatives. Suppose one wants to solve the differential equation. In that case, the basic and classical approach is to approximate the derivatives using the classical finite difference approximation, and this will do in this section and see what will be the result and how to deal with this result numerically [46-50].

Let us write Taylor's expansion for a given function as follows:

$$f(x+\Delta x)=f(x)+(\Delta x)\frac{\partial f}{\partial x}+\frac{(\Delta x)^2}{2!}\frac{\partial^2 f}{\partial x^2}+\frac{(\Delta x)^3}{3!}\frac{\partial^3 f}{\partial x^3}+\ldots \quad \mathbf{(16)}$$

Equation (1) can be re-written in the following compact form:

$$f(x+\Delta x)=f(x)+\sum_{n=1}^{\infty}\frac{(\Delta x)^n}{n!}\frac{\partial^n f}{\partial x^n} \tag{17}$$

Re-arrange equations (17) as in the following manner:

$$\frac{f(x+\Delta x)-f(x)}{\Delta x}=\frac{1}{\Delta x}\sum_{n=1}^{\infty}\frac{(\Delta x)^n}{n!}\frac{\partial^n f}{\partial x^n}$$

$$\Rightarrow$$

$$\frac{f(x+\Delta x)-f(x)}{\Delta x}=\left[\sum_{n=1}^{\infty}\frac{(\Delta x)^{n-1}}{n!}\frac{\partial^n f}{\partial x^n}\right]=\frac{\partial f}{\partial x}+\frac{(\Delta x)}{2!}\frac{\partial^2 f}{\partial x^2}+\frac{(\Delta x)^2}{3!}\frac{\partial^3 f}{\partial x^3}+.... \tag{18}$$

Cancel all terms containing the higher order of (Δx), one can get the following approximation:

$$\frac{\partial f}{\partial x}=\left(\frac{f(x+\Delta x)-f(x)}{\Delta x}\right)+O(\Delta x) \tag{19}$$

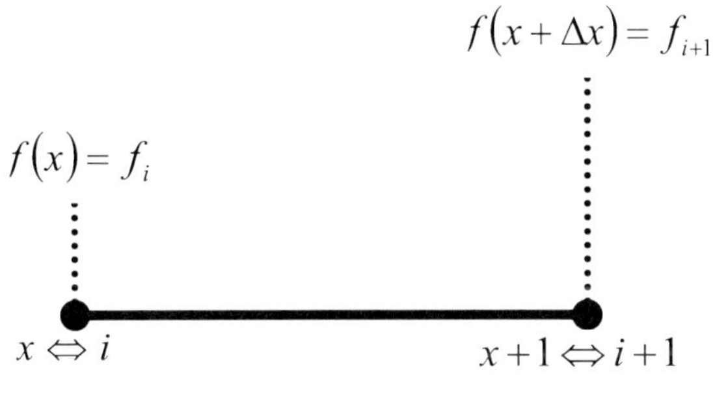

Fig.(4): Successive two nodes.

Equation (19) can be re-cast in a more general form:

$$\left(\frac{\partial f}{\partial x}\right)_i=\left(\frac{f_{i+1}-f_i}{\Delta x}\right)+O(\Delta x) \tag{20}$$

Equation (20) is called the first ***forward*** approximation for the first derivative and of order Δx. Assume that one wrote Taylor's expansion as follows:

$$f(x-\Delta x)=f(x)-(\Delta x)\frac{\partial f}{\partial x}+\frac{(\Delta x)^2}{2!}\frac{\partial^2 f}{\partial x^2}-\frac{(\Delta x)^3}{3!}\frac{\partial^3 f}{\partial x^3}+-.... \quad \textbf{(21)}$$

Again, equation (21) can be re-written in the following compact form:

$$f(x-\Delta x)=f(x)+\sum_{n=1}^{\infty}\left(\pm\frac{(\Delta x)^n}{n!}\frac{\partial^n f}{\partial x^n}\right) \quad \begin{cases}(+) & for\, even\,(n)\\ (-) & for\, odd\,(n)\end{cases} \quad \textbf{(22)}$$

Re-arrange equation (22), leads to:

$$\frac{f(x-\Delta x)-f(x)}{\Delta x}=\sum_{n=1}^{\infty}\left(\pm\frac{(\Delta x)^{n-1}}{n!}\frac{\partial^n f}{\partial x^n}\right) \quad \begin{cases}(+) & for\, even\,(n)\\ (-) & for\, odd\,(n)\end{cases} \quad \textbf{(23)}$$

Similarly, as done before, one can get another formula for the first approximation for the first derivative as follows:

$$\frac{\partial f}{\partial x}=\left(\frac{f(x)-f(x-\Delta x)}{\Delta x}\right)+O(\Delta x) \quad \textbf{(24)}$$

Equation (24) in terms of index notations can be re-written as:

$$\left(\frac{\partial f}{\partial x}\right)_i=\left(\frac{f_i-f_{i-1}}{\Delta x}\right)+O(\Delta x) \quad \textbf{(25)}$$

Equation (25) is called the first ***backward*** approximation for the first derivative and of order Δx. Now we derived the first approximation for the derivative of the first order, we will turn our attention to the derivation of approximating the second derivatives so as we can be able to solve simple partial differential equations.

Central Finite Difference Approximation for 1st Derivative

Let us have two Taylor's expansion of the following forms:

$$f(x+\Delta x)=f(x)+(\Delta x)\frac{\partial f}{\partial x}+\frac{(\Delta x)^2}{2!}\frac{\partial^2 f}{\partial x^2}+\frac{(\Delta x)^3}{3!}\frac{\partial^3 f}{\partial x^3}+.... \quad \textbf{(26)}$$

And

$$f(x-\Delta x)=f(x)-(\Delta x)\frac{\partial f}{\partial x}+\frac{(\Delta x)^2}{2!}\frac{\partial^2 f}{\partial x^2}-\frac{(\Delta x)^3}{3!}\frac{\partial^3 f}{\partial x^3}+-.... \quad \textbf{(27)}$$

Subtract equation (27) from equation (26), leads to:

$$f(x+\Delta x)-f(x-\Delta x)=2(\Delta x)\frac{\partial f}{\partial x}+2\frac{(\Delta x)^3}{3!}\frac{\partial^3 f}{\partial x^3}+.... \quad \textbf{(28)}$$

Neglect all terms starting from the term containing $(\Delta x)^3$, this gives:

$$\frac{\partial f}{\partial x}=\left(\frac{f(x+\Delta x)-f(x-\Delta x)}{2(\Delta x)}\right)+O(\Delta x)^2 \quad \textbf{(29)}$$

Equation (29), can be written in terms of index notations as follows:

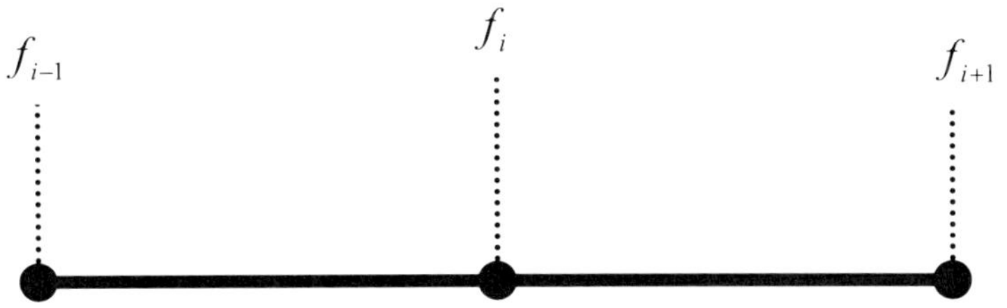

Fig. (5). Successive three nodes.

$$\left(\frac{\partial f}{\partial x}\right)_i = \left(\frac{f_{i+1} - f_{i-1}}{2(\Delta x)}\right) + O(\Delta x)^2 \qquad \textbf{(30)}$$

Equation (30) is called the ***Central difference approximation*** for the first derivative and of order $(\Delta x)^2$.

Forward and Backward Finite Difference Approximation for 2nd derivative

In the last three sections, we derived the approximation of the first derivative using forward, backward and central differences. In this section, we will derive the approximation of the second derivatives based on Taylor's expansion [51-60]. Assume that we have two Taylor expansions of the following successive forms:

$$f(x + \Delta x) = f(x) + (\Delta x)\frac{\partial f}{\partial x} + \frac{(\Delta x)^2}{2!}\frac{\partial^2 f}{\partial x^2} + \frac{(\Delta x)^3}{3!}\frac{\partial^3 f}{\partial x^3} + ... \qquad \textbf{(31)}$$

And

$$f(x + 2\Delta x) = f(x) + (2\Delta x)\frac{\partial f}{\partial x} + \frac{(2\Delta x)^2}{2!}\frac{\partial^2 f}{\partial x^2} + \frac{(2\Delta x)^3}{3!}\frac{\partial^3 f}{\partial x^3} + ... \qquad \textbf{(32)}$$

Multiply equation (31) by 2 and subtract the result from equation (32), leads to:

$$-2f(x + \Delta x) + f(x + 2\Delta x) = -f(x) + (\Delta x)^2\frac{\partial^2 f}{\partial x^2} + (\Delta x)^3\frac{\partial^3 f}{\partial x^3} + ... \qquad \textbf{(33)}$$

Neglect all terms starting from the term containing   , this gives:

$$\frac{\partial^2 f}{\partial x^2} = \left(\frac{f(x + 2\Delta x) - 2f(x + \Delta x) + f(x)}{(\Delta x)^2}\right) + O(\Delta x) \qquad \textbf{(34)}$$

Equation (34), can be written in terms of index notations as follows:

$$\left(\frac{\partial^2 f}{\partial x^2}\right)_i = \left(\frac{f_{i+2} - 2f_{i+1} + f_i}{(\Delta x)^2}\right) + O(\Delta x) \tag{35}$$

Equation (35) is called the ***forward difference approximation*** for the first derivative and of order $(\Delta x)^2$.

$$\left(\frac{\partial^2 f}{\partial x^2}\right)_i = \left(\frac{f_i - 2f_{i-1} + f_{i-2}}{(\Delta x)^2}\right) + O(\Delta x) \tag{36}$$

Now we are ready to start the numerical solution of different types of partial differential equations. The major and basic idea is to replace the differential equation with its associated boundary and initial conditions with a linear system of algebraic equations, using than any numerical method to solve the resultant linear system of algebraic equations.

Numerical Solution of Elliptic P.D.E.

The direct example of elliptic partial differential equation is Laplace equation in two and three dimensions; the state equation in both two and three dimensions takes the following form:

Case (1): 2-D State Equation

$$\nabla^2 u = \frac{\partial^2 u}{\partial x^2} + \frac{\partial^2 u}{\partial y^2} = 0 \tag{37}$$

Case (2): 3-D State Equation

$$\nabla^2 u = \frac{\partial^2 u}{\partial x^2} + \frac{\partial^2 u}{\partial y^2} + \frac{\partial^2 u}{\partial z^2} = 0 \tag{38}$$

Before continuing, let us remain the reader to the notation ∇^2 that is called Laplacian operator, and it takes two different forms in Cartesian coordinates, depending on the dimension of the problem under hand.

Case (1): Laplacian in 2-D

$$\nabla^2 = \left(\frac{\partial^2}{\partial x^2} + \frac{\partial^2}{\partial y^2} \right) \tag{39}$$

Case (2): Laplacian in 3-D

$$\nabla^2 = \left(\frac{\partial^2}{\partial x^2} + \frac{\partial^2}{\partial y^2} + \frac{\partial^2}{\partial z^2} \right) \tag{40}$$

Basic Idea of Solution

Suppose that it is required to solve Laplace equation over a domain bounded by a fixed boundary. Over the boundary, boundary conditions are associated. The main idea of solution can be summarized as follows:

1- Divide the whole domain by vertical and horizontal lines, which are usually called – Grid – the intersection between any vertical and horizontal line is called a node.

2- Replace all derivatives that appear in the given P.D.E. into the corresponding difference equation.

3- Substituting by differences in the P.D.E., will lead to an algebraic equation.

4- Apply the resultant algebraic equation in step 3 to all nodes over the grid; we will get a number of algebraic equations equal the number of nodes.

5- Using any numerical method to solve the resultant system of equations.

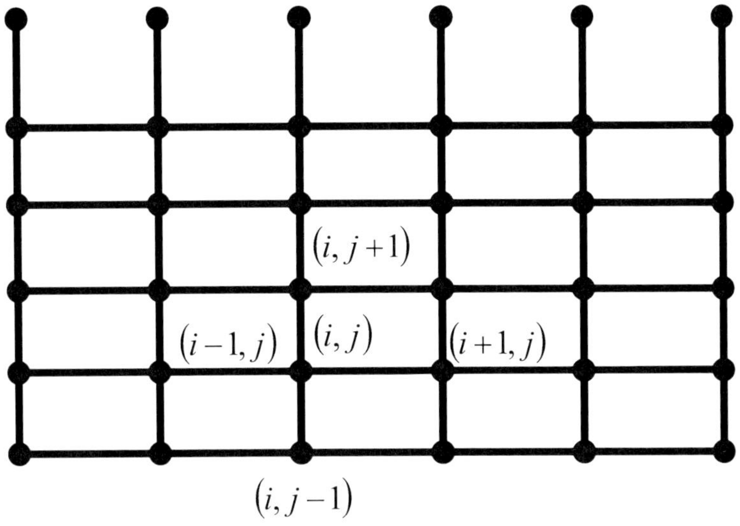

Fig. (6). Grids and nodes.

Important Notes

The index i represents the node number in the horizontal direction.

The index J represents the node number in the vertical direction.

Now let us write the difference approximation for both second derivatives, respectively:

$$\frac{\partial^2 u}{\partial x^2} = \frac{u_{i+1,j} - 2u_{i,j} + u_{i-1,j}}{(\Delta x)^2} \tag{41}$$

$$\frac{\partial^2 u}{\partial y^2} = \frac{u_{i,j+1} - 2u_{i,j} + u_{i,j-1}}{(\Delta y)^2} \tag{42}$$

By making use of equations (41) and (42) into Laplace equation, leads to:

$$\frac{\partial^2 u}{\partial x^2} + \frac{\partial^2 u}{\partial y^2} = \frac{u_{i+1,j} - 2u_{i,j} + u_{i-1,j}}{(\Delta x)^2} + \frac{u_{i,j+1} - 2u_{i,j} + u_{i,j-1}}{(\Delta y)^2} = 0 \tag{43}$$

Equation (43) is the difference equation that corresponds to Laplace equation; this equation, in case of unequal space grid distance in both directions, takes the following form:

General Form $\Delta x \neq \Delta y$

$$\frac{u_{i+1,j} - 2u_{i,j} + u_{i-1,j}}{(\Delta x)^2} + \frac{u_{i,j+1} - 2u_{i,j} + u_{i,j-1}}{(\Delta y)^2} = 0 \tag{44}$$

Special Form $\Delta x = \Delta y$

$$u_{i+1,j} + u_{i-1,j} + u_{i,j+1} + u_{i,j-1} - 4u_{i,j} = 0 \tag{45}$$

Before continuing and dealing with a numerical example, let us take a tour to different types of boundary conditions. Boundary conditions have three different types they are, Neumann, Dirichlet and Robin.

First Kind Dirichlet

In this type, the potential is known at all points of the boundary. The word potential refers mainly to the known, whatever its type, may be temperature, twisting angle, or something like that.

Example

Find the steady state temperature inside a square plate with pre-scribed boundary conditions shown in the attached (Fig. **7**) taking $a = b = 1$.

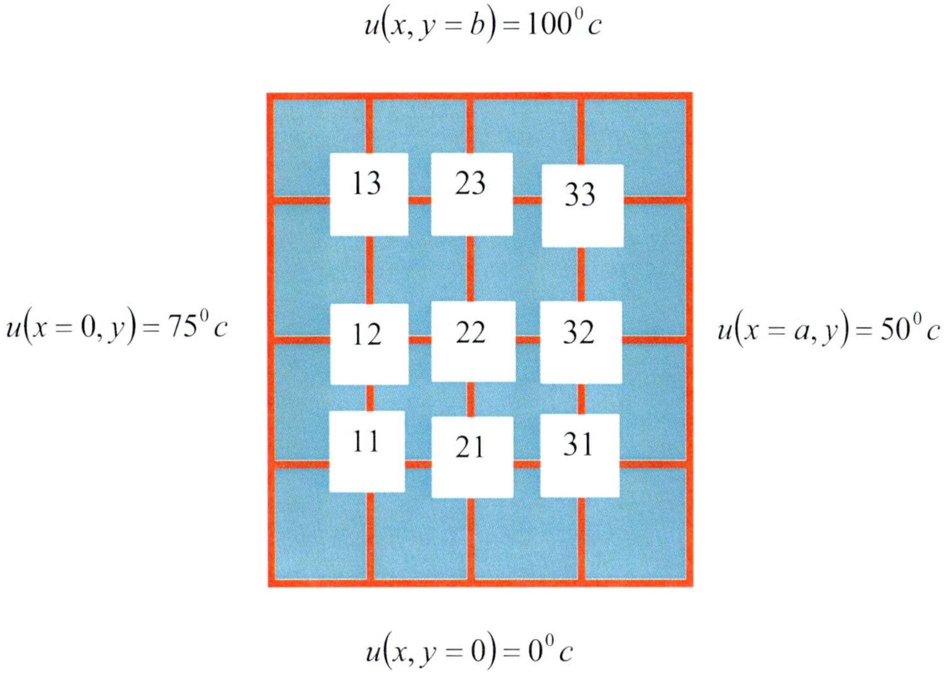

Fig.(7). Domain Problem and Grid distribution.

Solution

The domain is a square of side length equals unity, and the grid is constructed in such a way that equal space distance in both horizontal and vertical directions, $i.e., \Delta x = \Delta y = 0.25$. The next step of the solution procedure and due to equal grid size, we will make use of the differential equation given by equation (9). The next step is to number the nodes; in our example, we are in a vertical direction, as shown in the above Fig. (**4**). Then apply difference equation at each node separately, as follows.

At the point $(1,1)$

$$u_{i+1,j} + u_{i-1,j} + u_{i,j+1} + u_{i,j-1} - 4u_{i,j} = 0 \qquad \textbf{(a)}$$

At the point $(2,1)$

$$u_{11} + u_{13} + 0 + u_{22} - 4u_{21} = 0 \quad \textbf{(b)}$$

At the point $(3,1)$

$$u_{21} + 50 + 0 + u_{32} - 4u_{31} = 0 \quad \textbf{(c)}$$

At the point $(1,2)$

$$75 + u_{22} + u_{13} + u_{11} - 4u_{12} = 0 \quad \textbf{(d)}$$

By continuing, one can get nine equations in nine unknowns, then using Gauss or Gauss-Jordan or any other method, and then the final results can be summarized as follows:

$$\begin{aligned} u_{11} &= 43.00061 \\ u_{12} &= 63.21152 \\ u_{13} &= 78.58718 \\ u_{21} &= 33.29755 \\ u_{22} &= 56.11238 \\ u_{23} &= 76.06402 \\ u_{31} &= 33.88506 \\ u_{32} &= 52.33999 \\ u_{33} &= 69.71050 \end{aligned} \quad \textbf{(e)}$$

Second Kind Neumann

In this type, the potential derivative is known at all points of the boundary. The word potential derivative refers mainly to the known, whether its type, temperature gradient, twisting angle gradient, or something like that.

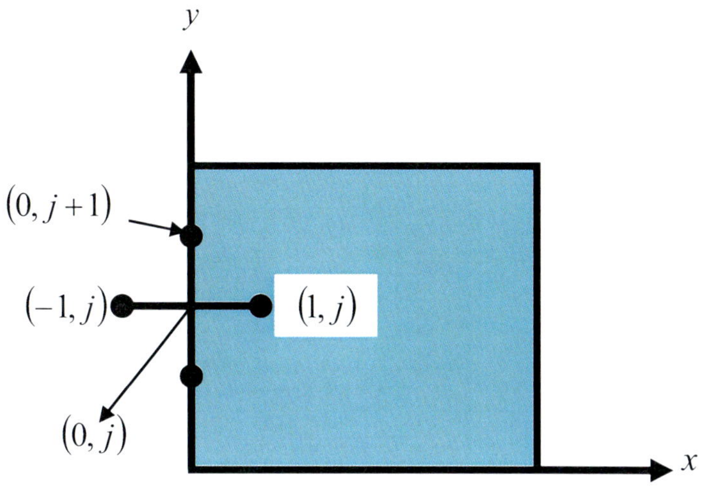

Fig. (8). 2nd kind Neumann.

Question will be raised if we want to solve the same example before, but with a little bit different is that the left side of the square will be subjected to boundary condition of Neumann's type, then the question will be, how can we do in this case? The answer is so simple, is that we take any node on that side and let us denote it as $(0, j)$, and assume the surrounding nodes will be as shown in Fig. (**5**), so if we put $i = 0$ in the differential equation, then it will take the following form:

$$u_{1,j} + u_{-1,j} + u_{0,j+1} + u_{0,j-1} - 4u_{0,j} = 0 \tag{46}$$

One can see an imaginary term and imaginary here does not mean complex, but it means it does not exist, but we imagine its existence and this term is $(u_{-1,j})$. Now the boundary condition of Neumann's type at the point $(0, j)$ can be written as:

$$\frac{\partial u_{0,j}}{\partial x} = \frac{u_{1,j} - u_{-1,j}}{2(\Delta x)} \tag{47}$$

The left-hand side is known in advance as a boundary condition; therefore, we can find the imaginary value $(u_{-1,j})$ as follows:

$$u_{-1,j} = u_{1,j} - 2(\Delta x)\frac{\partial u_{0,j}}{\partial x} \quad \textbf{(48)}$$

By making use of equation (48) into equation (47), then the later will take the following new form:

$$u_{1,j} + \left[u_{1,j} - 2(\Delta x)\frac{\partial u_{0,j}}{\partial x} \right] + u_{0,j+1} + u_{0,j-1} - 4u_{0,j} = 0 \quad \textbf{(49)}$$

Simplifying equation (49) will lead to the difference equation in the case of Neumann's type boundary condition, taking into consideration that we assumed equal grid size in both directions.

$$2u_{1,j} - 2(\Delta x)\frac{\partial u_{0,j}}{\partial x} + u_{0,j+1} + u_{0,j-1} - 4u_{0,j} = 0 \quad \textbf{(50)}$$

Example

Find the steady state temperature inside a square plate with pre-scribed boundary conditions shown in the attached Fig. taking $a = b = 1$.

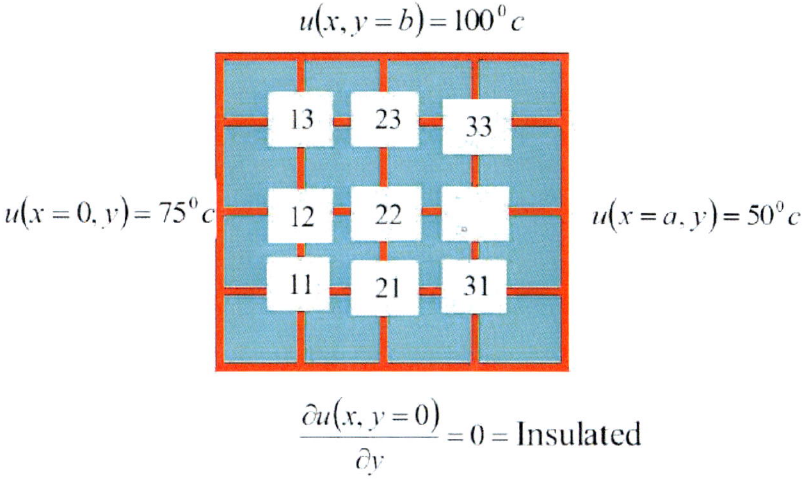

Fig. (9). Problem domain and discretization.

Solution

As we see, the lower side is perfectly insulated, so that, the difference equation that will be suitable for the solution is as follows:

$$u_{i+1,0} + u_{i-1,0} + 2u_{i,1} - 2(\Delta y)\frac{\partial u}{\partial y} - 4u_{i,0} = 0 \qquad \textbf{(a)}$$

Equation (a) is in its general form, but the equation that will be applied at the lower side takes the following form:

$$u_{i+1,0} + u_{i-1,0} + 2u_{i,1} - 4u_{i,0} = 0 \qquad \textbf{(b)}$$

Apply the difference equations (a) at all 6 nodes except the nodes adjacent to the lower side of the square. For the 3 nodes adjacent to the lower side of the square, apply equation (b). Finally, we get 9 equations in 9 unknowns, solving using Gauss or Gauss-Jordan; one can get the following results:

$$u_{10} = 71.91000 \quad u_{20} = 67.01000 \quad u_{30} = 59.54000$$
$$u_{11} = 72.81000 \quad u_{21} = 68.31000 \quad u_{31} = 60.57000 \qquad \textbf{(c)}$$
$$u_{12} = 76.01000 \quad u_{22} = 72.84000 \quad u_{32} = 64.42000$$
$$u_{13} = 83.41000 \quad u_{23} = 82.63000 \quad u_{33} = 74.26000$$

Irregular Boundaries

First of all, it is not usually we deal with regular boundaries, but in a wide range of applications, the boundaries of the domain are not straight lines but take different curvatures, then a question will arise, how can we deal with such boundaries?. To have an answer to such a question, let us have a curved boundary, see Fig. **(6)**. Suppose that the nearest node to the curved boundary, we see that there are two nodes intersecting with the curved boundary.

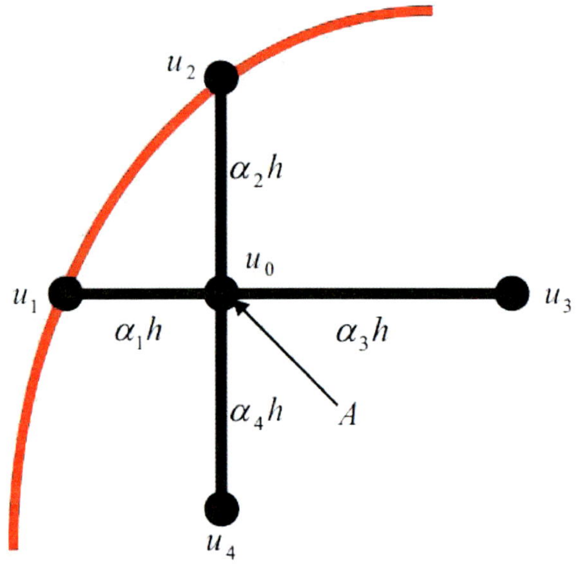

Fig. (10). Curved boundary.

Suppose that node A, like that node, is called ***unequal-armed star***; let us define the following:

$h = \Delta x = \Delta y$: Arm length and it is assumed equal in both directions

: Ratio of the arm length

$u_i, i = 0,1,2,3,4$: The potential (temperature) over the grid points

On the straight line joining the points over which the temperatures u_1, u_0 and u_3, we can write the potential derivative using the following formulae:

$$\left(\frac{\partial u}{\partial x}\right)_{1,0} \cong \frac{u_0 - u_1}{\alpha_1 h} \tag{51}$$

$$\left(\frac{\partial u}{\partial x}\right)_{0,3} \cong \frac{u_3 - u_0}{\alpha_3 h} \tag{52}$$

One may ask what we want to do. The answer is so simple; we want to derive a finite difference formula for the point near the curved boundary. To do that, let us start with the second derivative in each direction, separately, as follows:

$$\frac{\partial^2 u}{\partial x^2} = \frac{\partial}{\partial x}\left(\frac{\partial u}{\partial x}\right) \tag{53}$$

Making use of equations (51) and (52) into equation (53), therefore and after some mathematical steps, one can get the following approximation for the second derivative in x-direction, as follows:

$$\frac{\partial^2 u}{\partial x^2} = \frac{\left(\dfrac{u_3 - u_0}{\alpha_3 h}\right) - \left(\dfrac{u_0 - u_1}{\alpha_3 h}\right)}{\left(\dfrac{\alpha_1 + \alpha_3}{2h}\right)} \tag{54}$$

Equation (54) can be re-written in a simplified form:

$$\frac{\partial^2 u}{\partial x^2} = \frac{2}{h^2}\left[\frac{u_1 - u_0}{\alpha_1(\alpha_1 + \alpha_3)} + \frac{u_3 - u_0}{\alpha_3(\alpha_1 + \alpha_3)}\right] \quad \textbf{(55)}$$

By a similar procedure, we can derive the finite difference equation for the second derivative in the transverse y-direction as follows:

$$\frac{\partial^2 u}{\partial y^2} = \frac{2}{h^2}\left[\frac{u_2 - u_0}{\alpha_2(\alpha_2 + \alpha_4)} + \frac{u_4 - u_0}{\alpha_4(\alpha_2 + \alpha_4)}\right] \quad \textbf{(56)}$$

By making use of equations (55) and (56) into Laplace's equation, we get the final finite difference equation corresponding to Laplace's equation over curved boundaries.

$$\frac{2}{h^2}\left\{\left[\frac{u_1 - u_0}{\alpha_1(\alpha_1 + \alpha_3)} + \frac{u_3 - u_0}{\alpha_3(\alpha_1 + \alpha_3)}\right] + \left[\frac{u_2 - u_0}{\alpha_2(\alpha_2 + \alpha_4)} + \frac{u_4 - u_0}{\alpha_4(\alpha_2 + \alpha_4)}\right]\right\} = 0 \quad \textbf{(57)}$$

$$\left\{\left[\frac{u_1 - u_0}{\alpha_1(\alpha_1 + \alpha_3)} + \frac{u_3 - u_0}{\alpha_3(\alpha_1 + \alpha_3)}\right] + \left[\frac{u_2 - u_0}{\alpha_2(\alpha_2 + \alpha_4)} + \frac{u_4 - u_0}{\alpha_4(\alpha_2 + \alpha_4)}\right]\right\} = 0 \quad \textbf{(58)}$$

Therefore, equation (58) is the difference equation for Laplace's equation, applied at all points near the curved boundaries.

Example

Find the steady state temperature at the indicated grid points on the plate shown in the following Fig. (**11**) and its associated boundary conditions.

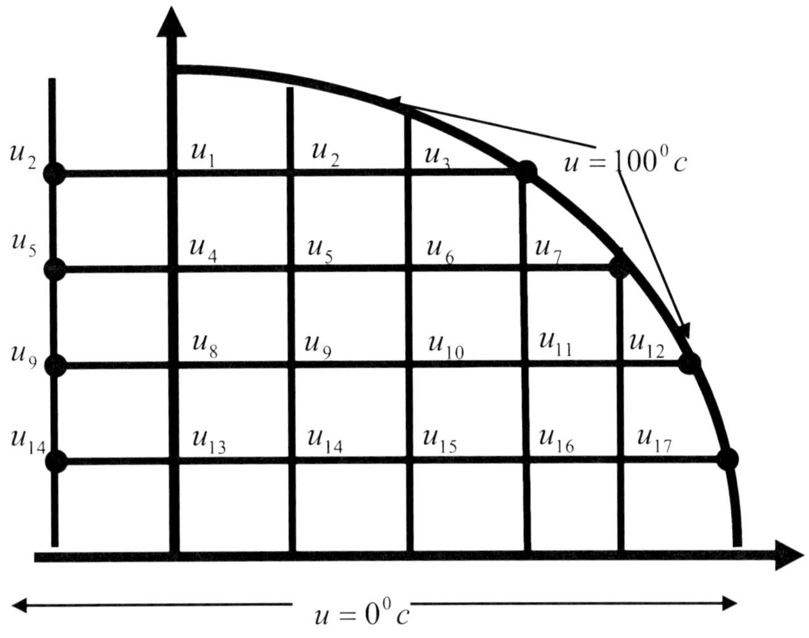

Fig. (11). Problem configuration for curved boundaries.

Solution

First of all, it is important to remember that the whole domain is the upper half of semi-circle, and because of symmetry, we took only the quarter in the first quadrant. Secondly, the points of temperatures u_2, u_3, u_{12} and u_{17} are points of unequal arms, therefore, the ratios $\alpha_i, i = 1,2,3,4$ should be determined. From the analytical geometry, the arm length of u_2 and u_{17} equals $0.899h$ and for the points of u_3 and u_{12} equals $0.5826h$ and the rest of all points are equal and equals to $h = 0.2$. We will deal with the aforementioned four points in what fellow, using equation (7). The final results will be listed in the next table.

Table

x	y	Point Number	Finite difference results	Analytical results
0	0.8	1	86.053	85.906
0.2	0.8	2	87.548	87.417
0.4	0.8	3	92.124	92.094
0	0.6	4	69.116	68.807
0.2	0.6	5	70.773	70.482
0.4	0.6	6	75.994	75.772
0.6	0.6	7	85.471	85.405
0	0.4	8	48.864	48.448
0.2	0.4	9	50.436	50.000
0.4	0.4	10	55.606	55.151
0.6	0.4	11	65.891	65.593
0.8	0.4	12	84.189	84.195
0	0.2	13	25.466	25.133
0.2	0.2	14	26.501	26.109
0.4	0.2	15	30.102	29.527
0.6	0.2	16	38.300	37.436

0.8	0.2	17	57.206	57.006

Numerical Solution of Parabolic P.D.E.

The second type of partial differential equation we will deal with in a numeric way is the parabolic partial differential equation. This type of partial differential equation is found in a wide range of engineering and industrial applications. A major criterion of this type of equation is its time-dependent, so a care should be taken into consideration, especially when dealing with time. The direct example of this type of partial differential equations is the diffusion problem, which is apparently found in many displaces like food, ceramic, steel, and others. Let us start with the simplest diffusion case problem in fixed domain problems, and so the state equation takes the following form [61-65]:

$$\frac{\partial^2 u}{\partial x^2} = \frac{c\rho}{k}\frac{\partial u}{\partial t} \tag{59}$$

In the grid corresponding to the diffusion equation, we will have vertical lines corresponding to different times, and the horizontal lines will correspond to the independent variable x. The Index: *i* will be the variable x and Index: *i* for the time. The same idea for the solution procedure is that we find the finite difference corresponds to each derivative that appears in the differential equation, and then replace these derivatives with the corresponding differences; we finally get the differential equation. The mesh or grid that corresponds to the one-dimensional heat diffusion equation is shown in Fig. **(7)**.

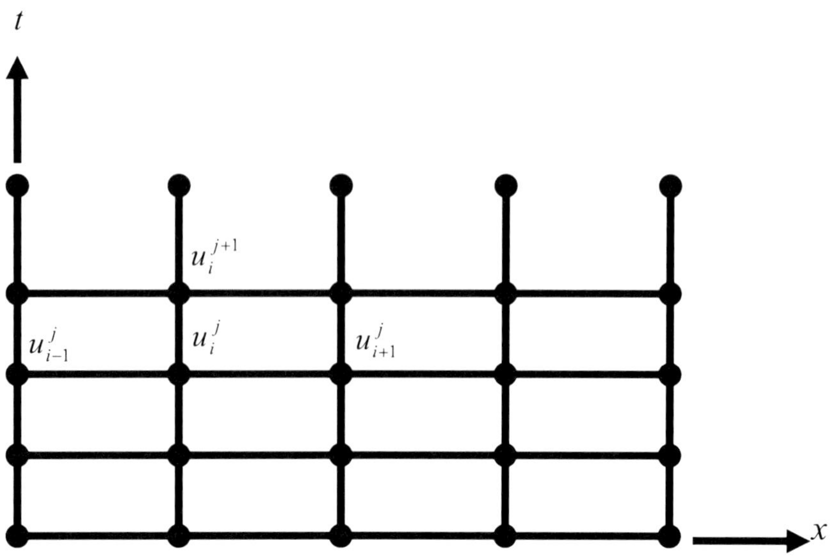

Fig. (12). Grid for 1-D heat diffusion.

The difference equation that corresponds to the second derivative is as follows:

$$\left(\frac{\partial^2 u}{\partial x^2}\right)_{\substack{x=x_i \\ t=t_J}} = \frac{u_{i+1}^J - 2u_i^J + u_{i-1}^J}{(\Delta x)^2} + O(\Delta x)^2 \tag{60}$$

The difference equation that corresponds to the first derivative is as follows:

$$\left(\frac{\partial u}{\partial t}\right)_{\substack{x=x_i \\ t=t_J}} = \frac{u_i^{J+1} - u_i^J}{(\Delta t)} + O(\Delta t) \tag{61}$$

By making use of equations (60) and (61) into equation (59), will lead to the difference equation for one-dimensional heat diffusion equation Fig. (**12**):

$$u_i^{J+1} = \left(\frac{k(\Delta t)}{c\rho(\Delta x)^2}\right)\left[u_{i+1}^J + u_{i-1}^J\right] + \left[1 - \frac{2k(\Delta t)}{c\rho(\Delta x)^2}\right]u_i^J \tag{62}$$

Notes

u_i^{J+1} Temperature at the point (i) and time $(J+1)$..Basic unknown

u_i^J Temperature at the point (i) and time (J)..Known

u_{i-1}^J Temperature at the point $(i-1)$ and time (J)..Known

u_{i+1}^J Temperature at the point $(i+1)$ and time (J)..Known

$\left(\dfrac{k(\Delta t)}{c\rho(\Delta x)^2}\right)$ is a ration plays an important play in the results, as we will see later on. This ratio may be one of the following three cases:

Case (I)

$\left(\dfrac{k(\Delta t)}{c\rho(\Delta x)^2}\right) = \dfrac{1}{2}$ in this case equation (4) will take the following simplified form:

$$u_i^{J+1} = \left(\frac{1}{2}\right)\left[u_{i+1}^J + u_{i-1}^J\right] \tag{63}$$

This simplification does have any direct effect on the accuracy of the results; just it makes the difference equation quite simple.

Case (II)

$\left(\dfrac{k(\Delta t)}{c\rho(\Delta x)^2}\right) < \dfrac{1}{2}$, in this case, we expect an improvement in the obtained, and there is a very important note herein, the size step in x-direction has a direct effect on the accuracy of the solution.

Case (III)

$\left(\frac{k(\Delta t)}{c\rho(\Delta x)^2}\right) > \frac{1}{2}$, this case only decreases the solution steps but on the other hand, we cannot ensure high expected improvement in the obtained results.

Example

A large flat steel plate is 2 cm thickness if the initial temperature within the plate is given as a function of the distance from one face and is defined as follows:

$$u(x,0) = \begin{cases} 100x & 0 \le x \le 1 \\ 100(2-x) & 1 \le x \le 2 \end{cases}$$

Find the temperature distribution as a function of both space variables and time if both faces are kept at zero temperature. In this problem, the material properties of the steel are taken as follow:

$$k = 0.13\text{cal}/\sec \cdot \text{cm} \cdot {}^0\text{C}$$
$$c = 0.11\text{cal}/\text{g} \cdot {}^0\text{C}$$
$$\rho = 0.11\text{g}/\text{cm}^3$$

Solution

In this case problem, we will solve one-dimensional heat diffusion problem Fig. (**13**).

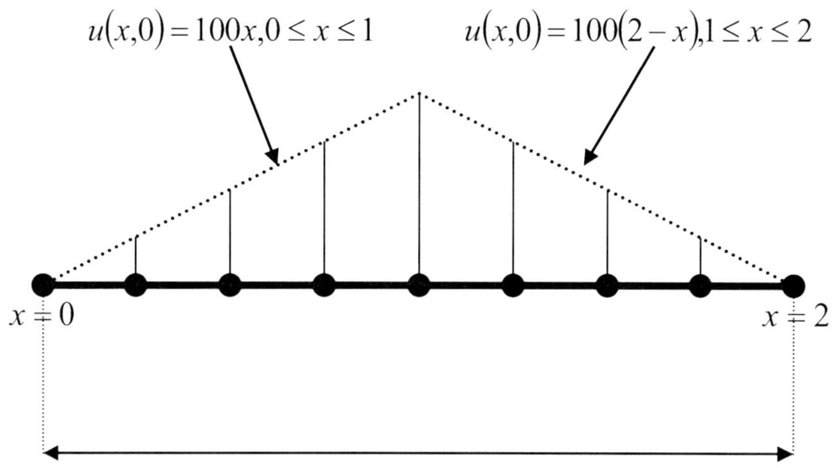

Fig. (13). Problem configuration.

The problem configuration is shown in Fig. (**9**). The next step is to determine the time step assuming that the step size in x-direction $\Delta x = 0.25$, as follows:

$$\left(\frac{k(\Delta t)}{c\rho(\Delta x)^2}\right) = \frac{1}{2}$$

$$\Rightarrow$$

$$\Delta t = \frac{(0.11)(7.8)(0.25)^2}{(2)(0.13)} = 0.206\,\text{second} \qquad \textbf{(a)}$$

Due to the above assumption regarding time step size, the difference equation that will be used is as follow:

$$u_i^{J+1} = \left(\frac{1}{2}\right)\left[u_{i+1}^J + u_{i-1}^J\right] \qquad \textbf{(b)}$$

Apply the difference equation (b) to all internal nodes, shown in Fig. (**9**), and the summary of the computed results will be clearly shown in the following table:

Time	Calculated temperature					
	$x = 0.0$	$x = 0.25$	$x = 0.50$	$x = 0.75$	$x = 1.0$	$x = 1.25$
0.000	0	25	50	75	100	75
0.206	0	25	50	75	75	75
0.412	0	25	50	62.5	75	62.5
0.619	0	25	43.75	62.5	62.5	62.5
0.825	0	21.88	43.75	53.12	62.5	53.12
1.031	0	21.88	37.5	53.12	53.12	53.12
1.238	0	18.75	37.5	45.31	53.12	45.31
1.444	0	18.75	32.03	45.31	45.31	45.13
1.650	0	16.16	32.03	38.67	45.31	38.67
1.856	0	16.16	27.34	38.67	38.67	38.67
2.062	0	13.67	27.34	33.01	38.67	33.01

The analytical solution corresponding to the given problem is as follows:

$$u(x,t) = 800\sum_{n=0}^{\infty}\frac{1}{\pi^2(2n+1)^2}\cos\frac{(2n+1)\pi(x-1)}{2}e^{-0.3738(2n+1)^2 t} \tag{c}$$

Crank-Nicolson Method

To understand the Crank-Nicolson method, let us re-write the heat diffusion again as follows:

$$\frac{\partial^2 u}{\partial x^2} = \frac{c\rho}{k}\frac{\partial u}{\partial t} \tag{64}$$

Now the difference equation corresponds to time, will be:

$$\frac{\partial u}{\partial t} = \frac{u_i^{j+1} - u_i^j}{\Delta t} \tag{65}$$

And the difference equation that corresponds to the second derivative will be:

$$\frac{\partial^2 u}{\partial x^2} = \frac{1}{2}\left\{\left(\frac{u_{i+1}^j - 2u_i^j + u_{i-1}^j}{(\Delta x)^2}\right) + \left(\frac{u_{i+1}^{j+1} - 2u_i^{j+1} + u_{i-1}^{j+1}}{(\Delta x)^2}\right)\right\} \tag{66}$$

By making use of equations (65) and (66) into the heat equation (64), one will get the difference equation and it will take the following form:

$$\frac{1}{2}\left\{\left(\frac{u_{i+1}^j - 2u_i^j + u_{i-1}^j}{(\Delta x)^2}\right) + \left(\frac{u_{i+1}^{j+1} - 2u_i^{j+1} + u_{i-1}^{j+1}}{(\Delta x)^2}\right)\right\} = \left[\frac{c\rho}{k}\right]\left(\frac{u_i^{j+1} - u_i^j}{\Delta t}\right) \tag{67}$$

Notes

1- The differences that appear on the right-hand side are central differences at the mid time step, *i.e.*, at $t_{j+\frac{1}{2}}$.

2- The first term on the left hand side represents the central difference at time t_{j+1}

3- The second term on the left hand side represents the central difference at time t_j

Let us now make the following simplification:

(68)

By making use of equation (68) into (67), and re-arrange the later, we will get the following equation:

$$-ru_{i-1}^{j+1}+(2+2r)u_i^{j+1}-ru_{i+1}^{j+1}=ru_{i-1}^{j}+(2-2r)u_i^{j}+ru_{i+1}^{j} \qquad \textbf{(69)}$$

Equation (69) is called Crank-Nicolson' formula for heat diffusion equation, and this method has important criteria that it is stable for numerical computations for any value of the parameter r, taking into consideration that by decreasing the value of the parameter r, butter results are obtained with good stability.

Example

Solve

$$\frac{\partial^2 u}{\partial x^2}=\frac{1}{D}\frac{\partial u}{\partial t} \qquad \textbf{(a)}$$

$D=0.119\text{cm}^2/\sec$

With the following boundary and initial conditions:

$$u(x,0)=20 \qquad \textbf{(b)}$$

$$u(0,t)=0 \qquad \textbf{(c)}$$

$$u(20,t)=10 \qquad \textbf{(d)}$$

Solution

To solve the example, we first must do some preparations, they are:

1- Determination of the domain length and this can be determined from the boundary conditions; it is 20 units

2- Determination of the step size in x-direction

3- Determination of the time step

We do graph the grid of the solution, as shown in Fig. (**10**).

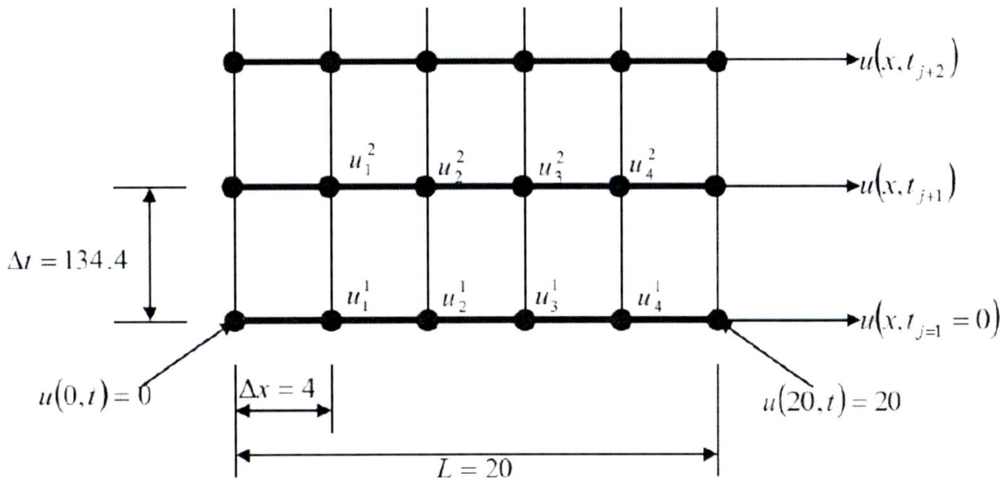

Fig. (14). Problem configuration.

Next let us re-write the relation of the parameter r:

$$r = \frac{k(\Delta t)}{c\rho(\Delta x)^2} \qquad \textbf{(e)}$$

Now, and before determining the time step, it is important to choose a suitable value for r, then we will be able then to determine the time step. If one assumes that $r = 1$,then:

$$r = \frac{k(\Delta t)}{c\rho(\Delta x)^2} \tag{f}$$

$$\frac{\partial^2 u}{\partial x^2} = \frac{c\rho}{k}\frac{\partial u}{\partial t}$$

&

$$\frac{\partial^2 u}{\partial x^2} = \frac{1}{D}\frac{\partial u}{\partial t} \tag{g}$$

$$\Rightarrow \frac{1}{D} = \frac{c\rho}{k}$$

$$\Rightarrow D = \frac{c\rho}{k} = 0.119$$

Substituting by above numeric values, one can get a time step equals 134.4 sec

$$-ru_{i-1}^{j+1} + (2+2r)u_i^{j+1} - ru_{i+1}^{j+1} = ru_{i-1}^{j} + (2-2r)u_i^{j} + ru_{i+1}^{j} \tag{h}$$

Put $r = 1$ into equation (8), then it will take the following form:

$$-u_{i-1}^{j+1} + 4u_i^{j+1} - u_{i+1}^{j+1} = u_{i-1}^{j} + ru_{i+1}^{j} \tag{i}$$

Now we are ready to compute the temperature at the next time step taking into consideration that we are now at the time step $t_{j+1} = t_2 = 0 + 134.4 = 134.4$, then apply equation (i) to the points 1,2,3, and 4, respectively, as follows:

$$-0.0 + 4u_1^2 - u_2^2 = 0.0 + 2.0 \tag{j}$$

$$-u_1^2 + 4u_2^2 - u_3^2 = 2.0 + 2.0 \tag{k}$$

$$-u_2^2 + 4u_3^2 - u_4^2 = 2.0 + 2.0 \tag{l}$$

$$-u_3^2 + 4u_4^2 - 10 = 2.0 + 10.0 \tag{m}$$

The same procedure repeat itself at each time step four equations like equations (10-14) will be derived; then by any numerical method, the solution will be obtained.

Time in seconds	**Calculated concentrations**					
	$x=0$	$x=4$	$x=8$	$x=12$	$x=16$	$x=20$
0.0	0.000	2.000	2.000	2.000	2.000	10.000
134.4	0.000	0.980	2.019	3.072	5.992	10.000
268.8	0.000	1.070	2.363	4.305	6.555	10.000
403.2	0.000	1.276	2.861	4.762	6.962	10.000
537.6	0.000	1.471	3.065	5.115	7.159	10.000

Time in seconds	**Analytical solution**	
	$x=4$	$x=12$
134.4	1.078	3.191
268.8	1.108	4.272
403.2	1.340	4.873
537.6	1.543	5.248

Parabolic P.D.E. in Two-dimensions

The heat diffusion equation is of parabolic type. Let us now start writing the heat equation in two-dimensions, and it will have the following form:

$$\frac{\partial u}{\partial t}=\frac{k}{c\rho}\left(\frac{\partial^2 u}{\partial x^2}+\frac{\partial^2 u}{\partial y^2}\right) \tag{1=70}$$

The difference equation solution will not differ herein than that for one-dimension, except here an additional term will be added and the grid will be in three-dimensions, two of which represents the xy-plane and the third one for the time, as shown in Fig. **(15)**.

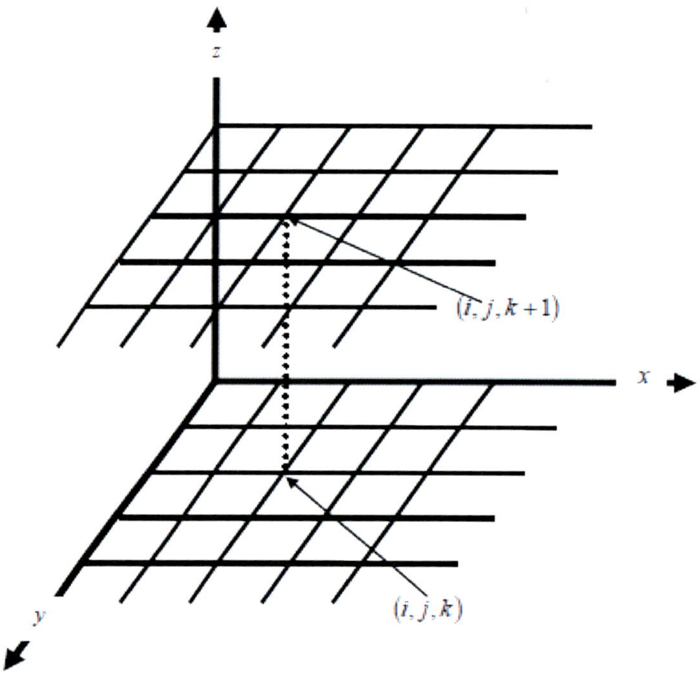

Fig. 15: Grid in 3-D problems.

Notes

1- The index i will be x-direction

2- The index j will be y-direction

3- The index k will be t-direction

The next step is to represent the derivatives in terms of differences, taking into consideration, that we will assume equal space grid in both x and y directions, *i.e.*, $\Delta x = \Delta y$. Another point should be taken into consideration is the parameter r, defined previously as:

$$r = \frac{k(\Delta t)}{c\rho(\Delta x)^2} \tag{71}$$

But for 2-D problems, equation (71) will take another form as follows:

$$r = \frac{k(\Delta t)}{c\rho\{(\Delta x)^2 + (\Delta y)^2\}} \tag{72}$$

Now then, the general difference equation for 2-D heat diffusion equation will be of the form:

$$u_{i,j}^{k+1} - u_{i,j}^{k} = r\left[u_{i+1,j}^{k} - 2u_{i,j}^{k} + u_{i-1,j}^{k} + u_{i,j+1}^{k} - 2u_{i,j}^{k} + u_{i,j-1}^{k}\right] \tag{73}$$

Equation (73), can take the following form:

$$u_{i,j}^{k+1} = r\left[u_{i+1,j}^{k} + +u_{i-1,j}^{k} + u_{i,j+1}^{k} + u_{i,j-1}^{k}\right] + (1-4r)u_{i,j}^{k} \tag{74}$$

Notes

1- In case of constant boundary conditions, the maximum allowable value for the parameter r is $r = \frac{1}{4}$

2- In case of un-equal space grid size $\Delta x \neq \Delta y$, the following relation should be satisfied, that is:

$$\frac{k(\Delta t)}{c\rho\left\{(\Delta x)^2+(\Delta y)^2\right\}} \leq \frac{1}{8} \tag{75}$$

The general Crank-Nicolson formula for two-dimensional heat equation can take the following formula:

$$u_{i,j}^{k+1}-u_{i,j}^{k}=\left(\frac{r}{2}\right)\begin{bmatrix}\left(u_{i+1,j}^{k+1}-2u_{i,j}^{k+1}+u_{i-1,j}^{k+1}\right)+\left(u_{i+1,j}^{k}-2u_{i,j}^{k}+u_{i-1,j}^{k}\right)\\ \left(u_{i,j+1}^{k+1}-2u_{i,j}^{k+1}+u_{i,j-1}^{k+1}\right)+\left(u_{i,j+1}^{k}-2u_{i,j}^{k}+u_{i,j-1}^{k}\right)\end{bmatrix} \tag{76}$$

This means we have to solve system of $M \times N$ linear algebraic equations at each time step, where $M \times N$ represents the number of unknowns in x-direction and $M \times N$ represents the number of unknowns in y-direction.

Heritage of Moving Boundary Problem

The topic free and moving boundary problems started since 1890, when Stefan followed the melting of ice cap. He noticed that by growing up the time, the ice change his state from solid state to liquid state. Moving boundary problems play important rules in Engineering and Industry. Any researcher cannot distinguish the important applications in Engineering and industry. For long time, researchers treat such problems as sharp interface problems, *i.e.*, the moving interface between phases may be point, line or area according to the dimension of the problem underhand.In fact, this assumption is not true but the interface may be line, area or volume according to the dimension, these geometries are called *mushy zones*, have specific properties and contain certain conditions and may or may not have partial differential equations within it [65-70].

Practical Applications

- Metal casting [71]
- Ground freezing problems [72]
- Thermal storage [73]
- Vaporization [74]
- Crystallization [75]
- Ceramic industry [76]
- Food industry [77]
- Ablation problems and can be found in [78]
- Material treatment [79]
- Thermal protection of re-entry vehicles in space technology [80]
- Laser drilling and cutting in manufacturing [81]

Free Boundary Problems

The term "***Free Boundary Problems***" is used to describe the case when the boundary is stationary and a steady-state problem exists. Consequently, free boundary problems require the solution of elliptic partial differential equations. As an example of practical application for such a type of free boundary problem is the cavitations phenomena. Free Boundary Problems is used to describe the case when the boundary is stationary and steady-state problem exist. Consequently, free boundary problems require the solution of elliptic partial differential equations, such as Laplace's equation. Practical example: The cavitations phenomena [82] Fig. (**16**).

Fig. (16). Cavitations around Hydrofoil.

Moving Boundary Problems

Definition

Moving boundary problems can be defined as set of partial differential equations with associated boundary and initial conditions over a domain which has one or more of its boundaries vary with time [83].

- The term "***Moving Boundary Problems***" is used to describe the case when the position of the boundary has to be determined as a function of time and space.

The determination of the movement of the moving boundaries is very important and is considered as an integral part of the solution. Moving boundary problems are usually referred to as Stefan problems. The difficulty in solving moving boundary problems refers mainly to the non-linearity caused by the presence of nonlinear boundary condition at the moving boundary. This difficulty makes it difficult in most cases to find an analytical solution, except for very simple problems. Moving boundary problems has a wide range of engineering and industrial applications such as metal casting; ground freezing problems and thermal storage. Ablation problems are another type of such problems and can be found in different engineering and industrial applications such as material treatment and processing, thermal protection of re-entry vehicles in space technology, laser drilling and cutting in manufacturing, among others. When dealing with different types of Stefan problems, which originally the basis of moving boundary problems, it should be taken into consideration the criteria of the type underhand [84-90].

- ***There is a major difference between Stefan and ablation problem is that the overall domain in Stefan problem remains fixed in space while the domain in the ablation problem is variable and diminishes in size with time.***

- ***Vaporization is another type of Stefan problems, in such a case, three phases are appearing throughout the process under the assumption of sharp interface between phases.***

In case of existence of mushy zone between solid and liquid, the situation is then changed completely. Since the early stages of modeling and simulation of Stefan problems, different restrictions had been taken into consideration such as the geometry. Therefore, very few analytical solutions are available and only for simple cases due to the high non-linearity at the moving boundaries. In recent years, efforts concentrate on the use of numerical techniques due to rapid development in computer technology and its high performance. Boundary element method becomes one of the most popular numerical methods, which were applied to a wide range of engineering and industrial applications. All these applications and some others have one thing in common, that is the formation of a moving interface during the process of heat and mass transfer. This interface may be defined by a sharp boundary or by a region, depending on the application considered [91-95].

Important Note

In both free and moving boundary problems, two conditions are needed on the free or moving boundary, one to determine the boundary itself and the other to complete definition of the solution of the differential equation (Fig. **17**). Suitable conditions on the fixed boundaries and, where appropriate, an initial condition is also prescribed as usual.

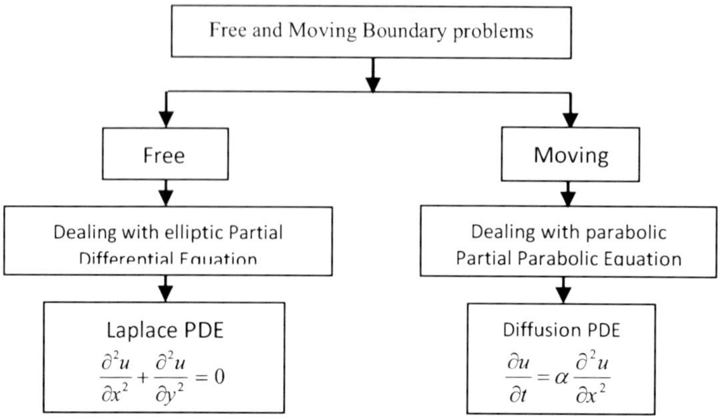

Fig. (17). Free and moving boundary problems.

Mathematical Description of Moving Boundary Problems

Moving boundary problems can be defined as set of partial differential equations with associated boundary and initial conditions over domain has one or more of its boundaries vary with time [96].

Detailed Description

From now on, we would like to learn something extremely important how to describe Phase change problems with moving boundaries when writing a scientific thesis or a paper that will be published in a journal, whether this journal is local or international or will be presented at an international conference. How can you express phase change problem with moving boundary (boundaries)?

Physical Description

Where a description of the problem that will be solved by starting with a physical description of the problem, what will be exposed to the boundary conditions, and what happens after the exposure of the domain to the aforementioned conditions.

Mathematical Formulation

Where from the outset the quality of the number of phases is determined through the physical description, then we start writing the mathematical equations for each phase with the inclusion of the boundary and initial conditions for each phase separately. There will be additional equations at the bound or moving boundaries represented in energy balance equations and other equations depending on the nature of the problem [97].

Difficulties in Moving Boundary Problems

The difficulty in solving moving boundary problems refers mainly to the non-linearity caused by the presence of nonlinear boundary condition at the moving boundary. When mathematically dealing with free and moving boundary problems, the student encounters a major problem, which is the non-linearity problem represented by the boundary conditions at the boundary or the set of moving limits, particularly in the energy equilibrium condition called the Stefan condition.

Characteristics of Sharp Interface

The sharp moving interface takes different forms, may be *point* in 1-D problems, *line* in 2-D problems or *plane* in 3-D problems, as shown in Fig. (**1**).

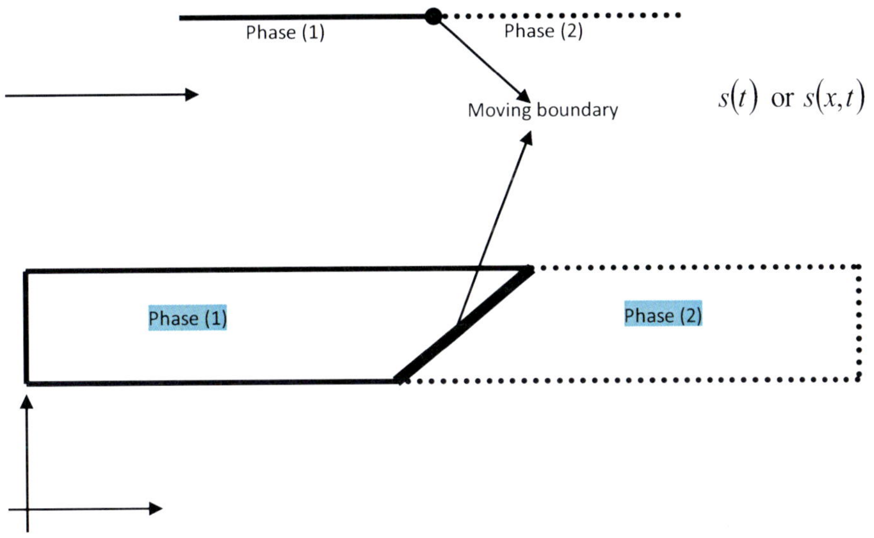

Fig. (18). Sharp moving interface in 1-D and 2-D problems.

From the previous Fig. (**18**), the student can conclude the following points:

1- In one-dimensional problems, we deal with the moving term as a point.

2- In the two-dimensional problems, we treat the moving term as a line or a Curve.

3- In three-dimensional problems, we treat the moving boundary as the Surface.

4- In all the previous cases there is one thing in common regardless of the dimensions of the problem under study, namely, how to track the movement of the moving boundary over time.

Moving Interface from Mathematical Point of View

- May or may not, according to the practical application, partial differential equation valid within it.

When dealing with the boundary or moving boundary as a sharp interface, we have two boundary conditions at this or these boundaries, but this speech will change to some extent when we deal with the boundary between any two areas as a region with special specifications and equations.

- Moving interface usually defined by certain condition, and the well known condition is called "Energy balance", this condition takes the following general formula.

$$\text{Input flux} - \text{output flux} = \pm \rho L V_n$$

In the above equation:

$\pm$ Signs represents the case problem (Melting or Solidification), respectively.

Dirichlet boundary condition also can found.

Simple Stefan Melting Problem

One-phase, one-dimensional melting/solidification problem

A semi-infinite sheet of ice, initially at the melting temperature, taken to be zero, the surface of which is raised at time, $t = 0$ to a temperature above zero, at which it is subsequently maintained, a boundary surface or interface on which melting occurs, moves from the surface into the sheet and separates a region of water from one of ice at zero temperature. The location of the moving boundary – interface- $x = s(t)$ in the $x - t$ plane is drawn below, where $x = s(t)$ denotes the thickness of the water phase at time t and x is the space coordinate measured from the outer surface of the sheet, $x = 0$ [98] (Fig. **19**).

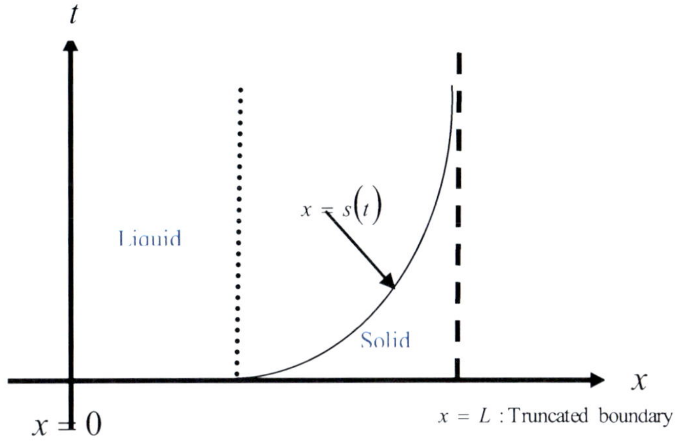

Fig. (19). Problem configuration.

Mathematical Formulation

If $u(x,t)$ represents the temperature distribution in the liquid phase, at any point and any time, then the mathematical formulation will contains the following:

Governing parabolic partial differential equation, its solution will represent the heat flow in the liquid phase,

$$\frac{\partial u}{\partial t} = \alpha \frac{\partial^2 u}{\partial x^2} \qquad 0 \leq \text{x} \leq s(t)$$

Two boundary conditions at the end points of the liquid phase,

$$u(x=0,t)=U_0 \quad ,t>0$$

Since the other term is the same as the moving term, the auxiliary condition that is used to find the complete solution to the differential equation will be the same as one of the two conditions related to the moving boundary, and therefore the condition will be temperature at the moving boundary, while the other condition will be related to the heat balance on the same Limit and based on the above, the condition related to temperature takes the following form:

$$u(x=s(t),t)=0 \quad ,t>0$$

An initial condition(s) when we have solid phase only before starting the melting process.

With regard to the time condition, it differs from one issue to another, but in our case we have two conditions related to the temporal moment, which is the time zero, that is, before the start of the melting process, the first condition will relate to the temperature of the solid substance (snow) before the consolidation process and the second condition relates to the position of the moving boundary before the beginning of the process itself, *i.e.* The fusion process, and mathematically these two conditions take the following form:

$$u(x,t=0)=0 \quad ,x\geq 0$$
$$s(t=0)=0$$

Usually, two boundary conditions at the moving boundary, one of which is used to find the complete solution of the governing PDE, while the second one represents the energy balance at the that moving boundary. The last condition is named as Stefan's condition.We mentioned that the conditions related to the moving limit in our case are two conditions, and we said that one of these conditions is an additional condition that helps in finding a complete solution to the differential equation, and the second condition concerns the thermal equilibrium or energy balance at this limit, and accordingly we will write the equilibrium thermal equilibrium or energy balance from the formula. The general condition for Stefan is as follows:

$$\text{Input flux} - \text{output flux} = \pm \rho L V_n$$

Very Important Considerations

1- The present problem has one dimension and one phase

2- Based on the first item, we find that there is no Input flux only the Output flux

3- Since we indicated in the first item that we are dealing with a one-dimensional issue, speed takes shape $V_n = \dfrac{ds(t)}{dt}$

Based on the foregoing considerations, the Stefan condition takes the following picture:

$$-K\frac{\partial u(x,t)}{\partial x} = \rho L\frac{ds(t)}{dt}$$

The mathematical modeling fulfills all the physical description and thus the set of equations with their attachments are all ready for solution. With regard to the methods of the solution, we will devote several chapters to it according to the type of method. In what follows I will generalize the previous problem, a study that becomes a One-Dimensional, Two-Phase Melting/Solidification problem.

Simple Stefan Melting Problem

Two-phase, One-dimensional Melting/solidification Problem

A semi-infinite sheet of ice, initially at temperature below the melting temperature, the melting temperature, not necessarily uniform, and heat flow occurs in both liquid (water) and solid (ice) phases, respectively. In this problem, we have three major unknowns; they are respectively, temperature in both liquid and solid phases as well as tracing the moving interface (Fig. **20**).

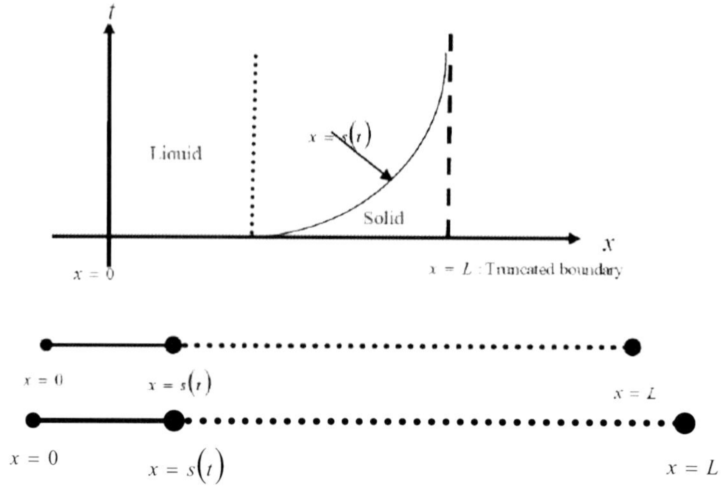

Fig. (20). Problem Configuration.

Mathematical Formulation

If $u_\ell(x,t), u_s(x,t)$ represent the temperature distribution in both liquid and solid phases, at any point and any time, then the mathematical formulation will contains the following:

Governing parabolic partial differential equations for both phases, their solution will represent the heat flow in the two phases,

For Liquid phase;

$$\frac{\partial u_\ell(x,t)}{\partial t} = \alpha_\ell \frac{\partial^2 u_\ell(x,t)}{\partial x^2} \qquad 0 \le \mathrm{x} \le s(t)$$

For Solid phase;

$$\frac{\partial u_s(x,t)}{\partial t} = \alpha_s \frac{\partial^2 u_s(x,t)}{\partial x^2} \qquad s(t) < \mathrm{x} \le L$$

Two boundary conditions for each phase, at the end points of phase underhand,

$$u(x=0,t)=U_0 \quad ,t>0$$

$$u_\ell(x=s(t),t)=U_m \quad ,t>0$$
$$u_s(x=s(t),t)=U_m \quad ,t>0$$

$$u_s(x=L,t)=U_i \quad ,t>0$$

$$u(x=s(t),t)=0 \quad ,t>0$$

An initial condition(s) when we have solid phase only before starting the melting process,

$$u(x,t=0)=U_i \quad ,x\geq 0$$
$$s(t=0)=0$$

Stefan's condition

$$\text{Input flux} - \text{output flux} = \pm\rho L V_n$$

$$K_s\frac{\partial u_s(x,t)}{\partial x}-K_\ell\frac{\partial u_\ell(x,t)}{\partial x}=\rho L\frac{ds(t)}{dt}$$

Multi-phase-phase, one-Dimensional Melting/solidification Problem

In what follows, we will show another more complex model, and we will explain that the method used in modeling is almost unchanged, in which the equations and conditions can increase, but the concept does not change. The current example is a research published by the author in 1998 in the Journal of the Faculty of Engineering, Cairo University, and the title of the research as it was published as follows [99]:

One-dimensional Multiphase Moving Boundary Problems

Problem Description and Formulation

Consider a solid material occupied a mould of length, ℓ, subject to a variable source of heat Q(t) at the extreme point x = 0, and in contact with a fluid at a constant temperature u_i at the other end x = ℓ. Assuming that the heat transfer is one-dimensional and that the material thermal properties such as conductivity and heat capacity are constant within each phase, but differ from one phase to another. The material density is assumed the same within each phase and from one to another. In this problem three different phases occurs, they are pre-melting process, melting process and the vaporization process [100] (Fig. **21**).

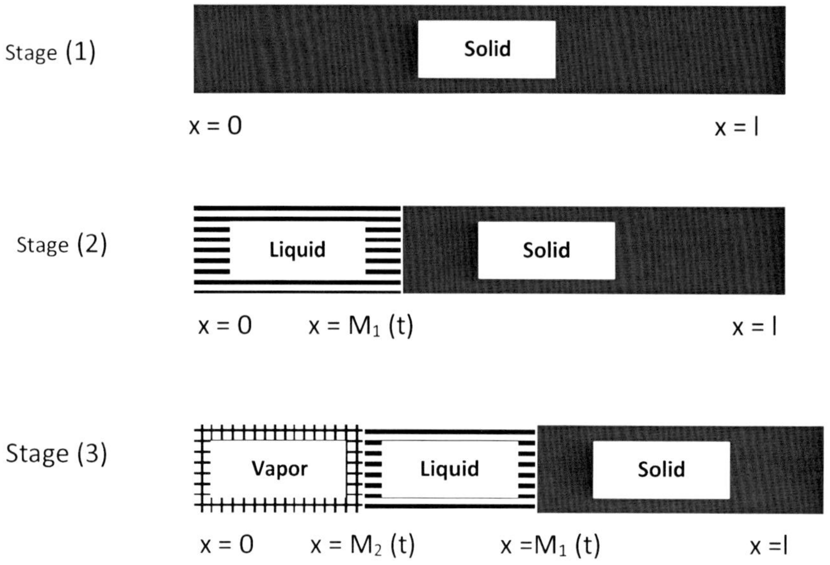

Fig. (21). Multi-Phase Problem configuration.

Stage (1): Pre-Melting Process

The first stage: the pre-melting stage

We find that the solid is subjected to a temperature as a boundary condition, taking into account that with increasing the temperature to a certain extent, the degree of this substance rises, but it will not reach the melting point like any other material that does not convert directly by heating from the solid state to the liquid state.

Physically, what does this mean?

It simply means that there is a so-called latent heat of fusion, which raises the temperature of the solid particles.

Mathematically, how do we express such a situation and expose it to a boundary condition?

Three equations express this:

1- The partial differential equation represented by the thermal diffusion equation.

2- Boundary condition at the left side

3- Boundary condition at the right end

For any researcher who reads this paper, he finds that the author wrote an introduction of three lines before listing the equations and conditions, as follows:

"*Due to the positive heat, input flux, the temperature throughout the solid increases with no change of phase, i.e. there is latent heat of fusion. The governing equations for this stage are:*"

$$\frac{\partial u_s}{\partial t} = \alpha_s \frac{\partial^2 u_s}{\partial x^2} \qquad 0 \leq x \leq \ell \qquad \textbf{(77)}$$

$$-K_s \frac{\partial u_s}{\partial t} = Q(t) \qquad x = 0 \qquad \textbf{(78)}$$

$$-K_s\frac{\partial u_s}{\partial t}=\mathrm{h}(\mathrm{u}_s(\mathrm{x},\mathrm{t})-\mathrm{u}_i) \qquad \mathrm{x}=\ell \tag{79}$$

Stage (2): Melting stage

The second stage: the stage of initiation of fusion

This stage begins when the boundary temperature reaches the melting temperature, and at this moment another phase begins, other than the solid phase, and thus the problem becomes two phase. Each phase has its equations and its boundary and temporal conditions. The researcher now has to see how the author of the research model the mathematically formulating this stage, as explained below. *when the temperature at the boundary* $\mathrm{x}=0$ *reaches the melting temperature,* u_m*, the solid starts melting and two phases start appearing, liquid and solid phase respectively. The governing equations are as follows:*

For liquid-phase

$$\frac{\partial u_\ell}{\partial t}=\alpha_\ell\frac{\partial^2 u_\ell}{\partial x^2} \qquad 0\le x\le M_1(\mathrm{t}) \tag{80}$$

$$-K_\ell\frac{\partial u_\ell}{\partial t}=Q(\mathrm{t}) \qquad \mathrm{x}=0 \tag{81}$$

$$\rho L_m\frac{dM_1}{dt}=K_s\frac{\partial \mathrm{u}_s}{\partial \mathrm{x}}-K_\ell\frac{\partial \mathrm{u}_\ell}{\partial \mathrm{x}} \qquad \text{at } \mathrm{x}=\mathrm{M}_1(t) \tag{82}$$

For solid-phase:

$$\frac{\partial u_s}{\partial t}=\alpha_s\frac{\partial^2 u_s}{\partial x^2} \qquad \mathrm{M}_1(\mathrm{t})\le \mathrm{x}\le \ell \tag{83}$$

$$-K_s\frac{\partial u_s}{\partial t}=\mathrm{h}\left(\mathrm{u}_s(x,t)-u_i\right) \qquad \mathrm{x}=\ell \tag{84}$$

$$u_s(x,t) = u_m \qquad x = M_1(t) \tag{85}$$

With the following initial conditions:

$$M_1(t) = 0 \qquad \text{at} \quad t \le t_m \tag{86}$$

$$u(x,0) = u_i \qquad 0 \le x \le \ell \tag{87}$$

Stage (3): Vaporization Stage

when the temperature at the boundary $x = 0$ *reaches the vaporization temperature* u_v*, a second moving boundary appears which separates the liquid from the vapor. Assuming the vapor is removed as soon as it appears, and the liquid region occupies the region* $M_2(t) \le x \le M_1(t)$*, and the solid occupies the region* $M_2(t) \le x \le \ell$*. This stage is finished when the moving boundary* $M_1(t)$ *reaches the extreme* $x = \ell$*. This stage can be formulated as follows:*

$$\rho_s C_s \frac{\partial u_s}{\partial t} = K_s \frac{\partial^2 u_s}{\partial x^2} \qquad M_1(t) \le x \le \ell \tag{88}$$

$$-K_s \frac{\partial u_s}{\partial x} = h(u_s(x,t) - u_i) \qquad \text{at} \qquad x = \ell \tag{89}$$

$$\rho_\ell C_\ell \frac{\partial u_\ell}{\partial t} = K_\ell \frac{\partial^2 u_\ell}{\partial x^2} \qquad M_2(t) \le x \le M_1(t) \tag{90}$$

$$\rho L_m \frac{dM_1(t)}{dt} = K_s\left(\frac{\partial u_s}{\partial x}\right) - K_\ell\left(\frac{\partial u_\ell}{\partial x}\right) \quad \text{at} \quad x = M_1(t) \tag{91}$$

$$\rho L_v \frac{dM_2(t)}{dt} = K_\ell\left(\frac{\partial u_\ell}{\partial x}\right) - Q(t) \qquad \text{at} \quad x = M_2(t) \tag{92}$$

In addition to the above boundary conditions, the following conditions apply for all three stages:

$$M_1(t) = 0 \qquad t \le t_m \tag{93}$$

$$M_2(t) = 0 \qquad t \le t_v \tag{94}$$

$$u(x, t) = u_m \qquad x = M_1(t),\ t \ge t_m \tag{95}$$

$$u(x, t) = u_v \qquad x = M_2(t),\ t \le t_m \tag{96}$$

$$u(x,0) = u_i \qquad 0 \le x \le \ell \tag{97}$$

Mushy Zone

Historical Background

- In 1974, Atthey started defining mushy zone when discussing the spot welding of two sheets of metals [101].

- In 1982, Rubinstein expressed concern about the precise physical significance of the term mushy zone, which is considered to be at a uniform temperature but has phase changes occurring within it [102].

- Mushy zones are mainly found in melting and solidification applications. Solidification processes play very important role in metallurgy and in many cases completely determine the physical and mechanical properties of solid alloy.

- Sometimes, it is possible to describe crystallization in a standard way by means of the well-known Stefan thermo-diffusion model with a planar front. However, often the plane front of solidification is broken by the constitutional super-cooling arising under certain circumstances ahead of the front. For this reason, some parts of the front grow in more favorable conditions than others, *i.e.* the front becomes morphologically unstable.

- The evolution of such an instability leads to the appearance of a meta-stable zone between solid and liquid phases, which is called the two-phase zone or mushy region. Thus, after a lapse of time, solidification is divided into three parts: solid, mushy and liquid phases.

Definition

Mushy zones are the regions that appear in the multi-component melt as porous layers of dendritic crystals that commonly form during solidification of multi-component melts [103].

Characteristics of Mushy Zones

The mushy zone takes different forms, may be *line* in 1-D problems, *area* in 2-D problems or *volume* in 3-D problems, as shown in Fig. (**22**).

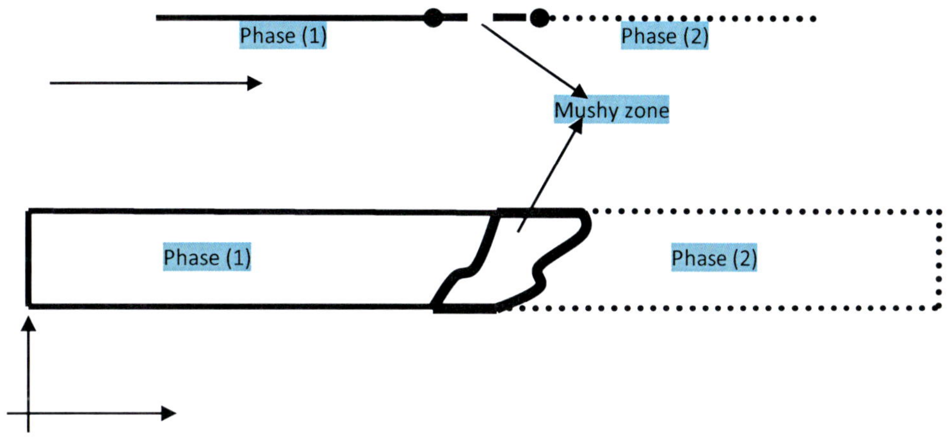

Fig. (22). Mushy zone in 1-D and 2-D problems.

Mushy zone from mathematical point of view

- Mushy zone usually defined by specified condition that relates the velocity of the moving boundaries bounding the mushy zone and the flux on both sides. This is the same as Stefan condition "Energy balance".

- Also there is an alternative condition relates the flux, velocity and the thickness of the mushy zone.

- Finally, we can say that according to the model assumed for any practical problem under investigation, one can add one or more conditions in the mushy zone.

Stefan Problems with and Without Mushy Zone

Mathematical analyses of solidification or melting are complicated by the need to apply boundaries at the solid/liquid interfaces which are evolving with time and whose positions must be found as part of the calculation. The case of a pure melt being cooled by conduction of heat to its boundaries is relatively straightforward since the geometry of the solidification front is similar to that of the boundary walls, such so-called "***Stefan***" problems [104]. In a general case of Stefan problems there are two or more phases, according to problem underhand, appear throughout the process, between these phases there are moving boundaries. In the past, all researchers assumed sharp interfaces, but nowadays, mushy zones are considered. In the above definition of Stefan problems, if a pure melt is super-cooled, so that latent heat is conducted away from the solidification front through the liquid, then the solidification front becomes extremely convoluted and forms intricate branching patterns. Snowflakes provide a common example of this phenomenon. When the liquid is an alloy such behavior is commonplace even when the liquid is not initially super-cooled. Analytical techniques can not follow the evolution of such convolutions far beyond initial perturbations from a flat interface through important results have been obtained through the use of high-speed computers. The matrix, or region of mixed phase, has been termed a "***Mush or Mushy zone***" by many previous authors. Hills, Loper & Roberts (1983) [105] developed a full set of thermodynamic equations for a mushy zone, and solved mush-reduced set of them approximately for the constrained growth of a binary alloy. Huppert & Worster (1986) formulated a simple mathematical model of mushy layer based on considerations of global conservation relationships. Its predictions agreed well with their observations of ice growing at a plane boundary from aqueous solutions of various salts. Several other treatments such as Burns

(1992) [106] led to models with solid phase-mushy zone and one phase. Tarzia (1995) [107] used the boundary integral method to solve a phase change problem.While Pavel (1999),[108]with other treatments, led to two regions consisting of the solid phase and mushy zone. S.G. Ahmed (2001) used the boundary integral method to solve a phase change problem and compared his results with Tarzia (1995) [60]. Huppert (1999)[109]classified the complex nature of the liquid-solid phase change processes in different systems. S. G Ahmed (2005)[110]used an approximate method to find the two main parameters for the mushy zone. Mekky and S. G. Ahmed (2009) [111] solved by Semi-analytical method a combined Stefan and ablation problem with mushy zone. Alexandrova (2009) [112] had investigated its behavior experimentally and analytically. By treating the mushy phase as a new single phase, and the macroscopic envelope of the convoluted solid as a phase boundary, it is necessary to follow the evolution of the two phase boundaries: the solid/mush and the mush/liquid interface. Kunle (2011) [112] studied numerically the effect of magnetic field upon an alloy with mushy zone.

Ablation Problems With And Without Mushy Zones

When a body is exposed to heat flux, the surface of the body changes phase. If the changed phase is immediately removed upon formation, this referred to ablation problem. Ablation problems can be found in different engineering and industrial applications such as *material treatment and processing*, *thermal protection of re-entry vehicles in space technology*, *laser drilling* and *cutting in manufacturing*. There is a major difference between Stefan and ablation problem is that the overall domain in Stefan problem remains fixed in space while the domain in the ablation problem is variable and diminishes in size with time. The difference in ablation problem with mushy zone and that without mushy zone appears in the mathematical model and formulation as well as the applications. Indeed the more practical cases appear in the case of ablation with mushy zone that represents the reality very well.

Approximate Treatment of Mushy zone Problems

For a long time, all researchers solved phase change problems with moving boundaries and sharp interfaces between the boundaries that appear throughout the whole process. In reality, this assumption is wrong since a transition zone between, for example solid and liquid (solidification problem), which have certain properties and is called ***mushy zone***.

Physical Description

A solid domain Ω exposed to high input heat flux; therefore, after time t_m, melting starting occurs in an inward direction. At any time before the completion of the process there exist, three phases appear throughout. The three phases are as follows liquid, mushy zone, and solid, respectively [113]. The configuration of the whole process at any instant of time is shown in the Fig. (**23**).

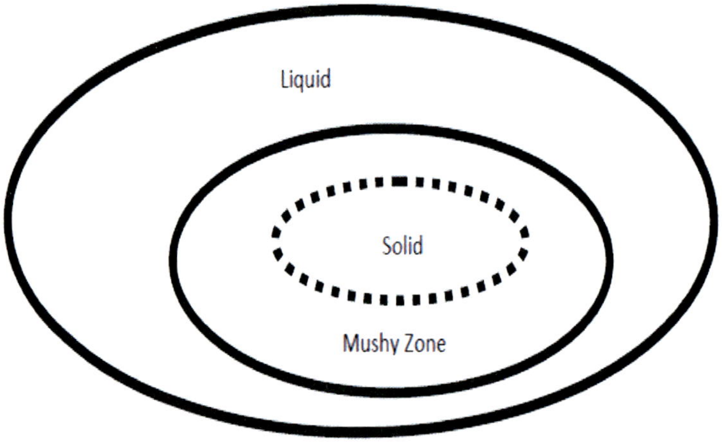

Fig. (23). 2-D problem configuration.

General Problem Configuration

For the sake of simplicity, let us start the proposed method by one-dimensional case problem; in this manner, the problem can be described as follow:

A finite solid initially at a uniform temperature, the boundary $x = 0$ subjected to a high input heat flux; therefore, three phases occur. In the present paper, we mainly consider the third stage of this problem in which the vapor is removed upon formation; therefore, the problem remains two phase problem with mushy zone between liquid and solid, see Fig. (**24**).

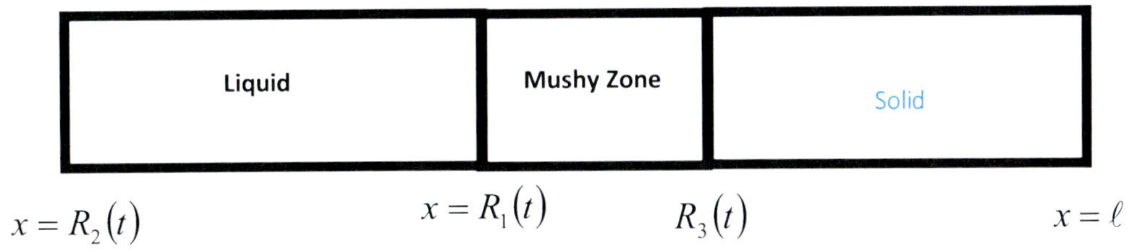

Fig. (24). 1-D problem configuration.

Mathematical Formulation

$$k_\ell \frac{\partial^2 u_\ell}{\partial x^2} = \frac{\partial u_\ell}{\partial t} \qquad R_2(t) < x < R_1(t) \tag{98}$$

$$k_S \frac{\partial^2 u_S}{\partial x^2} = \frac{\partial u_S}{\partial t} \qquad R_3(t) < x < \ell \tag{99}$$

$$u_S(\ell, t) = U_i \tag{100}$$

$$u_\ell(R_2(t), t) = U_V \tag{101}$$

$$\frac{\partial u_\ell}{\partial x} = \frac{\gamma}{[R_3(t) - R_1(t)]} \tag{102}$$

$$\left(K_\ell \frac{\partial u_\ell}{\partial x} - K_S \frac{\partial u_S}{\partial x} \right) = \rho L \left[\varepsilon \frac{dR_1(t)}{dt} + (1 - \varepsilon) \frac{dR_3(t)}{dt} \right] \tag{103}$$

In this formulation, we have five unknowns, the temperature distribution in the liquid and solid phases and three moving boundaries $R_2(t), R_1(t)$ and $R_3(t)$. In S. G. Ahmed's papers [70], a tabulated relation between the two parameters γ and

ε was found, and in the present thesis, we will use these data as input data files when computation is carried out.

Derivation of the Proposed Method

Step: 1

The first step in classical Goodman's method is to assume temperature profiles for the phases that appear throughout the whole process:

$$u_\ell(x,t)=\left[\left(\frac{U_m}{R_1-R_2}\right)+B_2(x-R_1)^2\right](x-R_2)+\left[\left(\frac{U_v}{R_2-R_1}\right)+B_2(x-R_2)^2\right](x-R_1) \quad \textbf{(104)}$$

$$u_S(x,t)=\left[\frac{U_m}{R_3-\ell}+B_3(x-R_3)^2\right](x-\ell)+\left[\frac{U_i}{R_3-\ell}+B_3(x-\ell)^2\right](R_3-x) \quad \textbf{(105)}$$

Step: 2

Integrate equation (98) w. r. t. the space variable x from $x=R_2$ to $x=R_1$, and equation (99) w. r. t. the space variable x from $x=R_3$ to $x=\ell$ yields:

$$k_\ell\left\{\left(\frac{\partial u_\ell}{\partial x}\right)_{x=R_1}-\left(\frac{\partial u_\ell}{\partial x}\right)_{x=R_2}\right\}=\frac{d}{dt}\left\{\int_{x=R_2}^{x=R_1}u_\ell(x,t)dx\right\} \\ +u_\ell(R_1,t)\frac{dR_1}{dt}-u_\ell(R_2,t)\frac{dR_2}{dt} \quad \textbf{(106)}$$

$$k_s\left\{\left(\frac{\partial u_s}{\partial x}\right)_{x=\ell}-\left(\frac{\partial u_s}{\partial x}\right)_{x=R_3}\right\}=\frac{d}{dt}\left\{\int_{x=R_3}^{x=\ell}u_s(x,t)dx\right\}-u_s(R_3,t)\frac{dR_3}{dt} \quad \textbf{(107)}$$

Step: 3

To find relation between moving boundary velocities and potential derivative, it is known that the total derivative of temperature at the moving boundaries is zero;

$$d(u_\ell(R_2,t)) = 0 \tag{108}$$

Equation (108) can be written in an expanded form as follow:

$$\frac{\partial u_\ell(R_2,t)}{\partial x}\frac{dR_2}{dt} + \frac{\partial u_\ell(R_2,t)}{\partial t} = 0 \tag{109}$$

Therefore; the velocities at R_3 and R_1 can be written as follow:

$$\left(\frac{dR_2}{dt}\right)_{x=R_2} = -k_\ell\left(\frac{\partial^2 u_\ell / \partial x^2}{\partial u_\ell / \partial x}\right)_{x=R_2} \tag{110}$$

$$\left(\frac{dR_1}{dt}\right)_{x=R_1} = -k_\ell\left(\frac{\partial^2 u_\ell / \partial x^2}{\partial u_\ell / \partial x}\right)_{x=R_1} \tag{111}$$

Similarly for solid phase;

$$\left(\frac{dR_3}{dt}\right)_{x=R_3} = -k_s\left(\frac{\partial^2 u_s / \partial x^2}{\partial u_s / \partial x}\right)_{x=R_3} \tag{112}$$

Step: 4

Analytical treatment of equation (9), this step started by differentiating assumed potential profile for liquid phase, then, substituting into the right-hand side, one can ensure that this side will equal to zero. The first term on the right-hand side is integrated first based on the assumed profile, then differentiating the result w. r. t. time, meanwhile, make use of equations (13) and (14), then after long mathematical manipulation, one can obtain the following equation:

$$\left(\frac{U_m+U_v}{2}\right)\left[\begin{array}{c}\dfrac{-6k_\ell B_2(R_1-R_2)}{\left(\dfrac{U_m-U_v}{R_1-R_2}\right)+B_2(R_1-R_2)^2}\\+\dfrac{6k_\ell B_2(R_2-R_1)}{\left(\dfrac{U_m-U_v}{R_1-R_2}\right)+B_2(R_2-R_1)^2}\end{array}\right]= \tag{113}$$
$$\left\{\frac{\eta_1(t)B_2}{\eta_2(t)+\eta_3(t)B_2}+\frac{\varsigma_1(t)B_2}{\eta_2(t)+\varsigma_2(t)B_2}\right\}$$

Where,

$$\eta_1(t)=-6k_\ell U_v(R_2-R_1)$$

$$\eta_2(t)=\left(\frac{U_v-U_m}{R_2-R_1}\right)$$

$$\eta_3(t)=(R_2-R_1)^2$$

$$\varsigma_1(t)=-6k_\ell U_m(R_2-R_1)$$

$$\varsigma_2(t)=(R_1-R_2)^2$$

Step: 5

Similar procedure is carried out again, but on equation (107), the result for this step is as follow:

$$\left[(\ell-R_3)^2+(R_3-\ell)^2\right]\left[\left(\frac{U_m-U_i}{R_3-\ell}\right)-B_3(R_3-\ell)\right]= \tag{114}$$
$$2(\ell-R_3)\left[\frac{5}{3}B_3R_3^3-5B_3\ell R_3^2+5B_3\ell^2R_3+\frac{1}{3}B_3\ell^2-\frac{1}{2}U_m+U_i\right]$$

It is clear from equation (113), that it contains three unknowns. One of them is the unknown function in the assumed profile, given by equation (104). Meanwhile, equation (114) contains two, one of them is the unknown function in the assumed profile given by equation (105), and therefore the procedure for the solution will be illustrated in the next section.

Solution Procedure

(1) Specify initial input data for both mushy zone parameters, ε and γ

(2) Assume initial position for both R_1 and R_3 (taking into consideration the satisfaction of heat balance equation, boundary conditions around the mushy zone with a prescribed tolerance, and previous studies to avoid divergence.

(3) From equation (113) evaluate B_2.

(4) Evaluate the error in equation (102) making use of equation (104).

(5) Repeat steps (1) to (4) till satisfying step (4) with prescribed tolerance.

(6) From equation (17) evaluate B_3 and the error in equation (103) but taking into consideration the last and accurate value for the first term in the left-hand side. If the error is not within a prescribed error, update the moving boundaries location; if it is satisfied, then go to the next time step.

Numerical Results and Discussion

Aluminum occupies semi-infinite domain initially at a uniform temperature u_i. An input heat flux is applied at $x = 0$; therefore, three phases appear. The vapor is removed upon formation and so the problem is still two-phase with mushy zone separating liquid and solid. The thermo-physical is shown in Table (**1**).

Table 1. Thermo-physical Properties

u_i	$932\,K$
u_V	$2543\ K$
L	$376560\ J/Kg$
K	$200\ W/mK$
ρ	$2710\ Kg/m^3$
C	$1200\ J/Kg\,K$

The first and main target herein is to trace the moving boundaries around the mushy zone to test the validity of the proposed method by comparing the present results with available previous results obtained by S. G. Ahmed (2005) [71]. We compared our results with the afro-mentioned results because the same definite domain was used in the two methods. The present results start with a comparison between the moving boundary separating the liquid and mushy zone, respectively, and denoted as the right side of the liquid phase, as shown in Fig. (**25**). From that figure, one can see that the error between the two results is comparatively good, especially no analytical solutions available for such problems in the finite domain.

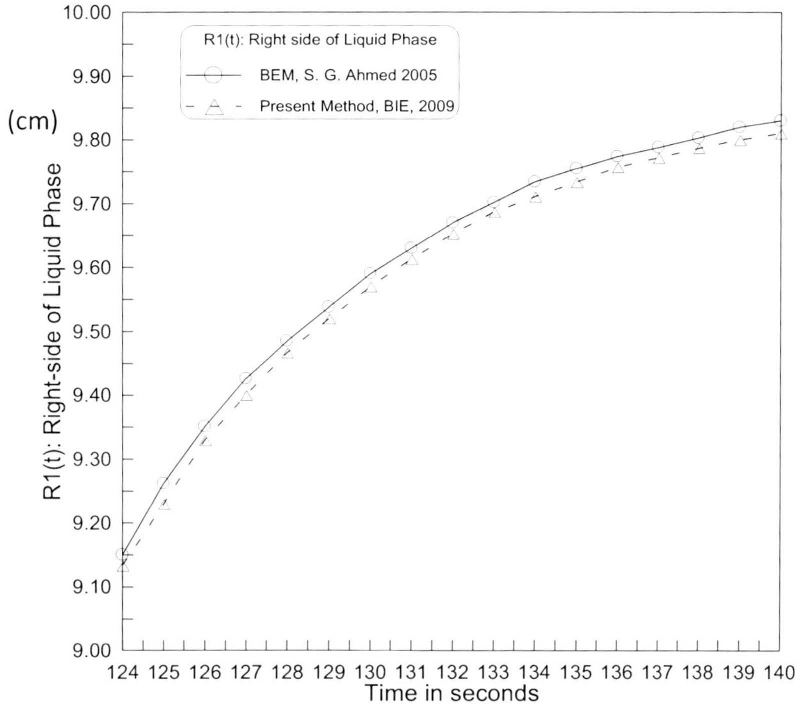

Fig.(25). $R_1(t)$ location of Right side of liquid phase over $0.1 < \varepsilon < 09$.

From the computation, it was found a very important result that is the moving boundary separating the mushy zone and the solid phase is affected directly by the parameter ε; therefore a comparison between the present results and S. G. Ahmed (2005) is made at two distinct values of that parameter, and this comparison is shown in two successive Figs. (**26** and **27**).

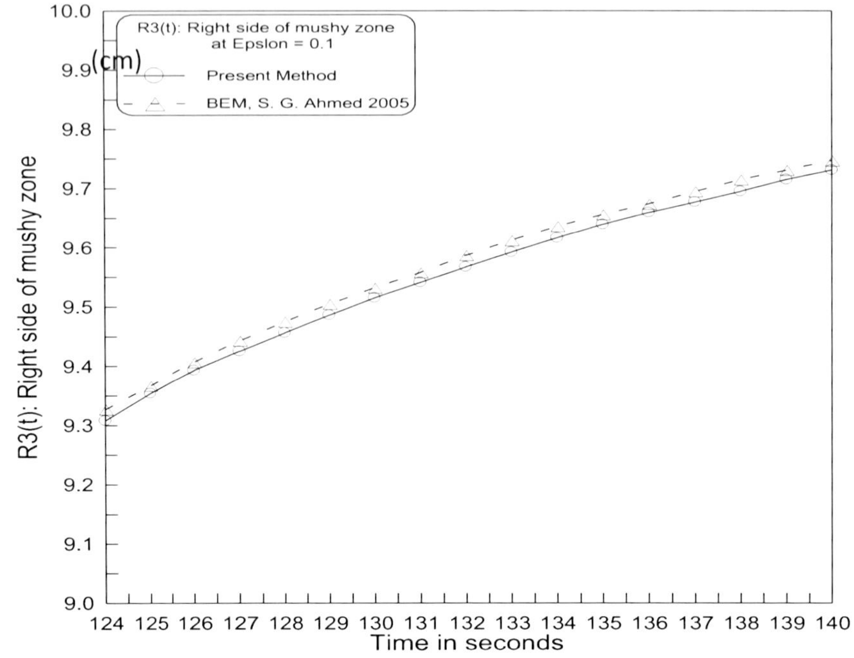

Fig. (26). Location of R3(t) right side of mushy zone at Epslon = 0.1.

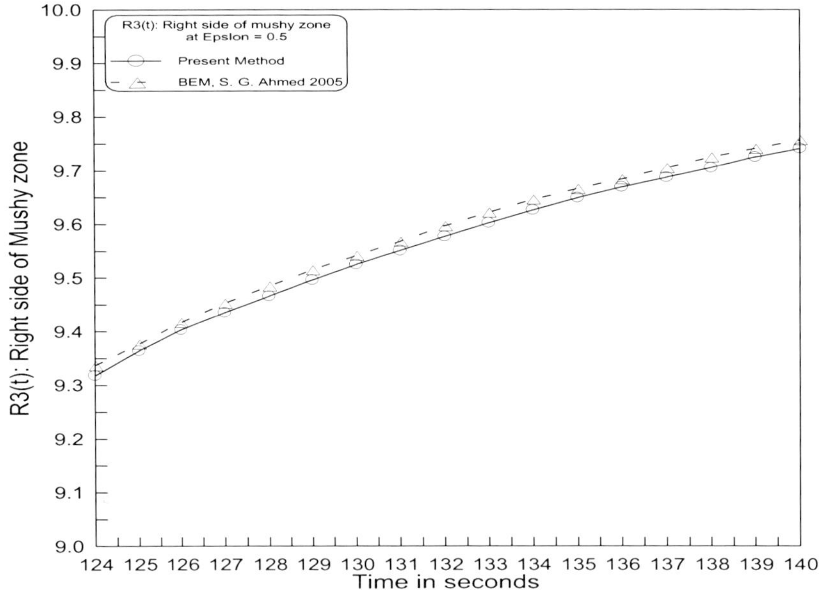

Fig. (27). Location of R3(t) right side of mushy zone at Epslon = 0.5.

Follow up the computation procedure, the temperature variations in both solid and liquid phases based on the proposed method only are shown in Figs. (**28** and **29**) As seen, the temperature variations are very well describing the physics of such a practical problem.

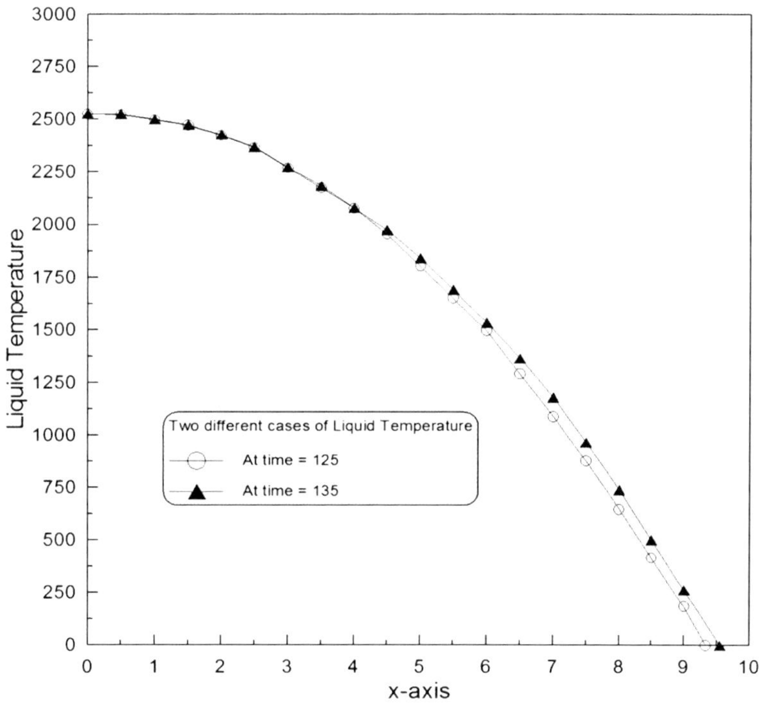

Fig. (28). Liquid temperature.

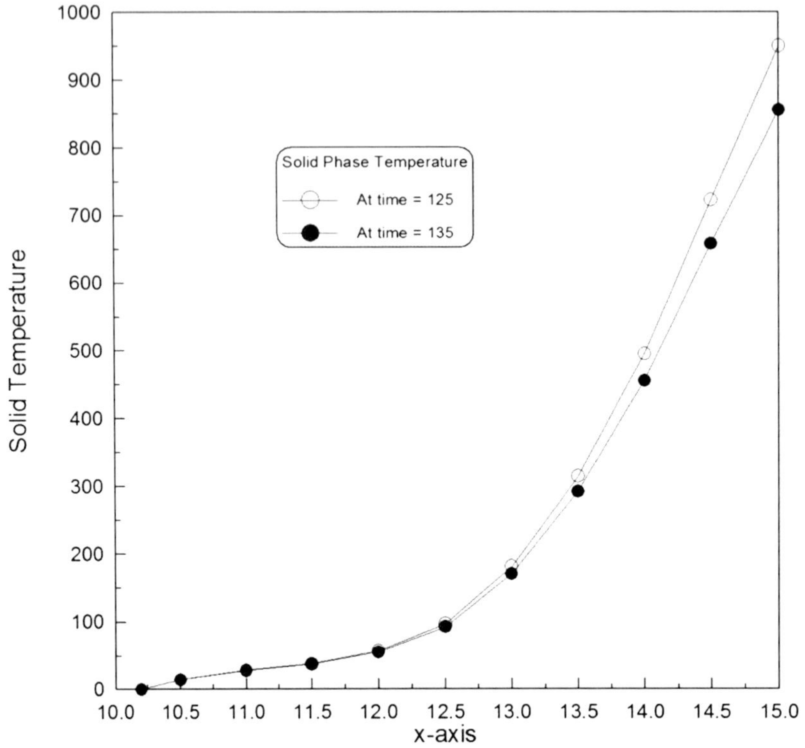

Fig. (29). Solid Phase Temperature at two different times.

Finally, and from the previous analysis one can summarize the overall results as follow:

(1) The heat balance integral method is an old method refers back to Goodman. This method was developed an approximate method for solving classical Stefan melting or solidification problem.

(2) In classical method the moving interface between solid and liquid was assumed sharp interface, but this is not actual.

(3) The present method is a ***new view of Goodman's method***; the new view comes from two ways.

(3.1) The first one is that the moving interface is not sharp but region called mushy region.

(3.2) While second one is the mathematical manipulation, *i.e.*; in classical Goodman's method the mathematics leads to three nonlinear differential equations, but in the present method we get only two equations and their numerical solution iteratively leads to the complete list of unknowns.

REFERENCES

[1] J. Crank, *Free and Moving Boundary Problems.* Clarendo Press: Oxford, 1984.

[2] D. Hegen, "Element-free Galerkin methods in combination with finite element approaches", *Comput. Methods Appl. Mech. Eng.,* vol. 135, no. 1-2, pp. 143-166, 1996. [http://dx.doi.org/10.1016/0045-7825(96)00994-2]

[3] I. Boztosun, A. Charafi, M. Zerroukat, and K. Djidjeli, "Thin-Plate Spline Radial Basis Function Scheme for Advection-Diffusion Problems", *Electronic Journal of Boundary Elements,* vol. BETEQ (2001), no. 2, pp. 267-282, 2002.

[4] A. La Rocca, A. Hernandez Rosales, and H. Power, "Radial basis function Hermite collocation approach for the solution of time dependent convection diffusion problems", *Eng. Anal. Bound. Elem.,* vol. 294, no. 4, pp. 359-370, 2005. [http://dx.doi.org/10.1016/j.enganabound.2004.06.005]

[5] D. Buddhi, "A simplified method of determining the latent heat of fusion", *Indian science congress,* 1991.

[6] C. J. Ho, and C. H. Chu, "Numerical Simulation of Melting Process Within a Square Enclosure With Oscillatory Surface Temperature", *HTD- Heat Transfer in Phase Change,* pp. 205-9, 1992.

[7] T.R. Goodman, "The Heat Balance Integral and its Application to Problems Involving a change of Phase", *Trans. Am. Soc. Mech. Eng.,* vol. 80, no. 2, pp. 335-342, 1958. [http://dx.doi.org/10.1115/1.4012364]

[8] J. Mennig, and M.N. Ozisik, "Coupled Integral Equation Approach for solving Melting or Solidification", *Int. J. Heat Mass Transf.,* vol. 28, no. 8, pp. 1481-1485, 1985. [http://dx.doi.org/10.1016/0017-9310(85)90250-9]

[9] S.G. Ahmed, "Numerical solution of ablation problem using modified constraint integral method", *J. Eng. Appl. Sci. (Asian Res. Publ. Netw.),* vol. 47, no. 3, 2000.

[10] E.J. Kansa, "Multiquadraic- a scattered data approximation scheme with applications to computational fluid dynamics.: Surface approximations and partial derivative estimates", *Comput. Math. Appl.,* vol. 19, no. 8-9, pp. 127-145, 1990. [http://dx.doi.org/10.1016/0898-1221(90)90270-T]

[11] E.J. Kansa, "Multiquadraic: A scattered data approximation scheme with applications to computation fluid-dynamics-II: solution to parabolic, hyperbolic and elliptic partial differential equations", *Comput. Math. Appl.,* vol. 19, no. 8-9, pp. 147-161, 1990. [http://dx.doi.org/10.1016/0898-1221(90)90271-K]

[12] R.L. Hardy, "Multiquadraic equations of topography and other irregular surfaces", *J. Geophys. Res.,* vol. 76, no. 8, pp. 1905-1915, 1971. [http://dx.doi.org/10.1029/JB076i008p01905]

[13] S.G.R. Brown, J.A. Spittle, D.J. Jarvis, and R. Walden-Bevan, "Numerical determination of liquid flow permeabilities for equiaxed dendritic structures", *Acta Mater.,* vol. 50, no. 6, pp. 1559-1569, 2002. [http://dx.doi.org/10.1016/S1359-6454(02)00014-9]

[14] Z Wu, and YC Hon, "Convergence error estimate in solving free boundary diffusion problem by radial basis functions method", *Eng Anal Bound Elements 27,* vol. 27, pp. 73-9, 2003. [http://dx.doi.org/10.1016/S0955-7997(02)00083-8]

[15] Z Wu, and YC Hon, *Convergence error estimate in solving free boundary diffusion problem by radial basis functions method.,* vol. 27, pp. 73-9, 20032003.*Eng Anal Bound Elements 27,* vol. 27, pp. 73-9,

20032003.
[http://dx.doi.org/10.1016/S0955-7997(02)00083-8]

[16] G.E. Fasshauer, "Solving partial differential equations by collocation with radial basis functions", In: *Surface Fitting and Multiresolution Methods.*, A. Le Mehaute, C. Rabut, L.L. Schumaker, Eds., Vanderbilt University Press: Nashville, 1997, pp. 131-138.

[17] Y. Wu, and C. Shu, "Development of RBF-DQ method for derivative approximation and its application to simulate natural convection in concentric annuli", *Comput. Mech.*, vol. 29, no. 6, pp. 477-485, 2002.
[http://dx.doi.org/10.1007/s00466-002-0357-4]

[18] A. Jardy, D. Ablitzer, B. Campin, and X. Robbe, "The use of a mathematical model to simulate a triple VAR processing route", *Proceedings of the 1999 Vacuum Metallurgy Conference,* 1999.

[19] K. Djidjeli, P.P. Chinchapatnam, P.B. Nair, and W.G. Price, "Global and compact meshless schemes for the unsteady convection-diffusion equation",

[20] X. Liu, G.R. Liu, K. Tai, and K.Y. Lam, "Radial Point Interpolation Collocation Method (RPICM) for Partial Differential Equations", *Comput. Math. Appl.*, vol. 50, no. 8, pp. 1425-1442, 2005.
[http://dx.doi.org/10.1016/j.camwa.2005.02.019]

[21] H. Wenland, "Piecewise polynomial, positive definite and compactly supported radial basis functions of minimal degree", *Adv. Comput. Math.*, p. 4, 1995.

[22] W. Shyy, Y. Pang, G.B. Hunter, D.Y. Wei, and M.H. Chen, "Modeling of turbulent transport and solidification during continuous ingot casting", *Int. J. Heat Mass Transf.*, vol. 35, no. 5, pp. 1229-1245, 1992.
[http://dx.doi.org/10.1016/0017-9310(92)90181-Q]

[23] G. Wright, and B. Fornberg, "scattered node compact finite difference type formulas generated from radial basis functions", *J. Comput. Phys.*, vol. 212, no. 1, pp. 99-123, 2006.
[http://dx.doi.org/10.1016/j.jcp.2005.05.030]

[24] C. A. Brebbia, "Dominguez", In: *Boundary Element Methods-An Introductory Course* 2nd addition. Comp. Mech. Publications and Mc Graw Hill: Southampton and Boston, 1982.

[25] P.J. Roache, *Computational Fluid Dynamics.* Hermosa Publishers: Albuquerque, NM, 1972.

[26] T.F. Russell, and M.F. Wheeler, "Finite Element and Finite Difference Methods for Continuous Flows in Porous Media", In: *The Mathematics of Reservoir Simulation, Frontiers in Applied Mathematics.*, R.E. Ewing, Ed., SIAM: Philadelphia, 1983.
[http://dx.doi.org/10.1137/1.9781611971071.ch2]

[27] William Ames, Numerical Methods for Partial Differential Equations, Academic Press, Inc, 199.

[28] B.J. Noye, and H.H. Tan, "Finite difference methods for the two-dimensional advection–diffusion equation", *Int. J. Numer. Methods Fluids,* vol. 9, no. 1, pp. 75-98, 1989.
[http://dx.doi.org/10.1002/fld.1650090107]

[29] D. Braess, *Finite Elements: Theory, Fast Solvers, and Applications in Solid Mechanics.* Cambridge University Press, 2001.

[30] N.L. Gibson, Solving Fluid-plate Interaction Problem Using Finite Element Methods with Domain Decomposition Strategies., 2001.

[31] C. Johnson, *Numerical solution of partial differential equations by the Finite element method.* Cambridge University Press New York, 1987

[32] C.T. Kelley, "Iterative Methods for Linear and Nonlinear Equations", *Society for Industrial & Applied Mathematics,* 199.

[33] C.A. Brebbia, *The Boundary element Method for Engineers.* Pentech Press, 1978.

[34] C.A. Brebbia, and S. Walker, *Boundary Element Techniques in Engineering.* Newnes-Butterworths,

1980.

[35] Alok Gray Sutradhar, J. Leonard, and Glaucio H. Paulino, *Symmetric Galerkin Boundary Element Method* Springer-Verlag Berlin Heidelberg Jahr: Verla, 2008.

[36] S. Mukherjee, and Y.X. Mukherjee, *Boundary Methods: Elements, Contours, and Nodes.* CRC Press: Boca Raton, FL, 2005.
[http://dx.doi.org/10.1201/9781420027860]

[37] S. Mukherjee, and Y.X. Mukherjee, "The hyper-singular boundary contour method for three-dimensional linear elasticity", *J. Appl. Mech.,* vol. 65, no. 2, pp. 300-309, 1998.
[http://dx.doi.org/10.1115/1.2789055]

[38] L.B. Lucy, "A numerical Approach to Testing the Fission Hypothesis", *Astron. J.,* vol. 82, no. 12, pp. 1013-1924, 1977.
[http://dx.doi.org/10.1086/112164]

[39] R. Gingold, and J. Monaghan, "Smoothed particle hydrodynamics: theory and application to non-spherical stars", *Mon. Not. R. Astron. Soc.,* vol. 181, no. 3, pp. 375-389, 1977.
[http://dx.doi.org/10.1093/mnras/181.3.375]

[40] J.J. Monaghan, "Why Particle Methods work", *SIAM Sci. Stat. Comp.,* vol. 3, no. 4, pp. 423-433, 1982.

[41] J.W. Swegle, D.L. Hicks, and S.W. Attaway, "Smoothed particle hydrodynamics stability analysis", *J. Comput. Phys.,* vol. 116, no. 1, pp. 123-134, 1995.
[http://dx.doi.org/10.1006/jcph.1995.1010]

[42] S.W. Attaway, M.W. Heinstein, and J.W. Swegle, "Coupling of smooth particle hydrodynamics with the finite element method", *Nucl. Eng. Des.,* vol. 150, no. 2-3, pp. 199-205, 1994.
[http://dx.doi.org/10.1016/0029-5493(94)90136-8]

[43] W. Liu, S. Jun, and Y. Zhang, "Reproducing kernel particle methods", *Int. J. Numer. Methods Fluids,* vol. 20, no. 8-9, pp. 1081-1106, 1995.
[http://dx.doi.org/10.1002/fld.1650200824]

[44] J. Bonet, and S. Kulasegaram, "Correction and stabilization of smooth particle hydrodynamics methods with application in metal forming simulation", *Int. J. Numer. Methods Eng.,* vol. 47, no. 6, pp. 1189-1214, 1999.
[http://dx.doi.org/10.1002/(SICI)1097-0207(20000228)47:6<1189::AID-NME830>3.0.CO;2-I]

[45] P. Lancaster, and K. Salkauskas, "Surfaces generated by moving least squares methods", *Math. Comput.,* vol. 37, no. 155, pp. 141-158, 1981.
[http://dx.doi.org/10.1090/S0025-5718-1981-0616367-1]

[46] D. Levin, "The approximation power of moving least-squares", *Math. Comput.,* vol. 67, no. 224, pp. 1517-1531, 1998.
[http://dx.doi.org/10.1090/S0025-5718-98-00974-0]

[47] S.G. Ahmed, "A numerical method based boundary integral equation to solve multiphase moving boundary problems", *British Journal of Mathematics & Computer Science,* vol. 15, no. 1, pp. 1-10, 2016.

[48] D.W. Tripp, and A. Mitchell, "Thermal regime in an EB hearth", In: *In Proceedings of the Conference on Electron Beam Melting and Refining – State of the Art 1985: Part Two* Bakish Materials Corp: Reno, Nevada, 1985.

[49] S. G. Ahmed, "An integral method to solve phase-change problems with/without mushy zones", *American Journal of Applied Mathematics,* vol. II, no. I, 2015.

[50] W. Liu, Y. Chen, S. Jun, J-S. Chen, T. Belytschko, C. Pan, R. Uras, and C. Chang, "Overview and application of the reproducing kernel particle methods", *Arch. Comput. Methods Eng.,* vol. 3, no. 1, pp. 3-80, 1996.
[http://dx.doi.org/10.1007/BF02736130]

[51] D. Grandemange, Y. Combres, B. Champin, A. Jardy, S. Hans, and D. Ablitzer, "The modeling of heat, momentum and solute transfers to optimize the melting parameters of VAR titanium ingots: an application to the "water" practice", *Proceedings of the 1997 Vacuum Metallurgy Conference,* 1997

[52] Pengbo Zhao, Renkun Dai, Wei Li, Qiuwang Wang, and Min Zeng, *Modeling the mushy zone during the melting process under Neumann boundary condition using the improved enthalpy-porosity method.,* vol. 78, no. 8, pp. 423-442, 2020. [http://dx.doi.org/10.1080/10407782.2020.1793540]

[53] R.N. Hills, D.E. Loper, P.H. Roberts, A thermodynamically consistent model of a mushy zone, The Quarterly Journal of Mechanics and Applied Mathematics, vol. 36, no. 4, pp. 505 540, 1983. [http://dx.doi.org/10.1093/qjmam/36.4.505]

[54] AS Burns, and LA Stickler, *Stewart Jr, solidification of an aqueous salt solution in a circular cylinder,* vol. 114, 1992.

[55] D.A. Tarzia, *Conference of Free and Moving Boundary Problems,* 1995 Bled, Slovenia.

[56] P.M. Pavel, "Thesis", *Bucharest,* 1999.

[57] H.E. Huppert, and M.G. Worster, "Dynamic solidification of a binary melt", *Nature,* vol. 314, no. 6013, pp. 703-707, 1986. [http://dx.doi.org/10.1038/314703a0]

[58] S.G. Ahmed, "An approximate method for evaporation problem with mushy zone", *Eng. Anal. Bound. Elem.,* vol. 29, no. 2, pp. 161-165, 2005. [http://dx.doi.org/10.1016/j.enganabound.2004.10.002]

[59] S.G. Ahmed, and M.L. Mekey, "A collocation and Cartesian grid methods using new radial basis function to solve class of partial differential equations", *Int. J. Comput. Math.,* pp. 1-14, 2009.

[60] I.V. Alexandrova, D.V. Alexandrov, D.L. Aseev, and S.V. Bulitcheva, "Mushy layer formation during solidification of binary alloys from a cooled wall: The role of boundary conditions", *Acta Phys. Pol. A,* vol. 115, no. 4, pp. 791-794, 2009. [http://dx.doi.org/10.12693/APhysPolA.115.791]

[61] M. Kunle, Entwicklung und untersuchung von moving least square verfahren zur numerischen simulation hydrodynamischer gleichungen., 2001.

[62] A. Waleed, *Mohamed, 'Advanced Mathematical Analysis for Phase Change Problems with and without Mushy Zones', PhD Thesis.* Zagazig University: Egypt, 2012.http://www.zu.edu.eg

[63] B. Nayroles, G. Touzot, and P. Villon, "Generalizing the finite element method: Diffuse approximation and diffuse elements", *Comput. Mech.,* vol. 10, no. 5, pp. 301-318, 1992. [http://dx.doi.org/10.1007/BF00364252]

[64] T. Belytschko, Y.Y. Lu, and L. Gu, "Element-free Galerkin methods", *Int. J. Numer. Methods Eng.,* vol. 37, no. 2, pp. 229-256, 1994. [http://dx.doi.org/10.1002/nme.1620370205]

[65] N. Sukumar, B. Moran, T. Black, and T. Belytschko, "An element-free Galerkin method for three-dimensional fracture mechanics", *Comput. Mech.,* vol. 20, no. 1-2, pp. 170-175, 1997. [http://dx.doi.org/10.1007/s004660050235]

[66] T. Belytschko, Y. Krongauz, D. Organ, M. Fleming, and P. Krysl, "Meshless methods: an overview and recent developments", *Comput. Methods Appl. Mech. Eng.,* vol. 139, no. 1-4, pp. 3-47, 1996. [http://dx.doi.org/10.1016/S0045-7825(96)01078-X]

[67] W. Liu, S. Jun, S. Li, J. Adee, and T. Belytschko, "Reproducing kernel particle methods for structural dynamics", *Int. J. Numer. Methods Eng.,* vol. 38, no. 10, pp. 1655-1679, 1995. [http://dx.doi.org/10.1002/nme.1620381005]

[68] *Using compactly supported radial basis functions to solve partial differential equations.,* 1999.www.witpress.com

[69] W. Liu, S. Li, and T. Belytschko, "Moving least square reproducing kernel methods: (i) Methodology and Convergence", *Comput. Methods Appl. Mech. Eng.,* 1996. [http://dx.doi.org/10.1016/S0045-7825(96)01082-1]

[70] J.M. Melenk, and I. Babuska, "The partition of unity finite element method: basic theory and applications", *Comput. Methods Appl. Mech. Eng.,* vol. 139, no. 1-4, pp. 289-314, 1996. [http://dx.doi.org/10.1016/S0045-7825(96)01087-0]

[71] J. Dolbow, N. Moes, and T. Belytschko, "Discontinuous enrichment in finite elements with a partition of unity method", *Finite Elem. Anal. Des.,* vol. 36, no. 3-4, pp. 235-160, 2000. [http://dx.doi.org/10.1016/S0168-874X(00)00035-4]

[72] I. Babuška, and J.M. Melenk, "The partition of unity method", *Int. J. Numer. Methods Eng.,* vol. 40, no. 4, pp. 727-758, 1997. [http://dx.doi.org/10.1002/(SICI)1097-0207(19970228)40:4<727::AID-NME86>3.0.CO;2-N]

[73] C.A. Duarte, and J.T. Oden, "An hp- adaptive method using clouds", *Comput. Methods Appl. Mech. Eng.,* vol. 139, no. 1-4, pp. 237-262, 1996. [http://dx.doi.org/10.1016/S0045-7825(96)01085-7]

[74] C.A. Duarte, J.T. Oden, and O.C. Zienkiewicz, "A new cloud based hp finite element method", *Comput. Methods Appl. Mech. Eng.,* vol. 153, no. 1-2, pp. 117-126, 1998. [http://dx.doi.org/10.1016/S0045-7825(97)00039-X]

[75] G.R. Liu, and Y.T. Gu, "A truly meshless method based on the strong-weak form, in Advances in Meshfree and X-FEM Methods", *Proceeding of the 1st Asian Workshop in Meshfree Methods,* 2002pp. 259-261

[76] T.P. Fries, and T. Belytschko, "The intrinsic XFEM: A method for arbitrary discontinuities without additional unknowns", *Int. J. Numer. Methods Eng.,* vol. 68, no. 13, pp. 1358-1385, 2006. [http://dx.doi.org/10.1002/nme.1761]

[77] T. Zhu, J. Zhang, and S. Atluri, "A local boundary integral equation (LBIE) method in computational mechanics and a meshless discretization approach", *Comput. Mech.,* vol. 21, no. 3, pp. 223-235, 1998. [http://dx.doi.org/10.1007/s004660050297]

[78] E.J. Sellountos, and D. Polyzos, "A meshless local boundary integral equation method for solving transient electrodynamic problems", *Comput. Mech.,* vol. 35, no. 4, pp. 265-276, 2004. [http://dx.doi.org/10.1007/s00466-004-0584-y]

[79] S.N. Atluri, and T.L. Zhu, "A new meshless local Petrov–Galerkin (MLPG) approach in computational mechanics", *Comput Mech,* vol. 22, no. 2, pp. 117-127, 1998.

[80] S. Long, and S.N. Atluri, "A meshless local Petrov–Galerkin (MLPG) formulation for solving the bending problem of a thin plate", *CMES: Comput Model Eng Sci,* vol. 3, no. 1, pp. 53-63, 2002.

[81] G.R. Liu, and Y.T. Gu, "Meshless local Petrov–Galerkin (MLPG) method in combination with finite element and boundary element approaches", *Comput. Mech.,* vol. 26, no. 6, pp. 536-546, 2000. [http://dx.doi.org/10.1007/s004660000203]

[82] H. Liu, and S.N. Atluri, "The meshless local Petrov–Galerkin (MLPG) method for solving convection-diffusion problem", *CMES: Comput Model Eng Sci,* vol. 1, no. 2, pp. 45-60, 2000.

[83] H. Liu, and S.N. Atluri, "The meshless local Petrov–Galerkin (MLPG) method for solving incompressible Navier–Stokes equations", *CMES: Comput Model Eng Sci,* vol. 2, no. 2, pp. 117-142, 2000.

[84] P. Jensen, "Finite difference techniques for variable grids", *Comput. Struc.,* vol. 2, no. 1-2, pp. 17-29, 1972. [http://dx.doi.org/10.1016/0045-7949(72)90020-X]

[85] N. Perrone, and R. Kao, "A general finite difference method for arbitrary meshes", *Comput. Struc.,* vol. 5, no. 1, pp. 45-58, 1975.

[http://dx.doi.org/10.1016/0045-7949(75)90018-8]

[86] J. Bonet, and S. Kulasegaram, "Correction and stabilization of smooth particle hydrodynamics methods with applications in metal forming simulation", *Int. J. Numer. Methods Eng.*, vol. 47, no. 6, pp. 1189-1214, 1999.
[http://dx.doi.org/10.1002/(SICI)1097-0207(20000228)47:6<1189::AID-NME830>3.0.CO;2-I]

[87] T. Liszka, "An interpolation method for an irregular net of nodes", *Int. J. Numer. Methods Eng.*, vol. 20, no. 9, pp. 1599-1612, 1984.
[http://dx.doi.org/10.1002/nme.1620200905]

[88] W. Schonauer, "Generation of difference and error formulae of arbitrary consistency order on an unstructured grid", *Z. Angew. Math. Mech.*, vol. 78, no. S3, pp. S1061-S1062, 1998.
[http://dx.doi.org/10.1002/zamm.19980781599]

[89] E. Onate, S. Idelsohn, O.C. Zienkiewicz, R.L. Taylor, and C. Sacco, "A stabilized finite point method for analysis of fluid mechanics problems", *Comput. Methods Appl. Mech. Eng.*, vol. 139, no. 1-4, pp. 315-346, 1996. a
[http://dx.doi.org/10.1016/S0045-7825(96)01088-2]

[90] E. Onate, S. Idelsohn, O.C. Zienkiewicz, and R.L. Taylor, "A finite point method in computational mechanics – applications to convective transport and fluid flow", *Int. J. Numer. Methods Eng.*, vol. 139, no. 22, pp. 3839-3866, 1996. b
[http://dx.doi.org/10.1002/(SICI)1097-0207(19961130)39:22<3839::AID-NME27>3.0.CO;2-R]

[91] E. Onate, C. Sacco, and S. Idelsohn, "A finite point method for incompressible flow problems", *Comput. Vis. Sci.*, vol. 2, no. 1-2, pp. 67-75, 2000.
[http://dx.doi.org/10.1007/s007910050053]

[92] E. Onate, and S. Idelsohn, "A mesh free finite point method for Advective diffusive transport and fluid flow problems", *Comput. Mech.*, vol. 21, no. 4-5, pp. 283-292, 1998.
[http://dx.doi.org/10.1007/s004660050304]

[93] E. Onate, F. Perazzo, and J. Miquel, "A finite point method for elasticity problems", *Comput. Struc.*, vol. 79, no. 22-25, pp. 2151-2163, 2001.
[http://dx.doi.org/10.1016/S0045-7949(01)00067-0]

[94] "Kee, B. A regularized strong-form Meshfree method for adaptive analysis. In Computational mechanics", In: *Computational Methods in Applied Sciences.*, C. Mota Soares, J. Martins, H. Rodrigues, J.e. Ambrósio, Eds., vol. Vol. 6. Springer, 2006, pp. 393-412.

[95] G.R. Liu, B.B.T. Kee, and L. Chun, "A Stabilized Least-Squares Radial Point Collocation Method (LS-RPCM)for Adaptive Analysis", *Comput. Methods Appl. Mech. Eng.*, vol. 195, no. 37-40, pp. 4843-4861, 2006.
[http://dx.doi.org/10.1016/j.cma.2005.11.015]

[96] G.R. Liu, J. Zhang, H. Li, K.Y. Lam, and B.B.T. Kee, "LI, H., Lam, K., AND Kee, B. Radial point interpolation based finite difference method for mechanics problems", *Int. J. Numer. Methods Eng.*, vol. 68, no. 7, pp. 728-754, 2006.
[http://dx.doi.org/10.1002/nme.1733]

[97] D. Wan Kim, and A.N.D. Liu, "W. K., Maximum principle and convergence analysis for the Meshfree point collocation method", *SIAM J. Numer. Anal.*, vol. 44, no. 2, pp. 515-539, 2006.
[http://dx.doi.org/10.1137/04060809X]

[98] E.J. Kansa, "Multiquadrics- a scattered data approximation scheme with applications to computational fluid dynamics.: Surface approximations and partial derivative estimates", *Comput. Math. Appl.*, vol. 19, no. 8-9, pp. 127-145, 1990.
[http://dx.doi.org/10.1016/0898-1221(90)90270-T]

[99] E.J. Kansa, "Multiquadrics: A scattered data approximation scheme with applications to computation fluid-dynamics-II: solution to parabolic, hyperbolic and elliptic partial differential equations", *Comput.*

Math. Appl., vol. 19, no. 8-9, pp. 147-161, 1990.
[http://dx.doi.org/10.1016/0898-1221(90)90271-K]

[100] R.L. Hardy, "Multiquadrics equations of topography and other irregular surfaces", *J. Geophys. Res.,* vol. 76, no. 8, pp. 1905-1915, 1971.
[http://dx.doi.org/10.1029/JB076i008p01905]

[101] S.L. Semiatin, V.G. Ivanchenko, S.V. Akhonin, and O.M. Ivasishin, "Diffusion models for evaporation losses during electron-beam melting of alpha/beta-titanium alloys", *Metall. Mater. Trans., B, Process Metall. Mater. Proc. Sci.,* vol. 35B, no. 2, pp. 235-245, 2004.
[http://dx.doi.org/10.1007/s11663-004-0025-5]

[102] Z Wu, and YC Hon, "Convergence error estimate in solving free boundary diffusion problem by radial basis functions method", *Eng Anal Bound Elements,* vol. 27, pp. 73-9, 2003.
[http://dx.doi.org/10.1016/S0955-7997(02)00083-8]

[103] Z Wu, and YC Hon, "Convergence error estimate in solving free boundary diffusion problem by radial basis functions method", *Eng Anal Bound Elements,* vol. 27, pp. 73-9, 20032003.
[http://dx.doi.org/10.1016/S0955-7997(02)00083-8]

[104] G.E. Fasshauer, "Solving partial differential equations by collocation with radial basis functions", In: *Surface Fitting and Multiresolution Methods.,* A. Le Mehaute, C. Rabut, L.L. Schumaker, Eds., Vanderbilt University Press: Nashville, 1997, pp. 131-138.

[105] W. Shi, W. Shyy, and R. Mei, Numer. Heat Transfer, Part B 40,

[106] H. Nakamura, and A. Mitchell, "Effect of beam oscillation rate on Al evaporation from a Ti-6Al-4V alloy in the electron beam melting process", *ISIJ Int.,* vol. 32, no. 5, pp. 583-592, 1992.
[http://dx.doi.org/10.2355/isijinternational.32.583]

[107] C-Y. Ku, J-E. Xiao, and C-Y. Liu, "A Novel Meshfree Approach with a Radial Polynomial for Solving Nonhomogeneous Partial Differential Equations", *Mathematics,* vol. 8, no. 2, p. 270, 2020.
[http://dx.doi.org/10.3390/math8020270]

[108] C-Y. Ku, J-E. Xiao, and C-Y. Liu, "Space–Time Radial Basis Function–Based Meshless Approach for Solving Convection–Diffusion Equations", *Mathematics,* vol. 8, no. 10, p. 1735, 2020.
[http://dx.doi.org/10.3390/math8101735]

[109] M. Ritchie, S.L. Cockroft, P.D. Lee, A. Mitchell, and T. Wang, "X-ray based measurement of composition during electron beam melting of AISI 316 stainless steel: Part ii. Evaporative processes and simulation", *Metall. Mater. Trans., A Phys. Metall. Mater. Sci.,* vol. 34A, no. 13, pp. 863-877, 2003.
[http://dx.doi.org/10.1007/s11661-003-1013-0]

[110] A. Powell, U. Pal, J. van den Avyle, B. Damkroger, and J. Szekely, "Analysis of multicomponent evaporation in electron beam melting and refining of titanium alloys", *Metall. Mater. Trans., B, Process Metall. Mater. Proc. Sci.,* vol. 28B, no. 6, pp. 1227-1234, 1997.
[http://dx.doi.org/10.1007/s11663-997-0078-3]

[111] S.A. Bawazeer, S.S. Baakeem, and A. Mohamad, "A New Radial Basis Function Approach Based on Hermite Expansion with Respect to the Shape Parameter", *Mathematics,* vol. 7, no. 10, p. 979, 2019.
[http://dx.doi.org/10.3390/math7100979]

[112] G. Fasshaur, "Approximate moving least squares approximation for time-dependent PDEs", *In Fifth World Congress on Computational Mechanics (WCCM V),* 2002.

[113] G.E. Fasshaur, "Matrix-free multilevel moving least-squares methods", In Approximation Theory X (St. Louis, MO, 2001), C. K. Chui, L. L. Schumaker, and J. Stockler (eds.), Innov. Appl. Math., Vanderbilt Univ. Press, Nashville, TN, USA.

CHAPTER 2

Boundary Elements for Potential Problems

Abstract: The present chapter focuses on the elementary development and features of one of the most important numerical techniques that were started by Brebbia CA in 1970. The chapter contains the basic strategy and development of these methods and techniques. The chapter focuses mainly on the boundary element methods and its potential problems.

Keywords: Phase change problem, boundary element methods.

INTRODUCTION

Since 1970, boundary element methods have started solving a wide range of practical, engineering, and industrial problems. Simply, the boundary element methods transfer the governing equation and its associated boundary conditions to a boundary integral equation. This equation is valid for the boundary in addition to the internal points of the domain. The basic idea to derive the integral equation is to minimize the error and deal only with the boundaries. There are two directions when deriving the integral equation, the *direct* and *indirect* approaches. In the present thesis, the first approach will be used for the derivation of the integral equation. The direct approach of formulation consists of transforming the partial differential equation describing the behavior of the unknown inside and on the boundary of the domain into an integral equation relating only boundary values and then finding out the numerical solution of the integral equation [1-20].

1. Basic Integral equation based on:

(a) Weighted residuals

(b) Betti's reciprocal theorem

(c) Green's third identity

(d) Fundamental principles

Consider Laplace's equation two or three-dimensional case; see Fig. **(1)**.

S. G. Ahmed

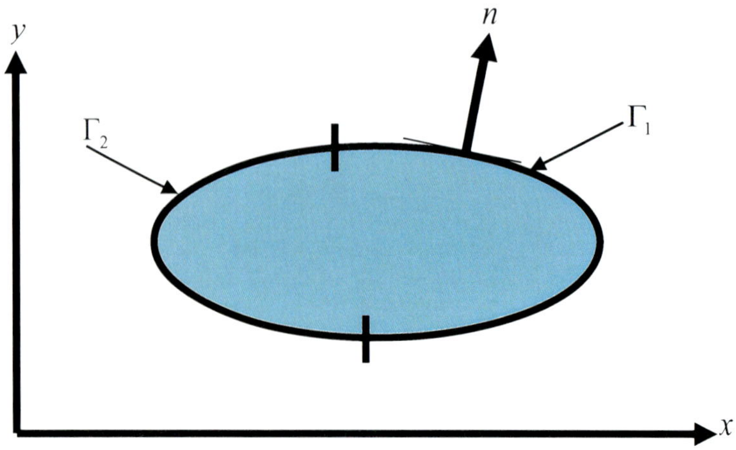

Fig. (1). Problem configuration for 2-D potential problem.

Governing Equation

$$\nabla^2 u = 0 \tag{1}$$

Essential boundary condition

$$u = \bar{u} \text{ over } \Gamma_1 \tag{2}$$

Natural boundary condition

$$q = \bar{q} \text{ over } \Gamma_2 \tag{3}$$

Note that:

$\Gamma_1 + \Gamma_2 = \Gamma$, represent the boundary of the domain of the interest.

The domain may be subjected to more complex boundary conditions, called

Robin – Mixed – boundary condition

$$\alpha u + \beta q = \gamma \tag{4}$$

In equation (4), the parameters α, β, γ are known.

As long as u is not the exact solution, this means that there exist errors, therefore, re-write the Laplace's equation and its associated boundary conditions as follow:

$$R = \nabla^2 u \neq 0 \tag{5}$$

$$R_1 = u - \bar{u} \neq 0 \tag{6}$$

$$R_2 = q - \bar{q} \neq 0 \tag{7}$$

In equations (5-7), both u and q are approximate values. Weighting the governing equation as follow:

$$\int_{\Omega} R u^* d\Omega = \int_{\Gamma_2} R_2 u^* d\Gamma - \int_{\Gamma_2} R_1 q^* d\Gamma \tag{8}$$

Substituting by equations (5-7) into equation (8) leads to:

$$\int_{\Omega} \left(\nabla^2 u\right) u^* d\Omega = \int_{\Gamma_2} \left(q - \bar{q} \right) u^* d\Gamma - \int_{\Gamma_2} \left(u - \bar{u} \right) q^* d\Gamma \tag{9}$$

Integrating by parts the L.H.S. of equation (9), one can get:

$$\int_{\Omega} \left(\nabla^2 u\right) u^* d\Omega = \left[u^* \frac{\partial u}{\partial n} \right]_{\Omega} - \int_{\Omega} \left(\frac{\partial u}{\partial n} \frac{\partial u^*}{\partial n} \right) d\Omega \tag{10}$$

Note the following:

$$\left[u^* \frac{\partial u}{\partial n} \right]_{\Omega} \equiv \int_{\Gamma} \left(u^* \frac{\partial u}{\partial n} \right) d\Gamma = \int_{\Gamma_2} \left(u^* q \right) d\Gamma \tag{11}$$

Then;

$$\int_{\Gamma_2}\left(u^* q\right)d\Gamma - \int_{\Omega}\left(\frac{\partial u}{\partial n}\frac{\partial u^*}{\partial n}\right)d\Omega = \int_{\Gamma_2}\left(q-\bar{q}\right)u^* d\Gamma - \int_{\Gamma_2}\left(u-\bar{u}\right)q^* d\Gamma \tag{12}$$

Re-arrange equation (12) leads to:

$$\begin{aligned}&-\int_{\Omega}\left(\frac{\partial u}{\partial n}\frac{\partial u^*}{\partial n}\right)d\Omega = \int_{\Gamma_2}\left(q-\bar{q}\right)u^* d\Gamma - \int_{\Gamma_2}\left(u-\bar{u}\right)q^* d\Gamma - \int_{\Gamma_2}u^* q d\Gamma\\ &\Rightarrow\\ &-\int_{\Omega}\left(\frac{\partial u}{\partial n}\frac{\partial u^*}{\partial n}\right)d\Omega = -\int_{\Gamma_2}\bar{q}u^* d\Gamma - \int_{\Gamma_2}uq^* d\Gamma + \int_{\Gamma_2}\bar{u}q^* d\Gamma\end{aligned} \tag{13}$$

The L.H.S. of equation (13) can be evaluated as follows:

$$-\int_{\Omega}\left(\frac{\partial u}{\partial n}\frac{\partial u^*}{\partial n}\right)d\Omega = -\left(u\frac{\partial u^*}{\partial n}\right)_{\Omega} + \int_{\Omega}\left(u\frac{\partial^2 u^*}{\partial n^2}\right)d\Omega \tag{14}$$

But the first term in the R.H.S. of equation (14) is equivalent to the integration of a part of the domain, *i.e.*,

$$-\int_{\Omega}\left(\frac{\partial u}{\partial n}\frac{\partial u^*}{\partial n}\right)d\Omega = -\int_{\Gamma_1}\bar{u}q^* d\Gamma + \int_{\Omega}\left(u\frac{\partial^2 u^*}{\partial n^2}\right)d\Omega \tag{15}$$

Equating the R.H.S. of equations (13) and (15), one can get:

$$-\int_{\Gamma_2}\bar{q}u^* d\Gamma - \int_{\Gamma_2}uq^* d\Gamma + \int_{\Gamma_2}\bar{u}q^* d\Gamma = -\int_{\Gamma_1}\bar{u}q^* d\Gamma + \int_{\Omega}\left(u\frac{\partial^2 u^*}{\partial n^2}\right)d\Omega \tag{16}$$

Re-arrange equation (16) leads to:

$$\int_{\Omega}\left(u\frac{\partial^2 u^*}{\partial n^2}\right)d\Omega = -\int_{\Gamma_2}u^* q d\Gamma - \int_{\Gamma_2}u^*\bar{q}\, d\Gamma + \int_{\Gamma_2}q^* u d\Gamma + \int_{\Gamma_2}q^*\bar{u}\, d\Gamma \tag{17}$$

Equation (17) is the starting point for the application of the boundary element method.

Fundamental Solution

First of all, let's ask our-self: what does it mean by the fundamental solution? The answer is quite simple. It represents the field generated by a concentrated unit charge at a point "i". The effect of this charge is propagated from that point to infinity without any consideration of boundary conditions. For Laplace's equation, its fundamental solution u^*, see Fig. **(2)**, is defined as [21-29]:

$$\nabla^2 u^* + \Delta^i = 0 \qquad \textbf{(18)}$$

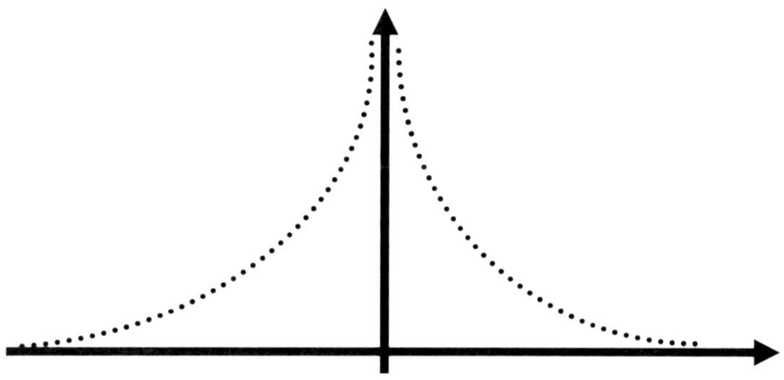

Fig. (2). Dirac Delta Function.

In equation (18)

$$\Delta^i = \text{Dirac Delta Function} = \begin{cases} \infty & \text{at } x = x_i \\ 0 & \text{elesewhere} \end{cases} \qquad \textbf{(19)}$$

The solution of equation (19) is as follows:

- For 2-D problems

$$u^{*} = \frac{1}{2\pi}\ln\left(\frac{1}{r}\right) \qquad \textbf{(20-a)}$$

- For 3-D problems

$$u^{*} = \frac{1}{4\pi r} \qquad \textbf{(20-b)}$$

In equations (20), r represents the distance between the source and field points. An important property for Dirac delta function is that the integral of Δ^{i} multiplied by any other function is equal to the value of the latter at that point, mathematically:

$$\int_{\Omega} u\left(\nabla^{2}u^{*}\right)d\Omega = \int_{\Omega} u\left(-\Delta^{i}\right)d\Omega = -u^{i} \qquad \textbf{(21)}$$

Substituting from equation (21) into the left hand side of equation (17), then the later one takes the following form:

$$-u^{i} = -\int_{\Gamma_2} u^{*}q\,d\Gamma - \int_{\Gamma_2} u^{*}\bar{q}\,d\Gamma + \int_{\Gamma_2} q^{*}u\,d\Gamma + \int_{\Gamma_2} q^{*}\bar{u}\,d\Gamma \qquad \textbf{(22)}$$

Note:

The first and the last terms of the R.H.S. of equation (22) are known, while the other two terms are unknown. By re-arranging equation (22), one can get the following form:

$$c^{i}u^{i} + \int_{\Gamma_2} q^{*}u\,d\Gamma + \int_{\Gamma_2} q^{*}\bar{u}\,d\Gamma = \int_{\Gamma_2} u^{*}q\,d\Gamma + \int_{\Gamma_2} u^{*}\bar{q}\,d\Gamma \qquad \textbf{(23)}$$

In equation (23), the parameter c^i depends mainly on the point under consideration; it usually varies between three general values according to the following relation:

$$c_i = \frac{\alpha}{2\pi} = \begin{cases} 1 & i \in \Omega \\ \frac{1}{2} & i \in \Gamma \\ 0 & i \in elsewhere \end{cases} \tag{24}$$

In equation (24) α is the angle between two adjacent elements.

For smooth boundary, equation (23) takes the following general form:

$$c^i u^i + \int_{\Gamma_2} q^* u d\Gamma + \int_{\Gamma_2} q^* \bar{u} d\Gamma = \int_{\Gamma_2} u^* q d\Gamma + \int_{\Gamma_2} u^* \bar{q} d\Gamma \tag{25}$$

TYPES OF BOUNDARY ELEMENTS

Constant Element

In the case of the constant element, the boundary is divided into N elements in such a way that both potential and potential derivatives are assumed to be constant over each element, and each element has only one node centered at the midpoint of the element, see Fig. (**3**).

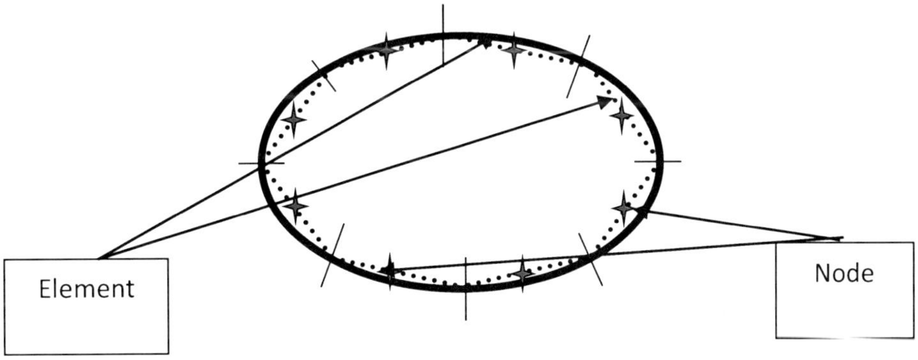

Fig. (3). Constant Element.

Linear Element

In the case of linear elements, the boundary is divided into N elements in such a way that both potential and potential derivatives are assumed to vary linearly over each element, and each element has two nodes at the end points of the element, see Fig. **(4)**.

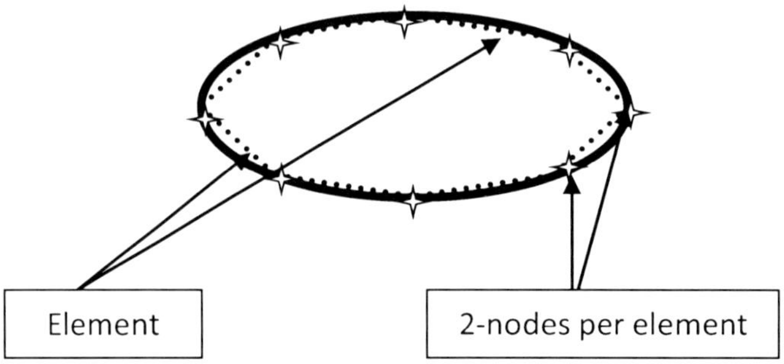

Fig. (4). Linear Element.

Quadratic Element

In the case of a quadratic element, the boundary is divided into N elements in such a way that both potential and potential derivatives are assumed to vary in a quadratic behavior over each element, and each element has three nodes two at the end points of the element and one at the midpoint of the element, see Fig. **(5)**.

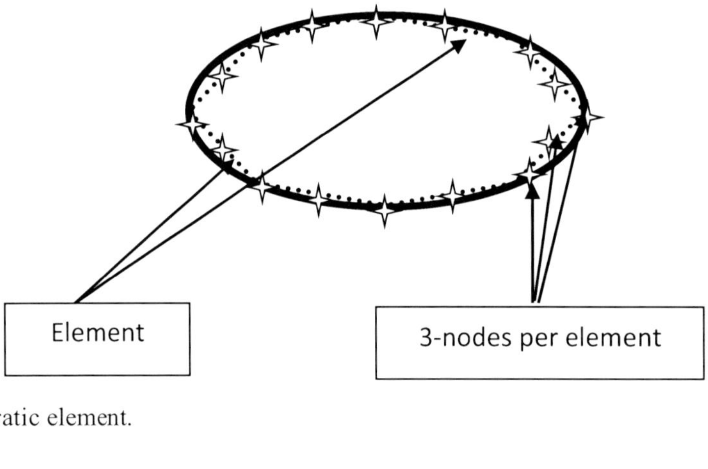

Fig. (5). Quadratic element.

DISCRETIZATION OF THE BOUNDARY INTEGRAL EQUATION

The boundary integral equation (25) is discretized to N elements, whatever its type, constant, linear or quadratic according to the nature of the problem under investigation. For the purpose of understanding the boundary element method, let us start discretization using constant elements. Therefore, equation (25) takes the following form:

$$c^i u^i + \sum_{j=1}^{N} \int_{\Gamma_2} q^* u d\Gamma + \sum_{j=1}^{N} \int_{\Gamma_2} q^* \bar{u} d\Gamma = \sum_{j=1}^{N} \int_{\Gamma_2} u^* q d\Gamma + \sum_{j=1}^{N} \int_{\Gamma_2} u^* \bar{q} d\Gamma \tag{26}$$

As mentioned before, in case of using constant element, u and q are taken constants over the element and defined at the mid-point of the element at a point called the node of the element. Therefore, equation (26) takes the form:

$$c^i u^i + \sum_{j=1}^{N} \left(\int_{\Gamma_2} q^* d\Gamma \right) u^j + \sum_{j=1}^{N} \left(\int_{\Gamma_2} q^* \bar{u} d\Gamma \right) = \sum_{j=1}^{N} \left(\int_{\Gamma_2} u^* d\Gamma \right) q^j + \sum_{j=1}^{N} \left(\int_{\Gamma_2} u^* d\Gamma \right) \bar{q}^j \tag{27}$$

Equation (28) in a general form is as follows:

$$c^i u^i + \sum_{j=1}^{N} \left(\int_{\Gamma_j} q^* d\Gamma \right) u^j = \sum_{j=1}^{N} \left(\int_{\Gamma_j} u^* d\Gamma \right) q^j \tag{28}$$

$$c^i u^i = \sum_{j=1}^{N} G^{ij} q^j - \sum_{j=1}^{N} H^{ij} u^j \tag{29}$$

Equation (28) in some details as follows:

Source point at $\xi = 1$

$$\begin{aligned} c^1 u^1 &= G^{11} q^1 + G^{12} q^2 + G^{13} q^3 + \dots\dots\dots + G^{1N} q^N \\ &\quad - \overline{H}^{11} u^1 - \overline{H}^{12} u^2 - \overline{H}^{13} u^3 - \dots\dots\dots - \overline{H}^{1N} u^N \end{aligned} \tag{30}$$

Source point at $\xi = 2$

$$c^2u^2 = G^{21}q^1 + G^{22}q^2 + G^{23}q^3 + \dots\dots\dots + G^{2N}q^N - \overline{H}^{21}u^1 - \overline{H}^{22}u^2 - \overline{H}^{23}u^3 - \dots\dots\dots - \overline{H}^{2N}u^N \tag{31}$$

Source point at $\xi = N$

$$c^Nu^N = G^{N1}q^1 + G^{N2}q^2 + G^{N3}q^3 + \dots\dots + G^{NN}q^N - \overline{H}^{N1}u^1 - \overline{H}^{N2}u^2 - \overline{H}^{N3}u^3 - \dots\dots - \overline{H}^{NN}u^N \tag{32}$$

Therefore, one obtains the following system in a matrix form:

$$\begin{bmatrix} c^1 & c^2 & c^3 & \dots\dots\dots c^N \end{bmatrix}_{(1\times N)} \begin{bmatrix} u^1 \\ u^2 \\ u^3 \\ \dots\dots \\ u^N \end{bmatrix}_{(N\times 1)} = \begin{bmatrix} G^{11} & G^{12} & G^{13} & \dots\dots G^{1N} \\ G^{21} & G^{22} & G^{23} & \dots\dots G^{2N} \\ G^{31} & G^{32} & G^{33} & \dots\dots G^{3N} \\ \dots\dots & & & \\ G^{N1} & G^{N2} & G^{N3} & \dots\dots G^{NN} \end{bmatrix}_{(N\times N)} \times$$

$$\times \begin{bmatrix} q^1 \\ q^2 \\ q^3 \\ \dots\dots \\ q^N \end{bmatrix}_{(1\times N)} - \begin{bmatrix} \overline{H}^{11} & \overline{H}^{12} & \overline{H}^{13} & \dots\dots \overline{H}^{1N} \\ \overline{H}^{21} & \overline{H}^{22} & \overline{H}^{23} & \dots\dots \overline{H}^{2N} \\ \overline{H}^{31} & \overline{H}^{32} & \overline{H}^{33} & \dots\dots \overline{H}^{3N} \\ \dots\dots & & & \\ \overline{H}^{N1} & \overline{H}^{N2} & \overline{H}^{N3} & \dots\dots \overline{H}^{NN} \end{bmatrix}_{(N\times N)} \begin{bmatrix} u^1 \\ u^2 \\ u^3 \\ \dots\dots \\ u^N \end{bmatrix}_{(N\times 1)} \tag{33}$$

Re-arrange equation (33) to take the following form:

$$\begin{bmatrix} c^1 & c^2 & c^3 & .....c^N \end{bmatrix}_{(1\times N)} \begin{bmatrix} u^1 \\ u^2 \\ u^3 \\ ..... \\ u^N \end{bmatrix}_{(N\times 1)} + \begin{bmatrix} \overline{H}^{11} & \overline{H}^{12} & \overline{H}^{13} & ....\overline{H}^{1N} \\ \overline{H}^{21} & \overline{H}^{22} & \overline{H}^{23} & ....\overline{H}^{2N} \\ \overline{H}^{31} & \overline{H}^{32} & \overline{H}^{33} & ....\overline{H}^{3N} \\ .... \\ \overline{H}^{N1} & \overline{H}^{N2} & \overline{H}^{N3} & ....\overline{H}^{NN} \end{bmatrix}_{(N\times N)} \begin{bmatrix} u^1 \\ u^2 \\ u^3 \\ ..... \\ u^N \end{bmatrix}_{(N\times 1)} =$$

$$= \begin{bmatrix} G^{11} & G^{12} & G^{13} & ....G^{1N} \\ G^{21} & G^{22} & G^{23} & ....G^{2N} \\ G^{31} & G^{32} & G^{33} & ....G^{3N} \\ .... \\ G^{N1} & G^{N2} & G^{N3} & ....G^{NN} \end{bmatrix}_{(N\times N)} \begin{bmatrix} q^1 \\ q^2 \\ q^3 \\ ..... \\ q^N \end{bmatrix}_{(1\times N)} \quad \textbf{(34)}$$

The L.H.S. of equation (34) can be treated mathematically such that it can take the following form:

$$\begin{bmatrix} \underbrace{H^{11}+c^1} & H^{12} & H^{13} & ....H^{1N} \\ H^{21} & \underbrace{H^{22}+c^2} & H^{23} & ....H^{2N} \\ H^{31} & H^{32} & \underbrace{H^{33}+c^3} & ....H^{3N} \\ .... \\ H^{N1} & H^{N2} & H^{N3} & ....\underbrace{H^{NN}+c^N} \end{bmatrix}_{(N\times N)} \begin{bmatrix} u^1 \\ u^2 \\ u^3 \\ ..... \\ u^N \end{bmatrix}_{(N\times 1)}$$

$$= \begin{bmatrix} G^{11} & G^{12} & G^{13} & ....G^{1N} \\ G^{21} & G^{22} & G^{23} & ....G^{2N} \\ G^{31} & G^{32} & G^{33} & ....G^{3N} \\ .... \\ G^{N1} & G^{N2} & G^{N3} & ....G^{NN} \end{bmatrix}_{(N\times N)} \begin{bmatrix} q^1 \\ q^2 \\ q^3 \\ ..... \\ q^N \end{bmatrix}_{(1\times N)} \quad \textbf{(35)}$$

Influence Matrices

In equation (29), there are two types of general integrals, over all elements on the boundary they form what is called influence matrices. These matrices take the following forms [30-41]:

$$H^{ij} = \int_{\Gamma_j} q^* d\Gamma \tag{36}$$

$$G^{ij} = \int_{\Gamma_j} u^* d\Gamma \tag{37}$$

In equations (36) and (37)

i: Boundary node number

j: Boundary element number

Taking equations (36) and (37) into equation (29), the later takes the form:

$$H^{ij} u^j = G^{ij} q^j \tag{38}$$

Consider "i" varies from "1" to "N"; therefore, the matrix H^{ij} will take two different values according to the relation between "i" and "j", as follows:

$$H^{ij} = \begin{cases} \overline{H}^{ij} & i \neq j \\ \overline{H}^{ij} + \dfrac{1}{2} & i = j \end{cases} \tag{39}$$

Notes:

(1) When $i \neq j$, this type of integral can be evaluated using Gauss quadrature rule.

(2) When $i = j$, this type of integrals will have singularities, and so treatment of such integrals requires special treatment.

Elementary Example

To understand the main steps of boundary element, let us consider a square of 6 units' length a side. The vertical sides are subject to potential derivative boundary conditions, while the other horizontal sides are subject to potential boundary conditions. The problem configuration is shown in Fig. **(6)**. Follow up the procedure described before till we get the discretized system in matrix notations as follow:

$$H_{(N\times N)}U_{(N\times 1)} = G_{(N\times N)}Q_{(N\times 1)} \tag{a}$$

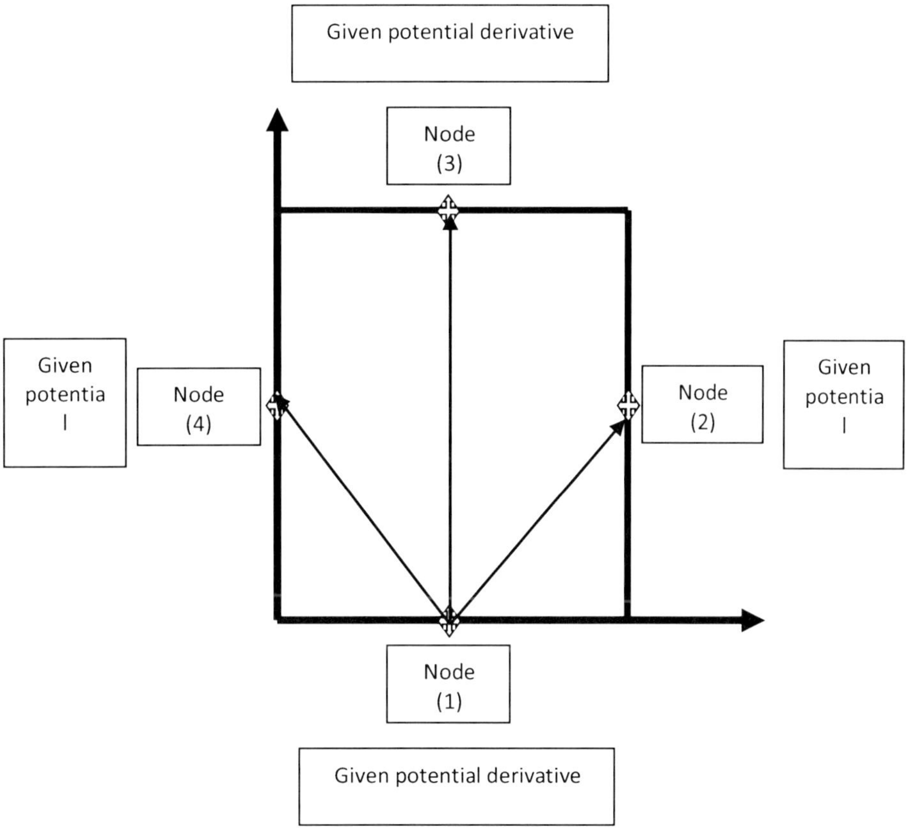

Fig. (6). Nodes distribution for constant element.

Procedure

(1) Stand at node (1) over elements 1-4 and goes anti-clockwise direction, mathematically, $i=1$ & $j=1 \to 4$, we get:

$$i=1 \text{ \& } j=1 \Rightarrow H^{11}U^{1} = G^{11}Q^{1} \quad \textbf{(b)}$$

$$i=1 \text{ \& } j=2 \Rightarrow H^{12}U^{2} = G^{12}Q^{2} \quad \textbf{(c)}$$

$$i=1 \text{ \& } j=3 \Rightarrow H^{13}U^{3} = G^{13}Q^{3} \quad \textbf{(d)}$$

$$i=1 \text{ \& } j=4 \Rightarrow H^{14}U^{4} = G^{14}Q^{4} \quad \textbf{(e)}$$

(2) Stand at node (2) over elements 1-4 and goes anti-clockwise direction, mathematically, $i=2$ & $j=1 \to 4$, we get:

$$i=2 \text{ \& } j=1 \Rightarrow H^{21}U^{1} = G^{21}Q^{1} \quad \textbf{(f)}$$

$$i=2 \text{ \& } j=2 \Rightarrow H^{22}U^{2} = G^{22}Q^{2} \quad \textbf{(g)}$$

$$i=2 \text{ \& } j=3 \Rightarrow H^{23}U^{3} = G^{23}Q^{3} \quad \textbf{(h)}$$

$$i=2 \text{ \& } j=4 \Rightarrow H^{24}U^{4} = G^{24}Q^{4} \quad \textbf{(i)}$$

(3) Stand at node (3) over elements 1-4 and goes anti-clockwise direction, mathematically, $i=3$ & $j=1 \to 4$, we get:

$$i=3 \text{ \& } j=1 \Rightarrow H^{31}U^{1} = G^{31}Q^{1} \quad \textbf{(j)}$$

$$i=3 \text{ \& } j=2 \Rightarrow H^{32}U^{2} = G^{32}Q^{2} \quad \textbf{(k)}$$

$$i=3 \text{ \& } j=3 \Rightarrow H^{33}U^{3} = G^{33}Q^{3} \quad \textbf{(l)}$$

$$i=3 \text{ \& } j=4 \Rightarrow H^{34}U^{4} = G^{34}Q^{4} \quad \textbf{(m)}$$

(4) Stand at node (4) over elements 1-4 and goes anti-clockwise direction, mathematically, $i=3$ & $j=1 \to 4$, we get:

$$i=4 \text{ \& } j=1 \Rightarrow H^{41}U^{1} = G^{41}Q^{1} \quad \textbf{(n)}$$

$$i = 4 \ \& \ j = 2 \Rightarrow H^{42}U^2 = G^{42}Q^2 \quad \textbf{(o)}$$

$$i = 4 \ \& \ j = 3 \Rightarrow H^{43}U^3 = G^{43}Q^3 \quad \textbf{(p)}$$

$$i = 4 \ \& \ j = 4 \Rightarrow H^{44}U^4 = G^{44}Q^4 \quad \textbf{(q)}$$

Next is to form the matrix system taking into consideration the boundary conditions over each element as described before:

$$\begin{bmatrix} H^{11} & H^{12} & H^{13} & H^{14} \\ H^{21} & H^{22} & H^{23} & H^{24} \\ H^{31} & H^{32} & H^{33} & H^{34} \\ H^{41} & H^{42} & H^{43} & H^{44} \end{bmatrix} \begin{bmatrix} U^1 \\ \overline{U}^2 \\ U^3 \\ \overline{U}^4 \end{bmatrix} = \begin{bmatrix} G^{11} & G^{12} & G^{13} & G^{14} \\ G^{21} & G^{22} & G^{23} & G^{24} \\ G^{31} & G^{32} & G^{33} & G^{34} \\ G^{41} & G^{42} & G^{43} & G^{44} \end{bmatrix} \begin{bmatrix} \overline{Q}^1 \\ Q^2 \\ \overline{Q}^3 \\ Q^4 \end{bmatrix} \quad \textbf{(r)}$$

Before continuity, let us make the following summary (Fig. 7):

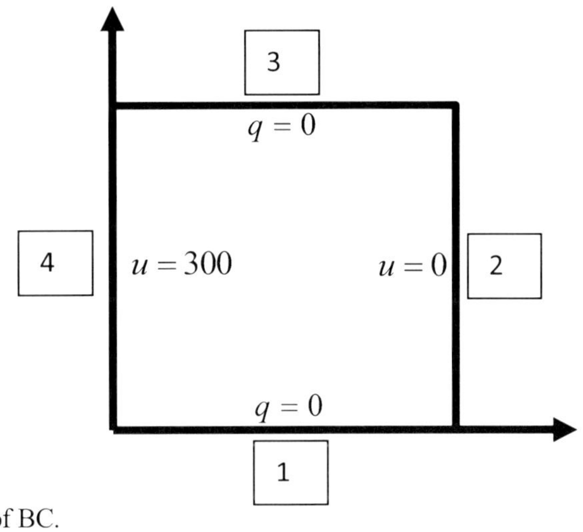

Fig (7). Distribution of BC.

Over element (1)
U^1..................Unknown
Q^1..................Known

Over element (2)
U^2..................Known
Q^2..................Unknown

Over element (3) U^3...................Unknown
Q^3...................Known

Over element (4) U^4...................known
Q^4...................Unknown

Hence, it is essential to re-arrange the above matrix system to collect known values on one side and the unknown values on the other side. To do that, let us take the first equation from the above matrix system before re-arrangement as follows:

Before re-arrangement

$$H^{11}U^1 + H^{12}\overline{U}^2 + H^{13}U^3 + H^{14}\overline{U}^3 = \\ = G^{11}\overline{Q}^1 + G^{12}Q^2 + G^{13}\overline{Q}^3 + G^{14}Q^4 \quad \textbf{(s)}$$

After re-arrangement

$$H^{11}U^1 - G^{12}Q^2 + H^{13}U^3 - G^{14}Q^4 = \\ = G^{11}\overline{Q}^1 - H^{12}\overline{U}^2 + G^{13}\overline{Q}^3 - H^{14}\overline{U}^3 \quad \textbf{(t)}$$

Similar procedure can be done for the other three equations, the resulting system matrix can be re-written as follows:

$$\begin{bmatrix} H^{11} & -G^{12} & H^{13} & -G^{14} \\ H^{21} & -G^{22} & H^{23} & -G^{24} \\ H^{31} & -G^{32} & H^{33} & -G^{34} \\ H^{41} & -G^{42} & H^{43} & -G^{44} \end{bmatrix} \begin{bmatrix} U^1 \\ Q^2 \\ U^3 \\ Q^4 \end{bmatrix} = \begin{bmatrix} G^{11} & -H^{12} & G^{13} & -H^{14} \\ G^{21} & -H^{22} & G^{23} & -H^{24} \\ G^{31} & -H^{32} & G^{33} & -H^{34} \\ G^{41} & -H^{42} & G^{43} & -H^{44} \end{bmatrix} \begin{bmatrix} \overline{Q}^1 \\ \overline{U}^2 \\ \overline{Q}^3 \\ \overline{U}^4 \end{bmatrix} \quad \textbf{(u)}$$

Note the right hand side for the above system is now known, therefore, the system can now be written in a general matrix notation as:

$$AX = b \quad \textbf{(v)}$$

Internal Points

Once the boundary unknowns are calculated, then the unknown at the internal points can be obtained in a very simple manner using the same boundary element equation as described before but with constant c equals one, i.e.;

$$u^i = \int_\Gamma q u^* d\Gamma - \int_\Gamma u q^* d\Gamma \quad \textbf{(w)}$$

Fundamental solution considered acting at any internal "i", therefore, the boundary integral equation (23) after discretization takes the following form:

$$u^i = \sum_{j=1}^{N} G^{ij} q^j - \sum_{j=1}^{N} H^{ij} u^j \quad \textbf{(x)}$$

The coefficients G^{ij} & H^{ij} are calculated at each internal point.

It is recommended when dealing with internal points to evaluate the potential derivative in both directions. Evaluating the potential derivative in both directions comes from differentiating equation (23) at any internal point as follows:

$$\left(q_{x_1}\right)^i = \left(\frac{\partial u}{\partial x_1}\right)^i = \int_\Gamma q \left(\frac{\partial u^*}{\partial x_1}\right)^i d\Gamma - \int_\Gamma u \left(\frac{\partial q^*}{\partial x_1}\right)^i d\Gamma$$

&

$$\left(q_{x_2}\right)^i = \left(\frac{\partial u}{\partial x_2}\right)^i = \int_\Gamma q \left(\frac{\partial u^*}{\partial x_2}\right)^i d\Gamma - \int_\Gamma u \left(\frac{\partial q^*}{\partial x_2}\right)^i d\Gamma \quad \textbf{(y)}$$

It is also important to remember that all integrations for internal points are carried out numerically.

Calculating the Integrals

When evaluating the integrals, one encounters two and different major types of integrals. These two types depend mainly on the relation between source and field points, i.e.; "i" and "j".

Case (1) When $i \neq j$

In this case, there will be no singularity, therefore, the integration is carried out using Gauss quadrature.

Case (2) When $i = j$

In this case, there will be singularity; therefore, the integration requires special treatments such as logarithmic methods.

It is important to remember that the constant element is treated in a very special manner because the matrix $\overline{H}^{ij}$ has no singularity; the details are shown in the next section.

Integrals Evaluation for Constant Element

Case when $i = j$

$$\overline{H}^{ij} = \int_{\Gamma_{i=j}} q^* d\Gamma = \int_{\Gamma_{i=j}} \frac{\partial q^*}{\partial n} d\Gamma = \int_{\Gamma_{i=j}} \left(\frac{\partial q^*}{\partial r} \frac{\partial r}{\partial n} \right) d\Gamma = 0 \qquad \textbf{(z)}$$

As we see, the final result of equation (26) equals zero because the element coordinate and the normal to the boundary are usually perpendicular, as shown in Fig. **(8)**.

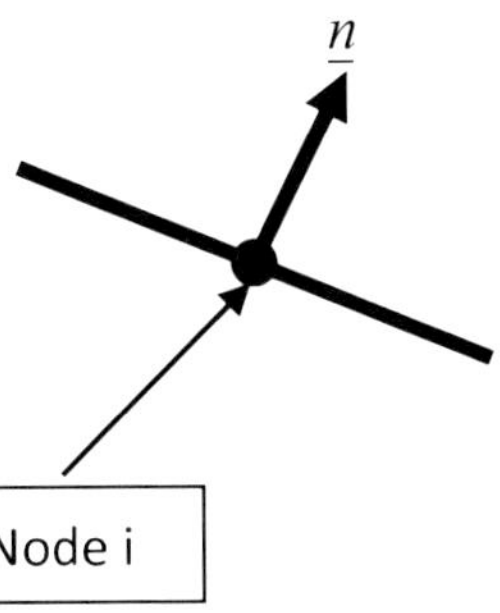

Fig. (8). Configuration of node (i).

G^{ij} **in Case** When $i = j$

$$G^{ij} = \int_{\Gamma_{i=j}} u^* d\Gamma \quad \textbf{(z-1)}$$

Making use of u^* in case of two-dimensional case problem, just as an example, one can get:

$$G^{ij} = \int_{\Gamma_{i=j}} u^* d\Gamma = \frac{1}{2\pi} \int_{\Gamma_{i=j}} \ln\left(\frac{1}{r}\right) dr \quad \textbf{(z-2)}$$

Refer to Fig. (**9**), one can get:

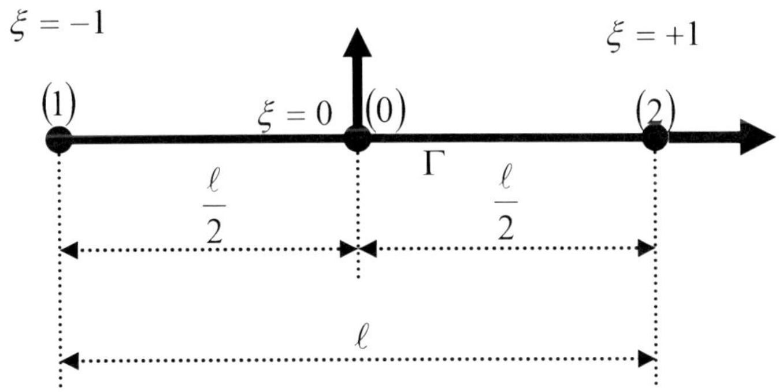

Fig. (9) Normalization of the element.

Let:

$$r = \left| \xi \times \frac{\ell}{2} \right| \tag{z-3}$$

In equation (29) ℓ represents the length of the element under consideration.

Now

$$G^{ij} = \frac{1}{2\pi} \int_{(1)}^{(2)} \ln\left(\frac{1}{r}\right) dr = \frac{1}{\pi} \int_{0}^{(2)} \ln\left(\frac{1}{r}\right) dr = \frac{1}{\pi} \int_{0}^{(2)} \ln\left(\frac{2}{\xi\ell}\right) d\left(r = \left|\xi \times \frac{\ell}{2}\right|\right) \tag{z-4}$$

Note:

$$d\xi = \left(\frac{2}{\ell}\right) dr \tag{z-5}$$

Substituting from equation (31) into the last term of equation (30) leads to:

$$\begin{aligned} G^{ij} &= \frac{1}{\pi}\frac{\ell}{2} \int_{0}^{(2)} \ln\left(\frac{2}{\xi\ell}\right) d\xi = \frac{\ell}{2\pi} \int_{0}^{(2)} \ln\left(\frac{1}{\left(\ell/2\right)} \times \frac{1}{\xi}\right) d\xi \\ &= \frac{\ell}{2\pi} \int_{0}^{1} \ln\left(\frac{1}{\left(\ell/2\right)}\right) + \ln\left(\frac{1}{\xi}\right) d\xi = \frac{\ell}{2\pi} \ln\left(\frac{1}{\left(\ell/2\right)}\right) + \frac{\ell}{2\pi} I_1 \end{aligned} \tag{z-6}$$

Now

$$I_1 = \int_{0}^{1} \ln\left(\frac{1}{\xi}\right) d\xi \tag{z-7}$$

Let

$$v = \ln\left(\frac{1}{\xi}\right) \Rightarrow \xi = e^{-v} \Rightarrow d\xi = -e^{-v}dv$$

New limits of integrations:

$\xi = 0 \Leftrightarrow v \to \infty$

&

$\xi = 1 \Leftrightarrow v = 0$

Then the integral in equation (33) takes the form:

$$I_1 = \int_{\infty}^{0} v\left(-e^{-v}\right)dv = 1 \qquad \textbf{(z-8)}$$

Substituting from (34) into equation (32), leads to:

$$G^{ij} = \frac{\ell}{2\pi}\left[\ln\left(\frac{1}{\left(\ell/2\right)}\right) + 1\right] \qquad \textbf{(z-9)}$$

Case when $i \neq j$

In this case, the integrals are calculated using Gaussian quadrature rules. There are two different cases for such integrals, one for singular integrals and the other for non-singular integrals. The details for both types will be presented after dealing with computer codes to show how these integrals can be calculated.

Computer Code for Potential Problems using Constant Elements

The following code takes the name "POCONBE" and it comes from:

PO : Potential

CON : Constant Element

BE : Boundary Element

This code is used for:

(1) Laplace type problems

(2) Valid for iso-tropic problems

(3) Can be run in any IMB/PC type or XT or AT or Compatible.

(4) It gives all unknown boundary values.

The POCONBE consists of six major parts, as follow:

Main Program

Define the maximum of the system of equations, input and output units.

Subroutine INPUTPC

Read all input data

Subroutine GHMATPC

- Form the matrices H and G
- Re-arrange them to matrix A in accordance with the boundary conditions
- Create the R.H.S. vector F in the system AX=F

Subroutine SLNPD

Solve the system AX=F using Gauss elimination method

Subroutine INTERPC

Compute the unknowns at the internal points

Subroutine OUTPTPC

Write the input data and the results

Some important symbols in the code, they are as follow:

Kode

One dimensional array indicating the type of boundary condition *e.g.*;

$\text{Kode}(J) = 0$: Potential is known

$\text{Kode}(J) = 1$: Potential derivative is known

X: One dimensional real array of x_1 coordinate extreme point of the point of boundary element, see Fig. (**10**).

Y: One dimensional real array of x_2 coordinate extreme point of the point of boundary element, see Fig. (**10**).

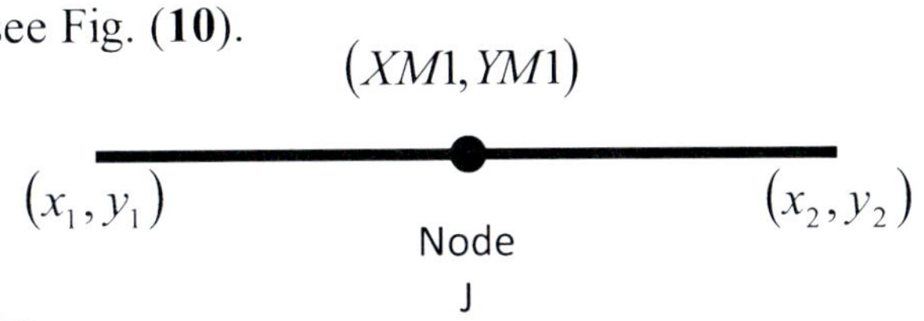

Fig. (10). Code and nodes

XM1, YM1: Coordinate of the node J

G & H: Matrices defined in the system HU=GQ

FI: Pre-scribed values of boundary conditions.

To understand how FI can be used, see the next Fig. (**11**).

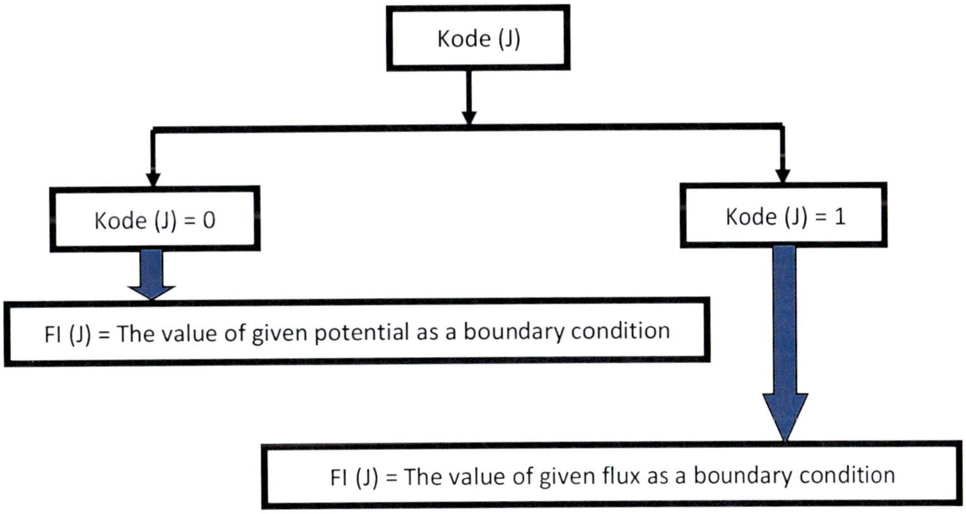

Fig. (11). Values of codes used through the BE Code.

DFI: Right hand side vector in the system AX=F

CX&CY: Coordinates of internal point

SOL: Vector of potential values for internal points

Evaluation of the Matrices H and G

As mentioned before, the subroutine EXTINPC is used to evaluate the two matrices H and G using Gaussian quadrature. To show it does that, let us take 4-Gaussian points as an example.

Case when $i = j$

In case of constant element,

$$\overline{H}^{ii} = 0 \Rightarrow H^{ii} = \frac{1}{2} \tag{40}$$

$$G^{ii} = \frac{1}{2\pi}\left[\ln\left(\frac{1}{\ell/2}\right)+1\right] \tag{41}$$

Case when $i \neq j$

$$H^{ij} = \sum_{k=1}^{4}\left(\frac{-1}{(RA)_k}\right)(\text{Dist})w_k\,\frac{\sqrt{(X1-X2)^2+(Y1-Y2)^2}}{2} \tag{42}$$

$$G^{ij} = \sum_{k=1}^{4}\ln\left(\frac{1}{(RA)_k}\right)w_k\,\frac{\sqrt{(X1-X2)^2+(Y1-Y2)^2}}{2} \tag{43}$$

Note

$$\frac{\sqrt{(X1-X2)^2+(Y1-Y2)^2}}{2} = \frac{\ell}{2}, \ell \text{ is the element length}$$

w_k is the weighting for each point

$(RA)_k$ is the distance from the source point to the collocation

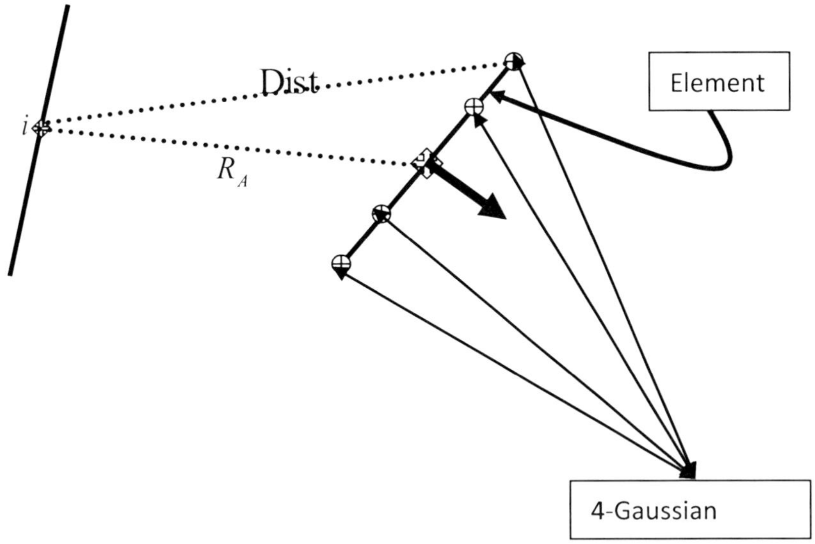

Fig. (12). Geometric computation and 4-Gaussian Points.

In what fellow, we will show how the distance $(RA)_k$ can be calculated; to do that, let us take the source point "i" at any element and the collocation point at any other element "j" as shown in Fig. **(13)**.

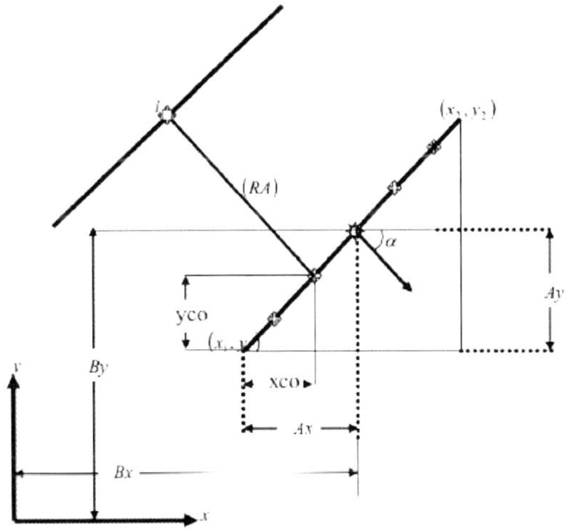

Fig. (13). Geometrical computational for nodes and element.

Referring to Fig. **(13)**,

$$\text{Element length} = \sqrt{(x_1 - x_2)^2 + (y_1 - y_2)^2}$$

Where:

$$Ax = (x_1 - x_2)$$
$$Ay = (y_1 - y_2)$$

And

$$r = (RA) = \sqrt{(\text{xco} - \text{xp})^2 + (\text{yco} - \text{yp})^2}$$

$$\frac{\text{xco}}{\text{Ax}} = \frac{\text{GI(I)}}{(\ell/2)}$$

$$\Rightarrow$$

$$\text{xco} = \text{Ax} \times \text{GI(I)} + \text{Bx}$$
$$\text{yco} = \text{Ay} \times \text{GI(I)} + \text{By}$$

xp & yp are given valu es

Therefore, we need the following in the ***computer code***:

$$Ax = \frac{(x_2 - x_1)}{2}$$

$$Ay = \frac{(y_2 - y_1)}{2}$$

$$Bx = \frac{(x_2 + x_1)}{2}$$

$$By = \frac{(y_2 + y_1)}{2}$$

$$xJ = SQRT(Ax**2 + Ay**2)$$

$$xco = Ax * GI(I) + Bx$$

$$yco = Ay * GI(I) + By$$

$$RA = SQRT((xco - xp)**2 + (yco - yp)**2)$$

$$ENX = -Ay / xJ$$

$$ENY = Ax / xJ$$

$$\Rightarrow$$

$$G = G + ALOG(1./RA) * OME(I) * xJ$$

$$H = H \mp (1./(RA)**2)((xco - xp) * ENX + (yco - yp) * ENY) * OME(I)**1$$

$-ve$ sign in case of clockwise direction

$+ve$ sign in case of anti - clockwise direction

The details of ENX and ENY are shown in Fig. (**14**).

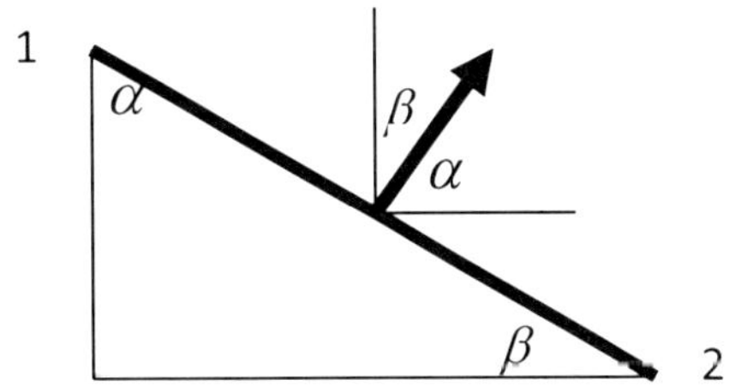

Fig. (14). Details of ENX and ENY.

$ENX = \cos\alpha$

$ENY = \cos\beta$

Refer back to our elementary example and let us show the details of computing the matrices H^{ij} & G^{ij}. As mentioned before, we will use four Gaussian points, and these points are shown in Figs. (**15** and **16**).

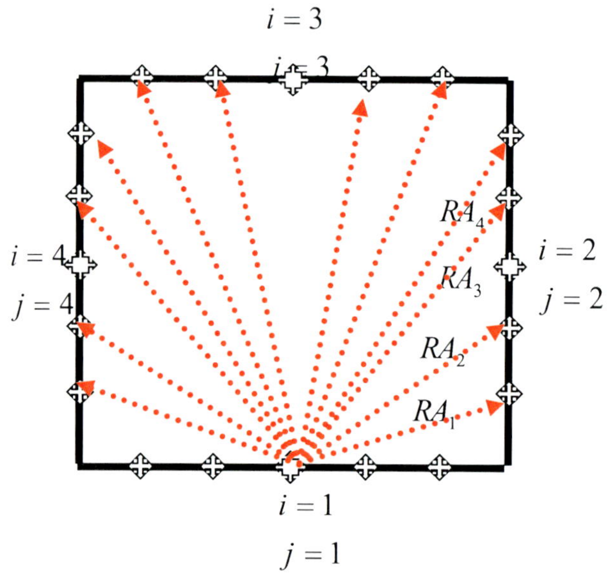

Fig. (15). i=1 source.

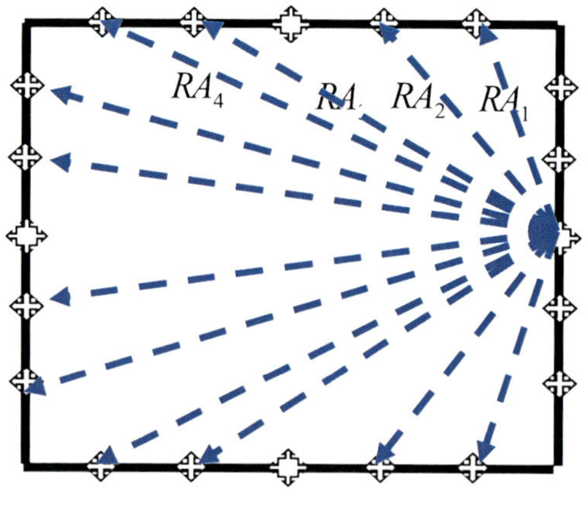

Fig. (16). Flow of computational from source to field points.

The Procedure is as Follows:

Numbering the elements anti-clockwise direction started from 1 to 4 (Fig. **17**).

The elements are numbered as $j = 1 \rightarrow 4$

Over each element, we have a node at the midpoint of the element, therefore $i = 1 \rightarrow 4$

We start at $i = 1, j = 1 \rightarrow 4$ as seen in Fig. **(12)**

Next stand at $i = 2$ and do the same $i = 2, j = 1 \rightarrow 4$ as seen in Fig. **(13)**

Repeat the previous two steps with $i = 3$ and $i = 4$

Note

The coordinates of extreme points for all elements is shown in Fig. **(14)**.

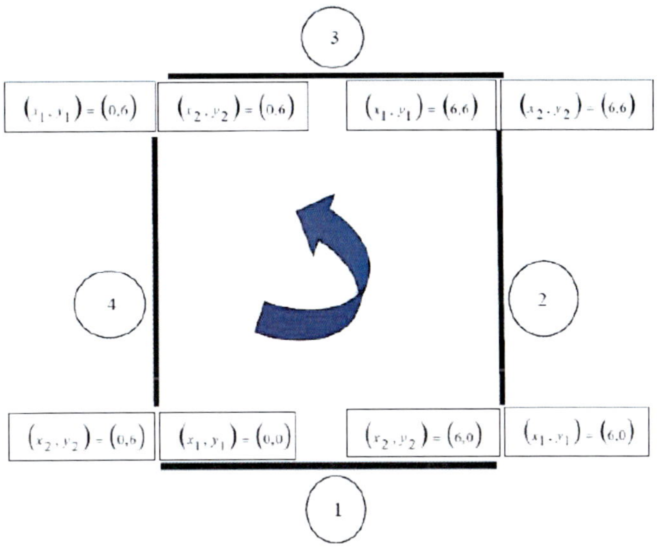

Fig. (17). Direction of computations.

Suppose that the source point is at $i = 1$ and the field point is at $j = 2$, in this case, both G^{ij} and H^{ij} can be calculated as follow:

$$G^{12} = \left\{ \ln\left(\frac{1}{(RA)_1}\right) w_1 + \ln\left(\frac{1}{(RA)_2}\right) w_2 + \ln\left(\frac{1}{(RA)_3}\right) w_3 + \ln\left(\frac{1}{(RA)_4}\right) w_4 \right\} \times \left\{ \frac{\sqrt{(x_1 - x_2)^2 + (y_1 - y_2)^2}}{2} \right\} \tag{44}$$

And

$$H^{12} = \sum_{k=1}^{4} \left[-\left(\frac{\text{Dist}}{(RA)_k^2}\right) w_k \left\{ \frac{\sqrt{(x_1 - x_2)^2 + (y_1 - y_2)^2}}{2} \right\} \right] \tag{45}$$

In equation (45), Dist is the perpendicular distance from the point of singularity (Collocation) to the element under consideration.

REFERENCES

[1] Delima-Silva Jnr and Wrobel, "BEM analysis of melting problems", In: *Boundary Element Technology IX* Computational Mechanics Publications: Southampton, 1994.

[2] S.G. Ahmed, "An approximate method for evaporation problem with mushy zone", *Eng. Anal. Bound. Elem.*, vol. 29, no. 2, pp. 161-165, 2005.
[http://dx.doi.org/10.1016/j.enganabound.2004.10.002]

[3] T. Zhu, J. Zhang, and S.N. Atluri, "A meshless numerical method based on the local boundary integral equation (LBIE) to solve linear and nonlinear boundary value problems", *Eng. Anal. Bound. Elem.*, vol. 23, no. 5-6, pp. 375-389, 1999.
[http://dx.doi.org/10.1016/S0955-7997(98)00096-4]

[4] K.M. Singh, and M. Tanaka, "On exponential variable transformation based boundary element formulation for advection-diffusion problems", *Eng. Anal. Bound. Elem.*, vol. 24, no. 3, pp. 225-235, 2000.
[http://dx.doi.org/10.1016/S0955-7997(00)00003-5]

[5] I. Boztosum, A. Charafi, M. Zerroukat, and K. Djidjeli, "Thin plate spline radial basis function scheme for Advection-diffusion problems", *Electronic Journal of Boundary Elements,* vol. 2, pp. 267-282, 2002.

[6] PANDAT (Software), Retrieved from: http://www.computherm.com

[7] S.G. Ahmed, "A collocation method using new radial basis functions of thin plate and Multiquadrics types", *Eng. Anal. Bound. Elem.,* vol. 30, no. 8, pp. 697-701, 2006.
[http://dx.doi.org/10.1016/j.enganabound.2006.03.001]

[8] X. Liu, and K. Tai, "Point interpolation collocation method for the solution of partial differential equations", *Eng. Anal. Bound. Elem.,* vol. 30, no. 7, pp. 598-609, 2006.
[http://dx.doi.org/10.1016/j.enganabound.2005.12.003]

[9] K.M. Liew, Y. Cheng, and S. Kitipornchai, "Boundary element-free method (BEFM) for two-dimensional elastodynamic analysis using laplace transforms", *Int. J. Numer. Methods Eng.,* vol. 64, no. 12, pp. 1610-1627, 2005.
[http://dx.doi.org/10.1002/nme.1417]

[10] A.E. Norhan, *Mohamd, ' Boundary Elements Methods and its Non-Linear Analysis of Hydrofoils', Supervisors: Prof. S. G. Ahmed and Prof. A. F. Abd-Elgawad, Zagazig University.* Zagazig University, 2002.

[11] T. Zhu, J. Zhang, and S.N. Atluri, "A meshless numerical method based on the local boundary integral equation (LBIE) to solve linear and non-linear boundary value problems", *Eng. Anal. Bound. Elem.,* vol. 23, no. 5-6, pp. 375-389, 1999.
[http://dx.doi.org/10.1016/S0955-7997(98)00096-4]

[12] R.P. Shaw, "' A boundary integral equation approach to the one-dimensional ablation problem", In: *Boundary Element Methods in Engineering.* Springer-Verlag: Berlin, 1982.
[http://dx.doi.org/10.1007/978-3-662-11273-1_10]

[13] L.C. Wrobel, "' A boundary element solution to Stefan's problem", In: *Boundary Elements V.* Springer-Verlag: Berlin, 1983.

[14] M. Heinlein, S. Mukherjee, and C. Richmond, "' A boundary element method analysis of temperature fields and stress during solidification'", *Acta Mech.,* vol. 59, no. 1-2, pp. 59-81, 1986.
[http://dx.doi.org/10.1007/BF01177060]

[15] C.K. Hsieh, C-Y. Choi, and A.J. Kassab, "' Solution of Stefan problems by a boundary element method", In: *Boundary Element Technology VII.* Computational Mechanics Publication, Southampton and Elsevier: London, 1992.
[http://dx.doi.org/10.1007/978-94-011-2872-8_32]

[16] M. Zerroukat, and C.R. Chatwin, "' An explicit variable time step method for implicit moving boundary problems'", *Commun. Numer. Methods Eng.*, vol. 10, no. 3, pp. 227-235, 1994.
[http://dx.doi.org/10.1002/cnm.1640100306]

[17] C.A. Brebbia, J.C.F. Telles, and L.C. Wrobel, ' *Boundary Element Techniques.* Springer-Verlag: Berlin, Heidelberg, New York, Tokyo, 1984.
[http://dx.doi.org/10.1007/978-3-642-48860-3]

[18] C.A. Brebbia, and P. Skerget, "Nonlinear time-dependent potential problem using BEM", In: *Boundary Elements VI* Computational Mechanics Publications, Southampton and Springer-Verlag: Berlin, 1984.

[19] W. Delmia-Silva Jr, and L.C. Wrobel, ' *A boundary element formulation for multi-dimensional ablation problems', Boundary Elements XIV, Computational Mechanics publication.* Southampton and Elsevier: London, 1993.

[20] R.P. Shaw, ' *A boundary integral approach to the one-dimensional ablation problem', Boundary Element Methods in Engineering.* Springer-Verlag: Berlin, 1982.

[21] W.Q. Dennis, E.O. Mark, and C.V. Donald, "The boundary element method applied to a moving free boundary problem,' Int. J. for Numer", *Int. J. Numer. Methods Eng.*, vol. 46, no. 8, pp. 1335-1346, 1999.
[http://dx.doi.org/10.1002/(SICI)1097-0207(19991120)46:8<1335::AID-NME757>3.0.CO;2-9]

[22] R. Pollandt, "Solving non-linear differential equations of mechanics with the boundary element method and radial basis functions", Int. J. for Numer, *Int. J. Numer. Methods Eng.*, vol. 40, no. 1, pp. 61-73, 1997.
[http://dx.doi.org/10.1002/(SICI)1097-0207(19970115)40:1<61::AID-NME50>3.0.CO;2-Q]

[23] G.R. Liu, and Y.T. Gu, "Boundary Meshfree methods based on the boundary point interpolation methods", *Eng. Anal. Bound. Elem.*, vol. 28, no. 5, pp. 475-487, 2004.
[http://dx.doi.org/10.1016/S0955-7997(03)00101-2]

[24] M. Yasko, "Boundary Element Method for a Hydrofoil near the Free Surface", J, *Eng. Anal. Bound. Elem.*, vol. 21, no. 2, pp. 191-194, 1998.
[http://dx.doi.org/10.1016/S0955-7997(98)00010-1]

[25] S.A. Kinnas, H. Lee, and Y.L. Young, ""Boundary Element Techniques for the Prediction of Sheet and Developed Tip Vortex Cavitation", Electronic Journal of Boundary Elements, Vol", *BETEQ*, vol. 2001, no. 2, pp. 151-178, 2002.

[26] S.G. Ahmed, and L.C. Wrobel, "An Alternative BEM Approach for Solving Cavity Flow Problems", *International Conference Boundary Element Technology XI,* 1996 Hawaii, USA.

[27] M. Yasko, "Boundary element method for a hydrofoil near the free surface', J", *Eng. Anal. Bound. Elem.*, vol. 21, no. 2, pp. 191-194, 1998.
[http://dx.doi.org/10.1016/S0955-7997(98)00010-1]

[28] S.G. Ahmed, and L.C. Wrobel, "An alternative BEM approach for solving cavity flow problems", *International Conference Boundary Element Technology XI,* 1996.

[29] L.C. Wrobel, "' A boundary element solution to Stefan's problem", In: *Boundary Elements V.* Springer-Verlag: Berlin, 1983.

[30] L. Gregory, "Young, Karen A. Mc Donald, Ahmet Polazoglu and William Ford,' Boundary element solutions for free boundary convection-diffusion problems'", *Numer. Heat Transf. A,* vol. 21, no. 3, pp. 299-311, 1992.

[http://dx.doi.org/10.1080/10407789208944878]

[31] A. Fic, A.J. Nowak, and R. Bialecki, "' Heat transfer analysis of the continuous casting process by the front tracking'", *Eng. Anal. Bound. Elem.,* vol. 24, no. 3, pp. 215-223, 2000. [http://dx.doi.org/10.1016/S0955-7997(00)00004-7]

[32] T. Zhu, J-D. Zhang, and S.N. Atluri, "A Local Boundary Integral Equation (LBIE) method in computational mechanics and a meshless discretization approach", *Comput. Mech.,* vol. 21, no. 3, pp. 223-235, 1998. [http://dx.doi.org/10.1007/s004660050297]

[33] P. Coppa, and A. Consorti, "Normal emissivity of samples surrounded by surfaces at diverse temperatures", *Measurement,* vol. 38, no. 2, pp. 124-131, 2005. [http://dx.doi.org/10.1016/j.measurement.2005.05.001]

[34] P.W. Partridge, C.A. Brebbia, and L.C. Wrobel, *The Dual Reciprocity Boundary Element Method.* Computational Mechanics Publications: Southampton, 1992.

[35] E. Perry-Debain, "Ter Morsche, H.G.,' B-Spline approximation and fast wavelet transform for an efficent evaluation of particular solutions for Poisson's equation'", *Eng. Anal. Bound. Elem.,* vol. 26, no. 1, pp. 1-13, 2002. [http://dx.doi.org/10.1016/S0955-7997(01)00080-7]

[36] P.A. Ramachandran, and K. Balakhrishnan, "' Radial basis functions as approximate particular solutions: review of recent progresses", *Eng. Anal. Bound. Elem.,* vol. 24, no. 7-8, pp. 575-582, 2002. [http://dx.doi.org/10.1016/S0955-7997(00)00037-0]

[37] S. Chantasiriwan, "' Cartesian grid methods using radial basis functions for solving Poisson, Helmholtz and diffusion-convection equations'", *Eng. Anal. Bound. Elem.,* vol. 28, no. 12, pp. 1417-1425, 2004. [http://dx.doi.org/10.1016/j.enganabound.2004.08.004]

[38] S.G. Ahmed, "A collocation method using new radial basis functions of thin plate and multiquadraic types'", *Eng. Anal. Bound. Elem.,* vol. 30, no. 8, pp. 697-701, 2006. [http://dx.doi.org/10.1016/j.enganabound.2006.03.001]

[39] A.J. Nowak, "Application of the multiple reciprocity method to potential problems", In: *Chapter 3, The Multiple Reciprocity Boundary Element Method,* A.J. Nowak, A.C Neves, Eds., Computational Mechanics Publications, 1994.

[40] L.C. Delima-Silva Jnr And Wrobel, "BEM analysis of melting problems", In: *Boundary Element Technology IX,* C.A. Brebbia, A.J.L Kassab, Eds., Computational Mechanics Publications: Southampton, 1994.

[41] N. Zabaras, and S. Mukherjee, "' An analysis of the solidification problems by the boundary element method'", *Int. J. Numer. Methods Eng.,* vol. 24, no. 10, pp. 1879-1900, 1987. [http://dx.doi.org/10.1002/nme.1620241006]

CHAPTER 3

Phase Change and Boundary Element Methods

Abstract: In the present chapter, the author focused on the direct and indirect application of the boundary element methods to solve phase change problems. The chapter starts with the basic principles of the boundary element methods with the complete derivation of the final system of the algebraic equations.

Keywords: Direct BEM, Indirect BEM, Phase change problems.

INTRODUCTION

Moving boundary problems can be defined as a set of partial differential equations with associated boundaries and initial conditions over a domain that has one or more of its boundaries varying with time. Determining the movement of the moving boundaries is very important and is considered an integral part of the solution. Moving boundary problems are usually referred to as Stefan problems [1-8]. The difficulty in solving moving boundary problems refers mainly to the non-linearity caused by the presence of non-linear boundary conditions at the moving boundary. Processes related to melting and solidification are widely encountered in many industrial applications, such as metal alloy casting, food freezing, ice forming, thermal energy storage, *etc*. Mathematical modeling of transport phenomena that occur during a phase change is becoming an important tool for predicting the state of the final product. Problems related to improper control of the transport mechanisms involve void formation, concentration variation of the dissolved component, and cracking due to thermal stresses. In the present chapter, the boundary element methods for phase change problems with fixed boundaries will be presented in detail. An integral representation is obtained starting from the governing equation for these problems, which is the Fourier unsteady heat conduction equation. To this end, Green's second identity is used together with the fundamental solution to the heat conduction equation [9-14] and its normal derivative. An elaborate integration in time is then proceeded taking advantage of the fundamental solution properties. The basic concepts of the boundary element methods are reviewed and a discussion about the discretization of the boundary integral representation and its numerical implementation, both in space and time, is ensued. The convolution time-marching scheme for a numerical solution is presented and discussed in detail. Since 1970, the boundary element

S. G. Ahmed

methods started solving a wide range of practical, engineering, and industrial problems. Simply, the boundary element methods transfer the governing equation and its associated boundary conditions to a boundary integral equation. This equation is valid for the boundary in addition to the internal points of the domain. The basic idea to derive the integral equation is to minimize the error and deal only with the boundaries. There are two directions when deriving the integral equation, the direct and indirect approaches. In the present thesis, the first approach will be used for the derivation of the integral equation. The direct approach of formulation consists of transforming the partial differential equation describing the behavior of the unknown inside and on the boundary of the domain into an integral equation relating only boundary values and then finding out the numerical solution of the integral equation [15-20].

Diffusion Problems with Fixed Boundaries

The governing equation for thermal diffusion in the absence of the source in the domain Ω and with constant material properties is as follows [21-27]:

$$\frac{\partial^2 u(x,y,t)}{\partial x^2}+\frac{\partial^2 u(x,y,t)}{\partial y^2}=\frac{1}{k}\frac{\partial u(x,y,t)}{\partial t} \qquad \forall(x,y)\in\Omega \tag{1}$$

With one, two or the three boundary conditions

Dirichlet boundary condition

$$u(x,y,t)=\bar{u}(x,y,t), \qquad (x,y)\in\Gamma_u \tag{2}$$

Neumann boundary condition

$$\frac{\partial u(x,y,t)}{\partial n}=\bar{q}_N(x,y,t), \qquad (x,y)\in\Gamma_{qN} \tag{3}$$

Robin boundary condition

$$\frac{\partial u(x,y,t)}{\partial n}=q_R(x,y,t), \qquad (x,y)\in\Gamma_{qR} \tag{4}$$

Where $\Gamma = \Gamma_u + \Gamma_{qN} + \Gamma_{qR}$ is the boundary of the domain Ω.

Initial condition

$$u(\mathrm{x}, \mathrm{y}, \mathrm{t}=0) = u_0(\mathrm{x}, \mathrm{y}), \qquad (x, y) \in \Omega \tag{5}$$

In these equations,

$k = \dfrac{K}{\rho c}$	Thermal diffusivity
K	Thermal conductivity
ρ	Density
c	Specific heat
h	Heat coefficient
u_a	Ambient temperature
$u(\mathrm{x}, \mathrm{y}, \mathrm{t})$	Temperature
$\dfrac{\partial u}{\partial n}$	Normal derivative of the temperature
$q = -K\dfrac{\partial u}{\partial n}$	Thermal flux
t	Time

Fundamental Solution

The time-dependent fundamental solution u^* is of the following form [28]:

$$u^*(p,t_F;s,t)=\frac{1}{\left[4\pi k(t_F-t)\right]^{\frac{m}{2}}}\exp\left[-\frac{r^2(p,s)}{4k(t_F-t)}\right]H(t_F-t) \quad \textbf{(6)}$$

In equation (2.6), m denotes the number of spatial dimensions and $r=|p-s|$, the distance between points p is the ***source*** and s is the field. The Heaviside function $H(t_F-t)$ is included to emphasize the fact that the solution is identically zero for the time $t>t_F$; this condition is called the ***causality condition*** [29].

Properties of the fundamental solution

The different properties of the fundamental solution are listed below [30]:

$$k\nabla u^*(p,t_F;s,t)-\frac{\partial u^*(p,t_F;s,t)}{\partial t}=-\delta(p,s)\delta(t_F,t) \quad \textbf{(7)}$$

$$\int_{t_0}^{t_F}dt\int_{\Omega}u(s,t)\delta(p,s)\delta(t_F,t)d\Omega=u(p,t_F) \quad \textbf{(8)}$$

$$\lim_{t\to t_F}u^*(p,t_F;s,t)=\delta(p,s) \quad \textbf{(9)}$$

$$\lim_{t\to t_F}\int_0^{t_F}u^*(p,t_F;s,t)dt=\delta(p,s) \quad \textbf{(10)}$$

Boundary Integral Representation of Diffusion Problems with Fixed Boundaries

The type of ***weighted residual statement*** commonly used in the BEM requires that the error introduced with discretization is forced to zero in the domain Ω

shown in Fig. (**1**) so that the governing equation is identically satisfied and permits it to be different from zero over the boundary Γ so that the boundary conditions can be approximately satisfied there.

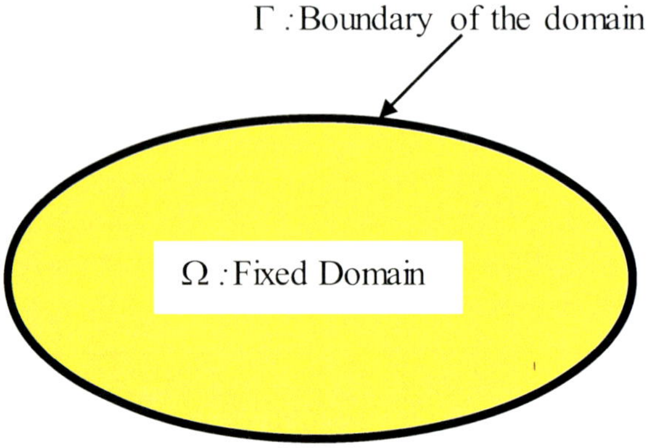

Fig. (1). Configuration of the fixed domain problem.

Taking this into consideration, the following weighted residual statement can be written for the governing equation (1), where the fundamental solution u^*, as given by equation (6), is the corresponding weight function:

$$\int_{t_0}^{t_F} dt \int_{\Omega} \left(k\nabla^2 u - \frac{\partial u}{\partial t} \right) u^* d\Omega = 0 \tag{11}$$

Green's second identity –see appendix A- can be written as:

$$\int_{\Omega} \left(u^* \nabla^2 u - u \nabla^2 u^* \right) d\Omega = \int_{\Gamma} \left(u^* \frac{\partial u}{\partial n} - u \frac{\partial u^*}{\partial n} \right) d\Gamma \tag{12}$$

From which, one may write:

$$\int_{\Omega}\left(u^{*}\nabla^{2}u\right)d\Omega=\int_{\Gamma}\left(u^{*}\frac{\partial u}{\partial n}-u\frac{\partial u^{*}}{\partial n}\right)d\Gamma+\int_{\Omega}\left(u\nabla^{2}u^{*}\right)d\Omega \tag{13}$$

If one also considers the time integration for equation (13), the integral on the left-hand side of this equation can be easily identified as the first integrand of equation (11), so that,

$$\int_{t_0}^{t_F}dt\left[\int_{\Omega}\left(u^{*}\frac{\partial u}{\partial t}-u\frac{\partial u^{*}}{\partial t}\right)d\Gamma+\int_{\Omega}\left(u\nabla^{2}u^{*}\right)d\Omega\right]-\int_{t_0}^{t_F}dt\int_{\Omega}\left(\frac{1}{k}u^{*}\frac{\partial u}{\partial t}\right)d\Omega=0 \tag{14}$$

One can operate on the last integral of equation (14) as follows:

$$\int_{t_0}^{t_F}dt\int_{\Omega}\left(\frac{1}{k}u^{*}\frac{\partial u}{\partial t}\right)d\Omega=\int_{t_0}^{t_F}dt\int_{\Omega}\frac{1}{k}\left[\frac{\partial}{\partial t}\left(u^{*}u\right)-u\frac{\partial u^{*}}{\partial t}\right]d\Omega \tag{15}$$

As the boundaries are fixed, *i.e.*, the domain under consideration is independent of time, one can invert the order of integration and perform it first in time and then in space. Taking this into consideration, the first integral on the right-hand side of equation (15) becomes:

$$\int_{t_0}^{t_F}dt\int_{\Omega}\left(\frac{1}{k}u^{*}\frac{\partial u}{\partial t}\right)d\Omega=\frac{1}{k}\int_{t_0}^{t_F}dt\frac{\partial}{\partial t}\left[\int_{\Omega}\left(u^{*}u\right)d\Omega\right]=\frac{1}{k}\left[\int_{\Omega}\left(u^{*}u\right)d\Omega\right]_{t=t_0}^{t=t_F} \tag{16}$$

Therefore, equation (15) can be re-written as:

$$\int_{t_0}^{t_F}dt\int_{\Omega}\frac{1}{k}\frac{\partial u}{\partial t}u^{*}d\Omega=\frac{1}{k}\left[\int_{\Omega}u^{*}ud\Omega\right]_{t=t_0}^{t=t_F}-\int_{t_0}^{t_F}dt\int_{\Omega}\frac{1}{k}\frac{\partial u^{*}}{\partial t}ud\Omega \tag{17}$$

Taking equation (17) into (14) and multiplying the whole equation by diffusivity, one obtains:

$$\int_{t_0}^{t_F} dt \int_{\Omega} u(s,t)\left(k\nabla^2 u^*(p,t_F;s,t) + \frac{\partial u^*(p,t_F;s,t)}{\partial t} \right) d\Omega - \left[\int_{\Omega} u^*(p,t_F;s,t) u(s,t) d\Omega \right]_{t=t_0}^{t=t_F} =$$

$$= -k \int_{t_0}^{t_F} dt \int_{\Gamma} \left(u^*(p,t_F;s,t) \frac{\partial u(s,t)}{\partial n} - u(s,t) \frac{\partial u^*(p,t_F;s,t)}{\partial n} \right) d\Gamma \tag{18}$$

At this point, one should investigate the singularity that occurs in the integrals of equation (18) at time $t = t_F$. The boundary integrals in equation (18), or at least one of them, will always be different from zero but for the trivial case of prescribed homogeneous boundary and initial conditions. The behavior of the ***domain integrals*** in equation (18) at a time $t = t_F$ will now be analyzed. To this end, it is convenient to name those integrals as:

$$I_1 = \int_{t_0}^{t_F} dt \int_{\Omega} u(s,t)\left(k\nabla^2 u^*(p,t_F;s,t) + \frac{\partial u^*(p,t_F;s,t)}{\partial t} \right) d\Omega \tag{19}$$

$$I_2 = -\int_{\Omega} u^*(p,t_F;s,t=t_F) u(s,t_F) d\Omega \tag{20}$$

$$I_3 = \int_{\Omega} u^*(p,t_F;s,t_0) u(s,t_0) d\Omega \tag{21}$$

The integral I_3 can be recognized as the initial condition one and can be left as it is for the moment, while further attention is focused on integrals I_1 and I_2. In order to avoid ending the integrations exactly at the peak of a Dirac delta function, one may subtract or add to the upper limit of integrals an arbitrarily small quantity ε. In the former case, integral I is identically zero for t in the range $0,\ t_F - \varepsilon$ because of equation (7) and so, taking the limit as $\varepsilon \to 0$ and accounting for condition given by equation (9), one has:

$$I_1 = 0 \tag{22}$$

And,

$$I_2 = -u(s,t_F) \tag{23}$$

Thus,

$$I_1 + I_2 + I_3 = -u(s,t_F) + \int_\Omega u^*(p,t_F;s,t_0)u(s,t_0)d\Omega \tag{24}$$

Taking equation (24) into (18), one finally arrives at:

$$u(s,t_F) = k\int_{t_0}^{t_F} dt \int_\Gamma \left(u^*(p,t_F;s,t)\frac{\partial u(s,t)}{\partial n} - u(s,t)\frac{\partial u^*(p,t_F;s,t)}{\partial t} \right) d\Gamma + \int_\Omega u^*(p,t_F;s,t_0)u(s,t_0)d\Omega \tag{25}$$

The same relation can be obtained by adding ε to the upper limit of the integrals in equation (18) then taking the limit as $\varepsilon \to 0$. In this case, $u^*(p,t_F;s,t_F+\varepsilon)$ is equal to zero due to the causality condition. Applying properties given by equations (7) and (8) to integrals I produces:

$$I_1 = -u(p,t_F) \tag{26}$$

And,

$$I_2 = 0 \tag{27}$$

The general form of equation (25), valid for any position at the point under consideration may assume, either in the domain or on the boundary, is:

$$c(s)u(s,t_F) = k\int_{t_0}^{t_F} dt \int_{\Gamma}\left(u^*(p,t_F;s,t)\frac{\partial u(s,t)}{\partial n} - u(s,t)\frac{\partial u^*(p,t_F;s,t)}{\partial t}\right)d\Gamma$$
$$+\int_{\Omega} u^*(p,t_F;s,t_0)u(s,t_0)d\Omega \quad \textbf{(28)}$$

It should be noticed that the first two integrals of equation (28) represent the effects of boundary conditions, while the third integral stands for the influence of the initial conditions. The coefficient $c(s)$ on the left-hand side of equation (28) is a function of the position of the point P and also of the smoothness of the boundary in the neighborhood of P. It may assume one of the following values [31]:

$$c(s) = \begin{cases} 0 & \text{P outside the domain } \Omega \\ 1 & \text{P inside the domain } \Omega \\ \dfrac{\theta}{\ell\pi} & \text{P on the boundary } \Gamma \end{cases} \quad \textbf{(29)}$$

In equation (29), θ is the solid angle of the boundary at the point P, finally, $c(s)$ is equal to $\frac{1}{2}$ for a point located on one of the two extremes (smooth boundaries).

Phase Change Problems with Moving Boundaries And Sharp Interface

In this section, the boundary integral representation for diffusion problems with moving boundaries and sharp interface, see Fig. (**2**) is obtained. In order to obtain the boundary integral representation for a moving boundary problem ***Reynolds's transport theorem*** – see Appendix C – has been used. The procedures for the simultaneous time-space integration are presented in detail, as well as their numerical implementation in the context of the boundary element method.

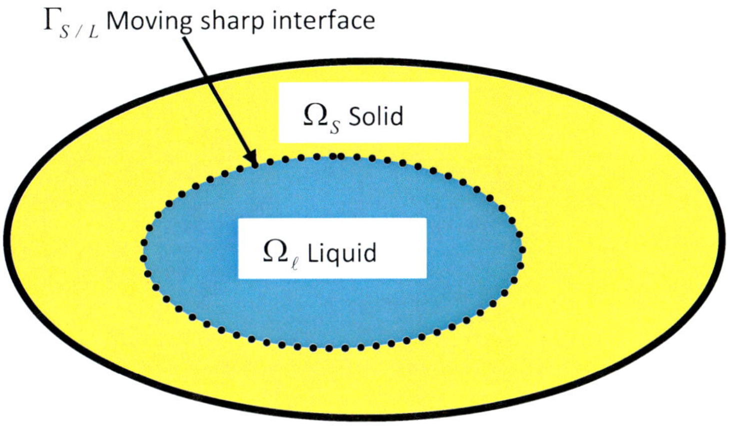

Fig. (2). Configuration of general two phase problems.

Governing Equation and Boundary Conditions

In the absence of heat source in the domain Ω and with constant material properties over each phase α, the diffusion equation with moving boundaries and the sharp interface is also governed by the same differential equation given by equation (2.1), but now written for each phase α, shown in Figs (**3** - **5**), respectively.

$$\frac{\partial^2 u_s(\mathrm{x},\mathrm{y},t)}{\partial x^2} + \frac{\partial^2 u_s(\mathrm{x},\mathrm{y},t)}{\partial y^2} = \frac{\mathbf{1}}{k_s}\frac{\partial u_s(\mathrm{x},\mathrm{y},t)}{\partial t} \qquad \forall(\mathrm{x},\mathrm{y}) \in \Omega_s \tag{30}$$

And

$$\frac{\partial^2 u_\ell(\mathrm{x},\mathrm{y},t)}{\partial x^2} + \frac{\partial^2 u_\ell(\mathrm{x},\mathrm{y},t)}{\partial y^2} = \frac{\mathbf{1}}{k_\ell}\frac{\partial u_\ell(\mathrm{x},\mathrm{y},t)}{\partial t} \qquad \forall(\mathrm{x},\mathrm{y}) \in \Omega_\ell \tag{31}$$

The same set of boundary and initial conditions given by equations (2-5) still apply here for the fixed boundaries, but now one has also to prescribe special boundary conditions on the ***solid-liquid interface***.

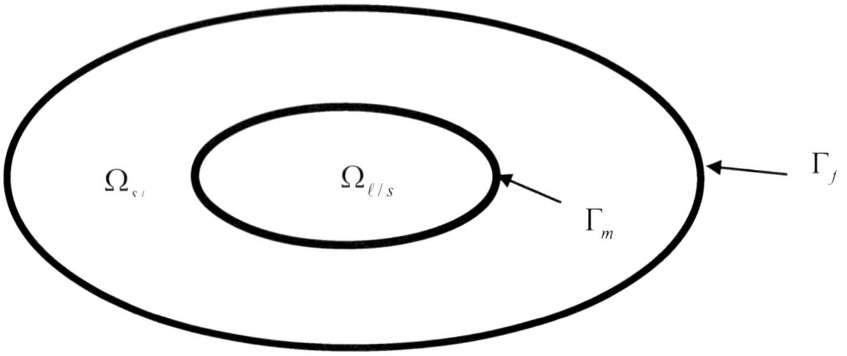

Fig. (3). Solidification or melting.

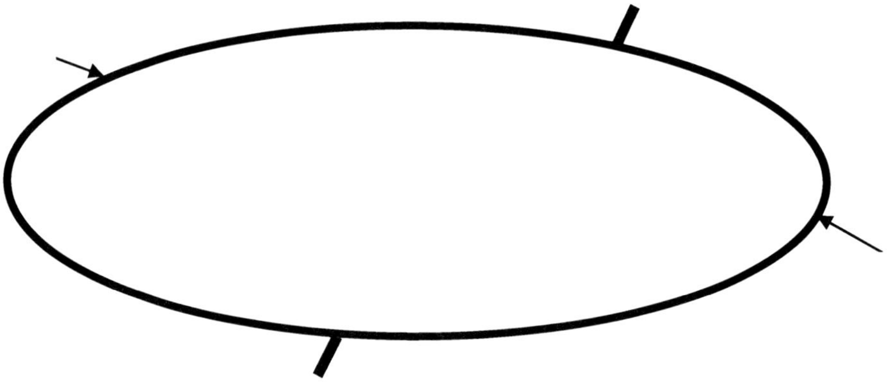

Fig. (4). Ablation.

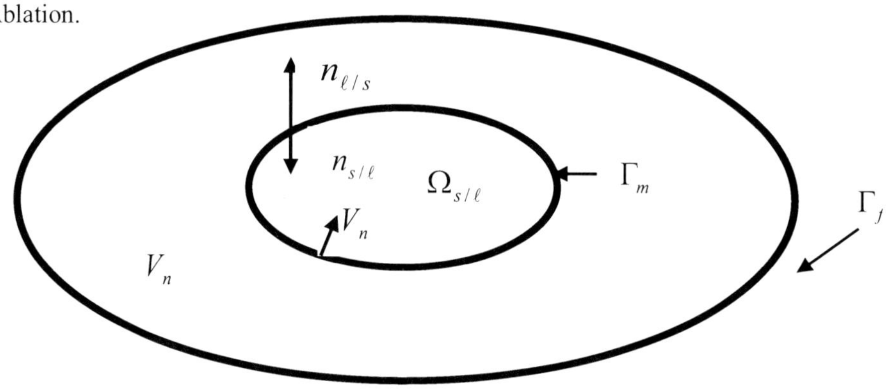

Fig. (5). Possible configuration of two phase (Solidification or melting).

These can be of the following types:

$$u = \bar{u} = u_m(t) \tag{32}$$

$$V_n(\mathrm{x},t) = \frac{1}{\rho_s L}\left[K_s \frac{\partial u_s}{\partial n} - K_\ell \frac{\partial u_\ell}{\partial n} \right] \tag{33}$$

$$V_n(\mathrm{x},t) = \frac{K}{\rho L}\left[\frac{\partial u(\mathrm{x},\mathrm{t})}{\partial n} \right] \tag{34}$$

$$Q(\mathrm{x},\mathrm{t}) = -K\left[\frac{\partial u(\mathrm{x},\mathrm{t})}{\partial n} \right] + \rho L V_n(\mathrm{x},t) \tag{35}$$

In equations (32-35) $x \in \Gamma_m(t)$, the subscripts s and ℓ stand for the solid and liquid phases, k_s, the thermal diffusivity of the solid phase is equal to $\frac{K_s}{\rho_s c_s}$ in terms of its conductivity K_s, density ρ_s and specific heat c_s, similar notation applies for k_ℓ, u_m is the melting temperature, L is the latent heat and V_n is the normal velocity of the interface or moving boundary given by:

$$V_n(\mathrm{x},t) = \frac{\partial R}{\partial t}.n_s = \frac{\partial X_i}{\partial t}.n_{si} \tag{36}$$

Where $i = 1,2,3$ corresponding to the x, y, z Cartesian axes, n is the unit outward normal to the interface or moving boundary at a point on $\Gamma_m(t)$ and;

$$R(t) = x(t)i + y(t)j + z(t)k \tag{37}$$

Equation (37) is a point on the interface or moving boundary $\Gamma_m(t)$, defined in terms of unit vectors (i, j, k) along the global Cartesian axes. The boundary conditions given by equations (32)-(35) are characteristic of the so-called classical approach, which is also the one adopted in the present work. In such an approach, the phase change is confined conceptually to an infinitesimally thin surface, across which the discontinuity of flux of sensible heat and the evolution of latent heat must balance. Phase change across the surface is regarded as instantaneous and complete, and the two contiguous phases occupy mutually exclusive zones. The melting temperature u_m is, in general, a function of time. This allows the analysis to be extended to cases of dissolution-melting into a binary mixture. In this kind of problem, the melting rate depends on particle diffusion (mass transfer) as well as heat transfer [32-37]. Equation (33) is known as the Stefan boundary condition and is used when heat conduction occurs in both the solid and liquid phases. Equation (34) can be seen as a particular case of (33) and is used when heat is conducted in only one of the phases, the other remaining at a constant temperature, characterizing a one-phase problem. Equations (33) and (34) are generally associated with Dirichlet boundary conditions on the fixed boundaries of the body under analysis. Finally, equation (35) is commonly found in one-phase problems such as melting and ablation in which thermal fluxes originated by Neumann or Robin boundary conditions are continuously applied to an initially sub-cooled body leading to its eventual melting. At least one extra initial condition, other than $u = u_0(x, y)$, is required in phase change problems. This condition contains information regarding the state of the domain $\Omega(t)$ at the instant t_m when phase change commences and generally assumes ***one of the following forms***:

$$\Omega_s(t_m) = 0 \tag{38}$$

$$\Omega_\ell(t_m) = 0 \tag{39}$$

$$u_{pre-melting}(x, t_m) = u_{post-melting}(x, t_m), \qquad x \in \Omega(t_m) \tag{40}$$

The initial conditions are given by equations (38) and (39) are associated with the solidification and melting processes, respectively, and commonly appear when Dirichlet boundary conditions are prescribed on the fixed boundaries. Equation (40) represents the initial condition required in melting and ablation processes which undergo a sub-cooled period until the heated boundaries reach the melting temperature.

Boundary Integral Representation of Diffusion Problems with Moving Boundaries and Sharp Interface

The steps for obtaining the boundary integral representation for diffusion problems with moving boundaries are similar to those presented before, where the fixed boundaries case was studied. The same fundamental solution and its properties still apply. One can now start discussing the formulation by re-writing equation (11) for the $\alpha-$phase under consideration (either solid or liquid):

$$\int_{t_0}^{t_F} dt \int_{\Omega_\alpha(t)} \left(\nabla^2 u_\alpha - \frac{1}{k_\alpha} \frac{\partial u_\alpha}{\partial t} \right) u_\alpha^* d\Omega = 0 \qquad \textbf{(41)}$$

A neater form of equation (41) is obtained by dropping most of the $\alpha-$subscripts and adopting the convention that the diffusivity and the variables u^*, u and their time and space derivatives are referred to as the $\alpha-$phase over which they are integrated. That is:

$$\int_{t_0}^{t_F} dt \int_{\Omega_\alpha(t)} \left(\nabla^2 u - \frac{1}{k} \frac{\partial u}{\partial t} \right) u^* d\Omega = 0 \qquad \textbf{(42)}$$

It should be emphasized at this point that the $\alpha-$phase spatial domain Ω_α is a function of time and is unknown a priori. It is mainly this fact that distinguishes the formulation for moving boundary problems from that for fixed boundaries. Proceeding as before, the Green's second identity is applied to the first term in the integrand of equation (42), producing:

$$\int_{t_0}^{t_F} dt \left\{ \int_{\Gamma_\alpha(t)} \left(u^* \frac{\partial u}{\partial n} - u \frac{\partial u^*}{\partial n} \right) d\Gamma + \int_{\Omega_\alpha(t)} \left(\nabla^2 u^* \right) u d\Omega \right\} - \int_{t_0}^{t_F} dt \int_{\Omega_\alpha(t)} \frac{1}{k} \left(\frac{\partial u}{\partial t} u^* \right) d\Omega = 0 \tag{43}$$

The last integral on the left-hand side of equation (43) is operated as follows:

$$\int_{t_0}^{t_F} dt \int_{\Omega_\alpha(t)} \frac{1}{k} \left(\frac{\partial u}{\partial t} u^* \right) d\Omega = \int_{t_0}^{t_F} dt \int_{\Omega_\alpha(t)} \frac{1}{k} \left(\frac{\partial}{\partial t} \left(u^* u \right) - u \frac{\partial u^*}{\partial t} \right) d\Omega \tag{44}$$

Applying Reynolds's transport theorem, see appendix C, to the first term in the right-hand side of equation (44), leads to:

$$\int_{t_0}^{t_F} dt \int_{\Omega_\alpha(t)} \frac{1}{k} \frac{\partial}{\partial t} \left(u^* u \right) d\Omega = \frac{1}{k} \int_{t_0}^{t_F} dt \frac{d}{dt} \int_{\Omega_\alpha(t)} \left(u^* u \right) d\Omega - \int_{t_0}^{t_F} dt \int_{\Gamma_\alpha(t)} \frac{1}{k} \left(u^* u \right) V_n d\Gamma \tag{45}$$

Therefore, equation (44) can be re-written as:

$$\begin{aligned} \int_{t_0}^{t_F} dt \int_{\Omega_\alpha(t)} \frac{1}{k} \left(\frac{\partial u}{\partial n} u^* \right) d\Omega = \frac{1}{k} \left[\int_{\Omega_\alpha(t)} \left(u^* u \right) d\Omega \right]_{t=t_0}^{t=t_F} \\ - \int_{t_0}^{t_F} dt \int_{\Gamma_\alpha(t)} \frac{1}{k} \left(u^* u \right) V_n d\Gamma - \int_{t_0}^{t_F} dt \int_{\Omega_\alpha(t)} \frac{1}{k} u \frac{\partial u^*}{\partial t} d\Omega \end{aligned} \tag{46}$$

Taking (46) into (44) and multiplying the whole equation by the diffusivity, one obtains:

$$\int_{t_0}^{t_F} dt \int_{\Omega_\alpha(t)} u(s,t)\left(k\nabla^2 u^*(p,t_F;s,t) + \frac{\partial u^*(p,t_F;s,t)}{\partial t}\right) d\Omega - \int_{\Omega_\alpha(t)} \left[u^*(p,t_F;s,t)u(s,t)\right]_{t=t_0}^{t=t_F} d\Omega$$

$$= -k\int_{t_0}^{t_F} dt \int_{\Gamma_\alpha(t)} \left(\begin{array}{l} u^*(p,t_F;s,t)\dfrac{\partial u(s,t)}{\partial n} - u(s,t)\dfrac{\partial u^*(p,t_F;s,t)}{\partial n} \\ +\dfrac{1}{k}u^*(p,t_F;s,t)u(s,t)V_n(s,t) \end{array}\right) d\Gamma$$

(47)

Therefore, the boundary integral representation for diffusion problems with moving boundaries takes the following form:

$$u(p,t_F) =$$

$$= k\int_{t_0}^{t_F} dt \int_{\Gamma_\alpha(t)} \left(u^*(p,t_F;s,t)\frac{\partial u(s,t)}{\partial n} - u(s,t)\frac{\partial u^*(p,t_F;s,t)}{\partial n} + \frac{1}{k}u^*(p,t_F;s,t)u(s,t)V_n(s,t)\right) d\Gamma$$

$$+ \int_{\Omega_\alpha(t_0)} u^*(p,t_F;s,t_0)u(s,t_0)d\Omega$$

(48)

Equation (48) gives the temperature $u(s,t_F)$ at the source point $P \in \Omega_\alpha(t)$ at time t_F in terms of the boundary temperature and their normal derivatives, as well as the initial temperature $u(s,t_0)$, being valid for any point in the domain $\Omega_\alpha(t_0)$. The general form of equation (48) valid for any position the point under consideration may assume, either in the domain or on the boundary, is:

$$c(p)u(p,t_F)=$$

$$= k\int_{t_0}^{t_F} dt \int_{\Gamma_\alpha(t)} \left(u^*(p,t_F;s,t)\frac{\partial u(s,t)}{\partial n} - u(s,t)\frac{\partial u^*(p,t_F;s,t)}{\partial n} + \frac{1}{k}u^*(p,t_F;s,t)u(s,t)V_n(s,t) \right) d\Gamma$$

$$+ \int_{\Omega_\alpha(t_0)} u^*(p,t_F;s,t_0)u(s,t_0)d\Omega$$

(49)

where the coefficient $c(p)$ on the left-hand side of equation (49) is a function of the position of the point P as well as of the smoothness of the boundary in its neighborhood. Equation (49) is the boundary integral representation of diffusion problems with moving boundaries.

Boundary Integral Representation of Phase Change Problems with Moving Boundaries and Mushy Zones

The steps for obtaining the boundary integral representation for diffusion problems with moving boundaries and mushy zones are similar to those presented before. The main difference herein is that each phase described by the diffusion equation will be bounded by two different types of boundaries; one is moving with unknown velocity should be determined as a major part of the solution, and the boundary will be fixed, see Fig. **(6)**.

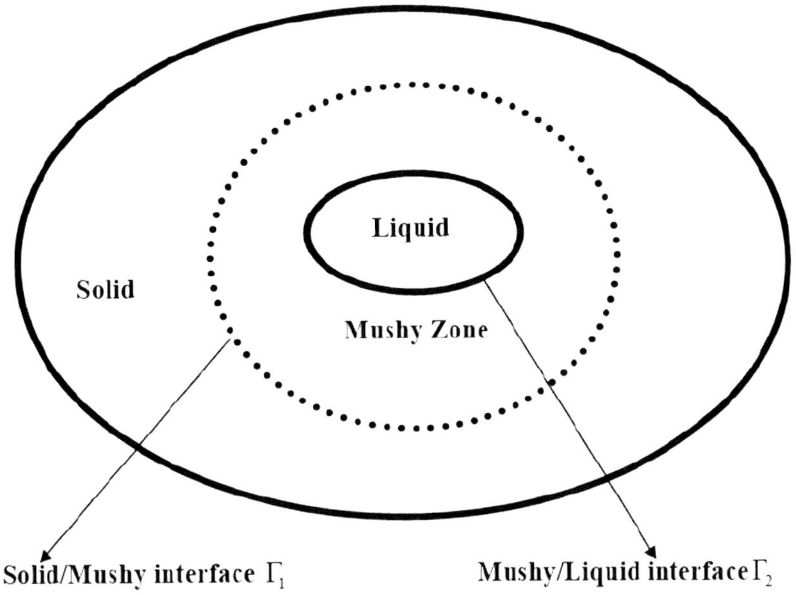

Fig. (6). Configuration of phase change problem with mushy zone.

Boundary Integral Equation (BIE) for Fixed-moving Phase

The general form of the boundary integral equation valid for any position the point under consideration may assume, either in the domain or on the boundary takes the following form:

$$
\begin{aligned}
&c(p)u(p,t_F) = \\
&= k\int_{t_0}^{t_F} dt \int_{\Gamma_\alpha(t)} \left(u^*(p,t_F;s,t)\frac{\partial u(s,t)}{\partial n} - u(s,t)\frac{\partial u^*(p,t_F;s,t)}{\partial n} + \frac{1}{k}u^*(p,t_F;s,t)u(s,t)V_n(s,t) \right) d\Gamma \\
&+ \int_{\Omega_\alpha(t_0)} u^*(p,t_F;s,t_0)u(s,t_0)d\Omega
\end{aligned}
\qquad (50)
$$

Boundary Integral Equation (BIE) for Moving-moving Phase

The general form of the boundary integral equation valid for any position the point under consideration may assume, either in the domain or on the boundary takes the following form:

$$
\begin{aligned}
&C(\zeta)u_\ell(\zeta,t) = \\
&= \alpha_\ell \int_{t_m+\Delta t}^{t_V} \left[u^*(\zeta,x,t_V,t)\frac{\partial u_\ell(x,t)}{\partial x} - u_\ell(x,t)\frac{\partial u^*(\zeta,x,t_V,t)}{\partial x} \right]_{x=R_2(t)}^{x=R_1(t)} dt \\
&+ \int_{t_m+\Delta t}^{t_e} \left[u_\ell(x,t)u^*(\zeta,x,t_V,t)\frac{dR_1(t)}{dt} \right]_{x=R_2(t)}^{x=R_1(t)} dt
\end{aligned}
\qquad \textbf{(51)}
$$

Boundary Element Method for Fixed Boundary Problems

In this section, attention is focused on the solution of equation (28) which relates boundary values of the temperature, its normal derivative and the initial values of the temperature over the domain. Since the analytical solution of this equation is only possible in the case of very simple geometries and boundary conditions, in general one must adopt a numerical procedure which reduces the integral equation to an algebraic form.

The Basic Steps of This Procedure, Which Constitute the Essence of the Boundary Element Method, can be Summarized as Follows:

(1) The original boundary is replaced by a finite number of boundary elements where the spatial and temporal variation of both temperature and its normal derivative are approximated using functions that ***locally*** interpolate the temperature and its normal derivative in terms of their values at nodal points of each element at each time step.

(2) The integral equation (28) is applied, in a discretized way, to each nodal point **p** of the boundary at the time $t = t_F$ and the integrals are computed over each element. Since the time variation of functions u and $\frac{\partial u}{\partial n}$ is not known a priori, a time-stepping technique has to be introduced. This produces a system of N algebraic equations involving N nodal values of u and $\frac{\partial u}{\partial n}$.

(3) The boundary conditions along the time history of the analysis are imposed at the nodes using the known values of either u or $\frac{\partial u}{\partial n}$. In this way, the number of unknowns is reduced to N and the algebraic system of linear equations can be solved for the remaining N unknowns at the time $t = t_F$.

(4) Once the boundary solution at time $t = t_F$ is found, the temperatures at arbitrary points at that time $t = t_F$ can be obtained by means of equation (25), also in a discretized form.

In the following sections, the above-mentioned steps for the numerical solution will be discussed in detail.

Boundary Discretization

In what follows, the discretization procedure for the solution of diffusion problems using boundary elements is described. There are essentially two sorts of approximation involved in the numerical solution of equation (28). The first one concerns the boundary geometry and the second the variation of the functions u *or* $\frac{\partial u}{\partial n}$ inside each element. In both cases, one can use interpolation functions in terms of nodal values to approximate the continuous variation of the geometry and unknowns by a discrete set of nodal values. These functions, usually written in terms of polynomials, are of the same sort as those used in finite elements, but using one dimension less. Thus, in three-dimensional problems they will be surface elements and in two-dimensional problems, they will be either straight line or curved segments. For each kind of domain two-dimension or three-dimension there will be different types of elements, depending on the function chosen to interpolate the geometry and the unknowns u *or* $\frac{\partial u}{\partial n}$ within each element [38-40].

Time Discretization

As diffusion problems have a transient character, and the time variation of functions representing the temperature and its normal derivative are not known a priori, a time-stepping technique has to be introduced for the numerical solution of equation (28). In the time-marching scheme employed here, which is a convolution-type one, the time integration process always restarts from the initial time and so, despite the increasing number of intermediate steps as time progresses, values of temperature at internal points need not be re-computed. In this scheme, one uses time-steps of constant size in fixed boundaries applications, only two new matrices for each node and time-step need to be computed and stored. This scheme also has the flexibility of allowing variable size time-steps and variable geometry. Furthermore, this scheme allows for the elimination of the domain initial conditions integral in two way common cases, transforming the formulation into a boundary-only one. The first case is that of a uniform initial temperature throughout the body. In this case, the domain integral is eliminated by shifting the temperatures scale by the value of the initial condition so that the initial temperature in the new scales equals zero. If the initial temperature is not constant but satisfies Laplace's equation, the domain integral in equation (28) can still be eliminated by transforming it into equivalent boundary integrals. The necessary procedures for the numerical implementation of the time-marching scheme here employed will be discussed.

Numerical treatment of the BIE of diffusion problem with fixed boundaries

The numerical solution of equation (28), the boundary Γ is discretized into a finite series of elements as was mentioned earlier. In order to do that, one chooses the type of boundary element among those described before. Furthermore, functions u and $\frac{\partial u}{\partial n}$ are assumed to vary within each element according to space and time interpolation functions such that:

$$u = \varphi^T \psi u^b \tag{52}$$

$$\frac{\partial u}{\partial n} = \varphi^T \psi \left(\frac{\partial u}{\partial n} \right)^b \tag{53}$$

Where φ and ψ are the space and the time interpolation functions, respectively, and the index b refers to the number of boundary nodes within each element to which nodal values of u and $\frac{\partial u}{\partial n}$ are associated, for instance, $b = 1$ for constant elements and $b = 2$ for linear elements, *etc.* When the original boundary Γ is approximated by a number N of elements Γ_j, the domain Ω subdivided into L cells, the time dimension subdivided into F time steps, due to of that equation (28) takes the following discretized form:

$$c_i u_F^i + k\sum_{j=1}^{N}\sum_{f=1}^{F}\left(\int_{\Gamma_j}\varphi^T\int_{t_{f-1}}^{t_F}\frac{\partial u^*}{\partial n}\psi\, dt d\Gamma\right)u^b = k\sum_{j=1}^{N}\sum_{f=1}^{F}\left(\int_{\Gamma_j}\varphi^T\int_{t_{f-1}}^{t_F}u^*\psi\, dt d\Gamma\right)\left(\frac{\partial u}{\partial n}\right)^b + \sum_{f=1}^{L}\int_{\Omega_j}u^* u_0 d\Omega \quad (54)$$

Linear Time Interpolation

Assuming a linear variation in time for functions u and $\frac{\partial u}{\partial n}$ within each time step, the following interpolation functions can be used:

$$\psi = \begin{Bmatrix}\psi_1 \\ \psi_2\end{Bmatrix} = \begin{Bmatrix}\dfrac{t_f - t}{\Delta t_f} \\ \dfrac{t - t_{f-1}}{\Delta t_f}\end{Bmatrix} \quad (55)$$

Where,

$$\Delta t_f = t_f - t_{f-1}$$

Applying equation (52) to all boundary nodes yields the following system of equations:

$$\sum_{f=1}^{F}\left(H_{fF}^1 u_{f-1} + H_{fF}^2 u_f\right) = \sum_{f=1}^{F}\left(G_{fF}^1\left(\frac{\partial u}{\partial n}\right)_{f-1} + G_{fF}^2\left(\frac{\partial u}{\partial n}\right)_f\right) + B_0\, \bar{u}_0 \quad (56)$$

Where the coefficients of the matrices involved are constituted of terms such as [41]:

$$h_{fF_{ij}}^{1,m} = \frac{k}{\Delta t_f}\int_{\Gamma_j}\varphi_m\int_{t_{f-1}}^{t_F}\left(t_f - t\right)\frac{\partial u^*}{\partial n}\, dt d\Gamma \quad (57)$$

$$h_{fF_{ij}}^{2,m} = \frac{k}{\Delta t_f} \int_{\Gamma_j} \varphi_m \int_{t_{f-1}}^{t_F} \left(t - t_{f-1}\right) \frac{\partial u^*}{\partial n} dt d\Gamma \tag{58}$$

$$g_{fF_{ij}}^{1,m} = \frac{k}{\Delta t_f} \int_{\Gamma_j} \varphi_m \int_{t_{f-1}}^{t_F} \left(t_f - t\right) u^* dt d\Gamma \tag{59}$$

$$g_{fF_{ij}}^{2,m} = \frac{k}{\Delta t_f} \int_{\Gamma_j} \varphi_m \int_{t_{f-1}}^{t_F} \left(t - t_{f-1}\right) u^* dt d\Gamma \tag{60}$$

Where the superscript m refers to the degree of variation of the function φ; *i.e.*, $m = 1$ if φ is constant, $m = 1,2$ if φ is linear, *etc.*, the $h_{fF_{ij}}^{1,m}$ and $g_{fF_{ij}}^{1,m}$ are influence coefficients defining the interaction between the point i under consideration at the time level t_F and a particular node i on an element j at the time t_f. When dealing with diffusion problems with fixed boundaries, one can invert the order of integration and perform the time integrals in equations (55-58) analytically.

Boundary Element Method for Moving Boundary Problems

In order to apply the boundary element method to solve equation (49), the general lines prescribed in section 7are still valid, but some additional procedures have to be adopted. The most striking difference between diffusion problems with fixed boundaries and the ones with moving boundaries is, of course, the presence of the moving boundary itself. As it has been said before, its position, which is unknown a priori, makes the problem non-linear, requiring iterations for its solution. In addition, as the boundary or interface moves, it assumes new positions during each time step (not only at the end of them), which makes the position a function of time. Thus, it is important to emphasize, once again, that although the explicit unknowns of equation (49) are the temperatures or their normal derivates and the velocity, the interface or boundary moves, so that the interface velocity enters this equation in an explicit as well as an implicit and non-linear manner through the fundamental solution. What this means is that, even within each time step, one can not invert the order of integration and precede the time integration analytically, as was done in a fixed boundaries problem, in reason of the time-dependence of the position. To overcome this difficulty, one has to use a new set of interpolation functions for the space, in which the position is expressed as a function of time, that is, $\phi = \phi(x(t))$ and to resort to simultaneous Gaussian quadrature in time and space. The procedures to this end will be presented in detail later. It should also be noticed that the position of the moving boundary is part of the solution and must be continuously updated in time. Another important aspect to be stressed at this point is related to the time-marching scheme, similar to the treatment given before, where the fixed boundary problems were studied, a convolution time-marching scheme is used to follow the time-history of the analysis. Therefore, when dealing with moving boundary problems, one must not forget that the fundamental solution is not only dependent on the time t_F and on time t, but also on the time-variable positions of the points P (source) and S (field). That is, despite the position of the source point being considered at a fixed instant in time t_F, the field point S assumes different positions throughout the time-history of the analysis.

Numerical Treatment of the BIE of Diffusion Problem with Moving Boundaries

As mentioned in the case of diffusion problems with fixed boundaries, linear sets of interpolation in time and space have been adopted. These sets are used for approximating the geometry (therefore, linear elements) and also for representing the variation of functions u, $\frac{\partial u}{\partial n}$ and V_n within each boundary element, such that:

$$u = \varphi^T \psi u^b \quad \textbf{(61)}$$

$$\frac{\partial u}{\partial n} = \varphi^T \psi \left(\frac{\partial u}{\partial n} \right)^b \quad \textbf{(62)}$$

Where, φ and ψ are the space and the time interpolation functions, respectively, and the index b refers to the number of boundary nodes within each element to which nodal values of u and $\frac{\partial u}{\partial n}$ are associated, and $b=2$ for linear. The two-dimensional fundamental solution for the thermal diffusion, u^b and its normal derivative $\frac{\partial u^b}{\partial n}$, for problems with either fixed or moving boundaries, are the same, is given by equations (59) and (60).Now then, one can write the discretized form of the boundary integral equation (49) as follows:

$$c_i u_F^i + k\sum_{f=1}^{F}\left(\int_{t_{f-1}}^{t_F} dt \sum_{j=1}^{N(t)} \int_{\Gamma_{\alpha j}(t)} \left[\frac{\partial u^*}{\partial n} - \frac{1}{k} u^* V_n\right] \varphi^T \psi \, d\Gamma\right) u^b = k\sum_{f=1}^{F}\left(\int_{t_{f-1}}^{t_F} dt \sum_{j=1}^{N(t)} \int_{\Gamma_{\alpha j}(t)} u^* \varphi^T \psi \, d\Gamma\right)\left(\frac{\partial u}{\partial n}\right)^b \quad \textbf{(63)}$$

As mentioned before, boundary discretization is a function of time. The reason is that as the phase change commences, elements on the moving boundary or interface undergo size variation, as their length are also functions of time, and eventually, nodes may come very close to each other, giving origin to numerical instabilities.

Linear Interpolation in Time and Space

Assuming linear variations for the functions representing the temperature and its normal derivative in space and within each time step, one has that the interpolation set for the time dimension is given by equation (53), but a new set must be defined for the space as a function of time, for the reasons explained above. Furthermore, both sets should also be expressed in terms of an intrinsic system of coordinates, as simultaneous Gaussian quadrature in time and space must be performed [42]. Applying equation (61) to all boundary nodes of each of the phases α yields the following system of equations:

$$\sum_{f=1}^{F}\left(H_{fF}^{1}-\frac{1}{k}G_{fF}^{1}V_{nf-1}\right)u_{f-1}+\left(H_{fF}^{2}-\frac{1}{k}G_{fF}^{2}V_{nf}\right)u_{f}= \\ =\sum_{f=1}^{F}\left(H_{fF}^{1}\left(\frac{\partial u}{\partial n}\right)_{f-1}+H_{fF}^{2}\left(\frac{\partial u}{\partial n}\right)_{f}\right) \tag{64}$$

Where, V_{nf} are diagonal matrices and the coefficients of the matrices H_{fF}^{1} and G_{fF}^{1} involved are constituted of terms such as:

$$h_{fF_{ij}}^{1,m}=k\int_{t_{f-1}}^{t_F}dt\int_{\Gamma_{\alpha j}}\Psi_1\varphi_m\frac{\partial u^*}{\partial n}d\Gamma \tag{65}$$

$$h_{fF_{ij}}^{2,m}=k\int_{t_{f-1}}^{t_F}dt\int_{\Gamma_{\alpha j}}\Psi_2\varphi_m\frac{\partial u^*}{\partial n}d\Gamma \tag{66}$$

$$g_{fF_{ij}}^{1,m}=k\int_{t_{f-1}}^{t_F}dt\int_{\Gamma_{\alpha j}}\Psi_1\varphi_m u^* d\Gamma \tag{67}$$

$$g_{fF_{ij}}^{2,m}=k\int_{t_{f-1}}^{t_F}dt\int_{\Gamma_{\alpha j}}\Psi_2\varphi_m u^* d\Gamma \tag{68}$$

In equations (63-66):

$$\phi=\begin{Bmatrix}\phi_1(x(t))\\ \phi_2(x(t))\end{Bmatrix}=\begin{Bmatrix}\dfrac{1-\zeta}{2}\\ \dfrac{1+\zeta}{2}\end{Bmatrix} \tag{69}$$

ζ is the local coordinate of any element as will be seen later in section (10) by equation (75).

And

$$\Psi = \begin{Bmatrix} \Psi_1(t) \\ \Psi_2(t) \end{Bmatrix} = \begin{Bmatrix} \dfrac{1-\eta}{2} \\ \dfrac{1+\eta}{2} \end{Bmatrix} \tag{70}$$

In equations (67) and (68):

$$\eta = \frac{2}{\Delta t_f} t - \frac{t_f + t_{f-1}}{\Delta t_f} \tag{71}$$

$$\Delta t_f = t_f - t_{f-1}$$

By substituting equations (69-70) into equations (65-68), equation (63) becomes ready for numerical evaluation for each phase.

INTEGRALS EVALUATION

Case when $i = j$

$$\overline{H}^{ij} = \int_{\Gamma_{i=j}} q^* d\Gamma = \int_{\Gamma_{i=j}} \frac{\partial u^*}{\partial n} d\Gamma = \int_{\Gamma_{i=j}} \left(\frac{\partial u^*}{\partial r} \frac{\partial r}{\partial n} \right) d\Gamma = 0 \tag{72}$$

As we see, the final result of equation (72) equals zero because the element coordinate and the normal to the boundary are usually perpendicular, as shown in Fig. **(7)**.

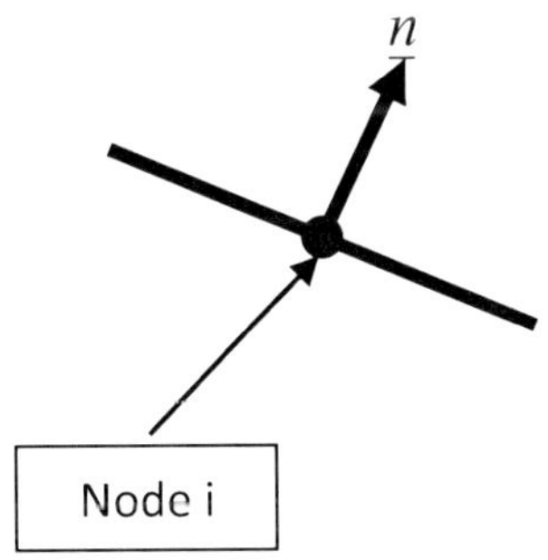

Fig. (7). Outward unit normal to the element.

G^{ij} **in Case** When $i = j$

$$G^{ij} = \int_{\Gamma_{i=j}} u^* d\Gamma \tag{73}$$

Making use of u^* in case of two-dimensional case problem, just as an example, one can get:

$$G^{ij} = \int_{\Gamma_{i=j}} u^* d\Gamma = \frac{1}{2\pi} \int_{\Gamma_{i=j}} \ln\left(\frac{1}{r}\right) dr \tag{74}$$

Refer to Fig. (**8**), one can get:

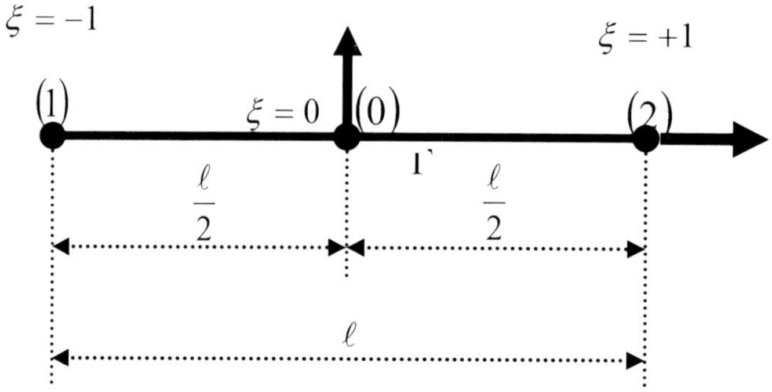

Fig. (8). Normalization of the element.

Let:

$$r = \left| \xi \times \frac{\ell}{2} \right| \tag{75}$$

In equation (75) ℓ represents the length of the element under consideration.

Now

$$G^{ij} = \frac{1}{2\pi}\int_{(1)}^{(2)} \ln\left(\frac{1}{r}\right)dr = \frac{1}{\pi}\int_{0}^{(2)} \ln\left(\frac{1}{r}\right)dr = \frac{1}{\pi}\int_{0}^{(2)} \ln\left(\frac{2}{\xi\ell}\right)d\left(r = \left|\xi \times \frac{\ell}{2}\right|\right) \quad \textbf{(76)}$$

Note:

$$d\xi = \left(\frac{2}{\ell}\right)dr \quad \textbf{(77)}$$

Substituting from equation (77) into the last term of equation (76) leads to:

$$\begin{aligned} G^{ij} &= \frac{1}{\pi}\frac{\ell}{2}\int_{0}^{(2)} \ln\left(\frac{2}{\xi\ell}\right)d\xi = \frac{\ell}{2\pi}\int_{0}^{(2)} \ln\left(\frac{1}{\left(\ell/2\right)} \times \frac{1}{\xi}\right)d\xi \\ &= \frac{\ell}{2\pi}\int_{0}^{1} \ln\left(\frac{1}{\left(\ell/2\right)}\right) + \ln\left(\frac{1}{\xi}\right)d\xi = \frac{\ell}{2\pi}\ln\left(\frac{1}{\left(\ell/2\right)}\right) + \frac{\ell}{2\pi}I_1 \end{aligned} \quad \textbf{(78)}$$

Now,

$$I_1 = \int_{0}^{1} \ln\left(\frac{1}{\xi}\right)d\xi \quad \textbf{(79)}$$

Let,

$$v = \ln\left(\frac{1}{\xi}\right) \Rightarrow \xi = e^{-v} \Rightarrow d\xi = -e^{-v}dv$$

New limits of integrations:

$\xi = 0 \Leftrightarrow v \to \infty$

&

$\xi = 1 \Leftrightarrow v = 0$

Then the integral in equation (79) takes the form:

$$I_1 = \int_{\infty}^{0} v\left(-e^{-v}\right)dv = 1 \tag{80}$$

Substituting from (80) into equation (78), leads to:

$$G^{ij} = \frac{\ell}{2\pi}\left[\ln\left(\frac{1}{\left(\ell/2\right)}\right) + 1\right] \tag{81}$$

Case when $i \neq j$

In this case, the integrals are calculated using Gaussian quadrature rules. There are two different cases for such integrals, one for singular integrals and the other for non-singular integrals [43-68]. The details for both types to show how these integrals can be calculated are very important for the computer codes.

COMPUTER CODE

Starting by the weighted residual technique of the diffusion equation as stated before in previous sections until reaching the boundary integral equation, which could be discretized in time and boundary and then transformed to several integrals in time along the boundary, which can be calculated by knowing the fundamental solution of the problem and boundary conditions to get a system of equations of the unknowns which may be the potential or the potential derivative according to the boundary conditions, and by solving this system of equations we can get the overall unknowns of the problem.

The code consists of the main program and nine subroutines, as follows:

Main Program

Define the maximum of the system of equations, input and output units and from which we call any subroutine according to the required mission.

Subroutine -1- INPUT

This subroutine is available and modified according to the requirements of the problem underhand it reads all input data.

Subroutine -2- GHMAT

Form the matrices H and G

Subroutine -3- TIMEINT

This subroutine evaluates the time integrals appear throughout any time step

Subroutine -4- SYSTEM

- Re-arrange the matrices H and G to matrix A in accordance with the boundary conditions
- Create the R.H.S. vector F in the system AX=F

Subroutine -5- SOLSYS

Solve the system AX=F using Gauss elimination method and in the present work, we use this subroutine as it is.

Subroutine -6- VERFIC

This subroutine is used to verify the energy balance at the moving interface by evaluating the right-hand side alone and left-hand side, then computes the difference and compares this difference with a prescribed tolerance.

Subroutine -7- UPDATING

This subroutine updates the location of the moving boundary based on the linear relation between time and position within each time step.

Subroutine -8- INTERP

Compute the unknowns at the internal points.

Subroutine -9- OUTPUT

Write the output data and the results in a file from which we can plot these results or present it as it is in a table.

TIME MARCHING SCHEME

Time dependent problems involve time integrals appearing throughout each time step; therefore, these integrals should be evaluated. There are two different worldwide computational schemes for such integrals.

Scheme (1)

This scheme treats each time step as a new problem and so, at the end of each step, it computes values of the function at a sufficient number of internal points to use them as pseudo-initial values for the next time step.

Scheme (2)

In this scheme, the time integration process always restarts from t_o and so, despite the increasing number of the intermediate steps as time progresses, values of u at the internal points ***need not be re-computed***. In the present work, all-time computations are carried out using the second time marching scheme.

Updating the moving boundaries

There are three major problems when dealing with moving boundary problems:

(1) How to guess an initial position for the moving boundary?

(2) How to update the wrong guess?

(3) What is the relation between the position at time step t_j and that at the next time step t_{j+1}

First of all, let us starts answering the 1st question, there is no specified method to try a correct guess for the moving boundary, it depends mainly on the nature of the problem and also by experience. For example, when dealing with melting/solidification problems, it was found that the moving interface takes exponential nature (increasing or decreasing) therefore, most researchers in this field found that the best guess, in the beginning is a horizontal surface in 2-D problems then by iterations this position will be corrected after satisfying the energy balance and the remaining boundary conditions at the moving boundary. The situation will be different when dealing with alloys problems involving impurities or dissolution problems, it was found from the experimental works that the initial guess will be a fraction from the total length of the mold. Secondly, the initial guess of the moving boundary sure will not be correct. It requires some iterations based on achieving mainly the heat or mass or both balance and the boundary condition (s) over it. To do that, a linear variation between the position and the time within time step is assumed, and this is the answer of question number three, see Fig. (**9**); *i.e.*,

$$X(t)^{J+1} = X(t)^{J} + V(t)^{J} \times \Delta t$$

In which:

$X(t)^{J}$: Position of the moving boundary at time t^{J}

$V(t)^{J}$: Velocity of the moving boundary at time t^{J}

$X(t)^{J+1}$: Position of the moving boundary at time t^{J+1}

Δt : Time step

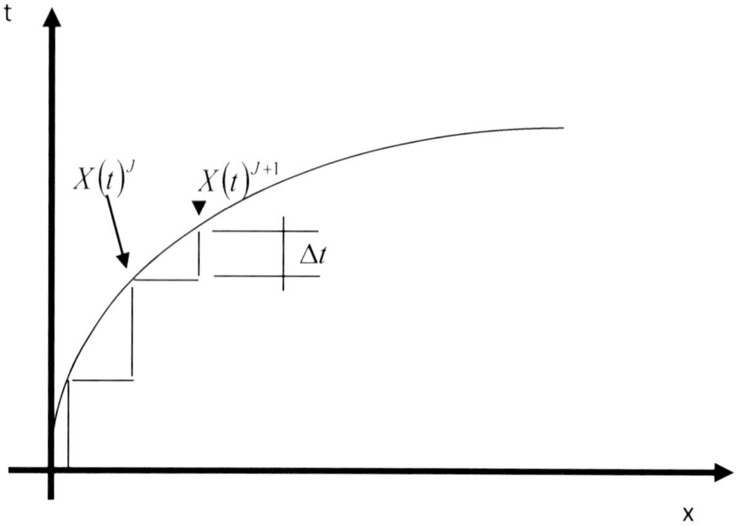

Fig. (9). Linear variation between the position and time.

REFERENCES

[1] M. Faghri, E.M. Sparrow, and A.T. Prata, "Finite difference solutions of convection-diffusion problems in irregular domains, using a non-orthogonal coordinates transformations", *Numer. Heat Transf.,* vol. 7, no. 2, pp. 183-209, 1984. [http://dx.doi.org/10.1080/01495728408961819]

[2] A.E. Croucher, and M.J. O' Sullivan, "Numerical methods for contaminant transport in rivers and estuaries", *Comput. Fluids,* vol. 27, no. 8, pp. 861-878, 1998. [http://dx.doi.org/10.1016/S0045-7930(98)00012-7]

[3] L. Gregory, "Young, Karen A. Mc Donald, Ahmet Polazoglu and William Ford,' Boundary element solutions for free boundary convection-diffusion problems", *Numer. Heat Transf. A,* vol. 21, no. 3, pp. 299-311, 1992. [http://dx.doi.org/10.1080/10407789208944878]

[4] A. Robert, "Brown and Do Hyun Kim,' Modeling of directional solidification: from Scheil to detailed numerical numerical simulation", *J. Cryst. Growth,* vol. 109, no. 1-4, pp. 50-65, 1991. [http://dx.doi.org/10.1016/0022-0248(91)90157-Z]

[5] E. Balsa-Canto, A.A. Alonso, and J.R. Banga, "Antonio A. Alonso and Julio R. Banga,' A novel efficient and reliable method for thermal process design and optimization. Part I: theory", *J. Food Eng.,* vol. 52, no. 3, pp. 227-234, 2002. [http://dx.doi.org/10.1016/S0260-8774(01)00110-8]

[6] V.A.F. Costa, and A.C.M. Sousa, "Modeling of flow and thermo kinetics during the cure of thick laminated composites", *Int. J. Therm. Sci.,* vol. 42, no. 1, pp. 15-22, 2005. [http://dx.doi.org/10.1016/S1290-0729(02)00003-0]

[7] S.M. Pimputkar, and S. Ostrach, "Convective effects in Crystals growth from the melt", *J. Cryst. Growth,* vol. 55, no. 3, pp. 614-646, 1981. [http://dx.doi.org/10.1016/0022-0248(81)90121-4]

[8] M. Yildiz, and S. Dost, "A continuum model for the liquid phase diffusion growth of bulk SiGe single crystals", *Int. J. Eng. Sci.,* vol. 43, no. 13-14, pp. 1059-1080, 2005. [http://dx.doi.org/10.1016/j.ijengsci.2005.06.001]

[9] E. James, "Simpson and Suresh V. Garimella,' The influence of gravity levels on the horizontal Bridgman crystal growth of an alloy", *Int. J. Heat Mass Transf.,* vol. 43, no. 11, pp. 1905-1923, 2000. [http://dx.doi.org/10.1016/S0017-9310(99)00275-6]

[10] D.W. Mackowski, V.R. Rao, D.G. Walker, and R.W. Knight, "Numerical investigation of the effects of thermal creep in physical vapor transport", *J. Cryst. Growth,* vol. 179, no. 1-2, pp. 297-308, 1997. [http://dx.doi.org/10.1016/S0022-0248(97)00087-0]

[11] R.W. Lewis, K. Morgan, H.R. Thomas, and Seetharamu, *The Finite Element Method in Heat Transfer Analysis.* Wiely: Chichester, UK, 1996.

[12] Y. Wu, J.M. Hill, and P.J. Flint, "A novel finite element method for heat transfer in the continuous caster", *J. Aust. Math. Soc. B Appl. Math.,* vol. 35, no. 3, pp. 263-288, 1994. [http://dx.doi.org/10.1017/S0334270000009292]

[13] D.J. Mcdaniel, and N. Zabaras, "A least-squares front tracking finite element method analysis of phase change with natural convection", *Int. J. Numer. Methods Eng.,* vol. 37, no. 16, pp. 2755-2777, 1994. [http://dx.doi.org/10.1002/nme.1620371605]

[14] A. Fic, A.J. Nowak, and R. Bialecki, "Heat transfer analysis of the continuous casting processthe front tracking'", *Eng. Anal. Bound. Elem.,* vol. 24, no. 3, pp. 215-223, 2000. [http://dx.doi.org/10.1016/S0955-7997(00)00004-7]

[15] R. Cholewa, A.J. Nowak, and L.C. Wrobel, "Application of BEM and sensitivity analysis to the solution of the governing diffusion-convection equation for a continuous casting process", *Eng. Anal. Bound. Elem.,* vol. 28, no. 4, pp. 389-403, 2004. [http://dx.doi.org/10.1016/S0955-7997(03)00086-9]

[16] R.A. Baialecki, A. Fic, A.J. Nowak, and B. Sarler, *BEM approaches to phase change system comparative studies of front tracking and enthalpy method scheme', Advanced Computational Methods in Heat Transfer V.* Computational Mechanics Publication: Southampton, UK, 1998, pp. 201-213.

[17] I. Nowak, A.J. Nowak, and L.C. Wrobel, "Tracking of phase change front for continuous casting-inverse BEM solution", In: *Proceedings of the ISIP,* M. Tanaka, G.S. Dulikravich, Eds., Elsevier: Nagano, Japan, Berlin, 2000, pp. 71-80. [http://dx.doi.org/10.1016/B978-008043693-7/50080-5]

[18] R. Cholewa, A.J. Nowak, and L.C. Wrobel, "Front tracking BEM solution of the heat transfer problem in the continuous casting process", *Fifth World Congress on Computational Mechanics* Viena, Austria3-9501554-0-6

[19] J. Crank, *Free and Moving Boundary Problems.* Clarendo Press: Oxford, 1984.

[20] I.V. Alexandrova, D.V. Alexandrov, D.L. Aseev, and S.V. Bulitcheva, "Mushy layer formation during solidification of binary alloys from a cooled wall: The role of boundary conditions", *Acta Phys. Pol. A,* vol. 115, no. 4, pp. 791-794, 2009. [http://dx.doi.org/10.12693/APhysPolA.115.791]

[21] D. Buddhi, A simplified method of determining the latent heat of fusion. Indian science congress: Indore, India, 1991.

[22] S.L. Mitchell, and M. Vynnycky, "Finite-difference methods with increased accuracy and correct initialization for one-dimensional Stefan problems", *Appl. Math. Comput.,* vol. 215, no. 4, pp. 1609-1621, 2009. [http://dx.doi.org/10.1016/j.amc.2009.07.054]

[23] T.R. Goodman, "The Heat Balance Integral and its Application to Problems Involving a change of Phase", *Trans. Am. Soc. Mech. Eng.,* vol. 80, no. 2, pp. 335-342, 1958. [http://dx.doi.org/10.1115/1.4012364]

[24] J. Mennig, and M.N. Ozisik, "Coupled Integral Equation Approach for solving Melting or Solidification", *Int. J. Heat Mass Transf.,* vol. 28, no. 8, pp. 1481-1485, 1985. [http://dx.doi.org/10.1016/0017-9310(85)90250-9]

[25] H. Ding, M.N.H. Gilani, and P.D.M. Spelt, "Sliding, pinch-off and detachment of a droplet on a wall in shear flow", *J. Fluid Mech.,* vol. 644, pp. 217-244, 2010. [http://dx.doi.org/10.1017/S0022112009992217]

[26] A. Saito, H.K. Hong, and O. Hirokane, "Heat transfer enhancement in the direct contact melting process", *Int. J. Heat Mass Transf.,* vol. 35, no. 2, pp. 295-305, 1992. [http://dx.doi.org/10.1016/0017-9310(92)90268-W]

[27] J.S. Wettlaufer, M.G. Worster, and H.E. Huppert, "Solidification of leads: Theory, experiment and field observations", *J. Geophys. Res.,* vol. 105, no. C1, pp. 1123-1134, 2000. [http://dx.doi.org/10.1029/1999JC900269]

[28] S.G. Ahmed, "Boundary integral formulation and mushy zone model for phase change problem", *Int. J. Numer. Methods Heat Fluid Flow,* vol. 11, no. 4, pp. 371-379, 2001. [http://dx.doi.org/10.1108/09615530110389207]

[29] S.G. Ahmed, "A New semi-analytical method for phase transformations in binary", *J. Comput. Appl. Math.,* vol. 206, no. 1, pp. 409-419, 2007. [http://dx.doi.org/10.1016/j.cam.2006.08.001]

[30] S.G. Ahmed, "' A New Numerical Algorithm for 2-D Moving Boundary Problems using Boundary Element Method', Computers and Mathematics with Applications", *Finally,* 2009.

[31] P. Lancaster, and K. Salkauskas, "Surfaces generated by moving least squares methods", *Math. Comput.,* vol. 37, no. 155, pp. 141-158, 1981. [http://dx.doi.org/10.1090/S0025-5718-1981-0616367-1]

[32] S.K.M. Chenoweth, J. Soria, and A. Ooi, "A singularity-avoiding moving least squares scheme for two-dimensional unstructured meshes", *J. Comput. Phys.,* vol. 228, no. 15, pp. 5592-5619, 2009. [http://dx.doi.org/10.1016/j.jcp.2009.04.036]

[33] Z.Q. Cheng, Y.Z. Wang, and B. Li, "A Survey of Methods for Moving Least Squares Surfaces", *IEEE/EG Symposium on Volume and Point-Based Graphics,* 2008 .

[34] D. Levin, "The approximation power of moving least-squares", *Math. Comput.,* vol. 67, no. 224, pp. 1517-1531, 1998. [http://dx.doi.org/10.1090/S0025-5718-98-00974-0]

[35] T. Belytschko, Y.Y. Lu, and L. Gu, "Element-free galerkin methods", *Int. J. Numer. Methods Eng.,* vol. 37, no. 2, pp. 229-256, 1994. [http://dx.doi.org/10.1002/nme.1620370205]

[36] S.N. Atluri, and T. Zhu, "A new meshless local Petrov–Galerkin (MLPG) approach in computational mechanics", *Comput. Mech.,* vol. 22, no. 2, pp. 117-127, 1998. [http://dx.doi.org/10.1007/s004660050346]

[37] K. Salkauskas, "Moving least squares interpolation with thin-plate splines and radial basis functions", *Comput. Math. Appl.,* vol. 24, no. 12, pp. 177-185, 1992. [http://dx.doi.org/10.1016/0898-1221(92)90178-K]

[38] M. Kunle, Entwicklung und untersuchung von moving least square verfahren zur numerischen simulation hydrodynamischer gleichungen., 2001.

[39] K.M. Liew, Y. Cheng, and S. Kitipornchai, "Boundary Element-free method (BEFM) for two-dimensional elastodynamic analysis using Laplace transforms", *Int. J. Numer. Methods Eng.,* vol. 64, no. 12, pp. 1610-1627, 2005. [http://dx.doi.org/10.1002/nme.1417]

[40] Z. Zhang, K.M. Liew Yumin Cheng, and Y.Y. Lee, "Analyzing 2D fracture problems with the improved element-free Galerkin method", *Eng. Anal. Bound. Elem.,* vol. 32, no. 3, pp. 241-250, 2008. [http://dx.doi.org/10.1016/j.enganabound.2007.08.012]

[41] Z. Zhang, P. Zhao, and K.M. Liew, "Improved element-free Galerkin method for two-dimensional potential problems", *Eng. Anal. Bound. Elem.,* vol. 33, no. 4, pp. 547-554, 2009. [http://dx.doi.org/10.1016/j.enganabound.2008.08.004]

[42] E.J. Kansa, "Multiquadrics- a scattered data approximation scheme with applications to computational fluid dynamics.: Surface approximations and partial derivative estimates", *Comput. Math. Appl.,* vol. 19, no. 8-9, pp. 127-145, 1990. [http://dx.doi.org/10.1016/0898-1221(90)90270-T]

[43] G.E. Fasshauer, "Solving partial differential equations by collocation with radial basis functions", In: *Surface Fitting and Multiresolution Methods.,* A. Le Mehaute, C. Rabut, L.L. Schumaker, Eds., Vanderbilt University Press: Nashville, 1997, pp. 131-138.

[44] X. Liu, and K. Tai, "Point Interpolation Collocation Method for the Solution of Partial Differential Equations", *Eng. Anal. Bound. Elem.,* vol. 30, no. 7, pp. 598-609, 2006. [http://dx.doi.org/10.1016/j.enganabound.2005.12.003]

[45] A. Intessar, and S.G.Ahmed. Hadi, "Numerical Solution for Advection-diffusion Equation using Collocation Based New Radial Basis Function", *Iraqi Journal for Mechanical and Material Engineering,* vol. 16, no. 2, 2016.

[46] V.S. Ajaev, "Spreading of thin volatile liquid droplets on uniformly heated surfaces", *J. Fluid Mech.,* vol. 528, pp. 279-296, 2005. [http://dx.doi.org/10.1017/S0022112005003320]

[47] J.G. Wang, and G.R. Liu, "A point interpolation meshless method based on radial basis functions", *Int. J. Numer. Methods Eng.,* vol. 54, no. 11, pp. 1623-1648, 2002. [http://dx.doi.org/10.1002/nme.489]

[48] H. Wenland, "Piecewise polynomial, positive definite and compactly supported radial basis functions of minimal degree", *Adv. Comput. Math.,* p. 4, 1995.

[49] S. David, K. Sefiane, and L. Tadrist, "Experimental investigation of the effect of thermal properties of the substrate in the wetting and evaporation of sessile drops", *Colloids Surf. A Physicochem. Eng. Asp.,* vol. 298, no. 1–2, pp. 108-114, 2007. [http://dx.doi.org/10.1016/j.colsurfa.2006.12.018]

[50] J.W. Cahn, and J.E. Hilliard, "Free energy of a nonuniform system. Part 3. Nucleation in a two-component incompressible fluid", *J. Chem. Phys.,* vol. 31, no. 3, pp. 688-699, 1959. [http://dx.doi.org/10.1063/1.1730447]

[51] J. Crank, *Free and Moving Boundary Problems.* Clarendo Press: Oxford, 1984.

[52] D. Buddhi, *A simplified method of determining the latent heat of fusion* Indian science congress: Indore. India, 1991, pp. 2-5.

[53] K. Ling, and W-Q. Tao, "A sharp-interface model coupling VOSET and IBM for simulations on melting and solidification", *Comput. Fluids,* vol. 178, no. 26, pp. 113-131, 2018. [http://dx.doi.org/10.1016/j.compfluid.2018.08.027]

[54] T.G. Myers, and S.L. Mitchell, "Application of the combined integral method to Stefan problems", *Appl. Math. Model.,* vol. 35, no. 9, pp. 4281-4294, 2011. [http://dx.doi.org/10.1016/j.apm.2011.02.049]

[55] E.J. Kansa, "Multiquadrics—A scattered data approximation scheme with applications to computational fluid-dynamics—I surface approximations and partial derivative estimates", *Comput. Math. Appl.,* vol. 19, no. 8–9, pp. 127-145, 1990. [http://dx.doi.org/10.1016/0898-1221(90)90270-T]

[56] H.Y. Erbil, "Evaporation of pure liquid sessile and spherical suspended drops: a review", *Adv. Colloid Interface Sci.,* vol. 170, no. 1-2, pp. 67-86, 2012. [http://dx.doi.org/10.1016/j.cis.2011.12.006] [PMID: 22277832]

[57] S.L. Mitchell, and T.G. Myers, "Application of standard and refined heat balance integral methods to one-dimensional Stefan problems", *SIAM Rev.,* vol. 52, no. 1, pp. 57-86, 2010. [http://dx.doi.org/10.1137/080733036]

[58] E.J. Kansa, "Multiquadrics: A scattered data approximation scheme with applications to computation fluid-dynamics-II: solution to parabolic, hyperbolic and elliptic partial differential equations", *Comput. Math. Appl.,* vol. 19, no. 8-9, pp. 147-161, 1990. [http://dx.doi.org/10.1016/0898-1221(90)90271-K]

[59] R.D. Deegan, O. Bakajin, T.F. Dupont, G. Huber, S.R. Nagel, and T.A. Witten, "Contact line deposits in an evaporating drop", *Phys. Rev. E Stat. Phys. Plasmas Fluids Relat. Interdiscip. Topics,* vol. 62, no. 1 Pt B, pp. 756-765, 2000. [http://dx.doi.org/10.1103/PhysRevE.62.756] [PMID: 11088531]

[60] Y.L. Wu, and C. Shu, "Development of RBF-DQ method for derivative approximation and its application to simulate natural convection in concentric annuli", *Comput. Mech.,* vol. 29, no. 6, pp. 477-485, 2002. [http://dx.doi.org/10.1007/s00466-002-0357-4]

[61] Z Wu, and YC Hon, "Convergence error estimate in solving free boundary diffusion problem by radial basis functions method", *Eng Anal Bound Elements 27,* vol. 27, pp. 73-9, 2003. [http://dx.doi.org/10.1016/S0955-7997(02)00083-8]

[62] G.E. Fasshauer, "Solving partial differential equations by collocation with radial basis functions", In: *Surface Fitting and Multiresolution Methods.,* A. Le Mehaute, C. Rabut, L.L. Schumaker, Eds., Vanderbilt University Press: Nashville, 1997, pp. 131-138.

[63] K. Djidjeli, P.P. Chinchapatnam, P.B. Nair, and W.G. Price, "Global and compact meshless schemes for the unsteady convection-diffusion equation", *International Symposium on Health Care and Biomedical Research Interaction,* 2004 Oujda, Morocco

[64] J.W. Cahn, "On spinodal decomposition", *Acta Metall.,* vol. 9, no. 9, pp. 795-801, 1961. [http://dx.doi.org/10.1016/0001-6160(61)90182-1]

[65] A. Carlson, M. Do-Quang, and G. Amberg, "Modeling of dynamic wetting far from equilibrium", *Phys. Fluids,* vol. 21, no. 12, p. 121701, 2009. [http://dx.doi.org/10.1063/1.3275853]

[66] Delima-Silva Jnr and Wrobel, "BEM analysis of melting problems", In: *Boundary Element Technology IX* Computational Mechanics Publications: Southampton, 1994.

[67] L.C. Wrobel, and C.A. Brebbia, "Computational Modelling of Free and Moving Boundary Problems", In: *xHeat Transfer.* vol. 2. Computational Mechanics Publications: Southampton and Walter de Grulyter Berlin, 1991.

[68] S.G. Ahmed, "A combined domain and integral techniques to solve planar oxidation problem", *Six International Conference of Free and Moving Boundary Problems,* 2001 Lemnos, Greece.

CHAPTER 4

Boundary Element Methods for Phase Change Over Domains with Fixed Boundaries

Abstract: In the present chapter, the principles of the boundary element methods for phase change problems over domains with fixed boundaries are developed briefly. The details are developed starting from the 1-D diffusion problems over domains with fixed boundaries subjected to different types of boundary and initial conditions. In these problems, a new criterion has emerged: the existence of a moving boundary, which requires special treatment when solving overall problems.

Keywords: Boundary elements methods, phase change problems, moving boundary problems, Stefan condition.

INTRODUCTION

In chapter one, the boundary element methods are introduced in detail to solve potential problems, followed by an analysis for different parameters that directly affect the accuracy of the output results. The Stefan problems over domains with fixed boundaries will be discussed with brief details so that the beginners of the boundary element methods in the branch of Stefan problems will not find any difficulties with this branch of engineering applications. Let us build this chapter on the solutions of the following general heat diffusion problems over the fixed domain, the governing equation with the associated boundary and initial conditions can be stated as follow [1-30]:

$$k\frac{\partial^2 u(x,t)}{\partial x^2}=\frac{\partial u(x,t)}{\partial t} \qquad 0<x<\ell \tag{1}$$

Boundary conditions

$$u=\bar{u} \qquad \text{Dirichlet} \tag{2}$$

$$q=\bar{q} \qquad \text{Neumann} \tag{3}$$

$$\frac{\partial u(x,t)}{\partial t}=\frac{h}{k}(u_a-u) \tag{4}$$

Initial conditions

S. G. Ahmed

$$u(x,0) = u_0(x) \tag{5}$$

In these equations:

k: Thermal conductivity

h: Heat transfer coefficients

u_a: Ambient temperature

BOUNDARY ELEMENT FORMULATION

Starting by the weighted residual statement for diffusion equation as follows [31-40]:

$$\int_0^\tau \int_\Omega \left(k \frac{\partial^2 u}{\partial x^2} - \frac{\partial u}{\partial t} \right) u^*(\xi, x; \tau, t) d\Omega dt = 0 \tag{6}$$

In equation **(6)** $u^*(\xi, x; \tau, t)$ is the fundamental solution for the diffusion equation and it is defined as [41-45]:

$$u^*(\xi, x; \tau, t) = \frac{1}{\sqrt{4\pi k(\tau - t)}} \exp\left[-\frac{r^2(\xi, x)}{4k(\tau - t)} \right]$$

In which, $r^2(\xi, x) = X(\xi) - X(x)$ and $X(\xi)$ is the source point while $X(x)$ is the field point.

Integrating equation **(6)** twice by parts leads to:

$$\int_0^\tau \int_\Omega \left(k \frac{\partial^2 u}{\partial x^2} - \frac{\partial u}{\partial t} \right) u^*(\xi, x; \tau, t) d\Omega dt$$

$$= \int_0^\tau \int_\Omega \left(k \frac{\partial^2 u^*(\xi, x; \tau, t)}{\partial x^2} + \frac{\partial u^*(\xi, x; \tau, t)}{\partial t} \right) u d\Omega dt \tag{7}$$

$$- k \int_0^\tau \int_\Gamma q^*(\xi, x; \tau, t) u d\Gamma dt = 0$$

Equation (**7**) can take the following simplified form:

$$\int_0^\tau \int_\Omega \left(k\frac{\partial^2 u}{\partial x^2} - \frac{\partial u}{\partial t} \right) u^*(\xi, x; \tau, 0)\, d\Omega\, dt = \\ = -k\int_0^\tau \int_\Omega u^*(\xi, x; \tau, 0)\, q\, d\Gamma\, dt - \int_0^\tau \int_\Gamma u(x,0)\, q^*(\xi, x; \tau, 0)\, d\Omega = 0 \tag{8}$$

For one-dimensional problems, equation (**8**) takes the following form:

$$\int_0^\tau \int_0^\ell \left(k\frac{\partial^2 u^*(\xi, x; \tau, t)}{\partial x^2} + \frac{\partial u^*(\xi, x; \tau, t)}{\partial t} \right) u\, dx\, dt \\ - k\int_0^\tau \int_0^\ell q^*(\xi, x; \tau, t)\, u\, dx\, dt = 0 \tag{9}$$

Note that in equation (**9**):

$$k\frac{\partial^2 u^*(\xi, x; \tau, t)}{\partial x^2} + \frac{\partial u^*(\xi, x; \tau, t)}{\partial t} = -\delta_i \tag{10}$$

And,

$$-\delta_i u^* = -u^* \tag{11}$$

Making use of equations (**10**) and (**11**) into equation (**9**), the last one takes the following form:

$$-k\int_0^\tau \int_0^\ell u^*(\xi, x; \tau, t)\, q\, dx\, dt - \int_0^\ell u(x,t)\, u^*(\xi, x; \tau, t)\, dx = 0 \tag{12}$$

For ***any point,*** the integral equation takes the following form:

$$u(\xi,\tau)=\int_0^\ell u(x,0)u^*(\xi,x;\tau,0)dx$$
$$+k\int_0^\tau\int_0^\ell u^*(\xi,x;\tau,t)q(x,t)dx\,dt-k\int_0^\tau\int_0^\ell q^*(\xi,x;\tau,t)u(x,t)dx\,dt \tag{13}$$

Note the following:

(1) The flux in $\int_0^\tau\int_0^\ell u^*(\xi,x;\tau,t)q(x,t)dx\,dt$ is a given boundary condition

(2) The potential in $\int_0^\tau\int_0^\ell q^*(\xi,x;\tau,t)u(x,t)dx\,dt$ is unknown

DISCRETIZATION OF TIME

Equation (1**3)** after the discretization of time takes the following form [71]:

$$u(\xi,t_n)=\int_0^\ell u(x,0)u^*(\xi,x;t_n,0)dx+k\sum_{i=1}^{n}\int_{t_{i-1}}^{t_i}\left[\frac{\partial u(x,t_i)}{\partial x}u^*(\xi,x;t_n,t)dt\right]_0^\ell$$
$$-k\sum_{i=1}^{n}\int_{t_{i-1}}^{t_i}\left[u(x,t_i)\frac{\partial u^*(\xi,x;t_n,t)}{\partial n}dt\right]_0^\ell \tag{14}$$

Assume that the potential u and the flux q are constant within each time step; therefore, the equation (4.14) can be written as:

$$u(\xi,t_n)=\int_0^{\ell}u(x,0)u^*(\xi,x;t_n,0)dx++k\sum_{i=1}^{n}\frac{\partial u(x,t_i)}{\partial x}\int_{t_{i-1}}^{t_i}\left[u^*(\xi,x;t_n,t)dt\right]_0^{\ell}$$
$$-k\sum_{i=1}^{n}\int_{t_{i-1}}^{t_i}u(x,t_i)\left[\frac{\partial u^*(\xi,x;t_n,t)}{\partial n}dt\right]_0^{\ell} \quad \textbf{(15)}$$

Equation (**15**) can be written in a more general form as follow:

$$u(\xi,t_n)=\int_0^{\ell}u(x,0)u^*(\xi,x;t_n,0)dx$$
$$+\int_0^{t_F}\left[q(x,t)u^*(\xi,x;t_n,t)-u(x,t)q^*(\xi,x;t_n,t)\right]dt \quad \textbf{(16)}$$

In equation (1**6**), we have two different time integrals, they are:

$$I_1=\int_0^{t_F}u^*(\xi,x;t_n,t)dt \quad \textbf{(17)}$$

And,

$$I_2=\int_0^{t_F}q^*(\xi,x;t_n,t)dt \quad \textbf{(18)}$$

EVALUATION TIME INTEGRALS

In this section, we will present briefly how to evaluate the integrals given by equations (17) and (18), respectively.

Integral given by equation (17)

$$I_1 = \int_0^{t_F} u^*(\xi, x; t_n, t)dt = \int_0^{t_F} \left\{ \frac{1}{\sqrt{4\pi k(t_n - t)}} \exp\left[-\frac{r^2(\xi, x)}{4k(t_n - t)} \right] \right\} dt$$
$$= \frac{1}{2\sqrt{\pi k}} \int_0^{t_F} \left\{ \frac{1}{\sqrt{(t_n - t)}} \exp\left[-\frac{r^2(\xi, x)}{4k(t_n - t)} \right] \right\} dt \quad \textbf{(19)}$$

Let,

$$z = \frac{r^2(\xi, x)}{4k(t_n - t)} \Rightarrow (t_n - t) = \frac{r^2(\xi, x)}{4kz} \Rightarrow t = t_n - \frac{r^2(\xi, x)}{4kz} \quad \textbf{(20)}$$

Therefore,

$$dt = -\frac{r^2}{4kz}\left(-\frac{dz}{z^2}\right)$$
$$\Rightarrow \quad \textbf{(21)}$$
$$dt = \frac{r^2}{4kz^2}dz$$

New limits of integrations:

At,

$$t = t_{i-1} \Leftrightarrow z = \frac{r^2}{4k(t_n - t_{i-1})} \quad \textbf{(22-1)}$$

And at,

$$t = t_i \Leftrightarrow z = \frac{r^2}{4k(t_n - t_i)} \tag{22-2}$$

Therefore, equation (**19**), takes the following new form:

$$I_1 = \int_0^{t_F} u^*(\xi, x; t_n, t)dt = \frac{1}{2\sqrt{\pi k}} \int_{\frac{r^2}{4k(t_n - t_{i-1})}}^{\frac{r^2}{4k(t_n - t_i)}} \left\{ \frac{1}{\sqrt{\left(\frac{r^2}{4kz}\right)}} \exp[-z] \right\} \left(\frac{r^2}{4kz^2} \right) dz \tag{23}$$

Let,

$$A = a_{i-1} = \frac{r^2}{4k(t_n - t_{i-1})}$$

&

$$B = a_i = \frac{r^2}{4k(t_n - t_i)} \tag{24}$$

Substituting from equation (**24**) into equation (**23**) leads to:

$$I_1 = \int_0^{t_F} u^*(\xi, x; t_n, t)dt = \frac{1}{2\sqrt{\pi k}} \int_A^B \left\{ \frac{1}{\sqrt{\left(\frac{r^2}{4kz}\right)}} \exp[-z] \right\} \left(\frac{r^2}{4kz^2} \right) dz \tag{25}$$

After simplification, equation (**25**) takes the following form:

$$I_1 = \frac{r}{2k\sqrt{\pi}} \int_A^B z^{-\frac{3}{2}} e^{-z} dz \tag{26}$$

Integrating by parts for the first time leads to:

$$I_1 = \frac{r}{4k\sqrt{\pi}}\left(-2e^{-B}B^{-\frac{1}{2}} + 2e^{-A}A^{-\frac{1}{2}}\right) - \frac{2r}{4k\sqrt{\pi}}\int_A^B z^{-\frac{1}{2}}e^{-z}dz \qquad \textbf{(27)}$$

Substituting by $A = a_{i-1}$ and $B = a_{i-2}$ into equation (**27**) and simplifying, we get:

$$I_1 = \frac{r}{2k\sqrt{\pi}}\left(-e^{-a_i}a_i^{-\frac{1}{2}} + e^{-a_{i-1}}a_{i-1}^{-\frac{1}{2}}\right) - \frac{2r}{4k\sqrt{\pi}}\int_A^B z^{-\frac{1}{2}}e^{-z}dz \qquad \textbf{(28)}$$

The next step is to evaluate the last integral that appeared in equation (2**8**). This will occurr as by denoting this integral by I_1^*, therefore it takes the form:

$$I_1^* = \int_A^B z^{-\frac{1}{2}}e^{-z}dz \qquad \textbf{(29)}$$

Let,

$$z = t^2 \Rightarrow dz = 2tdt \qquad \textbf{(30)}$$

New limits of integrations,

$$z = a_{i-1} \Leftrightarrow t = a_{i-1}^{\frac{1}{2}}$$

&

$$z = a_i \Leftrightarrow t = a_i^{\frac{1}{2}} \qquad \textbf{(31)}$$

Therefore, equation (4.2**9**) becomes:

$$I_1^* = 2\int_{\sqrt{a_{i-1}}}^{\sqrt{a_i}} e^{-t^2}dt \qquad \textbf{(32)}$$

The analytical solution for equation (3**2**) is:

$$I_1^* = \sqrt{\pi}\left\{erf\left(\sqrt{a_i}\right) - erf\left(\sqrt{a_{i-1}}\right)\right\} \quad \textbf{(33)}$$

Substutiting from equation (3**3)** into equation (2**8)**, we get:

$$I_1 = \frac{r}{2k\sqrt{\pi}}\left(a_{i-1}^{-\frac{1}{2}}e^{-a_{i-1}} - a_i^{-\frac{1}{2}}e^{-a_i}\right) + \sqrt{\pi}\left\{erf\left(\sqrt{a_{i-1}}\right) - erf\left(\sqrt{a_i}\right)\right\} \quad \textbf{(34)}$$

Integral given by equation (1**8)**

$$I_2 = \int_0^{t_F} \frac{\partial u^*(\xi, x; t_n, t)}{\partial x} dt = \int_{t_{i-1}}^{t_i} \frac{\partial}{\partial x}\left\{\frac{1}{\sqrt{4\pi k(t_n - t)}} \exp\left[-\frac{r^2(\xi,x)}{4k(t_n - t)}\right]\right\} dt$$
$$= \int_{t_{i-1}}^{t_i} \frac{\partial}{\partial x}\left\{\frac{1}{\sqrt{4\pi k(t_n - t)}} \frac{\partial}{\partial x}\left\{\exp\left[-\frac{r^2(\xi,x)}{4k(t_n - t)}\right]\right\}\right\} dt \quad \textbf{(35)}$$

As mentioned before, the distance $r(\xi, x)$ is defined as:

$$r(\xi, x) = \sqrt{X(\xi) - X(x)} = |\xi - x| = \pm(\xi - x) \quad \textbf{(36)}$$

Now equation (3**6)** takes the following form:

$$I_2 = \int_{t_{i-1}}^{t_i} \frac{\partial}{\partial x}\left\{\frac{1}{\sqrt{4\pi k(t_n - t)}} \exp\left[-\frac{r^2(\xi,x)}{4k(t_n - t)}\right] \frac{-1 \times 2r \times (\pm)}{4k(t_n - t)}\right\} dt \quad \textbf{(37)}$$

Equation (3**7)** after simplification becomes:

$$I_2 = (\pm)\frac{r}{4k^{\frac{3}{2}}\sqrt{\pi}} \int_{t_{i-1}}^{t_i} \left\{\frac{1}{(t_n - t)^{\frac{3}{2}}} \exp\left[-\frac{r^2(\xi,x)}{4k(t_n - t)}\right]\right\} dt \quad \textbf{(38)}$$

To evaluate the integral in equation (3**8)**, let us use the following substitution:

$$z = \frac{r^2(\xi,x)}{4k(t_n - t)} \Rightarrow t = t_n - \frac{r^2(\xi,x)}{4kz} \quad \textbf{(39)}$$

Therefore;

$$dt = \frac{r^2}{4kz^2}dz \quad \textbf{(40)}$$

New limits of integration:

$$t = t_{i-1} \Leftrightarrow z = a_{i-1} = \frac{r^2}{4k(t_n - t_{i-1})}$$

& **(41)**

$$t = t_i \Leftrightarrow z = a_i = \frac{r^2}{4k(t_n - t_i)}$$

Using equations (39-4**1)** into equation (4.3**8)**, we get:

$$I_2 = (\pm)\frac{r}{4k^{\frac{3}{2}}\sqrt{\pi}}\int_{a_{i-1}}^{a_i}\left(\frac{r^2}{4kz}\right)^{-\frac{3}{2}} e^{-z}\frac{r^2}{4k}\frac{1}{z^2}dz = $$
$$= (\pm)\frac{1}{2k\sqrt{\pi}}\int_{a_{i-1}}^{a_i} e^{-z} z^{-\frac{1}{2}}dz \quad \textbf{(42)}$$

The integral was evaluated before; therefore, the integral given by equation (1**8)** takes the following form:

$$I_2 = \int_{t_{i-1}}^{t_i} q^*(\xi,x;t_n,t)dt = \pm\frac{1}{2k}\left[erf\left(a_i^{\frac{1}{2}}\right) - erf\left(a_{i-1}^{\frac{1}{2}}\right)\right] \quad \textbf{(43)}$$

Note for the purpose of programming let us denote $\pm$ by $\text{sign}(r)$; therefore, one can write equation (43) as follows:

$$I_2 = \int_{t_{i-1}}^{t_i} q^*(\xi, x; t_n, t)dt = \text{sign}(r)\frac{1}{2k}\left[erf\left(a_i^{\frac{1}{2}}\right) - erf\left(a_{i-1}^{\frac{1}{2}}\right)\right] \tag{44}$$

After evaluating the time integrals, the boundary integral equation given by equation (15) can now take the following form:

$$\begin{aligned} u(\xi, t_n) &= \int_0^{\ell} u(x,0)u^*(\xi, x; t_n, 0)dx \\ &- \sum_{i=1}^{n}\left\{\frac{1}{2}\text{sign}(r)\left[erf\left(\sqrt{a_i}\right) - erf\left(\sqrt{a_{i-1}}\right)\right]u(x,t_i)\right\}_0^{\ell} \\ &+ \sum_{i=1}^{n}\left\{\frac{r}{2\sqrt{\pi}}\left\{\left(\frac{e^{-a_{i-1}}}{a_{i-1}^{-\frac{1}{2}}} - \frac{e^{-a_i}}{a_i^{-\frac{1}{2}}}\right) + \sqrt{\pi}\left[erf\left(\sqrt{a_{i-1}}\right) - erf\left(\sqrt{a_i}\right)\right]\right\}q(x,t_i)\right\}_0^{\ell} \end{aligned} \tag{45}$$

Evaluation of the domain integral

The next step is to evaluate the domain integral; as it is clear from equation (45), the domain integral is defined as [9-15]:

$$\text{Domain integral} = \int_0^{\ell} u(x,0)u^*(\xi, x; t_n, 0)dx \tag{46}$$

We know:

$$u^*(\xi, x; t_n, t) = \frac{1}{\sqrt{4\pi k(t_n - t)}}\exp\left[-\frac{r^2(\xi, x)}{4k(t_n - t)}\right] \tag{47}$$

Therefore

$$u^*(\xi, x; t_n, 0) = \frac{1}{\sqrt{4\pi k(t_n)}} \exp\left[-\frac{r^2(\xi, x)}{4kt_n}\right] \quad \textbf{(48)}$$

Substituting from (48) into (46) leads to:

$$\text{Domain integral} = \int_0^\ell u(x,0)\left\{\frac{1}{\sqrt{4\pi k(t_n)}} \exp\left[-\frac{r^2(\xi,x)}{4kt_n}\right]\right\}dx \quad \textbf{(49)}$$

Because it is assumed before that the initial temperature is constant, one can take the initial temperature out of the integral therefore, equation (49) takes the following form:

$$\text{Domain integral} = \int_0^\ell u(x,0)\left\{\frac{1}{\sqrt{4\pi k(t_n)}} \exp\left[-\frac{r^2(\xi,x)}{4kt_n}\right]\right\}dx \quad \textbf{(49a)}$$

$$\text{Domain integral} = u(x,0)\int_0^\ell \left\{\frac{1}{\sqrt{4\pi k(t_n)}} \exp\left[-\frac{r^2(\xi,x)}{4kt_n}\right]\right\}dx \quad \textbf{(50)}$$

Let,

$$z^2 = \frac{r^2(\xi,x)}{4kt_n} \quad \textbf{(51)}$$

But $r^2(\xi,x)$ defined before as

$$r^2(\xi,x) = (\xi - x)^2 \quad \textbf{(52)}$$

Therefore, equation (5**1)** becomes:

$$z^2 = \frac{(\xi - x)^2}{4k\,t_n} \Rightarrow z = \frac{(\xi - x)}{2\sqrt{k\,t_n}} \qquad \textbf{(53)}$$

From equation (4.5**3)**, one can get:

$$x = \xi - 2\sqrt{kt_n}\,z \Rightarrow dx = -2\sqrt{kt_n}\,dz \qquad \textbf{(54)}$$

New limits of integration:

$$x = 0 \Leftrightarrow z = \frac{\xi}{2\sqrt{kt_n}}$$

& **(55)**

$$x = \ell \Leftrightarrow z = \frac{\xi - \ell}{2\sqrt{kt_n}}$$

Using equations (53-5**5)** into equation (5**0)** and after evaluating the integral, one can get the following result for the domain integral:

$$\text{Domain integral} = (-)\frac{u(x,0)}{2}\left[erf\left(\frac{\xi - \ell}{2\sqrt{kt_n}}\right) - erf\left(\frac{\xi}{2\sqrt{kt_n}}\right)\right] \qquad \textbf{(56)}$$

Now the final form for the integral equation corresponding to the diffusion equation defined over fixed domain will be:

$$
\begin{aligned}
u(\xi,t_n) &= \frac{(-)u(x,0)}{2}\left[erf\left(\frac{\xi-\ell}{2\sqrt{kt_n}}\right)-erf\left(\frac{\xi}{2\sqrt{kt_n}}\right)\right] \\
&-\sum_{i=1}^{n}\left\{\frac{1}{2}\operatorname{sign}(r)\left[erf\left(\sqrt{a_i}\right)-erf\left(\sqrt{a_{i-1}}\right)\right]u(x,t_i)\right\}_0^{\ell} \\
&+\sum_{i=1}^{n}\left\{\frac{r}{2\sqrt{\pi}}\left\{\begin{array}{l}\left(\dfrac{e^{-a_{i-1}}}{a_{i-1}^{-\frac{1}{2}}}-\dfrac{e^{-a_i}}{a_i^{-\frac{1}{2}}}\right)\\ +\sqrt{\pi}\left[erf\left(\sqrt{a_{i-1}}\right)-erf\left(\sqrt{a_i}\right)\right]\end{array}\right\}q(x,t_i)\right\}_0^{\ell}
\end{aligned}
\tag{57}
$$

The corresponding general form for equation (57) is as follows:

$$
\begin{aligned}
u(\xi,t_n) &= \frac{u(x,0)}{2}\left[erf\left(\frac{\xi}{2\sqrt{kt_n}}\right)-erf\left(\frac{\xi-\ell}{2\sqrt{kt_n}}\right)\right] \\
&+\sum_{i=1}^{n}\left[q_{x=\ell}\int_{t_{i-1}}^{t_i}u^{*}\,dt-u_{x=\ell}\int_{t_{i-1}}^{t_i}q^{*}\,dt\right]_{x=\ell} \\
&-\sum_{i=1}^{n}\left[q_{x=0}\int_{t_{i-1}}^{t_i}u^{*}\,dt-u_{x=0}\int_{t_{i-1}}^{t_i}q^{*}\,dt\right]_{x=0}
\end{aligned}
\tag{58}
$$

ELEMENTARY EXAMPLE

ONE-DIMENSIONAL, ONE-PHASE DIFFUSION EQUATION

In this section, an elementary example will be introduced in detail. These details are the preliminary stage of designing code for these types of problems. Consider a bar of length, ℓ subjected to the uniform initial temperature $u_0 = 1.0$, at the time $t > 0$, the ends $x = 0$ and $x = \ell$ are kept at zero temperature. It is required to find flux at both ends using boundary integral equation (Fig. **1**) .

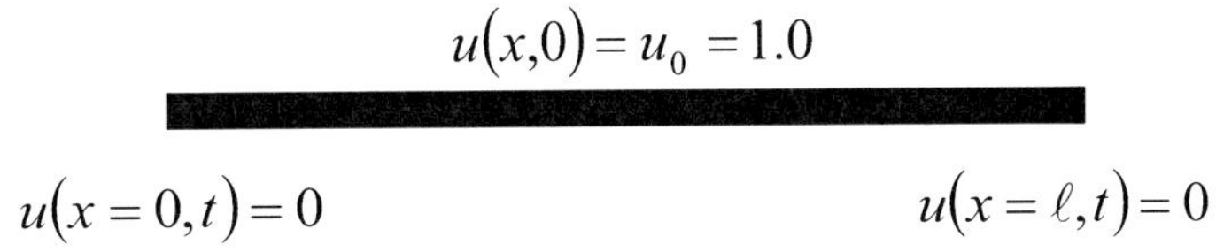

Fig. (1). Problem domain.

Let us start the solution by writing the general boundary integral equation as follows:

$$\begin{aligned} u(\xi,t_n) &= \frac{u(x,0)}{2}\left[erf\left(\frac{\xi}{2\sqrt{kt_n}}\right) - erf\left(\frac{\xi-\ell}{2\sqrt{kt_n}}\right)\right] \\ &+ \sum_{i=1}^{n}\left[q_{x=\ell}\int_{t_{i-1}}^{t_i} u^* \, dt - u_{x=\ell}\int_{t_{i-1}}^{t_i} q^* \, dt \right]_{x=\ell} \\ &- \sum_{i=1}^{n}\left[q_{x=0}\int_{t_{i-1}}^{t_i} u^* \, dt - u_{x=0}\int_{t_{i-1}}^{t_i} q^* \, dt \right]_{x=0} \end{aligned} \qquad \textbf{(Ex-1)}$$

The second step is to substitute by the numerical values of the associated boundary conditions; therefore, equation (Ex-**1)** takes the following simplified form:

$$\begin{aligned} u(\xi,t_n) &= \frac{u(x,0)}{2}\left[erf\left(\frac{\xi}{2\sqrt{kt_n}}\right) - erf\left(\frac{\xi-\ell}{2\sqrt{kt_n}}\right)\right] \\ &+ \sum_{i=1}^{n-1}\left[q_{x=\ell}\int_{t_{i-1}}^{t_i} u^* \, dt - q_{x=0}\int_{t_{i-1}}^{t_i} u^* \, dt \right] \end{aligned} \qquad \textbf{(Ex-2)}$$

Before introducing the details of the computation procedure, let us make a general overview of the equation (Ex-**2)** at three successive time steps to build a concept of computation for the readers.

At The First Time Step

$$u(\xi,t_1)=\frac{u_0}{2}\left[erf\left(\frac{\xi}{2\sqrt{kt_1}}\right)-erf\left(\frac{\xi-\ell}{2\sqrt{kt_1}}\right)\right]$$

$$+\left[q_{x=\ell}\int_{t_0}^{t_1}u^*\,dt-q_{x=0}\int_{t_0}^{t_1}u^*\,dt\right] \qquad \textbf{(Ex-3-1)}$$

At the Second Time Step

$$u(\xi,t_2)=\frac{u_0}{2}\left[erf\left(\frac{\xi}{2\sqrt{kt_2}}\right)-erf\left(\frac{\xi-\ell}{2\sqrt{kt_2}}\right)\right]+\left[q_{x=\ell}\int_{t_0}^{t_2}u^*\,dt-q_{x=0}\int_{t_0}^{t_2}u^*\,dt\right] \qquad \textbf{(Ex-3-2)}$$

Notes:

(1) The integral $\int_{t_0}^{t_2}u^*\,dt=\int_{t_0}^{t_1}u^*\,dt+\int_{t_1}^{t_2}u^*\,dt$

(2) Limits of integrations as follow:

$$t_0 \Leftrightarrow a_0=\frac{r^2}{4(t_2-t_0)}$$

$$t_1 \Leftrightarrow a_1=\frac{r^2}{4(t_2-t_1)}$$

$$t_2 \Leftrightarrow a_2=\frac{r^2}{4(t_2-t_2)} \Leftrightarrow \text{Time Singularit y}$$

At The Third Time Step

$$u(\xi,t_3)=\frac{u_0}{2}\left[erf\left(\frac{\xi}{2\sqrt{kt_3}}\right)-erf\left(\frac{\xi-\ell}{2\sqrt{kt_3}}\right)\right]+\left[q_{x=\ell}\int_{t_0}^{t_3}u^*dt-q_{x=0}\int_{t_0}^{t_3}u^*dt\right] \quad \textbf{(Ex-3-3)}$$

Notes:

(3) The integral $\int_{t_0}^{t_3}u^*dt=\int_{t_0}^{t_1}u^*dt+\int_{t_1}^{t_2}u^*dt+\int_{t_2}^{t_3}u^*dt$

(4) Limits of integrations as follow:

$$t_0 \Leftrightarrow a_0=\frac{r^2}{4(t_3-t_0)}$$

$$t_1 \Leftrightarrow a_1=\frac{r^2}{4(t_3-t_1)}$$

$$t_2 \Leftrightarrow a_2=\frac{r^2}{4(t_3-t_2)}$$

$$t_3 \Leftrightarrow a_3=\frac{r^2}{4(t_3-t_3)} \Leftrightarrow \text{Time Singularit y}$$

Numerical Computations

First Stage: Evaluation for the First Time Step

In this section, we will introduce a detailed numerical computation for the elementary example defined before. We will start the computation by evaluating the time integrals. To do so, let us start the computation by re-writing the boundary integral equation as follows:

$$u(\xi,t_n)=\frac{u(x,0)}{2}\left[erf\left(\frac{\xi}{2\sqrt{t_n}}\right)-erf\left(\frac{\xi-\ell}{2\sqrt{t_n}}\right)\right]$$
$$+\sum_{i=1}^{n-1}\left[q_{x=\ell}\int_{t_{i-1}}^{t_i}u^* dt-q_{x=0}\int_{t_{i-1}}^{t_i}u^* dt\right] \qquad \textbf{(NC-1)}$$

At the first time step $t_n = t_1$

The corresponding boundary integral equation takes the form:

$$u(\xi,t_1)=\frac{u_0}{2}\left[erf\left(\frac{\xi}{2\sqrt{t_1}}\right)-erf\left(\frac{\xi-\ell}{2\sqrt{t_1}}\right)\right]$$
$$+\left[q_{x=\ell}\int_{t_0}^{t_1}u^* dt-q_{x=0}\int_{t_0}^{t_1}u^* dt\right] \qquad \textbf{(NC-2)}$$

The limits of integration take the following form:

$$a_0=\frac{r^2}{4(t_1-t_0)}=\frac{r^2}{4\Delta t}$$
& **(NC-3)**
$$a_1=\frac{r^2}{4(t_1-t_i)}=\infty \quad \text{Time singularit y}$$

The next step is to apply the boundary integral equation (NC-**1)** at the two end points of the domain and each point will be considered a source point ξ .

Source point at $x=0$ **this corresponds to** $\xi=0$

The boundary integral equation (**1)** takes the form:

$$u(0,t_1)=\frac{u_0}{2}\left[erf(0)-erf\left(\frac{-\ell}{2\sqrt{t_1}}\right)\right]+\left[q_{x=\ell}\int_{t_0}^{t_1}u^*dt-q_{x=0}\int_{t_0}^{t_1}u^*dt\right] \quad \textbf{(NC-4)}$$

Note that $u(0,t_1)=0$ is already a boundary condition in our elementary example therefore equation (4) becomes:

$$-\frac{u_0}{2}erf\left(\frac{-\ell}{2\sqrt{t_1}}\right)=q_{x=\ell}\left(\int_{t_0}^{t_1}u^*dt\right)_{x=\ell}-q_{x=0}\left(\int_{t_0}^{t_1}u^*dt\right)_{x=0} \quad \textbf{(NC-5)}$$

Source point at $x=\ell$ **this corresponds to** $\xi=\ell$

The boundary integral equation **(1)** takes the form:

$$u(\ell,t_1)=\frac{u_0}{2}\left[erf\left(\frac{\ell}{2\sqrt{t_1}}\right)-erf(0)\right]+\left[q_{x=\ell}\int_{t_0}^{t_1}u^*dt-q_{x=0}\int_{t_0}^{t_1}u^*dt\right] \quad \textbf{(NC-6)}$$

Note that $u(\ell,t_1)=0$ is already a boundary condition in our elementary example therefore equation (**6**) becomes:

$$-\frac{u_0}{2}erf\left(\frac{\ell}{2\sqrt{t_1}}\right)=q_{x=\ell}\left(\int_{t_0}^{t_1}u^*dt\right)_{x=\ell}-q_{x=0}\left(\int_{t_0}^{t_1}u^*dt\right)_{x=0} \quad \textbf{(NC-7)}$$

Now equations (NC-**5)** and (NC-**7)** can be written in a matrix form as follows:

$$\begin{bmatrix}-u_{11} & u_{12}\\ -u_{21} & u_{22}\end{bmatrix}\begin{bmatrix}q_0\\ q_1\end{bmatrix}=\begin{bmatrix}b_1\\ b_2\end{bmatrix} \quad \textbf{(NC-8)}$$

Note:

The negative sing appear in the first column in the u_{ij} matrix comes from the sign of q_0 in equations (NC-**5)** and (NC-**7)**. The next step is the numerical calculation of the elements u_{ij}. Let us start the calculation by the element u_{11}.

$u_{11} \Leftrightarrow$ Source $\xi = 0$ and Field $x = 0$

In this case:

$r = 0$ because $r = \xi - x = 0$

$$a_0 = \frac{r^2}{4(t_1 - t_0)} = 0 \text{ because r equals zero} \quad \textbf{(NC-9)}$$

$$a_1 = \frac{r^2}{4(t_1 - t_1)} = \frac{0}{0}$$

$$u_{11} = \int_{t_0}^{t_1} u^*(\xi = 0, x = 0; t_1, t)dt =$$

$$= \frac{r}{2\sqrt{\pi}}\left\{\left(\frac{e^{-a_0}}{\sqrt{a_0}} - \frac{e^{-a_1}}{\sqrt{a_1}}\right) + \sqrt{\pi}\left(erf\left(\sqrt{a_0}\right) - erf\left(\sqrt{a_1}\right)\right)\right\} \quad \textbf{(NC-10)}$$

Substutiting from equation (NC-**9)** into equation (NC-1**0)**, we get:

$$u_{11} = \int_{t_0}^{t_1} u^*(\xi = 0, x = 0; t_1, t)dt =$$

$$= \frac{r}{2\sqrt{\pi}}\left\{\left(\frac{e^{-a_0}}{\sqrt{\frac{r^2}{4(t_1 - t_0)}}} - \frac{e^{-a_1}}{\sqrt{\frac{r^2}{4(t_1 - t_1)}}}\right)\right\} \qquad \textbf{(NC-11)}$$

$$+ \frac{r}{2\sqrt{\pi}}\left\{\sqrt{\pi}\left(erf\left(\sqrt{\frac{r^2}{4(t_1 - t_0)}}\right) - erf\left(\sqrt{\frac{r^2}{4(t_1 - t_1)}}\right)\right)\right\}$$

Simplify equation (NC-1**1)**, leads to:

$$u_{11} = \int_{t_0}^{t_1} u^*(\xi = 0, x = 0; t_1, t)dt =$$

$$= \frac{r}{2\sqrt{\pi}}\left\{\left(\frac{2e^{-a_0}\sqrt{(t_1 - t_0)}}{r} - \frac{2e^{-a_1}\sqrt{(t_1 - t_1)}}{r}\right)\right\} + \frac{r}{2}erf\left(\sqrt{a_1}\right) \qquad \textbf{(NC-12)}$$

Simplify equation (1**2)** once more, it becomes:

$$u_{11} = \int_{t_0}^{t_1} u^*(\xi = 0, x = 0; t_1, t)dt = \frac{1}{\sqrt{\pi}}\left\{\sqrt{(t_1 - t_0)}\right\} \qquad \textbf{(NC-13)}$$

$u_{12} \Leftrightarrow$ Source $\xi = 0$ and Field $x = \ell$

In this case:

$r \neq 0$

$$a_0 = \frac{r^2}{4(t_1 - t_0)} \qquad \textbf{(NC-14)}$$

$$a_1 = \frac{r^2}{4(t_1 - t_1)}$$

$$u_{12} = \int_{t_0}^{t_1} u^*(\xi = 0, x = \ell; t_1, t)dt =$$

$$= \frac{r}{2\sqrt{\pi}}\left\{\left(\frac{e^{-a_0}}{\sqrt{a_0}} - \frac{e^{-a_1}}{\sqrt{a_1}}\right) + \sqrt{\pi}\left(erf\left(\sqrt{a_0}\right) - erf\left(\sqrt{a_1}\right)\right)\right\} \qquad \textbf{(NC-15)}$$

Substituting from equation (NC-1**4)** into equation (NC-1**5)**, we get:

$$u_{12} = \int_{t_0}^{t_1} u^*(\xi = 0, x = \ell; t_1, t)dt =$$

$$= \frac{r}{2\sqrt{\pi}}\left\{\left(\frac{e^{-a_0}}{\sqrt{\frac{r^2}{4(t_1 - t_0)}}} - \frac{e^{-a_1}}{\sqrt{\frac{r^2}{4(t_1 - t_1)}}}\right)\right\} \qquad \textbf{(NC-16)}$$

$$+ \frac{r}{2\sqrt{\pi}}\left\{\sqrt{\pi}\left(erf\left(\sqrt{\frac{r^2}{4(t_1 - t_0)}}\right) - erf\left(\sqrt{\frac{r^2}{4(t_1 - t_1)}}\right)\right)\right\}$$

Simplify equation (NC-1**6)**, leads to:

$$u_{12} = \int_{t_0}^{t_1} u^*(\xi = 0, x = \ell; t_1, t)dt =$$

$$= \frac{1}{\sqrt{\pi}}\left\{e^{-a_0}\sqrt{(t_1 - t_0)}\right\} + \frac{r}{2}\left\{erf\left(\sqrt{a_0}\right) - erf\left(\sqrt{\infty}\right)\right\} \qquad \textbf{(NC-17)}$$

Simplify equation (NC-17) once more, it becomes:

$$u_{12} = \int_{t_0}^{t_1} u^*(\xi = 0, x = \ell; t_1, t)dt == \frac{1}{\sqrt{\pi}}\left\{e^{-a_0}\sqrt{(t_1 - t_0)}\right\} + \frac{r}{2}\left\{erf\left(\sqrt{a_0}\right) - 1\right\} \quad \textbf{(NC-18)}$$

$u_{21} \Leftrightarrow$ Source $\xi = \ell$ and Field $x = 0$

In this case:

$$r \neq 0$$

$$a_0 = \frac{r^2}{4(t_1 - t_0)} \quad \textbf{(NC-19)}$$

$$a_1 = \frac{r^2}{4(t_1 - t_1)}$$

$$u_{21} = \int_{t_0}^{t_1} u^*(\xi = \ell, x = 0; t_1, t)dt =$$

$$= \frac{r}{2\sqrt{\pi}}\left\{\left(\frac{e^{-a_0}}{\sqrt{a_0}} - \frac{e^{-a_1}}{\sqrt{a_1}}\right) + \sqrt{\pi}\left(erf\left(\sqrt{a_0}\right) - erf\left(\sqrt{a_1}\right)\right)\right\} \quad \textbf{(NC-20)}$$

Substutiting from equation (NC-19) into equation (NC-20), we get:

$$u_{22} = \int_{t_0}^{t_1} u^*(\xi = \ell, x = 0; t_1, t)dt =$$

$$= \frac{r}{2\sqrt{\pi}}\left\{\left(\frac{e^{-a_0}}{\sqrt{\dfrac{r^2}{4(t_1 - t_0)}}} - \frac{e^{-a_1}}{\sqrt{\dfrac{r^2}{4(t_1 - t_1)}}}\right)\right\}$$

$$+ \frac{r}{2\sqrt{\pi}}\left\{\sqrt{\pi}\left(erf\left(\sqrt{\frac{r^2}{4(t_1 - t_0)}}\right) - erf\left(\sqrt{\frac{r^2}{4(t_1 - t_1)}}\right)\right)\right\} \quad \textbf{(NC-21)}$$

Simplify equation (NC-2**1)**, leads to:

$$u_{21} = \int_{t_0}^{t_1} u^*(\xi = \ell, x = 0; t_1, t)dt$$

$$= \frac{1}{\sqrt{\pi}}\left\{e^{-a_0}\sqrt{(t_1 - t_0)}\right\} + \frac{r}{2}\left\{erf\left(\sqrt{a_0}\right) - erf\left(\sqrt{\infty}\right)\right\} \quad \textbf{(NC-22)}$$

Simplify equation (NC-2**2)** once more, it becomes:

$$u_{21} = \int_{t_0}^{t_1} u^*(\xi = 0, x = \ell; t_1, t)dt = \frac{1}{\sqrt{\pi}}\left\{e^{-a_0}\sqrt{(t_1 - t_0)}\right\} + \frac{r}{2}\left\{erf\left(\sqrt{a_0}\right) - 1\right\} \quad \textbf{(NC-23)}$$

$u_{22} \Leftrightarrow$ Source $\xi = \ell$ and Field $x = \ell$

In this case:

$$r = \xi - x = 0$$

$$a_0 = \frac{r^2}{4(t_1 - t_0)} = 0$$

$$a_1 = \frac{r^2}{4(t_1 - t_1)} = \frac{0}{0} \quad \textbf{(NC-24)}$$

$$u_{22}=\int_{t_0}^{t_1} u^{*}(\xi=\ell,x=0;t_1,t)dt=$$

$$=\frac{r}{2\sqrt{\pi}}\left\{\left(\frac{e^{-a_0}}{\sqrt{a_0}}-\frac{e^{-a_1}}{\sqrt{a_1}}\right)+\sqrt{\pi}\left(erf\left(\sqrt{a_0}\right)-erf\left(\sqrt{a_1}\right)\right)\right\} \quad \textbf{(NC-25)}$$

Substituting from equation (NC-24) into equation (NC-25), we get:

$$u_{22}=\int_{t_0}^{t_1} u^{*}(\xi=\ell,x=0;t_1,t)dt=$$

$$=\frac{r}{2\sqrt{\pi}}\left\{\left(\frac{e^{-a_0}}{\sqrt{\frac{r^2}{4(t_1-t_0)}}}-\frac{e^{-a_1}}{\sqrt{\frac{r^2}{4(t_1-t_1)}}}\right)\right\} \quad \textbf{(NC-26)}$$

$$+\frac{r}{2\sqrt{\pi}}\left\{\sqrt{\pi}\left(erf\left(\sqrt{\frac{r^2}{4(t_1-t_0)}}\right)-erf\left(\sqrt{\frac{r^2}{4(t_1-t_1)}}\right)\right)\right\}$$

Simplify equation (NC-26), leads to:

$$u_{22}=\int_{t_0}^{t_1} u^{*}(\xi=\ell,x=0;t_1,t)dt=$$

$$=\frac{1}{\sqrt{\pi}}\left\{e^{-a_0}\sqrt{(t_1-t_0)}\right\}+\frac{r}{2}\left\{erf\left(\sqrt{a_0}\right)-erf\left(\sqrt{\infty}\right)\right\} \quad \textbf{(NC-27)}$$

Simplify equation (NC-27) once more, it becomes:

$$u_{22}=\int_{t_0}^{t_1} u^{*}(\xi=0,x=\ell;t_1,t)dt=\frac{1}{\sqrt{\pi}}\left\{\sqrt{(t_1-t_0)}\right\} \quad \textbf{(NC-28)}$$

Now the system of equations for the first time step is defined by equations (NC-13), (NC-18), (NC-23), and (NC-28).

Numerical Computation

We will take the time step $\Delta t = 0.0025$ and the initial time equals zero. Now let us start computing the elements u_{11}, u_{12}, u_{21} and u_{22}.

Evaluation u_{11}

$$u_{11} = \int_{t_0}^{t_1} u^*(\xi = 0, x = 0; t_1, t)dt = \frac{1}{\sqrt{\pi}}\left\{\sqrt{(t_1 - t_0)}\right\}$$

$$u_{11} = \frac{1}{\sqrt{\pi}}\sqrt{0.0025 - 0} = -0.0282094$$

Evaluation u_{12}

$$u_{12} = \int_{t_0}^{t_1} u^*(\xi = 0, x = \ell; t_1, t)dt = \frac{1}{\sqrt{\pi}}\left\{e^{-a_0}\sqrt{(t_1 - t_0)}\right\} + \frac{r}{2}\left\{erf\left(\sqrt{a_0}\right) - 1\right\}$$

$$u_{12} = \frac{1}{2\sqrt{\pi}}\left[\frac{e^{-100}}{\sqrt{100}}\right] + \frac{1}{2}\left[erf\left(\sqrt{100}\right) - 1\right] = 1.04941405 \times 10^{-45}$$

Evaluation u_{21}

$$u_{21} = \int_{t_0}^{t_1} u^*(\xi = 0, x = \ell; t_1, t)dt = \frac{1}{\sqrt{\pi}}\left\{e^{-a_0}\sqrt{(t_1 - t_0)}\right\} + \frac{r}{2}\left\{erf\left(\sqrt{a_0}\right) - 1\right\}$$

$$u_{12} = 1.04941405 \times 10^{-45}$$

Evaluation u_{22}

$$u_{22} = \int_{t_0}^{t_1} u^*(\xi = 0, x = \ell; t_1, t)dt = \frac{1}{\sqrt{\pi}}\left\{\sqrt{(t_1 - t_0)}\right\}$$

$u_{22} = 0.0282094$

Evaluation b_1

$$b_1 = -\frac{u_0}{2} erf\left(\frac{-\ell}{2\sqrt{t_1}}\right) = -\frac{u_0}{2} erf\left(\frac{-\ell}{2\sqrt{t_0 + \Delta t}}\right) \cong -0.5$$

Evaluation b_2

$$b_1 = -\frac{u_0}{2} erf\left(\frac{-\ell}{2\sqrt{t_1}}\right) = -\frac{u_0}{2} erf\left(\frac{-\ell}{2\sqrt{t_0 + \Delta t}}\right) \cong -0.5$$

Now the system is given by equation **(8)** takes the following form:

$$\begin{bmatrix} -u_{11} & u_{12} \\ -u_{21} & u_{22} \end{bmatrix} \begin{bmatrix} q_0 \\ q_1 \end{bmatrix} = \begin{bmatrix} b_1 \\ b_2 \end{bmatrix}$$

$\Rightarrow$

$$\begin{bmatrix} -0.0282094 & 1.04941405 \times 10^{-45} \\ -1.04941405 \times 10^{-45} & 0.0282094 \end{bmatrix} \begin{bmatrix} q_0 \\ q_1 \end{bmatrix} = \begin{bmatrix} -0.5 \\ -0.5 \end{bmatrix}$$

Numerical solution leads to:

$q_0 = +17.72458933$

$q_\ell = -17.72458933$

Second Stage: Evaluation for the Second Time Step

The next stage is to extend the computation procedure to compute the flux for the elementary example at the second time step. Let us start this stage by re-writing the boundary integral equation in its general case for the elementary example:

$$u(\xi,t_n)=\frac{u(x,0)}{2}\left[erf\left(\frac{\xi}{2\sqrt{t_n}}\right)-erf\left(\frac{\xi-\ell}{2\sqrt{t_n}}\right)\right]$$
$$+\sum_{i=1}^{n-1}\left[\begin{array}{l} q_{x=\ell}\left(\int_{t_{i-1}}^{t_i}u^*(\xi,x;t_n,t)dt\right)_{x=\ell} \\ -q_{x=0}\left(\int_{t_{i-1}}^{t_i}u^*(\xi,x;t_n,t)dt\right)_{x=0}\end{array}\right] \qquad \textbf{(NEC-1)}$$

Equation **(NEC-1)** for the second time step takes the following form:

$$u(\xi,t_2)=\frac{u_0}{2}\left[erf\left(\frac{\xi}{2\sqrt{t_2}}\right)-erf\left(\frac{\xi-\ell}{2\sqrt{t_2}}\right)\right]$$
$$+\sum_{i=1}^{2}\left[q_{x=\ell}\left(\int_{t_0}^{t_2}u^*(\xi,\ell;t_2,t)dt\right)-q_{x=0}\left(\int_{t_0}^{t_2}u^*(\xi,0;t_2,t)dt\right)\right] \qquad \textbf{(NEC-2)}$$

Re-write equation **(NEC-2)** in some more details:

$$u(\xi,t_2)=\frac{u_0}{2}\left[erf\left(\frac{\xi}{2\sqrt{t_2}}\right)-erf\left(\frac{\xi-\ell}{2\sqrt{t_2}}\right)\right]$$
$$+q_{x=\ell}\left(\int_{t_0}^{t_1}u^*(\xi,\ell;t_2,t)dt\right)+q_{x=\ell}\left(\int_{t_1}^{t_2}u^*(\xi,\ell;t_2,t)dt\right) \qquad \textbf{(NEC-3)}$$
$$-q_{x=0}\left(\int_{t_0}^{t_1}u^*(\xi,0;t_2,t)dt\right)-q_{x=0}\left(\int_{t_1}^{t_2}u^*(\xi,0;t_2,t)dt\right)$$

It is known that ξ is the source point, and in our example, it takes two different values $\xi=0$ & $\xi=\ell$ and in both cases, we have zero boundary condition, therefore, $u(\xi,t)=0$ and so equation **(NEC-3)** takes the following form:

$$-\frac{u_0}{2}\left[erf\left(\frac{\xi}{2\sqrt{t_2}}\right)-erf\left(\frac{\xi-\ell}{2\sqrt{t_2}}\right)\right]$$

$$=q_{x=\ell}\left(\int_{t_0}^{t_1}u^*(\xi,\ell;t_2,t)dt\right)+q_{x=\ell}\left(\int_{t_1}^{t_2}u^*(\xi,\ell;t_2,t)dt\right) \qquad \textbf{(NEC-4)}$$

$$-q_{x=0}\left(\int_{t_0}^{t_1}u^*(\xi,0;t_2,t)dt\right)-q_{x=0}\left(\int_{t_1}^{t_2}u^*(\xi,0;t_2,t)dt\right)$$

Note that $\int_{t_0}^{t_1}u^*(\xi,\ell;t_2,t)dt$ and $\int_{t_0}^{t_1}u^*(\xi,0;t_2,t)dt$ are two known values from the first time step, and so the other two values are unknown and should be calculated at the second time step. The source point ξ in equation **(NEC-4)** will be applied for our elementary example at two different points $\xi=0$ and $\xi=\ell$. By applying equation **(NEC-4)** at these two source points, one gets the following two boundary integral equations:

$$-\frac{u_0}{2}\left[-erf\left(\frac{-\ell}{2\sqrt{t_2}}\right)\right]=q_{x=\ell}\left(\int_{t_0}^{t_1}u^*(\xi=0,x=\ell;t_2,t)dt\right)$$

$$+q_{x=\ell}\left(\int_{t_1}^{t_2}u^*(\xi=0,x=\ell;t_2,t)dt\right)-q_{x=0}\left(\int_{t_0}^{t_1}u^*(\xi=0,x=0;t_2,t)dt\right) \qquad \textbf{(NEC-5)}$$

$$-q_{x=0}\left(\int_{t_1}^{t_2}u^*(\xi=0,x=0;t_2,t)dt\right)$$

And

$$-\frac{u_0}{2}\left[erf\left(\frac{\ell}{2\sqrt{t_2}}\right)\right]=q_{x=\ell}\left(\int_{t_0}^{t_1}u^*(\xi=\ell,x=\ell;t_2,t)dt\right) \qquad \textbf{(NEC-6)}$$

$$+q_{x=\ell}\left(\int_{t_1}^{t_2}u^*(\xi=\ell,x=\ell;t_2,t)dt\right)-q_{x=0}\left(\int_{t_0}^{t_1}u^*(\xi=\ell,x=0;t_2,t)dt\right)$$

$$-q_{x=0}\left(\int_{t_1}^{t_2}u^*(\xi=\ell,x=0;t_2,t)dt\right)$$

Evaluation the time integrals

At ***first*** time step with a source at $\xi=0$ and field at $x=0$

$$(u_{11})_{t=t_1}=\int_{t_0}^{t_1}u^*(\xi=0,x=0;t_2,t)dt \qquad \textbf{(NEC-7)}$$

$$=\frac{r}{2\sqrt{\pi}}\left\{\left(\frac{e^{-a_0}}{\sqrt{a_0}}-\frac{e^{-a_1}}{\sqrt{a_1}}\right)+\sqrt{\pi}\left(erf\left(\sqrt{a_0}\right)-erf\left(\sqrt{a_1}\right)\right)\right\}$$

In this case:

$$r = \xi - x = 0$$

$$a_0 = \frac{r^2}{4(t_2 - t_0)} \qquad \textbf{(NEC-8)}$$

$$a_1 = \frac{r^2}{4(t_2 - t_1)}$$

Using equation **(NEC-8)** into equation **(NEC-7)** and simplify, we get:

$$(u_{11})_{t=t_1} = \int_{t_0}^{t_1} u^*(\xi = 0, x = 0; t_2, t)dt =$$

$$= \frac{1}{\sqrt{\pi}}\left\{\sqrt{(t_2 - t_0)} + \sqrt{(t_2 - t_1)}\right\} \qquad \textbf{(NEC-9)}$$

Numerical Computations

$$r = \xi - x = 0$$

$$a_0 = \frac{r^2}{4(t_2 - t_0)} = 0$$

$$a_1 = \frac{r^2}{4(t_2 - t_1)} = 0$$

$$(u_{11})_{t=t_1} = \frac{1}{\sqrt{\pi}}\left\{\sqrt{(t_2 - t_0)} + \sqrt{(t_2 - t_1)}\right\}$$

$$= \frac{1}{\sqrt{\pi}}\left\{\sqrt{(0.005 - 0)} + \sqrt{(0.005 - 0.0025)}\right\} = 0.01168475$$

At ***second*** time step with a source at and field at

$$(u_{11})_{t=t_2} = \int_{t_1}^{t_2} u^*(\xi = 0, x = 0; t_2, t)dt =$$

$$= \frac{r}{2\sqrt{\pi}}\left\{\left(\frac{e^{-a_1}}{\sqrt{a_1}} - \frac{e^{-a_2}}{\sqrt{a_2}}\right) + \sqrt{\pi}\left(erf\left(\sqrt{a_1}\right) - erf\left(\sqrt{a_2}\right)\right)\right\} \quad \textbf{(NEC-10)}$$

In this case:

$$r = \xi - x = 0$$

$$a_1 = \frac{r^2}{4(t_2 - t_1)} = 0 \quad \textbf{(NEC-11)}$$

$$a_2 = \frac{r^2}{4(t_2 - t_2)} = \frac{0}{0} \text{.............Singularit y}$$

Using equation **(NEC-11)** into equation **(NEC-10)** and simplify, we get:

$$(u_{11})_{t=t_2} = \int_{t_1}^{t_2} u^*(\xi = 0, x = 0; t_2, t)dt = \frac{1}{\sqrt{\pi}}\left\{\sqrt{(t_2 - t_1)}\right\} \quad \textbf{(NEC-12)}$$

Numerical Computations

$$(u_{11})_{t=t_2} = \frac{1}{\sqrt{\pi}}\left\{\sqrt{(t_2 - t_1)}\right\} = \frac{1}{\sqrt{\pi}}\left(\sqrt{0.005 - 0.0025}\right) = 0.02820947$$

At ***first*** time step with source at and field at $x = \ell$

$$(u_{12})_{t=t_1} = \int_{t_0}^{t_1} u^*(\xi = 0, x = \ell; t_2, t)dt =$$

$$= \frac{r}{2\sqrt{\pi}}\left\{\left(\frac{e^{-a_0}}{\sqrt{a_0}} - \frac{e^{-a_1}}{\sqrt{a_1}}\right) + \sqrt{\pi}\left(erf\left(\sqrt{a_0}\right)\ \ erf\left(\sqrt{a_1}\right)\right)\right\} \quad \textbf{(NEC-13)}$$

In this case:

$$r = \xi - x = -\ell$$

$$a_0 = \frac{r^2}{4(t_2 - t_0)} \qquad \textbf{(NEC-14)}$$

$$a_1 = \frac{r^2}{4(t_2 - t_1)}$$

Numerical Computations

$$r = \xi - x = -\ell$$

$$a_0 = \frac{r^2}{4(t_2 - t_0)} = \frac{1}{4(0.005 - 0.000)} = 50$$

$$a_1 = \frac{r^2}{4(t_2 - t_1)} = \frac{1}{4(0.005 - 0.0025)} = 100$$

$$(u_{12})_{t=t_1} = \int_{t_0}^{t_1} u^*(\xi = 0, x = \ell; t_2, t)dt =$$

$$= \frac{1}{2\sqrt{\pi}}\left\{\left(\frac{e^{-50}}{\sqrt{50}} - \frac{e^{-100}}{\sqrt{100}}\right) + \sqrt{\pi}\left(erf\left(\sqrt{50}\right) - erf\left(\sqrt{100}\right)\right)\right\}$$

$$= 7.694598628 \times 10^{-24}$$

At ***second*** time step with source at and field at $x = \ell$

$$(u_{12})_{t=t_2} = \int_{t_1}^{t_2} u^*(\xi = 0, x = \ell; t_2, t)dt =$$

$$= \frac{r}{2\sqrt{\pi}}\left\{\left(\frac{e^{-a_1}}{\sqrt{a_1}} - \frac{e^{-a_2}}{\sqrt{a_2}}\right) + \sqrt{\pi}\left(erf\left(\sqrt{a_1}\right) - erf\left(\sqrt{a_2}\right)\right)\right\} \qquad \textbf{(NEC-15)}$$

In this case:

$$r = \xi - x = -\ell$$

$$a_1 = \frac{r^2}{4(t_2 - t_1)} \qquad \textbf{(NEC-16)}$$

$$a_2 = \frac{r^2}{4(t_2 - t_2)} \text{.............Singularit y}$$

Using equation **(NEC-16)** into equation **(NEC-15)** and simplifying, we get:

$$(u_{12})_{t=t_2} = \int_{t_1}^{t_2} u^*(\xi = 0, x = \ell; t_2, t)dt =$$

$$= \frac{r}{2\sqrt{\pi}} \frac{e^{-a_1}}{\sqrt{a_1}} + \frac{r}{2}\left\{erf\left(\sqrt{a_1}\right) - 1\right\} \qquad \textbf{(NEC-17)}$$

Numerical Computations

$$(u_{12})_{t=t_2} = \int_{t_1}^{t_2} u^*(\xi = 0, x = \ell; t_2, t)dt =$$

$$= \frac{1}{2\sqrt{\pi}} \frac{e^{-100}}{\sqrt{100}} + \frac{1}{2}\left\{erf\left(\sqrt{100}\right) - 1\right\} = 1.04941405 \times 10^{-45}$$

At ***first*** time step with a source at $\xi = \ell$ and field at $x = 0$

$$(u_{21})_{t=t_1} = \int_{t_0}^{t_1} u^*(\xi = \ell, x = 0; t_2, t)dt =$$

$$= \frac{r}{2\sqrt{\pi}} \left\{ \left(\frac{e^{-a_0}}{\sqrt{a_0}} - \frac{e^{-a_1}}{\sqrt{a_1}} \right) + \sqrt{\pi}\left(erf\left(\sqrt{a_0}\right) - erf\left(\sqrt{a_1}\right)\right) \right\} \qquad \textbf{(NEC-18)}$$

In this case:

$$r = \xi - x = \ell$$

$$a_0 = \frac{r^2}{4(t_2 - t_0)} \qquad \textbf{(NEC-19)}$$

$$a_1 = \frac{r^2}{4(t_2 - t_1)}$$

Numerical Computations

$$r = \xi - x = \ell = 1$$

$$a_0 = \frac{r^2}{4(t_2 - t_0)} = \frac{1}{4(0.005 - 0.000)} = 50$$

$$a_1 = \frac{r^2}{4(t_2 - t_1)} = \frac{1}{4(0.005 - 0.0025)} = 100$$

$$(u_{21})_{t=t_1} = \int_{t_0}^{t_1} u^*(\xi = \ell, x = 0; t_2, t)dt =$$

$$= \frac{1}{2\sqrt{\pi}}\left\{\left(\frac{e^{-50}}{\sqrt{50}} - \frac{e^{-100}}{\sqrt{100}}\right) + \sqrt{\pi}\left(erf\left(\sqrt{50}\right) - erf\left(\sqrt{100}\right)\right)\right\}$$

$$= 7.694598628 \times 10^{-24}$$

At ***second*** time step with a source at $\xi = \ell$ and field at $x = 0$

$$(u_{21})_{t=t_2} = \int_{t_1}^{t_2} u^*(\xi = \ell, x = 0; t_2, t)dt =$$

$$= \frac{r}{2\sqrt{\pi}}\left\{\left(\frac{e^{-a_1}}{\sqrt{a_1}} - \frac{e^{-a_2}}{\sqrt{a_2}}\right) + \sqrt{\pi}\left(erf\left(\sqrt{a_1}\right) - erf\left(\sqrt{a_2}\right)\right)\right\} \qquad \textbf{(NEC-20)}$$

In this case:

$$r = \xi - x = \ell$$

$$a_1 = \frac{r^2}{4(t_2 - t_1)} \qquad \textbf{(NEC-21)}$$

$$a_2 = \frac{r^2}{4(t_2 - t_2)} \text{.............Singularit y}$$

Using equation **(NEC-21)** into equation **(NEC-20)** and simplify, we get:

$$(u_{21})_{t=t_2} = \int_{t_1}^{t_2} u^*(\xi = \ell, x = 0; t_2, t)dt =$$

$$= \frac{r}{2\sqrt{\pi}} \frac{e^{-a_1}}{\sqrt{a_1}} + \frac{r}{2}\left\{erf\left(\sqrt{a_1}\right) - 1\right\} \qquad \textbf{(NEC-22)}$$

Numerical Computations

$$r = \xi - x = \ell = 1$$

$$a_1 = \frac{r^2}{4(t_2 - t_1)} = \frac{1}{4(0.005 - 0.025)} = 100$$

$$a_2 = \frac{r^2}{4(t_2 - t_2)} = \frac{1}{4(0.005 - 0.005)} = \infty$$

$$(u_{21})_{t=t_2} = \int_{t_1}^{t_2} u^*(\xi = \ell, x = 0; t_2, t)dt =$$

$$= \frac{1}{2\sqrt{\pi}} \frac{e^{-100}}{\sqrt{100}} + \frac{r}{2}\left\{erf\left(\sqrt{100}\right) - 1\right\} = 1.04941405 \times 10^{-45}$$

At ***first*** time step with source at $\xi = \ell$ and field at $x = \ell$

$$(u_{22})_{t=t_1} = \int_{t_0}^{t_1} u^*(\xi = \ell, x = \ell; t_2, t)dt =$$

$$= \frac{r}{2\sqrt{\pi}}\left\{\left(\frac{e^{-a_0}}{\sqrt{a_0}} - \frac{e^{-a_1}}{\sqrt{a_1}}\right) + \sqrt{\pi}\left(erf\left(\sqrt{a_0}\right) - erf\left(\sqrt{a_1}\right)\right)\right\} \qquad \textbf{(NEC-23)}$$

In this case:

$$r = \xi - x = 0$$

$$a_0 = \frac{r^2}{4(t_2 - t_0)} \qquad \textbf{(NEC-24)}$$

$$a_1 = \frac{r^2}{4(t_2 - t_1)}$$

Using equation **(NEC-24)** into equation **(NEC-23)** and simplifying, we get:

$$(u_{22})_{t=t_1} = \int_{t_1}^{t_2} u^*(\xi = \ell, x = \ell; t_2, t)dt = \frac{1}{\sqrt{\pi}}\left\{\sqrt{t_2 - t_0} - \sqrt{t_2 - t_1}\right\} \qquad \textbf{(NEC-25)}$$

Numerical Computations

$$r = \xi - x = 0$$

$$a_0 = \frac{r^2}{4(t_2 - t_0)} = 0$$

$$a_1 = \frac{r^2}{4(t_2 - t_1)} = 0$$

$$(u_{22})_{t=t_1} = \int_{t_1}^{t_2} u^*(\xi = \ell, x = \ell; t_2, t)dt =$$

$$= \frac{1}{\sqrt{\pi}}\left\{\sqrt{0.005 - 0.000} - \sqrt{0.005 - 0.0025}\right\} = 0.01168475$$

At ***second*** time step with a source at $\xi = \ell$ and field at $x = \ell$

$$(u_{22})_{t=t_2} = \int_{t_1}^{t_2} u^*(\xi = \ell, x = \ell; t_2, t)dt =$$

$$= \frac{r}{2\sqrt{\pi}}\left\{\left(\frac{e^{-a_1}}{\sqrt{a_1}} - \frac{e^{-a_2}}{\sqrt{a_2}}\right) + \sqrt{\pi}\left(erf\left(\sqrt{a_1}\right) - erf\left(\sqrt{a_2}\right)\right)\right\} \quad \textbf{(NEC-26)}$$

In this case:

$$r = \xi - x = 0$$

$$a_1 = \frac{r^2}{4(t_2 - t_1)} \quad \textbf{(NEC-27)}$$

$$a_2 = \frac{r^2}{4(t_2 - t_2)} \text{.............Singularit y}$$

Using equation **(NEC-27)** into equation **(NEC-26)** and simplifying, we get:

$$(u_{22})_{t=t_2} = \int_{t_1}^{t_2} u^*(\xi = \ell, x = \ell; t_2, t)dt = \frac{1}{\sqrt{\pi}}\left(\sqrt{t_2 - t_1}\right) \quad \textbf{(NEC-28)}$$

Numerical Computations

$$r = \xi - x = 0$$

$$a_1 = \frac{r^2}{4(t_2 - t_1)} = 0$$

$$a_2 = \frac{r^2}{4(t_2 - t_2)} = \frac{0}{0}$$

$$(u_{22})_{t=t_2} = \int_{t_1}^{t_2} u^*(\xi = \ell, x = \ell; t_2, t)dt$$

$$- \frac{1}{\sqrt{\pi}}\left(\sqrt{0.005 - 0.0025}\right) = 0.02820947$$

Summary of the numerical results

$$q_0 = +17.72458933$$
$$q_\ell = -17.72458933$$

$$(u_{11})_{t=t_1} = \frac{1}{\sqrt{\pi}}\left\{\sqrt{(t_2-t_0)}+\sqrt{(t_2-t_1)}\right\}=$$
$$= \frac{1}{\sqrt{\pi}}\left\{\sqrt{(0.005-0)}+\sqrt{(0.005-0.0025)}\right\}= 0.01168475$$

$$(u_{11})_{t=t_2} = \frac{1}{\sqrt{\pi}}\left\{\sqrt{(t_2-t_1)}\right\}= \frac{1}{\sqrt{\pi}}\left(\sqrt{0.005-0.0025}\right)=$$
$$= 0.02820947$$

$$(u_{12})_{t=t_1} = \int_{t_0}^{t_1} u^*(\xi=0, x=\ell; t_2, t)dt =$$
$$= \frac{1}{2\sqrt{\pi}}\left\{\left(\frac{e^{-50}}{\sqrt{50}}-\frac{e^{-100}}{\sqrt{100}}\right)+\sqrt{\pi}\left(erf\left(\sqrt{50}\right)-erf\left(\sqrt{100}\right)\right)\right\}$$
$$= 7.694598628\times 10^{-24}$$

$$(u_{12})_{t=t_2} = \int_{t_1}^{t_2} u^*(\xi=0, x=\ell; t_2, t)dt =$$
$$= \frac{1}{2\sqrt{\pi}}\frac{e^{-100}}{\sqrt{100}}+\frac{1}{2}\left\{erf\left(\sqrt{100}\right)-1\right\}= 1.04941405\times 10^{-45}$$

$$(u_{21})_{t=t_1} = \int_{t_0}^{t_1} u^*(\xi=\ell, x=0; t_2, t)dt =$$
$$= \frac{1}{2\sqrt{\pi}}\left\{\left(\frac{e^{-50}}{\sqrt{50}}-\frac{e^{-100}}{\sqrt{100}}\right)+\sqrt{\pi}\left(erf\left(\sqrt{50}\right)-erf\left(\sqrt{100}\right)\right)\right\}$$
$$= 7.694598628\times 10^{-24}$$

$$(u_{21})_{t=t_2} = \int_{t_1}^{t_2} u^*(\xi = \ell, x = 0; t_2, t)dt =$$

$$= \frac{1}{2\sqrt{\pi}} \frac{e^{-100}}{\sqrt{100}} + \frac{r}{2}\left\{erf\left(\sqrt{100}\right) - 1\right\} = 1.04941405 \times 10^{-45}$$

$$(u_{22})_{t=t_1} = \int_{t_1}^{t_2} u^*(\xi = \ell, x = \ell; t_2, t)dt =$$

$$= \frac{1}{\sqrt{\pi}}\left\{\sqrt{0.005 - 0.000} - \sqrt{0.005 - 0.0025}\right\} = 0.01168475$$

$$(u_{22})_{t=t_2} = \int_{t_1}^{t_2} u^*(\xi = \ell, x = \ell; t_2, t)dt = \frac{1}{\sqrt{\pi}}\left(\sqrt{0.005 - 0.0025}\right) =$$

$$= 0.02820947$$

The next step is to re-write equations **(NEC-6)** and **(NEC-7)** in a compact form as follows:

$$-\frac{u_0}{2}\left[-erf\left(\frac{-\ell}{2\sqrt{t_2}}\right)\right] =$$
$$= \{[q_{x=\ell}][u_{12}]\}_{t_1} + \{[q_{x=\ell}][u_{12}]\}_{t_2} - \{[q_{x=0}][u_{11}]\}_{t_1} - \{[q_{x=0}][u_{11}]\}_{t_2} \quad \textbf{(NEC-29)}$$

Equation **(NEC-29)** can be re-arranged so that all known terms are moved to the left-hand side; therefore equation **(NEC-29)** becomes:

$$-\frac{u_0}{2}\left[-erf\left(\frac{-\ell}{2\sqrt{t_2}}\right)\right] - \{[q_{x=\ell}][u_{12}]\}_{t_1} + \{[q_{x=0}][u_{11}]\}_{t_1} =$$
$$= \{[q_{x=\ell}][u_{12}]\}_{t_2} - \{[q_{x=0}][u_{11}]\}_{t_2} \quad \textbf{(NEC-30)}$$

And,

$$-\frac{u_0}{2}\left[erf\left(\frac{\ell}{2\sqrt{t_2}}\right)\right]=$$
$$=\{[q_{x=\ell}][u_{22}]\}_{t_1}+\{[q_{x=\ell}][u_{22}]\}_{t_2}-\{[q_{x=0}][u_{21}]\}_{t_1}-\{[q_{x=0}][u_{21}]\}_{t_2} \qquad \textbf{(NEC-31)}$$

Similar procedure can be carried out on equation **(NEC-31)**, therefore it becomes:

$$-\frac{u_0}{2}\left[erf\left(\frac{\ell}{2\sqrt{t_2}}\right)\right]-\{[q_{x=\ell}][u_{22}]\}_{t_1}+\{[q_{x=0}][u_{21}]\}_{t_1}=$$
$$=\{[q_{x=\ell}][u_{22}]\}_{t_2}-\{[q_{x=0}][u_{21}]\}_{t_2} \qquad \textbf{(NEC-32)}$$

Substituting by the numerical results into equation **(NEC-30)**, we get:

$$-0.5-\{[-17.72458933][7.694598628\times10^{-24}]\}_{t_1}$$
$$+\{[+17.72458933][0.01168475]\}_{t_1}=$$
$$=\{[q_{x=\ell}][1.04941405\times10^{-45}]\}_{t_2}-\{[q_{x=0}][0.02820947]\}_{t_2} \qquad \textbf{(NEC-33)}$$

Substituting by the numerical results into equation **(NEC-31)**, we get:

$$-0.5-\{[-17.72458933][0.01168475]\}_{t_1}$$
$$+\{[+17.72458933][7.694598628\times10^{-24}]\}_{t_1}=$$
$$=\{[q_{x=\ell}][0.02820947]\}_{t_2}-\{[q_{x=0}][1.04941405\times10^{-45}]\}_{t_2} \qquad \textbf{(NEC-34)}$$

Solution of equations **(NEC-33)** and **(NEC-34)** leads to the values of both $q_{x=0}$, and $q_{x=\ell}$ at the next time step, these values are:

$$q_{x=0}=+10.382775$$

&

$$q_{x=\ell}=-10.382775$$

Third Stage: Evaluation for the Third Time Step

This stage is to extend the computation procedure to compute the flux for the elementary example at the third time step.

General Overview Notes For This Stage

(1) $$t_n = t_3 = \underbrace{\underbrace{\underbrace{t_0 + \Delta t}_{t_1} + 2\Delta t}_{t_2} + 2\Delta t}_{t_3}$$

(3) The limits of integrations corresponding to this time step will be:

$$a_0 = \frac{r^2}{4(t_3 - t_0)}$$

$$a_1 = \frac{r^2}{4(t_3 - t_1)}$$

$$a_2 = \frac{r^2}{4(t_3 - t_2)}$$

$$a_3 = \frac{r^2}{4(t_3 - t_3)} \text{.......Singularit y}$$

The boundary integral equation corresponding to this time step takes the form:

$$u(\xi, t_3) = \frac{u_0}{2}\left[erf\left(\frac{\xi}{2\sqrt{t_3}}\right) - erf\left(\frac{\xi - \ell}{2\sqrt{t_3}}\right)\right] + \sum_{i=1}^{3}\left[\begin{array}{l} q_{x=\ell}\left(\int_{t_0}^{t_3} u^*(\xi, x; t_3, t)dt\right)_{x=\ell} \\ -q_{x=0}\left(\int_{t_0}^{t_3} u^*(\xi, x; t_3, t)dt\right)_{x=0} \end{array}\right] \quad \textbf{(ETTH-1)}$$

Equation **(ETTH-1)** can be written in some details as follows:

$$u(\xi,t_3)=\frac{u_0}{2}\left[erf\left(\frac{\xi}{2\sqrt{t_3}}\right)-erf\left(\frac{\xi-\ell}{2\sqrt{t_3}}\right)\right]+$$
$$+q_{x=\ell}\left(\int_{t_0}^{t_1}u^*(\xi,x;t_3,t)dt+\int_{t_1}^{t_2}u^*(\xi,x;t_3,t)dt+\int_{t_2}^{t_3}u^*(\xi,x;t_3,t)dt\right)_{x=\ell} \qquad \textbf{(ETTH-2)}$$
$$-q_{x=0}\left(\int_{t_0}^{t_1}u^*(\xi,x;t_3,t)dt+\int_{t_1}^{t_2}u^*(\xi,x;t_3,t)dt+\int_{t_2}^{t_3}u^*(\xi,x;t_3,t)dt\right)_{x=0}$$

Equation **(ETTH-2)** still has a detailed form as follows:

$$u(\xi,t_3)=\frac{u_0}{2}\left[erf\left(\frac{\xi}{2\sqrt{t_3}}\right)-erf\left(\frac{\xi-\ell}{2\sqrt{t_3}}\right)\right]++[q_{x=\ell}]_{t_1}\int_{t_0}^{t_1}u^*(\xi,x=\ell;t_3,t)dt$$
$$+[q_{x=\ell}]_{t_2}\int_{t_1}^{t_2}u^*(\xi,x=\ell;t_3,t)dt+[q_{x=\ell}]_{t_3}\int_{t_2}^{t_3}u^*(\xi,x=\ell;t_3,t)dt$$
$$-[q_{x=0}]_{t_1}\int_{t_0}^{t_1}u^*(\xi,x=0;t_3,t)dt-[q_{x=0}]_{t_2}\int_{t_1}^{t_2}u^*(\xi,x=0;t_3,t)dt \qquad \textbf{(ETTH-3)}$$
$$-[q_{x=0}]_{t_3}\int_{t_2}^{t_3}u^*(\xi,x=0;t_3,t)dt$$

Important Notes

(1) In equation **(3)** both $[q_{x=0}]_{t_3}$ and $[q_{x=\ell}]_{t_3}$ are unknowns.

(2) All integrals are known except $\int_{t_2}^{t_3} u^*(\xi, x=\ell; t_3, t)dt$ and $\int_{t_2}^{t_3} u^*(\xi, x=0; t_3, t)dt$

(3) It is required to evaluate the integrals in note (**2**) and then form the system as done before for the first and second time steps to evaluate the unknown flux at the end points.

The next step is to apply equation **(ETTH-3)** at two different source points, they are $\xi = 0$ and $\xi = \ell$. Let us start by taking source point at $\xi = 0$, leads to:

$$
\begin{aligned}
u(\xi=0,t_3) &= \frac{u_0}{2}\left[erf\left(\frac{\ell}{2\sqrt{t_3}}\right)\right] + +[q_{x=\ell}]_{t_1}\int_{t_0}^{t_1} u^*(\xi=0,x=\ell;t_3,t)dt \\
&+[q_{x=\ell}]_{t_2}\int_{t_1}^{t_2} u^*(\xi=0,x=\ell;t_3,t)dt + [q_{x=\ell}]_{t_3}\int_{t_2}^{t_3} u^*(\xi=0,x=\ell;t_3,t)dt \\
&-[q_{x=0}]_{t_1}\int_{t_0}^{t_1} u^*(\xi=0,x=0;t_3,t)dt - [q_{x=0}]_{t_2}\int_{t_1}^{t_2} u^*(\xi=0,x=0;t_3,t)dt \\
&-[q_{x=0}]_{t_3}\int_{t_2}^{t_3} u^*(\xi=0,x=0;t_3,t)dt
\end{aligned}
\qquad \textbf{(ETTH-4)}
$$

Notes:

$$u_{11} = \int_{t_1}^{t_2} u^*(\xi=0,x=0;t_3,t)dt$$

& **(ETTH-5)**

$$u_{12} = \int_{t_1}^{t_2} u^*(\xi=0,x=\ell;t_3,t)dt$$

Now let us re-write equation **(ETTH-4)** by making use of equation **(ETTH-5)**, we get:

$$[u(\xi = 0, t_3) = 0] =$$

$$= \frac{u_0}{2}\left[erf\left(\frac{\ell}{2\sqrt{t_3}}\right)\right] + [q_{x=\ell}]_{t_1}[u_{12}]_{t_1} + [q_{x=\ell}]_{t_2}[u_{12}]_{t_2} + [q_{x=\ell}]_{t_3}[u_{12}]_{t_3} \qquad \textbf{(ETTH-6)}$$
$$-[q_{x=0}]_{t_1}[u_{11}]_{t_1} - [q_{x=0}]_{t_2}[u_{11}]_{t_2} - [q_{x=0}]_{t_3}[u_{11}]_{t_3}$$

$$-\frac{u_0}{2}\left[erf\left(\frac{\ell}{2\sqrt{t_3}}\right)\right] = [q_{x=\ell}]_{t_1}[u_{12}]_{t_1} + [q_{x=\ell}]_{t_2}[u_{12}]_{t_2} + [q_{x=\ell}]_{t_3}[u_{12}]_{t_3} \qquad \textbf{(ETTH-7)}$$
$$-[q_{x=0}]_{t_1}[u_{11}]_{t_1} - [q_{x=0}]_{t_2}[u_{11}]_{t_2} - [q_{x=0}]_{t_3}[u_{11}]_{t_3}$$

Notes:

(1) $[q_{x=\ell}]_{t_1}$ & $[q_{x=0}]_{t_1}$ Fluxes found from first time step

(2) $[u_{12}]_{t_1}$ & $[u_{11}]_{t_1}$ Evaluated from first time step

(3) $[q_{x=\ell}]_{t_2}$ & $[q_{x=0}]_{t_2}$ Fluxes found from second time step

(4) $[u_{12}]_{t_2}$ & $[u_{11}]_{t_2}$ Evaluated from first time step

(5) $[q_{x=\ell}]_{t_3}$ & $[q_{x=0}]_{t_3}$ Unknown to be found from third time step

(6) $[u_{12}]_{t_3}$ & $[u_{11}]_{t_3}$ Integral to be calculated in third time step

Now, re-arrange equation (7) so that all known values will be on one side and the unknowns in the other side, leads to:

$$-\frac{u_0}{2}\left[erf\left(\frac{\ell}{2\sqrt{t_3}}\right)\right]+[q_{x=0}]_{t_1}[u_{11}]_{t_1}+[q_{x=0}]_{t_2}[u_{11}]_{t_2}$$
$$-[q_{x=\ell}]_{t_1}[u_{12}]_{t_1}-[q_{x=\ell}]_{t_2}[u_{12}]_{t_2}=[q_{x=\ell}]_{t_3}[u_{12}]_{t_3}-[q_{x=0}]_{t_3}[u_{11}]_{t_3} \quad \textbf{(ETTH-8)}$$

Now, let us take source point at $\xi=\ell$, leads to:

$$u(\xi=\ell,t_3)=\frac{u_0}{2}\left[erf\left(\frac{\ell}{2\sqrt{t_3}}\right)\right]+[q_{x=\ell}]_{t_1}\int_{t_0}^{t_1}u^*(\xi=\ell,x=\ell;t_3,t)dt$$
$$+[q_{x=\ell}]_{t_2}\int_{t_1}^{t_2}u^*(\xi=\ell,x=\ell;t_3,t)dt+[q_{x=\ell}]_{t_3}\int_{t_2}^{t_3}u^*(\xi=\ell,x=\ell;t_3,t)dt$$
$$-[q_{x=0}]_{t_1}\int_{t_0}^{t_1}u^*(\xi=\ell,x=0;t_3,t)dt-[q_{x=0}]_{t_2}\int_{t_1}^{t_2}u^*(\xi=\ell,x=0;t_3,t)dt \quad \textbf{(ETTH-9)}$$
$$-[q_{x=0}]_{t_3}\int_{t_2}^{t_3}u^*(\xi=\ell,x=0;t_3,t)dt$$

Notes:

$$u_{21}=\int_{t_1}^{t_2}u^*(\xi=\ell,x=0;t_3,t)dt$$

& **(ETTH-10)**

$$u_{22}=\int_{t_1}^{t_2}u^*(\xi=\ell,x=\ell;t_3,t)dt$$

Now let us re-write equation **(ETTH-9)** by making use of equation **(ETTH-10)**, we get:

$$\left[u(\xi=\ell,t_3)=0\right]=\frac{u_0}{2}\left[erf\left(\frac{\ell}{2\sqrt{t_3}}\right)\right]+$$
$$+\left[q_{x=\ell}\right]_{t_1}\left[u_{22}\right]_{t_1}+\left[q_{x=\ell}\right]_{t_2}\left[u_{22}\right]_{t_2}+\left[q_{x=\ell}\right]_{t_3}\left[u_{22}\right]_{t_3} \qquad \textbf{(ETTH-11)}$$
$$-\left[q_{x=0}\right]_{t_1}\left[u_{21}\right]_{t_1}-\left[q_{x=0}\right]_{t_2}\left[u_{21}\right]_{t_2}-\left[q_{x=0}\right]_{t_3}\left[u_{21}\right]_{t_3}$$

$$-\frac{u_0}{2}\left[erf\left(\frac{\ell}{2\sqrt{t_3}}\right)\right]=\left[q_{x=\ell}\right]_{t_1}\left[u_{22}\right]_{t_1}+\left[q_{x=\ell}\right]_{t_2}\left[u_{22}\right]_{t_2}+\left[q_{x=\ell}\right]_{t_3}\left[u_{22}\right]_{t_3}$$
$$-\left[q_{x=0}\right]_{t_1}\left[u_{21}\right]_{t_1}-\left[q_{x=0}\right]_{t_2}\left[u_{21}\right]_{t_2}-\left[q_{x=0}\right]_{t_3}\left[u_{21}\right]_{t_3} \qquad \textbf{(ETTH-12)}$$

Notes:

(1) $\left[q_{x=\ell}\right]_{t_1}$ & $\left[q_{x=0}\right]_{t_1}$ Fluxes found from first time step

(2) $\left[u_{22}\right]_{t_1}$ & $\left[u_{21}\right]_{t_1}$ Evaluated from first time step

(3) $\left[q_{x=\ell}\right]_{t_2}$ & $\left[q_{x=0}\right]_{t_2}$ Fluxes found from second time step

(4) $\left[u_{22}\right]_{t_2}$ & $\left[u_{21}\right]_{t_2}$ Evaluated from first time step

(5) $\left[q_{x=\ell}\right]_{t_3}$ & $\left[q_{x=0}\right]_{t_3}$ Unknown to be found from third time step

(6) $\left[u_{22}\right]_{t_3}$ & $\left[u_{21}\right]_{t_3}$ Integral to be calculated in third time step

Now, re-arrange equation (1**2)** so that all known values will be on one side and the unknowns in the other side, leads to:

$$-\frac{u_0}{2}\left[erf\left(\frac{\ell}{2\sqrt{t_3}}\right)\right]+\left[q_{x=0}\right]_{t_1}\left[u_{21}\right]_{t_1}$$
$$+\left[q_{x=0}\right]_{t_2}\left[u_{21}\right]_{t_2}-\left[q_{x=\ell}\right]_{t_1}\left[u_{22}\right]_{t_1}-\left[q_{x=\ell}\right]_{t_2}\left[u_{22}\right]_{t_2}$$
$$=\left[q_{x=\ell}\right]_{t_3}\left[u_{22}\right]_{t_3}-\left[q_{x=0}\right]_{t_3}\left[u_{21}\right]_{t_3} \quad \textbf{(ETTH-13)}$$

Now, equations **(ETTH-8)** and **(ETTH-13)** can be re-written in a matrix form as follow:

$$AX=b \quad \textbf{(ETTH-14)}$$

Where

$$A=\begin{bmatrix}-u_{11} & u_{12}\\ -u_{21} & u_{22}\end{bmatrix} \quad \textbf{(ETTH-15)}$$

$$X=\begin{bmatrix}q_0\\ q_1\end{bmatrix} \quad \textbf{(ETTH-16)}$$

$$b=\begin{bmatrix}-\frac{u_0}{2}\left[erf\left(\frac{\ell}{2\sqrt{t_3}}\right)\right]+\left[q_{x=0}\right]_{t_1}\left[u_{11}\right]_{t_1}\\ +\left[q_{x=0}\right]_{t_2}\left[u_{11}\right]_{t_2}-\left[q_{x=\ell}\right]_{t_1}\left[u_{12}\right]_{t_1}-\left[q_{x=\ell}\right]_{t_2}\left[u_{12}\right]_{t_2}\\ -\frac{u_0}{2}\left[erf\left(\frac{\ell}{2\sqrt{t_3}}\right)\right]+\left[q_{x=0}\right]_{t_1}\left[u_{21}\right]_{t_1}\\ +\left[q_{x=0}\right]_{t_2}\left[u_{21}\right]_{t_2}-\left[q_{x=\ell}\right]_{t_1}\left[u_{22}\right]_{t_1}-\left[q_{x=\ell}\right]_{t_2}\left[u_{22}\right]_{t_2}\end{bmatrix} \quad \textbf{(ETTH-17)}$$

The next step is to evaluate the time integrals in equation **(ETTH-15)**; these integrals can be summarized as follow:

$$
\begin{aligned}
u_{11} &= \int_{t_1}^{t_2} u^*\left(\xi = 0, x = 0; t_3, t\right)dt \\
u_{12} &= \int_{t_1}^{t_2} u^*\left(\xi = 0, x = \ell; t_3, t\right)dt \\
u_{21} &= \int_{t_1}^{t_2} u^*\left(\xi = \ell, x = 0; t_3, t\right)dt \\
u_{22} &= \int_{t_1}^{t_2} u^*\left(\xi = \ell, x = \ell; t_3, t\right)dt
\end{aligned}
\qquad \textbf{(ETTH-18)}
$$

Let us now illustrate the number of integrals associated within the time steps up to the third one. To do that, refer to the following diagrammatic equation and have a look at the number of integrals associated with each time step.

$$
\int_{t_0}^{t_3} = \underbrace{\int_{t_0}^{t_1}}_{\substack{\text{Four integrals} \\ u_{11}, u_{12} \\ u_{21}, u_{22}}} + \underbrace{\int_{t_1}^{t_2}}_{\substack{\text{Four integrals} \\ u_{11}, u_{12} \\ u_{21}, u_{22}}} + \underbrace{\int_{t_2}^{t_3}}_{\substack{\text{Four integrals} \\ u_{11}, u_{12} \\ u_{21}, u_{22}}}
$$

Fig. (2): Overall time integration.

It is obvious from the Fig. (**2**) that we have 12 integrals that should be evaluated within this stage. What we are doing now is to evaluate these integrals one by one simultaneously.

For The First Time Step

At ***first*** time step with a source at $\xi = 0$ and field at $x = 0$

$$\{u_{11}\}_{t_1} = \int_{t_0}^{t_1} u^*(\xi = 0, x = 0; t_3, t)dt =$$

$$= \frac{r}{2\sqrt{\pi}}\left\{\left(\frac{e^{-a_0}}{\sqrt{a_0}} - \frac{e^{-a_1}}{\sqrt{a_1}}\right) + \sqrt{\pi}\left\{erf\sqrt{a_0} - erf\sqrt{a_1}\right\}\right\} \quad \textbf{(ETTH-19)}$$

In equation **(ETTH-19)**:

$$r = \xi - x = 0$$

$$a_0 = \frac{r^2}{4(t_3 - t_0)} \quad \textbf{(ETTH-20)}$$

$$a_1 = \frac{r^2}{4(t_3 - t_1)}$$

Therefore, equation **(ETTH-19)** takes the following final form:

$$\{u_{11}\}_{t_1} = \frac{1}{\sqrt{\pi}}\left(\sqrt{t_3 - t_0} - \sqrt{t_3 - t_1}\right) \quad \textbf{(ETTH-21)}$$

At ***first*** time step with a source at $\xi = 0$ and field at $x = \ell$

$$\{u_{12}\}_{t_1} = \int_{t_0}^{t_1} u^*(\xi = 0, x = \ell; t_3, t)dt =$$

$$= \frac{r}{2\sqrt{\pi}}\left\{\left(\frac{e^{-a_0}}{\sqrt{a_0}} - \frac{e^{-a_1}}{\sqrt{a_1}}\right) + \sqrt{\pi}\left\{erf\sqrt{a_0} - erf\sqrt{a_1}\right\}\right\} \quad \textbf{(ETTH-22)}$$

In equation **(ETTH-22)**:

$$r = \xi - x = -\ell$$

$$a_0 = \frac{r^2}{4(t_3 - t_0)} \qquad \textbf{(ETTH-23)}$$

$$a_1 = \frac{r^2}{4(t_3 - t_1)}$$

At ***first*** time step with source at $\xi = \ell$ and field at $x = 0$

$$\{u_{21}\}_{t_1} = \int_{t_0}^{t_1} u^*(\xi = \ell, x = 0; t_3, t)dt =$$

$$= \frac{r}{2\sqrt{\pi}}\left\{\left(\frac{e^{-a_0}}{\sqrt{a_0}} - \frac{e^{-a_1}}{\sqrt{a_1}}\right) + \sqrt{\pi}\left\{erf\sqrt{a_0} - erf\sqrt{a_1}\right\}\right\} \qquad \textbf{(ETTH-24)}$$

In equation **(ETTH-24)**:

$$r = \xi - x = \ell$$

$$a_0 = \frac{r^2}{4(t_3 - t_0)} \qquad \textbf{(ETTH-25)}$$

$$a_1 = \frac{r^2}{4(t_3 - t_1)}$$

At ***first*** time step with a source at $\xi = \ell$ and field at $x = \ell$

$$\{u_{22}\}_{t_1} = \int_{t_0}^{t_1} u^*(\xi = \ell, x = \ell; t_3, t)dt =$$

$$= \frac{r}{2\sqrt{\pi}}\left\{\left(\frac{e^{-a_0}}{\sqrt{a_0}} - \frac{e^{-a_1}}{\sqrt{a_1}}\right) + \sqrt{\pi}\left\{erf\sqrt{a_0} - erf\sqrt{a_1}\right\}\right\} \qquad \textbf{(ETTH-26)}$$

In equation **(ETTH-26)**:

$$r = \xi - x = 0$$

$$a_0 = \frac{r^2}{4(t_3 - t_0)} \qquad \textbf{(ETTH-27)}$$

$$a_1 = \frac{r^2}{4(t_3 - t_1)}$$

Therefore, equation **(ETTH-26)** becomes:

$$\{u_{22}\}_{t_1} = \frac{1}{\sqrt{\pi}}\left(\sqrt{t_3 - t_0} - \sqrt{t_3 - t_1}\right) \qquad \textbf{(ETTH-28)}$$

For The Second Time Step

At ***second*** time step with a source at $\xi = 0$ and field at $x = 0$

$$\{u_{11}\}_{t_2} = \int_{t_1}^{t_2} u^*(\xi = 0, x = 0; t_3, t)dt =$$

$$= \frac{r}{2\sqrt{\pi}}\left\{\left(\frac{e^{-a_1}}{\sqrt{a_1}} - \frac{e^{-a_2}}{\sqrt{a_2}}\right) + \sqrt{\pi}\left\{erf\sqrt{a_1} - erf\sqrt{a_2}\right\}\right\} \qquad \textbf{(ETTH-29)}$$

In equation **(ETTH-29)**:

$$r = \xi - x = 0$$

$$a_1 = \frac{r^2}{4(t_3 - t_1)} \qquad \textbf{(ETTH-30)}$$

$$a_2 = \frac{r^2}{4(t_3 - t_2)}$$

Therefore, equation **(ETTH-29)** takes the following final form:

$$\{u_{11}\}_{t_1} = \frac{1}{\sqrt{\pi}}\left(\sqrt{t_3 - t_1} - \sqrt{t_3 - t_2}\right) \qquad \textbf{(ETTH-31)}$$

At ***second*** time step with source at $\xi = 0$ and field at $x = \ell$

$$\{u_{12}\}_{t_2} = \int_{t_1}^{t_2} u^*(\xi = 0, x = \ell; t_3, t)dt =$$

$$= \frac{r}{2\sqrt{\pi}}\left\{\left(\frac{e^{-a_1}}{\sqrt{a_1}} - \frac{e^{-a_2}}{\sqrt{a_2}}\right) + \sqrt{\pi}\left\{erf\sqrt{a_1} - erf\sqrt{a_2}\right\}\right\} \quad \textbf{(ETTH-32)}$$

In equation (3**2**):

$$r = \xi - x = -\ell$$

$$a_1 = \frac{r^2}{4(t_3 - t_1)} \quad \textbf{(ETTH-33)}$$

$$a_2 = \frac{r^2}{4(t_3 - t_2)}$$

At ***second*** time step with source at $\xi = \ell$ and field at $x = 0$

$$\{u_{21}\}_{t_2} = \int_{t_1}^{t_2} u^*(\xi = \ell, x = 0; t_3, t)dt =$$

$$= \frac{r}{2\sqrt{\pi}}\left\{\left(\frac{e^{-a_1}}{\sqrt{a_1}} - \frac{e^{-a_2}}{\sqrt{a_2}}\right) + \sqrt{\pi}\left\{erf\sqrt{a_1} - erf\sqrt{a_2}\right\}\right\} \quad \textbf{(ETTH-34)}$$

In equation **(ETTH-34)**:

$$r = \xi - x = -\ell$$

$$a_1 = \frac{r^2}{4(t_3 - t_1)} \quad \textbf{(ETTH-35)}$$

$$a_2 = \frac{r^2}{4(t_3 - t_2)}$$

At ***first*** time step with source at $\xi = \ell$ and field at $x = \ell$

$$\{u_{22}\}_{t_2} = \int_{t_1}^{t_2} u^*(\xi = \ell, x = \ell; t_3, t)dt =$$

$$= \frac{r}{2\sqrt{\pi}}\left\{\left(\frac{e^{-a_1}}{\sqrt{a_1}} - \frac{e^{-a_2}}{\sqrt{a_2}}\right) + \sqrt{\pi}\left\{erf\sqrt{a_1} - erf\sqrt{a_2}\right\}\right\} \quad \textbf{(ETTH-36)}$$

In equation **(ETTH-36)**:

$$r = \xi - x = 0$$

$$a_1 = \frac{r^2}{4(t_3 - t_1)} \quad \textbf{(ETTH-37)}$$

$$a_2 = \frac{r^2}{4(t_3 - t_2)}$$

Therefore, equation (ETTH-36) becomes:

$$\{u_{22}\}_{t_2} = \frac{1}{\sqrt{\pi}}\left(\sqrt{t_3 - t_1} - \sqrt{t_3 - t_2}\right) \quad \textbf{(ETTH-38)}$$

For the third time step

At ***third*** time step with a source at $\xi = 0$ and field at $x = 0$

$$\{u_{11}\}_{t_3} = \int_{t_2}^{t_3} u^*(\xi = 0, x = 0; t_3, t)dt =$$

$$= \frac{r}{2\sqrt{\pi}}\left\{\left(\frac{e^{-a_2}}{\sqrt{a_2}} - \frac{e^{-a_3}}{\sqrt{a_3}}\right) + \sqrt{\pi}\left\{erf\sqrt{a_2} - erf\sqrt{a_3}\right\}\right\} \quad \textbf{(ETTH-39)}$$

In equation **(ETTH-39)**:

$$r = \xi - x = 0$$

$$a_2 = \frac{r^2}{4(t_3 - t_2)} \quad \text{(ETTH-40)}$$

$$a_1 = \frac{r^2}{4(t_3 - t_3)} \text{......Singularit y}$$

Therefore, equation **(ETTH-39)** takes the following final form:

$$\{u_{11}\}_{t_1} = \frac{1}{\sqrt{\pi}}\left(\sqrt{t_3 - t_2}\right) \quad \text{(ETTH-41)}$$

At ***third*** time step with source at $\xi = 0$ and field at $x = \ell$

$$\{u_{12}\}_{t_3} = \int_{t_2}^{t_3} u^*(\xi = 0, x = \ell; t_3, t)dt =$$

$$= \frac{r}{2\sqrt{\pi}}\left\{\left(\frac{e^{-a_2}}{\sqrt{a_2}} - \frac{e^{-a_3}}{\sqrt{a_3}}\right) + \sqrt{\pi}\left\{erf\sqrt{a_2} - erf\sqrt{a_3}\right\}\right\} \quad \text{(ETTH-42)}$$

In equation **(ETTH-42)**:

$$r = \xi - x = -\ell$$

$$a_2 = \frac{r^2}{4(t_3 - t_2)} \quad \text{(ETTH-43)}$$

$$a_3 = \frac{r^2}{4(t_3 - t_3)} = \infty$$

Therefore, equation **(ETTH-42)** becomes:

$$\{u_{12}\}_{t_3} = \frac{r}{2\sqrt{\pi}}\left\{\left(\frac{e^{-a_2}}{\sqrt{a_2}}\right) + \sqrt{\pi}\left\{erf\sqrt{a_2} - 1\right\}\right\} \quad \text{(ETTH-44)}$$

At ***third*** time step with source at $\xi = \ell$ and field at $x = 0$

$$\{u_{21}\}_{t_3} = \int_{t_2}^{t_3} u^*(\xi = \ell, x = 0; t_3, t)dt =$$

$$= \frac{r}{2\sqrt{\pi}}\left\{\left(\frac{e^{-a_2}}{\sqrt{a_2}} - \frac{e^{-a_3}}{\sqrt{a_3}}\right) + \sqrt{\pi}\left\{erf\sqrt{a_2} - erf\sqrt{a_3}\right\}\right\} \quad \textbf{(ETTH-45)}$$

In equation **(ETTH-45)**:

$$r = \xi - x = \ell$$

$$a_2 = \frac{r^2}{4(t_3 - t_2)} \quad \textbf{(ETTH-46)}$$

$$a_3 = \frac{r^2}{4(t_3 - t_3)}$$

At ***third*** time step with source at $\xi = \ell$ and field at $x = \ell$

$$\{u_{22}\}_{t_3} = \int_{t_2}^{t_3} u^*(\xi = \ell, x = \ell; t_3, t)dt =$$

$$= \frac{r}{2\sqrt{\pi}}\left\{\left(\frac{e^{-a_2}}{\sqrt{a_2}} - \frac{e^{-a_3}}{\sqrt{a_3}}\right) + \sqrt{\pi}\left\{erf\sqrt{a_2} - erf\sqrt{a_3}\right\}\right\} \quad \textbf{(ETTH-47)}$$

In equation **(ETTH-47)**:

$$r = \xi - x = 0$$

$$a_2 = \frac{r^2}{4(t_3 - t_2)} \quad \textbf{(ETTH-48)}$$

$$a_3 = \frac{r^2}{4(t_3 - t_3)}$$

Therefore, equation **(ETTH-48)** becomes:

$$\{u_{22}\}_{t_3} = \frac{1}{\sqrt{\pi}}\left(\sqrt{t_3 - t_2}\right) \quad \textbf{(ETTH-49)}$$

Preliminary Stage for Code Program

This section will prepare the general algorithm for solving the heat diffusion problem over a fixed domain. The first step of this algorithm is to create the time step, and this can be done using the following formula:

$$T(k) = T(k-1) + \Delta T, \;\; k = 1, N$$

Example

For

$$k = 1$$
$$T(1) = T(0) + \Delta T$$

$$k = 2$$
$$T(2) = T(1) + \Delta T$$

$$k = 3$$
$$T(3) = T(2) + \Delta T$$

Notes:

k :Counter
N :Total number of time steps
$T(0)$:Initial time

$kk = 1, k$: Do loop for evaluating time integrals at each time step

$$a(kk-1)=\frac{1}{4(T(k)-T(kk-1))}:$$

Do loop for evaluating limits of integrals at each time step

Check (1)

N :Total number of time steps $= 2$

$k = 1$

$$kk = 1,1 \Leftrightarrow a_0 = \frac{1}{4(t_1 - t_0)}$$

N :Total number of time steps $= 2$

$k = 2$

$$kk = 1 \Leftrightarrow a_0 = \frac{1}{4(t_2 - t_0)}$$

$$kk = 2 \Leftrightarrow a_1 = \frac{1}{4(t_2 - t_1)}$$

Check (2)

N :Total number of time steps $= 5$

$k = 1$

$$kk = 1,1 \Leftrightarrow a_0 = \frac{1}{4(t_1 - t_0)}$$

N :Total number of time steps $= 5$

$k = 2$

$$kk = 1 \Leftrightarrow a_0 = \frac{1}{4(t_2 - t_0)}$$

$$kk = 2 \Leftrightarrow a_1 = \frac{1}{4(t_2 - t_1)}$$

N : Total number of time steps = 5

$k = 3$

$$kk = 1 \Leftrightarrow a_0 = \frac{1}{4(t_3 - t_0)}$$

$$kk = 2 \Leftrightarrow a_1 = \frac{1}{4(t_3 - t_1)}$$

$$kk = 3 \Leftrightarrow a_3 = \frac{1}{4(t_3 - t_2)}$$

N : Total number of time steps = 5

$k = 4$

$$kk = 1 \Leftrightarrow a_0 = \frac{1}{4(t_4 - t_0)}$$

$$kk = 2 \Leftrightarrow a_1 = \frac{1}{4(t_4 - t_1)}$$

$$kk = 3 \Leftrightarrow a_3 = \frac{1}{4(t_4 - t_2)}$$

$$kk = 4 \Leftrightarrow a_4 = \frac{1}{4(t_4 - t_3)}$$

N : Total number of time steps = 5

$k = 5$

$$kk = 1 \Leftrightarrow a_0 = \frac{1}{4(t_5 - t_0)}$$

$$kk = 2 \Leftrightarrow a_1 = \frac{1}{4(t_5 - t_1)}$$

$$kk = 3 \Leftrightarrow a_3 = \frac{1}{4(t_5 - t_2)}$$

$$kk = 4 \Leftrightarrow a_4 = \frac{1}{4(t_5 - t_3)}$$

$$kk = 5 \Leftrightarrow a_5 = \frac{1}{4(t_5 - t_4)}$$

General Algorithm for the Computation of the First Time Step

$$t_n = t_1$$
$$t_1 = t_0 + \Delta t$$

$$a_0 = \frac{r^2}{4\left(\underbrace{t_1}_{\text{The index corresponds to "k"}} - \underbrace{t_0}_{\text{The index corresponds to "kk"}}\right)}$$

$$u_{11} = -\frac{1}{\sqrt{\pi}}\left(\sqrt{t_1 - t_0}\right)$$

$$u_{12} = \frac{r}{2\sqrt{\pi}}\left(\frac{e^{-a_0}}{\sqrt{a_0}}\right) + \frac{r}{2}\left\{erf\left(\sqrt{a_0}\right) - 1\right\}$$

$$u_{21} = -u_{12}$$

$$u_{22} = \frac{1}{\sqrt{\pi}}\left(\sqrt{t_1 - t_0}\right)$$

$$b_1 = b_1 = \frac{-u_0}{2}\left(erf\left[\frac{\ell}{2\sqrt{t_1}}\right]\right)$$

$$q_1 = \frac{\left(b_1 \times u_{22} - b_2 \times u_{12}\right)}{\left(u_{11} \times u_{22} - u_{21} \times u_{12}\right)}$$

$$q_2 = \frac{\left(b_1 \times u_{11} - b_2 \times u_{21}\right)}{\left(u_{11} \times u_{22} - u_{21} \times u_{12}\right)}$$

General Algorithm for the Computation of the Second Time Step

The following algorithm will consist of two major parts the first one is for the first time step while the second one is for the second time step.

$$t_n = t_2$$

$$t_1 = t_0 + \Delta t$$

$$t_2 = t_1 + \Delta t = t_0 + 2\Delta t$$

Fisrt part of the algorithm corresponds to the time integral in the first time step

$$a_0 = \frac{r^2}{4(t_2 - t_0)}$$

$$a_1 = \frac{r^2}{4(t_2 - t_1)}$$

$$u_{11} = \frac{1}{\sqrt{\pi}}\left(\sqrt{t_2 - t_0} - \sqrt{t_2 - t_1}\right)$$

$$u_{12} = \frac{r}{2\sqrt{\pi}}\left(\frac{e^{-a_0}}{\sqrt{a_0}} - \frac{e^{-a_1}}{\sqrt{a_1}}\right) + \sqrt{\pi}\left\{erf\left(\sqrt{a_0}\right) - erf\left(\sqrt{a_1}\right)\right\}$$

$$u_{21} = u_{12}$$

$$u_{22} = \frac{1}{\sqrt{\pi}}\left(\sqrt{t_2 - t_0} - \sqrt{t_2 - t_1}\right)$$

$$b_{11} = \frac{-u_0}{2}\left(erf\left[\frac{\ell}{2\sqrt{t_2}}\right]\right) +$$

$$+\left[q_1\right]\text{Evaluated at the first time step}\, u_{11} - \left[q_2\right]\text{Evaluated at the first time step}\, u_{12}$$

$$b_{22} = \frac{-u_0}{2}\left(erf\left[\frac{\ell}{2\sqrt{t_2}} \right] \right) +$$

$+[q_1]$ Evaluated at the first time step $u_{21} - [q_2]$ Evaluated at the first time step u_{22}

Second part of the algorithm corresponds to the time integral in the second time step

$$u_{11} = -\frac{1}{\sqrt{\pi}}\left(\sqrt{t_2 - t_1}\right)$$

$$u_{12} = \frac{r}{2\sqrt{\pi}}\left(\frac{e^{-a_1}}{\sqrt{a_1}}\right) + \frac{r}{2}\left\{erf\left(\sqrt{a_1}\right) - 1\right\}$$

$$u_{21} = -u_{12}$$

$$u_{22} = \frac{1}{\sqrt{\pi}}\left(\sqrt{t_2 - t_1}\right)$$

$$b_{11} = \frac{-u_0}{2}\left(erf\left[\frac{\ell}{2\sqrt{t_2}} \right] \right) +$$

$+[q_1]$Evaluated at the first time step $u_{11} - [q_2]$Evaluated at the first time step u_{12}

$$q_1 = \frac{(b_{11} \times u_{22} - b_{22} \times u_{12})}{(u_{11} \times u_{22} - u_{21} \times u_{12})}$$

$$q_2 = \frac{(b_{11} \times u_{11} - b_{22} \times u_{21})}{(u_{11} \times u_{22} - u_{21} \times u_{12})}$$

Now let us re-write the summary of the numerical results for the first and second time steps, so as we use these results for the computation for the third time step.

Summary for the First Time Step

$q_0 = +17.72458933$

&

$q_\ell = -17.72458933$

Summary for the Second Time Step

$q_{x=0} = +10.382775$

&

$q_{x=\ell} = -10.382775$

Numerical Computation for the Third Time Step

As described before, the computation for the third time step requires computations of the integrals from t_0 to t_1, t_1 to t_2 and t_2 to t_3. Each integral consists of four elements, and now the computations of these elements are as follows:

For the first time step

At ***first*** time step with a source at $\xi = 0$ and field at $x = 0$

$$\{u_{11}\}_{t_1} = \frac{1}{\sqrt{\pi}}\left(\sqrt{t_3 - t_0} - \sqrt{t_3 - t_1}\right) = 8.96602 \times 10^{-3}$$

At ***first*** time step with a source at $\xi = 0$ and field at $x = \ell$

$$\{u_{12}\}_{t_1} = \int_{t_0}^{t_1} u^*(\xi = 0, x = \ell; t_3, t)dt =$$

$$= \frac{r}{2\sqrt{\pi}}\left\{\left(\frac{e^{-a_0}}{\sqrt{a_0}} - \frac{e^{-a_1}}{\sqrt{a_1}}\right) + \sqrt{\pi}\left\{erf\sqrt{a_0} - erf\sqrt{a_1}\right\}\right\}$$

$$r = \xi - x = -\ell$$

$$a_0 = \frac{r^2}{4(t_3 - t_0)}$$

$$a_1 = \frac{r^2}{4(t_3 - t_1)}$$

$$\{u_{12}\}_{t_1} = 1.6316 \times 10^{-16}$$

At ***first*** time step with a source at $\xi = \ell$ and field at $x = 0$

$$\{u_{21}\}_{t_1} = \int_{t_0}^{t_1} u^*(\xi = \ell, x = 0; t_3, t)dt =$$

$$= \frac{r}{2\sqrt{\pi}}\left\{\left(\frac{e^{-a_0}}{\sqrt{a_0}} - \frac{e^{-a_1}}{\sqrt{a_1}}\right) + \sqrt{\pi}\left\{erf\sqrt{a_0} - erf\sqrt{a_1}\right\}\right\}$$

$$r = \xi - x = \ell$$

$$a_0 = \frac{r^2}{4(t_3 - t_0)}$$

$$a_1 = \frac{r^2}{4(t_3 - t_1)}$$

$$\{u_{12}\}_{t_1} = 1.6316 \times 10^{-16}$$

At ***first*** time step with a source at $\xi = \ell$ and field at $x = \ell$

$$\{u_{22}\}_{t_1} = \frac{1}{\sqrt{\pi}}\left(\sqrt{t_3 - t_0} - \sqrt{t_3 - t_1}\right) = 8.96602 \times 10^{-3}$$

For the second time step

At ***second*** time step with a source at $\xi = 0$ and field at $x = 0$

$$\{u_{11}\}_{t_1} = \frac{1}{\sqrt{\pi}}\left(\sqrt{t_3 - t_1} - \sqrt{t_3 - t_2}\right) = 0.0116848$$

At ***second*** time step with a source at $\xi = 0$ and field at $x = \ell$

$$\{u_{12}\}_{t_2} = \int_{t_1}^{t_2} u^*(\xi = 0, x = \ell; t_3, t)dt =$$

$$= \frac{r}{2\sqrt{\pi}}\left\{\left(\frac{e^{-a_1}}{\sqrt{a_1}} - \frac{e^{-a_2}}{\sqrt{a_2}}\right) + \sqrt{\pi}\left\{erf\sqrt{a_1} - erf\sqrt{a_2}\right\}\right\}$$

$$r = \xi - x = -\ell$$

$$a_1 = \frac{r^2}{4(t_3 - t_1)}$$

$$a_2 = \frac{r^2}{4(t_3 - t_2)}$$

$$\{u_{12}\}_{t_2} = 7.6946 \times 10^{-24}$$

At ***second*** time step with a source at $\xi = \ell$ and field at $x = 0$

$$\{u_{21}\}_{t_2} = \int_{t_1}^{t_2} u^*(\xi = \ell, x = 0; t_3, t)dt =$$

$$= \frac{r}{2\sqrt{\pi}}\left\{\left(\frac{e^{-a_1}}{\sqrt{a_1}} - \frac{e^{-a_2}}{\sqrt{a_2}}\right) + \sqrt{\pi}\left\{erf\sqrt{a_1} - erf\sqrt{a_2}\right\}\right\}$$

$$r = \xi - x = -\ell$$

$$a_1 = \frac{r^2}{4(t_3 - t_1)}$$

$$a_2 = \frac{r^2}{4(t_3 - t_2)}$$

$$\{u_{21}\}_{t_2} = 7.6946\times10^{-24}$$

At ***first*** time step with a source at $\xi = \ell$ and field at $x = \ell$

$$\{u_{22}\}_{t_2} = \frac{1}{\sqrt{\pi}}\left(\sqrt{t_3 - t_1} - \sqrt{t_3 - t_2}\right) = 0.0116848$$

For the third time step

At ***third*** time step with a source at $\xi = 0$ and field at $x = 0$

$$\{u_{11}\}_{t_3} = \frac{1}{\sqrt{\pi}}\left(\sqrt{t_3 - t_2}\right) = -.028209$$

At ***third*** time step with a source at $\xi = 0$ and field at $x = \ell$

$$\{u_{12}\}_{t_3} = \int_{t_2}^{t_3} u^*(\xi = 0, x = \ell; t_3, t)dt =$$

$$= \frac{r}{2\sqrt{\pi}}\left\{\left(\frac{e^{-a_2}}{\sqrt{a_2}} - \frac{e^{-a_3}}{\sqrt{a_3}}\right) + \sqrt{\pi}\left\{erf\sqrt{a_2} - erf\sqrt{a_3}\right\}\right\}$$

$$r = \xi - x = -\ell$$

$$a_2 = \frac{r^2}{4(t_3 - t_2)}$$

$$a_3 = \frac{r^2}{4(t_3 - t_3)} = \infty$$

$$\{u_{12}\}_{t_3} = \frac{r}{2\sqrt{\pi}}\left\{\left(\frac{e^{-a_2}}{\sqrt{a_2}}\right) + \sqrt{\pi}\left\{erf\sqrt{a_2} - 1\right\}\right\} = 1.049414\times10^{-45}$$

At ***third*** time step with a source at $\xi = \ell$ and field at $x = 0$

$$\{u_{21}\}_{t_3} = \int_{t_2}^{t_3} u^*(\xi = \ell, x = 0; t_3, t)dt =$$

$$= \frac{r}{2\sqrt{\pi}}\left\{\left(\frac{e^{-a_2}}{\sqrt{a_2}} - \frac{e^{-a_3}}{\sqrt{a_3}}\right) + \sqrt{\pi}\left\{erf\sqrt{a_2} - erf\sqrt{a_3}\right\}\right\}$$

$$r = \xi - x = \ell$$

$$a_2 = \frac{r^2}{4(t_3 - t_2)}$$

$$a_3 = \frac{r^2}{4(t_3 - t_3)}$$

$$\{u_{21}\}_{t_3} = -1.149414 \times 10^{-45}$$

At ***third*** time step with a source at $\xi = \ell$ and field at $x = \ell$

$$\{u_{22}\}_{t_3} = \frac{1}{\sqrt{\pi}}\left(\sqrt{t_3 - t_2}\right) = 0.028209$$

$$\{b_{11}\}_{t_3} = -0.5 - \left[-17.7246 \times 1.6316 \times 10^{-16}\right] + \left[17.7246 \times 8.96602 \times 10^{-3}\right]$$
$$-\left[-10.382775 \times 7.6946 \times 10^{-24}\right] + \left[10.382775 \times 0.0116848\right] = -0.220926$$

$$\{b_{22}\}_{t_3} = -0.5 - \left[-17.7246 \times 8.96602 \times 10^{-3}\right] + \left[17.7246 \times 17.7246 \times 10^{-16}\right]$$
$$-\left[-10.382775 \times 0.0116848\right] + \left[10.382775 \times 7.6946 \times 10^{-24}\right] = -0.220926$$

Now the system of equations for the third time step will be:

$$\begin{bmatrix} -0.028209 & 1.049414 \times 10^{-45} \\ 1.049414 \times 10^{-45} & 0.028209 \end{bmatrix}\begin{bmatrix} q_0 \\ q_\ell \end{bmatrix} = \begin{bmatrix} -0.220926 \\ -0.220926 \end{bmatrix}$$

The solution leads to the numerical values for q_0 and q_ℓ at the third time step. The flow chart, computer code and the results are shown below.

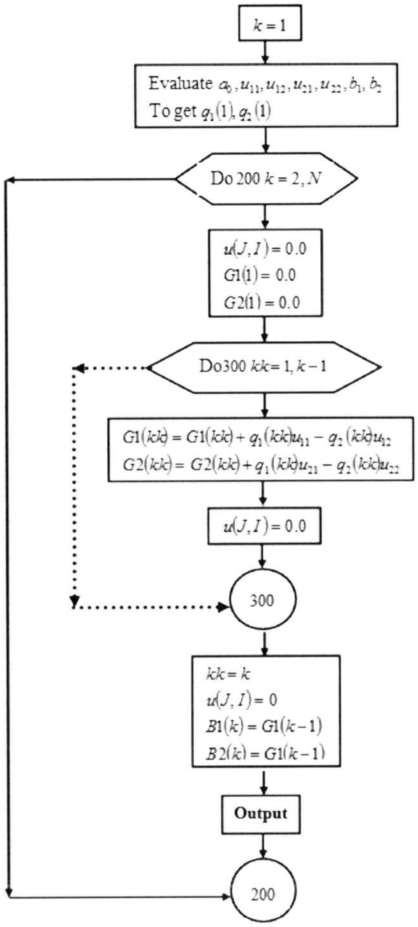

TREATMENT OF INTERNAL POINTS

Case (1): For the first time step only

This section will discuss the method of calculating the temperature at any internal point in the domain of interest. Let us start this discussion by re-writing the general boundary integral equation described before.

$$u(\xi,t_n)=\frac{u_0}{2}\left[erf\left(\frac{\xi}{2\sqrt{t_n}}\right)-erf\left(\frac{\xi-\ell}{2\sqrt{t_n}}\right)\right]$$
$$+\sum_{i=1}^{n}\left[q_{x=\ell}\left(\int_{t_{i-1}}^{t_i}u^*(\xi,x;t_n,t)dt\right)_{x=\ell}-q_{x=0}\left(\int_{t_{i-1}}^{t_i}u^*(\xi,x;t_n,t)dt\right)_{x=0}\right] \quad \textbf{(TIP-1)}$$

Refer to the domain in the present general one-dimensional heat diffusion equation in Fig. (**3**) and suppose we want to find the temperature at any internal point $\xi = x_1$.

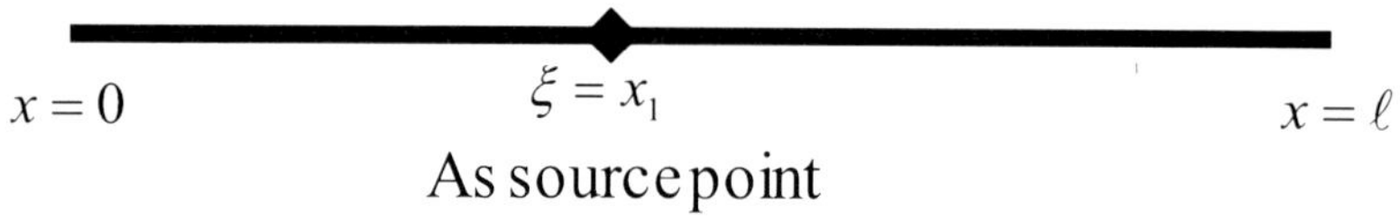

Fig. (3). Source point location.

Assume that all computations of the first time step are known; this means the flux at both ends of the domain are known; therefore, equation **(TIP-1)** will be of the form:

$$u(\xi,t_1)=\frac{u_0}{2}\left[erf\left(\frac{\xi}{2\sqrt{t_1}}\right)-erf\left(\frac{\xi-\ell}{2\sqrt{t_1}}\right)\right]$$
$$+q_{x=\ell}\int_{t_0}^{t_1}u^*(\xi,x=\ell;t_1,t_0)dt-q_{x=0}\int_{t_0}^{t_1}u^*(\xi,x=0;t_1,t_0)dt \quad \textbf{(TIP-2)}$$

In equation **(TIP-2)** both $q_{x=0}$ and $q_{x=\ell}$ are known and now the source point ξ can be applied at any interior point of the domain. In this case, the time integrals in the equation **(TIP-2)** should be evaluated analytically in a general way. Therefore, the next task is the analytical evaluation of the time integrals, which we are doing.

$$\int_{t_0}^{t_1} u^*(\xi, x; t_1, t_0)dt = \tag{TIP-3}$$
$$= \frac{r}{2\sqrt{\pi}}\left\{\left(\frac{e^{-a_0}}{\sqrt{a_0}} - \frac{e^{-a_1}}{\sqrt{a_1}}\right) + \sqrt{\pi}\left[erf\left(\sqrt{a_0}\right) - erf\left(\sqrt{a_1}\right)\right]\right\}$$

Note:

In equation **(TIP-3)** $r \neq 0$ but there exists a sign in a_1 should be taken into consideration, and it takes the following formula:

$$a_1 = \frac{r^2}{4(t_1 - t_1)} \tag{TIP-4}$$

Substituting from equation **(TIP-4)** into equation **(TIP-3)**, leads to:

$$\int_{t_0}^{t_1} u^*(\xi, x; t_1, t_0)dt = \frac{r}{2\sqrt{\pi}} \frac{e^{-a_0}}{\sqrt{a_0}} + \frac{r}{2}\left\{erf\left(\sqrt{a_0}\right) - 1\right\} \tag{TIP-5}$$

Now suppose that the source point $\xi = x_1$, and suppose that we want to apply the integral equation at that point then it takes the following form:

$$u(x_1, t_1) = \frac{u_0}{2}\left[erf\left(\frac{x_1}{2\sqrt{t_1}}\right) - erf\left(\frac{x_1 - \ell}{2\sqrt{t_1}}\right)\right] +$$
$$+ q_{x=\ell}\left[\frac{r_1}{2\sqrt{\pi}} \frac{e^{-a_0}}{\sqrt{a_0}} + \frac{r_1}{2}\left\{erf\left(\sqrt{a_0}\right) - 1\right\}\right] \tag{TIP-6}$$
$$+ q_{x=0}\left[\frac{r_2}{2\sqrt{\pi}} \frac{e^{-a_0}}{\sqrt{a_0}} + \frac{r_2}{2}\left\{erf\left(\sqrt{a_0}\right) - 1\right\}\right]$$

The distances r_1 and r_2 are shown in Fig. (**4**) and defined by:

$$r_1 = |x_1 - \ell|$$
$$r_2 = |x_1 - 0| \quad \textbf{(TIP-7)}$$

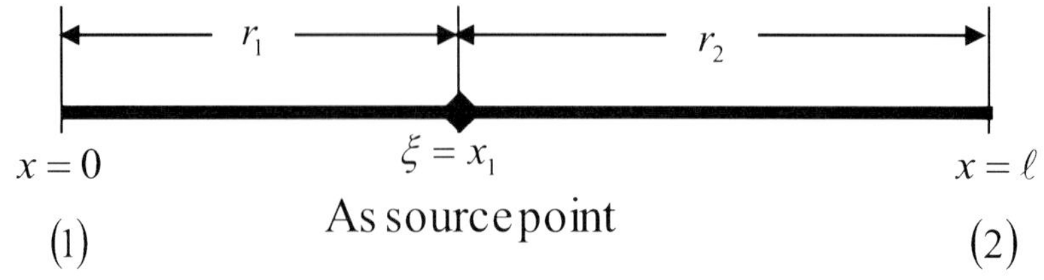

Fig. (4). Normalization.

The computer code defined before to evaluate the flux at the end points of the domain is now modified to evaluate the temperature at any internal point of the domain. We will check some of the modified loops in the designed code just to check its sequence.

Loop 141

$$I2 = 1$$

$$R1(1) = |0.1 - 0| = 0.1$$

$$RR(1) = \frac{0.1}{2\sqrt{0.005}} = 0.7071067812$$

$$RY(1)3.147102641$$

$$ERFRR(1) = 1 - \frac{1}{RY(1)} = 0.6822474148$$

$$DS = 0.1 - 0.0 = 0.1$$

$$BB(1) = 0 + 0.6822474148 = 0.6822474148$$

$$I2 = 2$$

$$R1(2) = |0.1 - 1| = 0.9$$

$$RR(2) = \frac{0.9}{2\sqrt{0.005}} = 6.363961031$$

$$RY(2) = 389340403.5$$

$$ERFRR(2) = 1 - \frac{1}{RY(2)} = -0.9999999$$

$$DS = 0.1 - 1.0 = -0.9$$

$$BB(2) = 0.6822474148 - 0.9999999 = -0.3177525826$$

Loop 550

$$I = 1$$

$$R(1) = |0.1 - 0| = 0.1$$

$$A1(0,1) = \frac{(0.1)^2}{4(t_1 - t_0)} = \frac{(0.1)^2}{4(0.005)} = 0.5$$

$$I = 2$$

$$R(2) = |0.1 - 1.0| = 0.9$$

$$A1(0,2) = \frac{(0.9)^2}{4(t_1 - t_0)} = \frac{(0.9)^2}{4(0.005)} = 40.5$$

Loop 551

$$I1 = 1$$

$$YS(1) = \sqrt{A1(0,1)} = \sqrt{0.5} = 0.7071067812$$

$$ERFYS(1) = 1.0$$

$$U1(1) = \frac{R(1)}{2\sqrt{\pi}}\left[\frac{e^{-A1(0,1)}}{YS(1)}\right] + \frac{R(1)}{2}\left[ERFYS(1) - 1.0\right] =$$

$$= \frac{0.1}{2\sqrt{\pi}}\left[\frac{e^{-0.5}}{0.7071067812}\right] + 0.0 = 0.02419707245$$

$$I1 = 2$$

$$YS(2) = \sqrt{A1(0,2)} = 6.363961031$$

$$ERFYS(2) = 1.0$$

$$U1(2) = \frac{R(2)}{2\sqrt{\pi}}\left[\frac{e^{-A1(0,2)}}{YS(2)}\right] + \frac{R(2)}{2}\left[ERFYS(2) - 1.0\right] =$$

$$= \frac{0.9}{2\sqrt{\pi}}\left[\frac{e^{-40.5}}{6.363961031}\right] + 0.0 = 1.027977357 \times 10^{-19}$$

TREATMENT OF INTERNAL POINTS

Case (2): For Second Time Step

In this section, we will extend the idea of how to evaluate the temperature at any internal point in the domain of interest. Let us start this discussion by re-writing the general boundary integral equation described before.

$$u(\xi, t_2) = \frac{u_0}{2}\left[erf\left(\frac{\xi}{2\sqrt{t_2}}\right) - erf\left(\frac{\xi - \ell}{2\sqrt{t_2}}\right)\right]$$
$$+ \sum_{i=1}^{2}\left[q_{x=\ell}\left(\int_{t_{i-1}}^{t_i} u^*(\xi, x; t_2, t)dt\right)_{x=\ell} - q_{x=0}\left(\int_{t_{i-1}}^{t_i} u^*(\xi, x; t_2, t)dt\right)_{x=0}\right] \quad \textbf{(TIP-C2-1)}$$

Equation **(TIP-C2-1)** can be re-written in an expanded form as follows:

$$u(\xi,t_2)=\frac{u_0}{2}\left[erf\left(\frac{\xi}{2\sqrt{t_2}}\right)-erf\left(\frac{\xi-\ell}{2\sqrt{t_2}}\right)\right]$$
$$+q_{x=\ell}\int_{t_0}^{t_1}u^*(\xi,\ell;t_2,t)dt+q_{x=\ell}\int_{t_1}^{t_2}u^*(\xi,\ell;t_2,t)dt \qquad \textbf{(TIP-C2-2)}$$
$$-q_{x=0}\int_{t_0}^{t_1}u^*(\xi,0;t_2,t)dt-q_{x=0}\int_{t_1}^{t_2}u^*(\xi,0;t_2,t)dt$$

Again equation **(TIP-C2-2)** can be re-written as follows:

$$u(\xi,t_2)=\frac{u_0}{2}\left[erf\left(\frac{\xi}{2\sqrt{t_2}}\right)-erf\left(\frac{\xi-\ell}{2\sqrt{t_2}}\right)\right]$$
$$+[q_{x=\ell}u_{12}]_{t_1}+[q_{x=\ell}u_{12}]_{t_2} \qquad \textbf{(TIP-C2-3)}$$
$$-[q_{x=0}u_{11}]_{t_1}-[q_{x=0}u_{11}]_{t_2}$$

The time integrals in equation **(TIP-C2-3)** are as follow:

$$[u_{11}]_{t_1}=\int_{t_0}^{t_1}u^*(\xi,0;t_2,t)dt=\left|\begin{array}{l}\frac{r_2}{2\sqrt{\pi}}\left(\frac{e^{-a_0}}{\sqrt{a_0}}-\frac{e^{-a_1}}{\sqrt{a_1}}\right)\\ +\sqrt{\pi}\left\{erf\left(\sqrt{a_0}\right)-erf\left(\sqrt{a_1}\right)\right\}\end{array}\right| \qquad \textbf{(TIP-C2-4)}$$

With

$$r_2=|x_1-0| \qquad a_0=\frac{r_2^2}{4(t_2-t_0)} \qquad a_1=\frac{r_2^2}{4(t_2-t_1)} \qquad \textbf{(TIP-C2-5)}$$

$$[u_{12}]_{t_1} = \int_{t_0}^{t_1} u^*(\xi,\ell;t_2,t)dt =$$

$$= \left[\frac{r_1}{2\sqrt{\pi}}\left(\frac{e^{-a_0}}{\sqrt{a_0}} - \frac{e^{-a_1}}{\sqrt{a_1}}\right) + \sqrt{\pi}\left\{erf\left(\sqrt{a_0}\right) - erf\left(\sqrt{a_1}\right)\right\}\right] \qquad \textbf{(TIP-C2-6)}$$

With

$$r_1 = |x_1 - \ell|$$

$$a_0 = \frac{r_1^2}{4(t_2 - t_0)}$$

$$a_1 = \frac{r_1^2}{4(t_2 - t_1)} \qquad \textbf{(TIP-C2-7)}$$

$$[u_{11}]_{t_2} = \int_{t_1}^{t_2} u^*(\xi,0;t_2,t)dt = \left[\frac{r_2}{2\sqrt{\pi}}\left(\frac{e^{-a_1}}{\sqrt{a_1}}\right) + \frac{r_2}{2}\left\{erf\left(\sqrt{a_1}\right) - 1\right\}\right] \qquad \textbf{(TIP-C2-8)}$$

With

$$r_2 = |x_1 - 0|$$

$$a_1 = \frac{r_2^2}{4(t_2 - t_1)} \qquad \textbf{(TIP-C2-9)}$$

$$a_2 = \frac{r_2^2}{4(t_2 - t_2)}$$

$$[u_{12}]_{t_2} = \int_{t_1}^{t_2} u^*(\xi,\ell;t_2,t)dt = \left[\frac{r_1}{2\sqrt{\pi}}\left(\frac{e^{-a_1}}{\sqrt{a_1}}\right) + \frac{r_1}{2}\left\{erf\left(\sqrt{a_1}\right) - 1\right\}\right] \qquad \textbf{(TIP-C2-10)}$$

With

$$r_1 = |x_1 - \ell|$$

$$a_1 = \frac{r_1^2}{4(t_2 - t_1)}$$

$$a_2 = \frac{r_1^2}{4(t_2 - t_2)}$$

REFERENCES

[1] J. Crank, *Free and Moving Boundary Problems.* Clarendon Press: Oxford, 1984.

[2] *Zerrokat, M and Chatwin, CR, Computational Moving Boundary Problems.* Research Studies Press Ltd, 1994.

[3] A. Bejan, and P.A. Tyvand, "The pressure melting of ice to an embedded cylinder", *J. Heat Transfer,* vol. 114, no. 2, pp. 532-535, 1992. [http://dx.doi.org/10.1115/1.2911311]

[4] M.K. Moallemi, B.W. Webb, and R. Viskanta, "An experimental and analytical study of close-contact melting", *J. Heat Transfer,* vol. 108, no. 4, pp. 894-899, 1986. [http://dx.doi.org/10.1115/1.3247030]

[5] A. Mehdi, and C.K. Hsieh, "Solution of ablation and combination of ablation and Stefan problems by a source and sink method", *Numer. Heat Transf. A,* vol. 26, no. 1, pp. 67-86, 1994. [http://dx.doi.org/10.1080/10407789408955981]

[6] S.G. Ahmed, "An approximate method for evaporation problem with mushy zone", *Eng. Anal. Bound. Elem.,* vol. 29, no. 2, pp. 161-165, 2005. [http://dx.doi.org/10.1016/j.enganabound.2004.10.002]

[7] H.G. Landau, "Heat Conduction in a Melting Solid", *Q. Appl. Math.,* vol. 8, no. 1, pp. 81-94, 1950. [http://dx.doi.org/10.1090/qam/33441]

[8] T.R. Goodman, "Application of Integral Methods to Transient Nonlinear Heat Transfer", *Adv. Heat Transf.,* vol. 1, pp. 55-122, 1964. [http://dx.doi.org/10.1016/S0065-2717(08)70097-2]

[9] E. Vallerani, "Integral Technique Solution to a Class of Simple Ablation Problems", *Meccanica,* vol. 9, no. 2, pp. 94-101, 1974. [http://dx.doi.org/10.1007/BF02209512]

[10] B.F. Blackwell, "Numerical prediction of one dimensional ablation using a finite control volume procedure with exponential differencing", *Numer. Heat Transf.,* vol. 14, no. 1, pp. 17-34, 1988. [http://dx.doi.org/10.1080/10407788808913631]

[11] M. Hogge, and P. Gerrekens, "Two dimensional finite element methods for surface ablation", *AIAA J.,* vol. 23, no. 3, pp. 465-472, 1985. [http://dx.doi.org/10.2514/3.8936]

[12] H. Abakians, and M.F. Modest, "Evaporative cutting of a semi-transparent body with a moving CW laser", *J. Heat Transfer,* vol. 110, no. 4a, pp. 924-930, 1988. [http://dx.doi.org/10.1115/1.3250594]

[13] L.C. Delima Silva Jnr And Wrobel, *BEM analysis of melting problems, in Boundary Element Technology IX.,* C.A. Brebbia, A.J.L Kassab, Eds., Computational Mechanics Publications: Southampton, 1994.

[14] R.N. Hills, D.E. Loper, and P.H. Roberts, "A thermodynamically consistent model of a mushy zone", *Q. J. Mech. Appl. Math.,* vol. 36, no. 4, pp. 505-539, 1983. [http://dx.doi.org/10.1093/qjmam/36.4.505]

[15] H.E. Huppert, and M.G. Worster, "Dynamic solidification of a binary melt", *Nature,* vol. 314, no. 6013, pp. 703-707, 1986. [http://dx.doi.org/10.1038/314703a0]

[16] AS Burns, LA Stickler, and Jr Stewart, "Solidification of an aqueous salt solution in a circular cylinder", vol.114 (1992).

[17] D.A. Tarzia, *Conference of Free and Moving Boundary Problems,* 1995 Bled, Slovenia.

[18] P.M. Pavel, Thesis, bucharest. 1999.

[19] S.G. Ahmed, "Boundary integral formulation and mushy zone model for phase change problem", *Int. J. Numer. Methods Heat Fluid Flow,* vol. 11, no. 4, pp. 371-379, 2001. [http://dx.doi.org/10.1108/09615530110389207]

[20] P.M. Pavel, and N. Ene, "Planar solidification solution with mushy zone", *Energy Convers. Manage.,* vol. 44, no. 18, pp. 2977-2989, 2003. [http://dx.doi.org/10.1016/S0196-8904(03)00052-9]

[21] M. Mekey, "L M. and S. G. Ahmed, Semi-analytical method for combined Stefan and ablation problems with mushy zones, Ain Shams University, Scientific Bulletin, Faculty of Engineering", *Part III,* vol. 42, no. 3, pp. 1241-1256, 2007.

[22] I.V. Alexandrova, D.V. Alexandrov, D.L. Aseev, and S.V. Bulitcheva, "Mushy layer formation during solidification of binary alloys from a cooled wall: The role of boundary conditions", *Acta Phys. Pol. A,* vol. 115, no. 4, pp. 791-794, 2009. [http://dx.doi.org/10.12693/APhysPolA.115.791]

[23] S. Nouri, M. Benzeghiba, and A. Benzeoui, "Numerical computation of magnetic field effect of an alloy with mushy zone", *ISESCO science and technology,* vol. 6, 2010.

[24] S.G. Ahmed, "Numerical solution of ablation problem using modified constraint integral method", *J. Eng. Appl. Sci. (Asian Res. Publ. Netw.),* vol. 47, no. 3, 2000.

[25] J. Mennig, and M.N. Ozisik, "Coupled Integral Equation Approach for solving Melting or Solidification", *Int. J. Heat Mass Transf.,* vol. 28, no. 8, pp. 1481-1485, 1985. [http://dx.doi.org/10.1016/0017-9310(85)90250-9]

[26] J.S. Langer, "Instabilities and pattern formation in crystal growth", *Rev. Mod. Phys.,* vol. 52, no. 1, pp. 1-28, 1980. [http://dx.doi.org/10.1103/RevModPhys.52.1]

[27] H. Lyle, "Cellular interface morphologies in directional solidification. IV. The formation of deep cells", In: *Phys. Rev* vol. B 31. , 1985, pp. 5931-5940.

[28] S.H. Davis, *Theory of Solidification.* Cambridge University Press: New York, 2001. [http://dx.doi.org/10.1017/CBO9780511546747]

[29] A. Saito, H.K. Hong, and O. Hirokane, "Heat transfer enhancement in the direct contact melting process", *Int. J. Heat Mass Transf.,* vol. 35, no. 295, pp. 295-305, 1992. [http://dx.doi.org/10.1016/0017-9310(92)90268-W]

[30] Y. Zhang, and A. Faghri, "Analysis of Freezing in an Eccentric Annulus", *J. Sol. Energy Eng.,* vol. 119, no. 3, pp. 237-241, 1997. [http://dx.doi.org/10.1115/1.2888025]

[31] E. Ben-Jacob, and J.S. Nigel Goldenfeld, "Langer, and Gerd Schön, Dynamics of Interfacial Pattern Formation", *Phys. Rev. Lett.,* vol. 51, no. 21, pp. 1930-1932, 1983. [http://dx.doi.org/10.1103/PhysRevLett.51.1930]

[32] E. Obryk, "Electroproduction from Nucleons in a Relativistic Quark Model", In: *Phys. Rev* vol. D 4. , 1971, pp. 2844-2848.

[33] C. Ross, "Kerr, Andrew W. Woods, M. Grae Worster, and Herbert E. Huppert, Solidification of an alloy cooled from above, Part 3. Compositional stratification within the solid", *J. Fluid Mech.,* vol. 218, no. -1, pp. 337-354, 1990. [http://dx.doi.org/10.1017/S0022112090001021]

[34] The Development of Crystal Growth Technology, *HANS J. SCHEEL CONSULTING, CH-8808 Pfaeffikon SZ.* John Wiley & Sons, Ltd.: Switzerland, 2003.

[35] L.C. Wrobel, "A boundary element solution to Stefan's problem", In: *Boundary Elements V.* Springer-Verlag: Berlin, 1983.

[36] R. Bonnerot, and P. Jamet, "A Third order accurate discontinuous finite element method for one-dimensional Stefan problem", *J. Comput. Phys.,* vol. 32, no. 2, pp. 109-128, 1979. [http://dx.doi.org/10.1016/0021-9991(79)90127-X]

[37] W.W. Mullins, and R.F. Sekerka, "The Stability of a Planar Interface During Solidification of a Dilute Binary Alloy", *J. Appl. Phys.,* vol. 35, no. 2, pp. 444-451, 1964. [http://dx.doi.org/10.1063/1.1713333]

[38] S.L. Braga, and R. Viskanta, "Solidification of a Binary Solution on a Cold Isothermal Surface", *Int. J. Heat Mass Transf.,* vol. 33, no. 4, pp. 745-754, 1990. [http://dx.doi.org/10.1016/0017-9310(90)90172-Q]

[39] W.Z. Cao, and D. Poulikakos, "Freezing of a Binary Alloy Saturating a Packed Bed of Spheres", *J. Thermophys. Heat Trans.,* vol. 5, no. 1, pp. 46-53, 1991. [http://dx.doi.org/10.2514/3.225]

[40] A. Faghri, Y. Zhang, and J.R. Howell, *Advanced Heat and Mass Transfer.* Global Digital Press: Columbia, MO, 2010.

[41] D. Juric, and G. Tryggvason, *A Front-Tracking Method for Liquid-Vapor Phase Change," in Advances in Numerical Modeling of Free Surface and Interface Fluid Dynamics, FED.* vol. Vol. 234. ASME, 1995, pp. 141-148.

[42] G. Segal, C. Vuik, and F. Vermolen, "A conserving discretization for the free boundary in a two-dimensional Stefan problem", *J. Comput. Phys.,* vol. 141, no. 1, pp. 1-21, 1998. [http://dx.doi.org/10.1006/jcph.1998.5900]

[43] W.D. Murray, and F. Landis, "Numerical and machine solutions of transient heat-conduction problems involving melting or freezing, Trans. ASME (C)", *J. Heat Transfer,* vol. 81, pp. 106-112, 1959. [http://dx.doi.org/10.1115/1.4008149]

[44] V.R. Voller, "An implicit enthalpy solution for phase change problems: with application to a binary alloy solidification", *Appl. Math. Model.,* vol. 11, no. 2, pp. 110-116, 1987. [http://dx.doi.org/10.1016/0307-904X(87)90154-5]

[45] S. Chen, B. Merriman, S. Osher, and P. Smereka, "A simple level set method for solving Stefan problems", *J. Comput. Phys.,* vol. 135, no. 1, pp. 8-29, 1997. [http://dx.doi.org/10.1006/jcph.1997.5721]

[46] S. Osher, and R. Fedkiw, *Level Set Methods and Dynamic Implicit Surfaces.* Springer: NewYork, 2003. [http://dx.doi.org/10.1007/b98879]

[47] M. Sussman, P. Smereka, and S. Osher, "A level set approach for computing solutions to incompressible two-phase flow", *J. Comput. Phys.,* vol. 114, no. 1, pp. 146-159, 1994. [http://dx.doi.org/10.1006/jcph.1994.1155]

[48] C.A. Brebbia, J.C.F. Telles, and L.C. Wrobel, *Boundary Element Techniques Theory and Applications in Engineering* Spring-Verlag: Berlin and New York, 1984.

[49] I. Fredholm, "Sur Une Classe d'equations Fonctionelles", *Acta Math.,* vol. 27, no. 0, pp. 365-390, 1903.
[http://dx.doi.org/10.1007/BF02421317]

[50] M.A. Jaswon, "Integral Equation Methods in Potential Theory, I", *Proc. Roy. Soc. Ser. A,* vol. 275, pp. 23-32, 1963.

[51] G.T. Symm, "Integral Equation Methods in Potential Theory, II", *Proc. Roy. Soc. Ser. A,* vol. 275, pp. 33-46, 1963.

[52] M.A. Jaswon, and A.R. Ponter, "An Integral Equation Solution of the Torsion Problem", *Proc. Roy. Soc. Ser. A,* vol. 273, pp. 237-246, 1963.

[53] J.L. Hess, and A.M.O. Smith, *Calculation of Potential Flow about Arbitrary Bodies, Progress in Aeronautical Science.,* D. Kuchemann, Ed., vol. Vol. 8. Pergamon Press: London, 1967.

[54] Y.K. Chuang, and J. Szekely, "On the use of Green's functions for solving melting or solidification problems", *Int. J. Heat Mass Transfer,* vol. 14, 1972.

[55] Y.K. Chuang, and J. Szekely, "The use of Green's functions for solving melting or solidification problems in a cylindrical coordinate system", *Int. J. Heat Mass Transfer,* vol. 15, 1972.

[56] R.P. Shaw, "A boundary integral equation approach to the one dimensional ablation problem", In: *Boundary Element Methods in Engineering.* Springer-Verlag: Berlin, 1982.
[http://dx.doi.org/10.1007/978-3-662-11273-1_10]

[57] C.J. Coleman, "a boundary integral formulation of the Stefan problem, appl", *Math. Model.,* vol. 10, 1986.

[58] C.J. Coleman, "a boundary integral approach to the solidification of dilute alloys", *Int. J. Heat Mass Transf.,* vol. 30, no. 8, pp. 1727-1732, 1987.
[http://dx.doi.org/10.1016/0017-9310(87)90319-X]

[59] K. O'Neill, "Boundary integral equation solution of moving boundary phase change problems", *Int. J. Num. Meth. Eng.,* vol. 19, 1983.

[60] *Heinlein, M., Mukherjee, S., and Richmond, A boundary element method analysis of temperature fields and stress during solidification.* vol. 59. Acta Mechanica, 1986.

[61] N. Zabaras, and S. Mukherjee, *An analysis of the solidification problems by the boundary element method, Int.* vol. Vol. 24. J. Num. Methods, 1987.

[62] D-S. Jnr, and L.C. Wrobel, "Boundary element formulation for multi-dimensional ablation problems", In: *Boundary Elements XIV* vol. 1. Computational Mechanics Publications, 1992.

[63] S.G. Ahmed, and L.C. Wrobel, "Numerical solution of solidification problems using integral equations", In: *Third Int. Conference of free and moving boundary problems.* Bled: Slovenia, 1995.

[64] H.S. Carslaw, and J.C. Jaejer, *Conduction of Heat in Solids.* 2nd ed. Clarendon Press: Oxford, 1959.

[65] J.S. Wettlaufer, M.G. Worster, and H.E. Huppert, "Solidification of leads: Theory, experiment and field observations", *J. Geophys. Res.,* vol. 105, no. C1, pp. 1123-1134, 2000.
[http://dx.doi.org/10.1029/1999JC900269]

[66] W.M. Abd-El Fatah, M.M. Helal, S.G. Ahmed, and M.I.M. Hessein, "Heat balance integral method applied to ablation problems with mushy zones", *Ninth International Congress of Fluid Dynamics & Propulsion* Alexandria, Egypt.

[67] W.M. Abd-el Fatah, M.M. Helal, M.I.M. Hessein, and S.G. Ahmed, "Integral solution of unidirectional solidification of binary melts from a cooled wall", In: *imbic*Culcutta, India, 2010.

[68] B.F. Zalewski, R.L. Mullen, and R.L. Muhanna, *Boundary Element Analysis of Systems using Interval Methods, authors.*U.S.A, 2006.

[69] W.M. Abd-El Fatah, and S.G. Ahmed, "Boundary Integral Formulation for Binary Alloys from a Cooling Solid Wall", *International Journal of Computational and Applied Mathematics,* vol. 5, no. 5, pp. 687-696, 2010.

[70] S.G. Ahmed, *On Diffusion and Hydrodynamics with Moving Boundaries, Ph. D., Chanel system between Portsmouth University, Uk.* And Zagazig University, 1996.

[71] P. Rao, and V.M.K. Sastri, "Efficient numerical method for two-dimensional phase change problems", *Int. J. Heat Mass Transf.,* vol. 27, no. 11, pp. 2077-2084, 1984. [http://dx.doi.org/10.1016/0017-9310(84)90194-7]

CHAPTER 5

Boundary Element Methods Stefan Problems over Domains with Moving Boundaries

Abstract: In the present chapter, the principles of the boundary element methods for phase change problems over domains with moving boundaries are developed in brief details. The details are developed starting from the 1-D diffusion problems over domains with fixed boundaries subjected to different types of boundary and initial conditions. In these problems, a new criterion has appeared: the existence of the moving boundary, which requires special treatment when solving overall problems.

Keywords: Boundary elements methods, moving boundary problems, Stefan condition.

INTRODUCTION

In the present chapter, the boundary element methods are introduced in detail to solve Stefan problems over domains with moving boundaries. Many physical and engineering processes can be modelled as moving boundary problems where a set of partial differential equations are to be solved for a domain whose boundaries are not known a *priori* but have to be determined as an integral part of the solution [1-15]. Let us build this chapter on solving the following general heat diffusion problems over moving domain, the governing equation with the associated boundary and initial conditions can be stated as follow [16-20]:

$$k\frac{\partial^2 u(x,t)}{\partial x^2}=\frac{\partial u(x,t)}{\partial t} \qquad 0<x<\ell \tag{1}$$

Boundary conditions

$$u=\bar{u} \qquad \text{Dirichlet} \tag{2}$$

S. G. Ahmed

$q = \bar{q}$ Neumann **(3)**

Initial conditions

$$u(x,0) = u_0(x) \quad \textbf{(4)}$$

Stefan condition

$$K_{\text{Input}} - K_{\text{output}} = \pm \rho L \frac{ds(t)}{dt} \quad \textbf{(5 and 6)}$$

In these equations:

K_{Input} : Thermal conductivity for initial material

K_{Output} : Thermal conductivity for output material

k : Thermal diffusivity

ρ : Density for the material assuming constant density

L : Latent heat for the material assuming constant Latent heat

$\frac{ds(t)}{dt}$: Velocity of the moving boundary for 1-D case problems

$\pm$: + For melting problems and – for solidification problems

BOUNDARY ELEMENT FORMULATION

Starting by the weighted residual statement for diffusion equation as follows [20-25]:

$$\int_0^{\tau} \int_{\Omega} \left(k \frac{\partial^2 u}{\partial x^2} - \frac{\partial u}{\partial t} \right) u^* (\xi, x; t_n, t) d\Omega dt = 0 \quad \textbf{(7)}$$

Equation (7) can take the following form:

$$\int_0^{\tau}\int_{\Omega} k\frac{\partial^2 u}{\partial x^2}u^*(\xi,x;t_n,t)d\Omega dt-\int_0^{\tau}\int_{\Omega}\frac{\partial u}{\partial t}u^*(\xi,x;t_n,t)d\Omega dt=0$$

$$\Leftrightarrow$$

$$I_1 - I_2 = 0 \tag{8}$$

In equations (7-8) $u^*(\xi,x;\tau,t)$ is the fundamental solution for diffusion equation, and it is defined as [26]:

$$u^*(\xi,x;\tau,t)=\frac{1}{\sqrt{4\pi k(t_n-t)}}\exp\left[-\frac{r^2(\xi,x)}{4k(t_n-t)}\right] \tag{9}$$

In which, $r^2(\xi,x)=X(\xi)-X(x)$ and $X(\xi)$ is the source point while $X(x)$ is the field point [27].

EVALUATION OF THE FIRST INTEGRAL

$$I_1=\int_0^{t_n}\int_{\Omega} k\frac{\partial^2 u}{\partial x^2}u^*(\xi,x;t_n,t)d\Omega dt \tag{10}$$

In one-dimensional problems $d\Omega = dx$; therefore, equation (10) becomes:

$$I_1=\int_0^{t_n}\int_0^{\ell} k\frac{\partial^2 u}{\partial x^2}u^*(\xi,x;t_n,t)dx\,dt \tag{11}$$

Now let us integrate equation (11) by parts, leading to:

$$I_1=\int_0^{t_n}\int_0^{\ell} k\frac{\partial^2 u}{\partial x^2}u^*dx\,dt=k\int_0^{t_n}\left\{\left[u^*\frac{\partial u}{\partial x}\right]_0^{\ell}-\int_0^{\ell}\frac{\partial u}{\partial x}\frac{\partial u^*}{\partial x}dx\right\}dt \tag{12}$$

Integration of the second integral once again by parts leads to:

$$I_1 = k\int_0^{t_n}\left[u^*\frac{\partial u}{\partial x}\right]_0^{\ell} - k\int_0^{t_n}\left\{\left[u\frac{\partial u^*}{\partial x}\right]_0^{\ell} - \int_0^{\ell} u\frac{\partial^2 u^*}{\partial x^2}dx\right\}dt \quad \textbf{(13)}$$

$$I_1 = k\int_0^{t_n}\left[u^*\frac{\partial u}{\partial x}\right]_0^{\ell} dt - k\int_0^{t_n}\left[u\frac{\partial u^*}{\partial x}\right]_0^{\ell} dt + k\int_0^{t_n}\int_0^{\ell} u\frac{\partial^2 u^*}{\partial x^2}dx\,dt \quad \textbf{(14)}$$

Therefore the first integral in equation (8) becomes:

$$I_1 = k\int_0^{t_n}\left[u^* q - uq^*\right]_0^{\ell} dt + k\int_0^{t_n}\int_0^{\ell} u\nabla^2 u^* dx\,dt \quad \textbf{(15)}$$

EVALUATION OF THE SECOND INTEGRAL

$$I_2 = \int_0^{t_n}\int_0^{\ell}\frac{\partial u}{\partial t}u^* dx\,dt \quad \textbf{(16)}$$

This integral can be re-written as:

$$I_2 = \int_0^{t_n}\int_0^{\ell}\left\{\frac{\partial}{\partial t}\left(uu^*\right) - u\frac{\partial u^*}{\partial t}\right\}dx\,dt \quad \textbf{(17)}$$

Leibniz's rule can take the following form:

$$\frac{\partial}{\partial t}\int_{\Omega} g\,dw = \int_{\Omega}\frac{\partial g}{\partial t}d\Omega + \int_{\Gamma} g\frac{ds}{dt}d\Gamma \quad \textbf{(18)}$$

Re-arrange equation (5.18) as follows:

$$\int_{\Omega}\frac{\partial g}{\partial t}d\Omega = \frac{\partial}{\partial t}\int_{\Omega} g\,dw - \int_{\Gamma} g\frac{ds}{dt}d\Gamma \quad \textbf{(19)}$$

Comparing equations (17) and (19), one can get:

$$I_2 = \int_0^{t_n} \frac{\partial}{\partial t} \int_0^{\ell} \left(uu^* \right) dw - \int_0^{t_n} \int_{\Gamma} uu^* \frac{ds}{dt} d\Gamma - \int_0^{t_n} \int_0^{\ell} u \frac{\partial u^*}{\partial t} dx\,dt \quad \textbf{(20)}$$

The differentiation in the first term cancels the integration process; therefore, equation (20) takes the following form:

$$I_2 = -\int_0^{\ell} u(x,0) u^* (\xi, x; t_n, 0) dx - \int_0^{t_n} \int_{\Gamma} uu^* \frac{ds}{dt} d\Gamma - \int_0^{t_n} \int_0^{\ell} u \frac{\partial u^*}{\partial t} dx\,dt \quad \textbf{(21)}$$

It is important to notify that the term $\frac{ds}{dt}$ appears in the time period t_m to t_n; therefore equation (21) can be written as:

$$\begin{aligned} I_2 &= -\int_0^{\ell} u(x,0) u^* (\xi, x; t_n, 0) dx \\ &- \int_{t_m}^{t_n} \int_{\ell}^{x(t)} u(x,t) u^* (\xi, x; t_n, t(x)) dx\,dt - \int_0^{t_n} \int_0^{\ell} u \frac{\partial u^*}{\partial t} dx\,dt \end{aligned} \quad \textbf{(22)}$$

As long as we express the time in the second integral as a function of space variable x, which represents the position of the moving boundary, hence we just integrate for space, then equation (22) can now take the following form:

$$\begin{aligned} I_2 &= -\int_0^{\ell} u(x,0) u^* (\xi, x; t_n, 0) dx \\ &- \int_{\ell}^{x(t)} u(x,t) u^* (\xi, x; t_n, t(x)) dx - \int_0^{t_n} \int_0^{\ell} u \frac{\partial u^*}{\partial t} dx\,dt \end{aligned} \quad \textbf{(23)}$$

Substituting from equations (15) and (23) into equation (8), one can get the following boundary integral equation corresponding to diffusion equation over moving boundary:

$$k\int_0^{t_n}\left[u^*q-uq^*\right]_0^{\ell}dt+k\int_0^{t_n}\int_0^{\ell}u\nabla^2u^*dx\,dt+\int_0^{\ell}u(x,0)u^*(\xi,x;t_n,0)dx$$
$$+\int_{\ell}^{x(t)}u(x,t)u^*(\xi,x;t_n,t(x))dx+\int_0^{t_n}\int_0^{\ell}u\frac{\partial u^*}{\partial t}dx\,dt=0 \quad \textbf{(24)}$$

Adding the double integrals in equation (24), we get:

$$\int_0^{t_n}\int_0^{\ell}\left(u\nabla^2u^*+u\frac{\partial u^*}{\partial t}\right)dx\,dt+k\int_0^{t_n}\left[u^*q-uq^*\right]_0^{\ell}dt$$
$$+\int_0^{\ell}u(x,0)u^*(\xi,x;t_n,0)dx+\int_{\ell}^{x(t)}u(x,t)u^*(\xi,x;t_n,t(x))dx=0 \quad \textbf{(25)}$$

But

$$\int_0^{t_n}\int_0^{\ell}u\left(\nabla^2u^*+\frac{\partial u^*}{\partial t}\right)dx\,dt=-c(\xi)u(\xi,t_n) \quad \textbf{(26)}$$

Substutiting from (26) into (25), we get:

$$-c(\xi)u(\xi,t_n)+k\int_0^{t_n}\left[u^*q-uq^*\right]_0^{\ell}dt+\int_0^{\ell}u(x,0)u^*(\xi,x;t_n,0)dx$$
$$+\int_{\ell}^{x(t)}u(x,t)u^*(\xi,x;t_n,t(x))dx=0 \quad \textbf{(27)}$$

Re-arrange equation (5.27), we get:

$$c(\xi)u(\xi,t_n) = k\int_0^{t_n}\left[u^* q - uq^*\right]_0^{\ell} dt + \int_0^{\ell} u(x,0)u^*(\xi,x;t_n,0)dx$$
$$+\int_{\ell}^{x(t)} u(x,t)u^*(\xi,x;t_n,t(x))dx = 0 \tag{28}$$

Important Note

The limits of integration of the first integral in the right-hand side are from to t_n, but this period is divided into two parts from 0 to t_m and from to. The space limits corresponding to these intervals are as follows: From to t_m correspond to to ℓ and from t to t correspond to 0 to $x(t)$ (Fig. **1**).

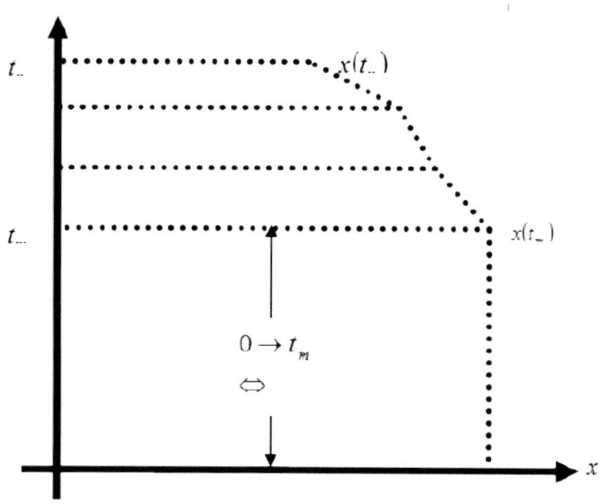

Fig. (1). Linear variation in time marching scheme.

Based on the above scheme, the boundary integral equation for the diffusion equation with moving boundary takes the following form:

$$c(\xi)u(\xi,t_n) = \int_0^{\ell} u(x,0)u^*(\xi,x;t_n,0)dx + \int_0^{t_m}\left[u^* q - uq^*\right]_0^{\ell} dt$$
$$+\int_{t_m}^{t_n}\left[u^* q - uq^*\right]_0^{x(t)} dt + \int_{\ell}^{x(t)} u(x,t)u^*(\xi,x;t_n,t(x))dx \tag{29}$$

Equation (29) can be written in more detail as follows:

$$
\begin{aligned}
c(\xi)u(\xi,t_n) &= \int_0^{\ell} u(x,0)u^*(\xi,x;t_n,0)dx \\
&+ \int_0^{t_m}\left[q(x,t)u^*(\xi,x;t_n,t)\right]_0^{\ell}dt - \int_0^{t_m}\left[u(x,t)q^*(\xi,x;t_n,t)\right]_0^{\ell}dt \\
&+ \int_{t_m}^{t_n}\left[q(x,t)u^*(\xi,x;t_n,t)\right]_0^{x(t)}dt - \int_{t_m}^{t_n}\left[u(x,t)q^*(\xi,x;t_n,t)\right]_0^{x(t)}dt \\
&+ \int_{\ell}^{x(t)} u(x,t)u^*(\xi,x;t_n,t(x))dx
\end{aligned}
\qquad (30)
$$

Once again, equation (5.30) can be written in more and brief details as follows:

$$
\begin{aligned}
&c(\xi)u(\xi,t_n)= \\
&= \underbrace{\int_0^{\ell} u(x,0)u^*(\xi,x;t_n,0)dx}_{\text{Domain integral within } 0 \text{ to } t_n} \\
&+ \int_0^{t_m} \left[q(x=\ell,t)u^*(\xi,x=\ell;t_n,t)\right]dt \\
&- \int_0^{t_m} \left[q(x=0,t)u^*(\xi,x=0;t_n,t)\right]dt \\
&- \int_0^{t_m} \left[u(x=\ell,t)q^*(\xi,x=\ell;t_n,t)\right]dt \\
&+ \int_0^{t_m} \left[u(x=0,t)q^*(\xi,x=0;t_n,t)\right]dt \\
&+ \int_{t_m}^{t_n} \left[q(x(t),t)u^*(\xi,x(t);t_n,t)\right]dt - \int_{t_m}^{t_n} \left[q(0,t)u^*(\xi,0;t_n,t)\right]dt \\
&- \int_{t_m}^{t_n} \left[u(x(t),t)q^*(\xi,x(t);t_n,t)\right]dt + \int_{t_m}^{t_n} \left[u(0,t)q^*(\xi,0;t_n,t)\right]dt \\
&+ \underbrace{\int_{\ell}^{x(t)} u(x,t)u^*(\xi,x;t_n,t(x))dx}_{\text{Domain integral within } t_m \text{to} t_m}
\end{aligned}
\quad \textbf{(31)}
$$

Equation (31) is now then applied to two end points of the phase of the domain bounded by the fixed end $x = 0$ and the moving boundary $x = x(t)$. Let us first take source point at $\xi = 0$ then equation (31) takes the following form:

$$
\begin{aligned}
\frac{1}{2}u(\xi=0,t_n) &= \int_0^{\ell} u(x,0)u^*(\xi=0,x;t_n,0)dx \\
&+ \int_0^{t_m}\left[q(x=\ell,t)u^*(\xi=0,x=\ell;t_n,t)\right]dt \\
&- \int_0^{t_m}\left[q(x=0,t)u^*(\xi=0,x=0;t_n,t)\right]dt \\
&- \int_0^{t_m}\left[u(x=\ell,t)q^*(\xi=0,x=\ell;t_n,t)\right]dt \\
&+ \int_0^{t_m}\left[u(x=0,t)q^*(\xi=0,x=0;t_n,t)\right]dt \\
&+ \int_{t_m}^{t_n}\left[q(x(t),t)u^*(\xi=0,x(t);t_n,t)\right]dt \\
&- \int_{t_m}^{t_n}\left[q(0,t)u^*(\xi=0,0;t_n,t)\right]dt \\
&- \int_{t_m}^{t_n}\left[u(x(t),t)q^*(\xi=0,x(t);t_n,t)\right]dt \\
&+ \int_{t_m}^{t_n}\left[u(0,t)q^*(\xi=0,0;t_n,t)\right]dt \\
&+ \int_{\ell}^{x(t)} u(x,t)u^*(\xi=0,x;t_n,t(x))dx
\end{aligned} \qquad \textbf{(32)}
$$

To prepare a computer code, let us denote the following integrals by the following names asthey will appear in the code.

$$
\mathrm{SMR1} = \int_0^{t_m}\left[q(x=0,t)u^*(\xi=0,x=0;t_n,t)\right]dt
$$

$$\mathrm{SMR2} = \int_{0}^{t_m} \left[q(x=\ell,t) u^*(\xi=0, x=\ell; t_n, t) \right] dt$$

$$\mathrm{SMR3} = \int_{0}^{t_m} \left[u(x=0,t) q^*(\xi=0, x=0; t_n, t) \right] dt$$

$$\mathrm{SMR4} = \int_{0}^{t_m} \left[u(x=\ell,t) q^*(\xi=0, x=\ell; t_n, t) \right] dt$$

$$\mathrm{SOLX1}(1) = \mathrm{SMR1} + \mathrm{SMR2} + \mathrm{SMR3} + \mathrm{SMR4}$$

$$\mathrm{COLYY1}(1) = \int_{t_m}^{t_n} \left[q(0,t) u^*(\xi=0, 0; t_n, t) \right] dt$$

$$\mathrm{COLYY2}(1) = \int_{t_m}^{t_n} \left[q(x(t),t) u^*(\xi=0, x(t); t_n, t) \right] dt$$

$$\mathrm{COLYY3}(1) = \int_{t_m}^{t_n} \left[u(0,t) q^*(\xi=0, 0; t_n, t) \right] dt$$

$$\mathrm{COLVV1}(1) = \int_{t_m}^{t_n} \left[u(x(t),t) q^*(\xi=0, x(t); t_n, t) \right] dt$$

$$\mathrm{COLVV2}(1) = \int_{\ell}^{x(t)} u(x,t) u^*(\xi=0, x; t_n, t(x)) dx$$

Multiply both sides of equation (32) by 2 to cancel $\frac{1}{2}$ from the left-hand side, it becomes:

$$
\begin{aligned}
u(\xi=0,t_n) &= 2\int_0^{\ell} u(x,0)u^*(\xi=0,x;t_n,0)dx \\
&+2\int_0^{t_m}\left[q(x=\ell,t)u^*(\xi=0,x=\ell;t_n,t)\right]dt \\
&-2\int_0^{t_m}\left[q(x=0,t)u^*(\xi=0,x=0;t_n,t)\right]dt \\
&-2\int_0^{t_m}\left[u(x=\ell,t)q^*(\xi=0,x=\ell;t_n,t)\right]dt \\
&+2\int_0^{t_m}\left[u(x=0,t)q^*(\xi=0,x=0;t_n,t)\right]dt \\
&+2\int_{t_m}^{t_n}\left[q(x(t),t)u^*(\xi=0,x(t);t_n,t)\right]dt \\
&-2\int_{t_m}^{t_n}\left[q(0,t)u^*(\xi=0,0;t_n,t)\right]dt \\
&-2\int_{t_m}^{t_n}\left[u(x(t),t)q^*(\xi=0,x(t);t_n,t)\right]dt \\
&+2\int_{t_m}^{t_n}\left[u(0,t)q^*(\xi=0,0;t_n,t)\right]dt \\
&+2\int_{\ell}^{x(t)} u(x,t)u^*(\xi=0,x;t_n,t(x))dx
\end{aligned}
\qquad \textbf{(33)}
$$

Equation (33) in symbolic and matrix forms as written in the computer code is as follow:

$$
\begin{bmatrix} a_{11} & a_{12} \end{bmatrix}\begin{bmatrix} u(0,t) \\ u(x(t),t) \end{bmatrix} = \begin{bmatrix} b_1 \end{bmatrix} \qquad \textbf{(34)}
$$

Where

$$a_{11} = 1 - \mathrm{COLYY3}(1)$$

$$a_{12} = \mathrm{COLVV1}(1) - \mathrm{COLVV2}(1)$$

$$b_1 = \mathrm{DINT}(1) + \mathrm{SOLX1}(1, \mathrm{N1}) + \mathrm{COLLY2}(1) + \mathrm{COLLY1}(1)$$

$$+ \sum \int_{t_m}^{t_{n-1}} \mathrm{COLVV2}(1)$$

$$+ \sum \int_{t_m}^{t_{n-1}} \begin{Bmatrix} \mathrm{COLYY2}(1) + \mathrm{COLYY1}(1) \\ + \mathrm{COLVV1}(1) + \mathrm{COLYY3}(1) \\ + \mathrm{COLVV2}(1) \end{Bmatrix}$$

Now let us take source point at $\xi = x(t)$ then equation (31) takes the following form:

$$\begin{aligned}
\frac{1}{2} u(\xi = x(t), t_n) &= \int_0^{\ell} u(x,0) u^*(\xi = x(t), x; t_n, 0) dx \\
&+ \int_0^{t_m} \left[q(x = \ell, t) u^*(\xi = x(t), x = \ell; t_n, t) \right] dt \\
&- \int_0^{t_m} \left[q(x = 0, t) u^*(\xi = x(t), x = 0; t_n, t) \right] dt \\
&- \int_0^{t_m} \left[u(x = \ell, t) q^*(\xi = x(t), x = \ell; t_n, t) \right] dt \\
&+ \int_0^{t_m} \left[u(x = 0, t) q^*(\xi = x(t), x = 0; t_n, t) \right] dt \\
&+ \int_{t_m}^{t_n} \left[q(x(t), t) u^*(\xi = x(t), x(t); t_n, t) \right] dt \\
&- \int_{t_m}^{t_n} \left[q(0, t) u^*(\xi = x(t), 0; t_n, t) \right] dt \\
&- \int_{t_m}^{t_n} \left[u(x(t), t) q^*(\xi = x(t), x(t); t_n, t) \right] dt \\
&+ \int_{t_m}^{t_n} \left[u(0, t) q^*(\xi = x(t), 0; t_n, t) \right] dt + \int_{\ell}^{x(t)} u(x, t) u^*(\xi = x(t), x; t_n, t(x)) dx
\end{aligned} \qquad \mathbf{(35)}$$

For the purpose of preparing a computer code, let us denote the following integrals by the following names as they will appear in the code.

$$\mathrm{SMR11} = \int_0^{t_m} \left[q(x=0,t) u^*\left(\xi = x(t), x=0; t_n, t\right) \right] dt$$

$$\mathrm{SMR22} = \int_0^{t_m} \left[q(x=\ell,t) u^*\left(\xi = x(t), x=\ell; t_n, t\right) \right] dt$$

$$\mathrm{SMR33} = \int_0^{t_m} \left[u(x=0,t) q^*\left(\xi = x(t), x=0; t_n, t\right) \right] dt$$

$$\mathrm{SMR44} = \int_0^{t_m} \left[u(x=\ell,t) q^*\left(\xi = x(t), x=\ell; t_n, t\right) \right] dt$$

$$\mathrm{SOLX1}(2) = \mathrm{SMR11} + \mathrm{SMR22} + \mathrm{SMR33} + \mathrm{SMR44}$$

$$\mathrm{COLYY1}(2) = \int_{t_m}^{t_n} \left[q(0,t) u^*\left(\xi = x(t), 0; t_n, t\right) \right] dt$$

$$\mathrm{COLYY2}(2) = \int_{t_m}^{t_n} \left[q(x(t),t) u^*\left(\xi = x(t), x(t); t_n, t\right) \right] dt$$

$$\mathrm{COLYY3}(2) = \int_{t_m}^{t_n} \left[u(0,t) q^*\left(\xi - x(t), 0; t_n, t\right) \right] dt$$

$$\mathrm{COLVV1}(2) = \int_{t_m}^{t_n} \left[u(x(t),t) q^*\left(\xi = x(t), x(t); t_n, t\right) \right] dt$$

$$\mathrm{COLVV2}(2) = \int_{\ell}^{x(t)} u(x,t) u^*\left(\xi = x(t), x; t_n, t(x)\right) dx$$

Multiply both sides of equation (5.35) by 2 to cancel $\frac{1}{2}$ from the left-hand side, it becomes:

$$
\begin{aligned}
u(\xi = x(t), t_n) &= 2\int_0^{\ell} u(x,0)u^*(\xi = x(t), x; t_n, 0)dx \\
&+ 2\int_0^{t_m} \left[q(x = \ell, t)u^*(\xi = x(t), x = \ell; t_n, t)\right]dt \\
&- 2\int_0^{t_m} \left[q(x = 0, t)u^*(\xi = x(t), x = 0; t_n, t)\right]dt \\
&- 2\int_0^{t_m} \left[u(x = \ell, t)q^*(\xi = x(t), x = \ell; t_n, t)\right]dt \\
&+ 2\int_0^{t_m} \left[u(x = 0, t)q^*(\xi = x(t), x = 0; t_n, t)\right]dt \\
&+ 2\int_{t_m}^{t_n} \left[q(x(t), t)u^*(\xi = x(t), x(t); t_n, t)\right]dt \\
&- 2\int_{t_m}^{t_n} \left[q(0, t)u^*(\xi = x(t), 0; t_n, t)\right]dt \\
&- 2\int_{t_m}^{t_n} \left[u(x(t), t)q^*(\xi = x(t), x(t); t_n, t)\right]dt \\
&+ 2\int_{t_m}^{t_n} \left[u(0, t)q^*(\xi = x(t), 0; t_n, t)\right]dt \\
&+ 2\int_{\ell}^{x(t)} u(x,t)u^*(\xi = x(t), x; t_n, t(x))dx
\end{aligned}
\qquad \textbf{(36)}
$$

Equation (5.36) in symbolic and matrix forms as written in the computer code is as follow:

$$\begin{bmatrix} a_{21} & a_{22} \end{bmatrix} \begin{bmatrix} u(0,t) \\ u(x(t),t) \end{bmatrix} = \begin{bmatrix} b_2 \end{bmatrix} \tag{37}$$

Where

$$a_{21} = -\text{COLYY3}(2)$$

$$a_{22} = 1 + \text{COLVV1}(2) - \text{COLVV2}(2)$$

$$b_2 = \text{DINT}(2) + \text{SOLX1}(2, \text{N1}) + \text{COLLY2}(2) + \text{COLLY1}(2)$$

$$+ \sum \int_{t_m}^{t_{n-1}} \text{COLVV2}(2)$$

$$+ \sum \int_{t_m}^{t_{n-1}} \begin{Bmatrix} \text{COLYY2}(2) + \text{COLYY1}(2) \\ + \text{COLVV1}(2) + \text{COLYY3}(2) \\ + \text{COLVV2}(2) \end{Bmatrix}$$

Finally imposing equations (34) and (37) to form the following matrix system:

$$\begin{bmatrix} a_{11} & a_{12} \\ a_{21} & a_{22} \end{bmatrix} \begin{bmatrix} u(0,t) \\ u(x(t),t) \end{bmatrix} = \begin{bmatrix} b_1 \\ b_1 \end{bmatrix} \tag{38}$$

The present problem is a one-dimensional, single-phase diffusion problem. This requires determining the time at which melting starts occurring and subsequently the appearance of the moving boundary.

TYPES OF INTEGRALS ENCOUNTERED

When solving one-dimensional, one phase moving boundary problem, one encountered different types of integrals. These integrals can be classified as follows [28-35]:

(1) Domain integrals from $x = 0$ to $x = \ell$ corresponds to the initial condition, this integral is of the form $\int_0^\ell u(x,0) u^*(\xi, x; t_n, 0) dx$

(2) Domain integrals from $x = \ell$ to $x = x(t)$ this integral is of the form $\int_{x=\ell}^{x=x(t)} u(x,t)u^*(\xi, x(t), t_n, t(x))dx$

(3) Time integrals from $t = 0$ to $t = t_m$

(4) Time integrals from $t = t_m$ to $t = t_{n-1}$, these integrals are called intermediate integrals.

(5) Time integrals from $t = t_{n-1}$ to $t = t_n$

In the section below, we will start to evaluate each type in brief detail.

DOMAIN INTEGRAL FROM $x = 0$ TO $x = \ell$

This integral corresponds to the initial condition associated with the diffusion problem with moving boundary [36-40]. In this case, we will consider uniform initial conditions over the whole domain.

The domain integral takes the form:

$$I = \int_0^\ell u(x,0)u^*(\xi, x; t_n, t)dx = \frac{u_0}{2}\left\{erf\left(\frac{\xi}{2\sqrt{t_n}}\right) - erf\left(\frac{\xi - \ell}{2\sqrt{t_n}}\right)\right\}$$

Notes

This integral will be calculated at each time step and u_0 is the initial temperature.

For the purpose of preparing a computer code, the domain integral will be denoted by $\mathrm{DINT}(\mathrm{i})$ such that:

$\mathrm{DINT}(1)$: Means source at $\xi = 0$

$\mathrm{DINT}(2)$: Means source at $\xi = \ell$

DOMAIN INTEGRAL FROM $x = \ell$ TO $x = x(t)$

This integral takes the form:

$$I = \int_{\ell}^{x(t)} u(x,t)u^*(\xi = x_i, x(t); t_n, t(x))dx$$

In Szekely 1970, it was considered that $u(x,t)$ inside this domain integral equal constant value, that is the melting temperature, but it was written the integral equation in a general form that was $u(x,t) = u_m(t_k)$, also in Wrobel 1983, also considered this temperature constant and equal to the melting temperature. In the present chapter, we considered this temperature has two different values, that is, it will equal u_m within t_m & t_{n-1} and will equal to u_{mov}, which is unknown and should be determined as a major part of the solution. Therefore, the analytical solution that corresponds to this integral takes the form:

$$I = \frac{1}{2}u_m \left\{ \begin{aligned} &\left[erf\left(\frac{D-A}{\sqrt{QA}}\right) - erf\left(\frac{G-A}{\sqrt{QG}}\right) \right] \\ &+ e^{\frac{4A}{Q}} \left[erf\left(\frac{D+A}{\sqrt{QD}}\right) - erf\left(\frac{G+A}{\sqrt{QG}}\right) \right] \end{aligned} \right\}$$

Where

$$A = (t_n - t_{i-1})S_i + (\xi - x_{i-1})$$
$$D = (t_n - t_{i-1})S_{i-1} + (x_i - x_{i-1})$$
$$F = (t_n - t_i)S_i$$
$$G = (t_n - t_{i-1})S_{i-1}$$
$$Q = \frac{4k}{S_i}$$

Note that the u_m still the same in the time period t_m & t_{n-1} and changed to u_{mov} an unknown one in the last time step.

TIME INTEGRALS FROM $t = 0$ TO $t = t_m$

In this case, we have four different time integrals as follow:

$$I_1 = q(\ell,t)\int_{t=0}^{t=t_m} u^*(\xi,\ell;t_n,t)dt$$

$$I_2 = q(0,t)\int_{t=0}^{t=t_m} u^*(\xi,0;t_n,t)dt$$

$$I_3 = u(\ell,t)\int_{t=0}^{t=t_m} q^*(\xi,\ell;t_n,t)dt$$

$$I_4 = u(0,t)\int_{t=0}^{t=t_m} q^*(\xi,0;t_n,t)dt$$

As we see in these integrals, the field point may be 0 or ℓ while the source point will be 0 or $x(t)$, therefore the analytical expressions for the above integrals are:

$$I = \int_{t_{i-1}}^{t_i} u^*(\xi,x;t_n,t)dt = \frac{r}{2\sqrt{\pi}}\left\{\begin{array}{l}\left[\dfrac{e^{-a_{i-1}}}{\sqrt{a_{i-1}}} - \dfrac{e^{-a_i}}{\sqrt{a_i}}\right] \\ +\sqrt{\pi}\left\{erf\sqrt{a_{i-1}} - erf\sqrt{a_i}\right\}\end{array}\right\}$$

$$I = \int_{t_{i-1}}^{t_i} q^*(\xi,x;t_n,t)dt = \frac{1}{2}sign(r)\times\left\{erf\sqrt{a_i} - erf\sqrt{a_{i-1}}\right\}$$

TIME INTEGRALS FROM $t = t_m$ TO $t = t_{n-1}$

In this case, the source point 0 and $x(t)$ and the field point 0 and $x(t)$. According to the elementary example herein, we have ten types of time integrals, they are as follow:

$$GG1 = \int_{t_m}^{t_{n-1}} u^*(\xi = x(t), 0; t_{n-1}, t)dt$$

$$GG2 = \int_{t_m}^{t_{n-1}} q^*(\xi = x(t), 0; t_{n-1}, t)dt$$

$$GG3 = \int_{t_m}^{t_{n-1}} u^*(\xi = 0, 0; t_{n-1}, t)dt$$

$$GG4 = \int_{t_m}^{t_{n-1}} q^*(\xi = 0, 0; t_{n-1}, t)dt$$

$$GG5 = \int_{t_m}^{t_{n-1}} u^*(\xi = x(t), x(t); t_{n-1}, t)dt$$

$$GG6 = \int_{t_m}^{t_{n-1}} q^*(\xi = x(t), x(t); t_{n-1}, t)dt$$

$$GG8 = \int_{t_m}^{t_{n-1}} u^*(\xi = 0, x(t); t_{n-1}, t)dt$$

$$GG9 = \int_{t_m}^{t_{n-1}} q^*(\xi = 0, x(t); t_{n-1}, t)dt$$

In addition to the above eight-time integrals, we still have two space integrals, they are:

$$\mathrm{GG7}(1) = \int_{\ell}^{x(t)} u_{\mathrm{melt}} u^{*}(\xi = 0, x; t_{n-1}, t(x))dt$$

$$\mathrm{GG7}(2) = \int_{\ell}^{x(t)} u_{\mathrm{melt}} u^{*}(\xi = x(t), x; t_{n-1}, t(x))dt$$

TREATMENT OF INTEGRALS IN WHICH FIELD POINT IS THE MOVING BOUNDARY

There are two different types of integrals in which the field point is located at the moving boundary, and these integrals are as follow:

$$I = \int_{t_m}^{t_{n-1}} q(x(t), t) u^{*}(\xi, x(t); t_{n-1}, t)dt =$$

$$= \left[\frac{q(x(t), t)}{2s_i}\right] \times \left\{ \begin{array}{l} \left[erf\left(\frac{G - A}{\sqrt{QG}}\right) - erf\left(\frac{F - A}{\sqrt{QF}}\right)\right] \\ + e^{\frac{4A}{Q}}\left[erf\left(\frac{G + A}{\sqrt{QG}}\right) - erf\left(\frac{F + A}{\sqrt{QF}}\right)\right] \end{array} \right\}$$

$$I = \int_{t_m}^{t_{n-1}} u(x(t), t) q^{*}(\xi, x(t); t_{n-1}, t)dt =$$

$$= \left[\frac{u(x(t), t)}{2}\right] \times \left\{ e^{\frac{4A}{Q}}\left[erf\left(\frac{E + A}{\sqrt{QF}}\right) - erf\left(\frac{G + A}{\sqrt{QG}}\right)\right]\right\}$$

TREATMENT OF TIME STEP OF SINGULARITY

For fixed boundary problems, we showed how we treat the singularity in both space and time integrals. The situation is different when dealing with moving boundary problems that appear different parameters in the expressions for the moving boundaries. Before dealing with such singularities, let us remind the expressions for different parameters that appeared in the analytical expressions;

$$A = (t_n - t_{i-1})S_i + (\xi - x_{i-1})$$
$$D = (t_n - t_{i-1})S_{i-1} + (x_i - x_{i-1})$$
$$F = (t_n - t_i)S_i$$
$$G = (t_n - t_{i-1})S_{i-1}$$
$$Q = \frac{4k}{S_i}$$
$$S_i = \frac{x_{i-1} - x_i}{t_i - t_{i-1}}$$

Substituting from S_i into the parameters $A \,\&\, D$, we get:

$$A = (\xi - x_i)$$
$$D = 0$$

Therefore, for the singular time step, the above parameters take the form:

$$A = (\xi - x_i)$$
$$D = 0$$
$$F = 0$$
$$G = (t_n - t_{i-1})S_{i-1}$$
$$Q = \frac{4}{S_i}$$
$$S_i = \frac{x_{i-1} - x_i}{t_i - t_{i-1}}$$

Because these parameters are used in evaluating the error functions in the integrals, the reader must make sure to take all possibilities for signing in the code.

PROBABILITIES OF DIFFERENT PARAMETERS

Different possibilities for signs can be viewed from the following table.

	$A=0$	$A=0$ $D=0$	$A=0$ $D=0$ $F=0$	$A=0$ $F=0$	$A<0$ $D=0$	$A<0$ $F=0$	$A<0$ $F=0$ $D=0$
$\frac{D-A}{\sqrt{QD}}$	$\sqrt{\frac{D}{Q}}$	0	0	$\sqrt{\frac{D}{Q}}$	∞	Has value	∞
$\frac{D+A}{\sqrt{QD}}$	$\sqrt{\frac{D}{Q}}$	0	0	$\sqrt{\frac{D}{Q}}$	$-\infty$	Has value	$-\infty$
$\frac{G-A}{\sqrt{QG}}$	$\sqrt{\frac{G}{Q}}$	$\sqrt{\frac{G}{Q}}$	$\sqrt{\frac{G}{Q}}$	$\sqrt{\frac{G}{Q}}$	Has value	Has value	Has value
$\frac{G+A}{\sqrt{QG}}$	$\sqrt{\frac{G}{Q}}$	$\sqrt{\frac{G}{Q}}$	$\sqrt{\frac{G}{Q}}$	$\sqrt{\frac{G}{Q}}$	Has value	Has value	Has value
$\frac{F-A}{\sqrt{QF}}$	$\sqrt{\frac{F}{Q}}$	$\sqrt{\frac{F}{Q}}$	0	0	Has value	∞	∞
$\frac{F+A}{\sqrt{QF}}$	$\sqrt{\frac{F}{Q}}$	$\sqrt{\frac{F}{Q}}$	0	0	Has value	$-\infty$	$-\infty$

ELEMENTARY EXAMPLE

Assume that we have a bar of length ℓ subjected to zero flux at the end $x = 0$ and a prescribed flux at the other end. This example was solved before by Wylder and Citron for the following parameters:

$$m = \frac{\sqrt{\pi}}{2} \frac{c(u_m - u_0)}{L} = 3.6675$$

$$r = k \frac{(u_m - u_0)}{\ell Q} = 0.75$$

$$\frac{m}{r} = 4.89$$

With unit values for ℓ, L, c, k and from the above quantities, $u_m = 4.13833$ & $Q = 5.51777$, and with the following restrictions for the purpose of calculations:

$$\Delta t = 0.01$$

$$\text{Exx} = \text{maximum tolerrance between } u_{mov} \text{ and } u_{melt} = 0.01$$

The following is an evaluation by hand for the first and second iterations to check the designated code.

For the first time step

$$X = x(42) = 0.9995$$

$$S = s(12) = 0.05$$

$$\frac{\partial u}{\partial x}(42) = 5.4677$$

$$u_{mov}(42) = 4.1374$$

Now guess new $X, S, \frac{\partial u}{\partial x}$

$$\left(\frac{\partial u}{\partial x}\right)^{(1)}_{(42)} = \left(u_a - u_{melt}\right) - S_{42} = (9.656 - 4.1383) - 0.05 = 5.4677$$

$$(x)^{(1)}_{(43)} = 0.9995 - (0.05 \times 0.01) = 0.999$$

INTEGRALS FROM t_m TO t_n, WITH SOURCE AT $\xi = 0$

The following calculation is for the first iteration in the second time step. Let us start with the first integral:

$$\mathrm{COLYY2}(1) = \int_{t_m}^{t_n} q(x(t),t)u^*(0,x(t),t_{43},t)dt$$

$$\mathrm{COLYY2}(1) = \int_{t_{41}}^{t_{43}} q(x(t),t)u^*(0,x(t),t_{43},t)dt$$

$$= \int_{t_{41}}^{t_{42}} q(x(t),t)u^*(0,x(t),t_{43},t)dt + \int_{t_{42}}^{t_{43}} q(x(t),t)u^*(0,x(t),t_{43},t)dt$$

Case (1): Non-singular time integral

$$I = \int_{t_{41}}^{t_{42}} q(x(t),t)u^*(0,x(t),t_{43},t)dt =$$

$$= \frac{q(x(t),t)}{2 \times S_i}\left\{ \begin{array}{l} e^{\frac{4A}{Q}}\left[erf\left(\dfrac{G+A}{\sqrt{QG}}\right) - erf\left(\dfrac{F+A}{\sqrt{QF}}\right)\right] \\ + \left[erf\left(\dfrac{G-A}{\sqrt{QG}}\right) - erf\left(\dfrac{F-A}{\sqrt{QF}}\right)\right] \end{array} \right\}$$

Where,

$$A = (t_n - t_{i-1})S_i + (\xi - x_{i-1})$$
$$F = (t_n - t_i)S_i$$
$$G = (t_n - t_{i-1})S_{i-1}$$
$$Q = \frac{4}{S_i}$$
$$i = m+1, m+2, \ldots\ldots$$

Note that, in the present calculations $i = m+1 = 42$ & $t_n = 43$, therefore:

$$A = (t_{43} - t_{41})S_{42} + (0 - x_{41}) = (0.43 - 0.41)\times 0.05 + (0 - 1.0) = -0.999$$
$$G = (t_{43} - t_{41})S_{42} = (0.43 - 0.41)\times 0.05 = 0.001$$
$$F = (t_{43} - t_{42})S_{42} = (0.43 - 0.42)\times 0.05 = 5\times 10^{-4}$$
$$Q = \frac{4}{S_{42}} = \frac{4}{0.05} = 80$$

Now,

$$\frac{G+A}{\sqrt{QG}} = -3.528462838$$

$$\Rightarrow erf\left(\frac{G+A}{\sqrt{QG}}\right) = -0.9999993683$$

$$\frac{G-A}{\sqrt{QG}} = 3.535533906$$

$$\Rightarrow erf\left(\frac{G-A}{\sqrt{QG}}\right) = 0.9999993996$$

$$\frac{F+A}{\sqrt{QF}} = -4.9925$$

$$\Rightarrow erf\left(\frac{F+A}{\sqrt{QF}}\right) = -1.0$$

$$\frac{F-A}{\sqrt{QF}}=4.9975$$

$$\Rightarrow erf\left(\frac{F-A}{\sqrt{QF}}\right)=1.0$$

Then

$$I=\int_{t_{41}}^{t_{42}} q(x(t),t)u^{*}(0,x(t),t_{43},t)dt=$$

$$=\left(\frac{5.4677}{2\times 0.05}\right)\left\{e^{\frac{4A}{Q}}[-0.9999993683+1]+[0.9999993996-1]\right\}$$

$$=5.704700892\times 10^{-8}=\mathrm{GG8}$$

Case (2): Singular time integral

$$I=\int_{t_{42}}^{t_{43}} q(x(t),t)u^{*}(0,x(t),t_{43},t)dt=$$

$$=\frac{q(x(t),t)}{S_i}\left\{\begin{array}{l} e^{\frac{4A}{Q}}\left[erf\left(\frac{G+A}{\sqrt{QG}}\right)-erf\left(\frac{F+A}{\sqrt{QF}}\right)\right] \\ +\left[erf\left(\frac{G-A}{\sqrt{QG}}\right)-erf\left(\frac{F-A}{\sqrt{QF}}\right)\right]\end{array}\right\}$$

Where,

$$A=(\xi-x_i)=(0-x_{43})=0.999$$

$$F=(t_n-t_i)S_i=(t_{43}-t_{43})S_{43}=(0.43-0.43)\times 0.05=0.00$$

$$G=(t_n-t_{i-1})S_i=(t_{43}-t_{42})S_{43}=(0.43-0.42)\times 0.05=5\times 10^{-4}$$

$$Q=\frac{4}{S_i}=80$$

Therefore,

$$\frac{G+A}{\sqrt{QG}} = -4.9925 \Rightarrow erf\left(\frac{G+A}{\sqrt{QG}}\right) = -1$$

$$\frac{G-A}{\sqrt{QG}} = 4.9975 \Rightarrow erf\left(\frac{G-A}{\sqrt{QG}}\right) = 1$$

$$\frac{F+A}{\sqrt{QF}} = -\infty \Rightarrow erf\left(\frac{F+A}{\sqrt{QF}}\right) = -1$$

$$\frac{F-A}{\sqrt{QF}} = \infty \Rightarrow erf\left(\frac{F-A}{\sqrt{QF}}\right) = 1.0$$

$$I = \int_{t_{42}}^{t_{43}} q(x(t),t)u^*(0,x(t),t_{43},t)dt = 0.0 = \mathrm{COLYY2}(1)$$

Evaluation flux integral

$$\mathrm{COLVV2}(1) = -\int_{t_{41}}^{t_{43}} u(x(t),t)q^*(0,x(t),t_{43},t)dt$$

$$= -\int_{t_{41}}^{t_{42}} u(x(t),t)q^*(0,x(t),t_{43},t)dt - \int_{t_{42}}^{t_{43}} u(x(t),t)q^*(0,x(t),t_{43},t)dt$$

Case (1): Non-singular time integral

$$I_1 = -\int_{t_{41}}^{t_{42}} u(x(t),t)q^*(0,x(t),t_{43},t)dt =$$

$$= -\frac{u(x(t),t)}{2}\left\{e^{\frac{4A}{Q}}\left[erf\left(\frac{F+A}{\sqrt{QF}}\right) - erf\left(\frac{G+A}{\sqrt{QG}}\right)\right]\right\}$$

$$erf\left(\frac{F+A}{\sqrt{QF}}\right)=-1$$

$$erf\left(\frac{G+A}{\sqrt{QG}}\right)=-0.9999993683$$

Therefore

$$I_1=-\int_{t_{41}}^{t_{42}} u(x(t),t)q^*(0,x(t),t_{43},t)dt=2.486794159\times10^{-6}=\text{GG9}$$

Case (2): Singular time integral

$$I_2=-\int_{t_{42}}^{t_{43}} u(x(t),t)q^*(0,x(t),t_{43},t)dt=-u(x(t),t)\int_{t_{42}}^{t_{43}} q^*(0,x(t),t_{43},t)dt$$

$$=-\frac{u(x(t),t)}{2}\left\{e^{\frac{4A}{Q}}\left[erf\left(\frac{F+A}{\sqrt{QF}}\right)-erf\left(\frac{G+A}{\sqrt{QG}}\right)\right]\right\}$$

Note that the temperature $u(x(t),t)$is taken out of the integral because it is unknown now the singular integral will be:

$$\int_{t_{42}}^{t_{43}} q^*(0,x(t),t_{43},t)dt=0$$

Because

$$erf\left(\frac{F+A}{\sqrt{QF}}\right)=-1$$

$$erf\left(\frac{G+A}{\sqrt{QG}}\right)=-1$$

$$\Rightarrow$$

$$\int_{t_{42}}^{t_{43}} q^*(0,x(t),t_{43},t)dt=0.0=\text{COLVV1}(1)$$

EVALUATION SINGULAR AND NON-SINGULAR DOMAIN INTEGRALS

In the next, we will evaluate the singular and non-singular domain integrals [41-68]. Let us start the evaluation by the following integral in which the source point will be at $\xi = 0$:

$$\mathrm{COLVV2}(1) = \int_{\ell}^{x(t)} u(x,t)u^{*}(0,x,t_{n},t)dt =$$

$$= \underbrace{\int_{\ell}^{x(t_{42})} u(x,t)u^{*}(0,x(t),t_{n},t)dt}_{\text{Non-singular integral}} + \underbrace{\int_{x(t_{42})}^{x(t_{43})} u(x,t)u^{*}(0,x(t),t_{n},t)dt}_{\text{Singular integral}}$$

The next step is to evaluate the non-singular and singular integrals. Let us start by evaluating the non-singular integral as follows:

$$\text{Non-singular integral} = \int_{\ell}^{x(t_{42})} u(x,t)u^{*}(0,x,t_{n},t)dt =$$

$$= u_{\text{melt}} \left\{ \begin{array}{l} e^{\frac{4A}{Q}} \left[erf\left(\dfrac{D+A}{\sqrt{QD}}\right) - erf\left(\dfrac{G+A}{\sqrt{QG}}\right) \right] \\ + \left[erf\left(\dfrac{D-A}{\sqrt{QD}}\right) - erf\left(\dfrac{G-A}{\sqrt{QG}}\right) \right] \end{array} \right\}$$

In this case, the different parameters are as follow:

$$A = (t_n - t_{i-1})S_i + (\xi - x_{i-1}) = (0.43 - 0.41) \times 0.05 + (0 - 1) = -0.999$$

$$G = 1 \times 10^{-3}$$

$$D = (t_n - t_{i-1})S_i + (x_i - x_{i-1}) = (0.43 - 0.41) \times 0.05 + (0.9995 - 1)$$

$$= 5 \times 10^{-4}$$

$$Q = \frac{4}{S_{42}} = 80$$

Therefore

$$\frac{D + A}{\sqrt{QD}} = -4.9925 \Rightarrow erf\left(\frac{D + A}{\sqrt{QD}}\right) = -1$$

$$\frac{D - A}{\sqrt{QD}} = 4.9975 \Rightarrow erf\left(\frac{G - A}{\sqrt{QG}}\right) = 1$$

$$\frac{G + A}{\sqrt{QG}} = -3.528462838 \Rightarrow erf\left(\frac{G + A}{\sqrt{QG}}\right)$$

$$= -0.9999993683$$

$$\frac{F - A}{\sqrt{QF}} = 3.535533906 \Rightarrow erf\left(\frac{F - A}{\sqrt{QF}}\right)$$

$$= -0.9999993996$$

Hence

$$\text{Non} - \text{singular integral} = \int_{\ell}^{x(t_{42})} u(x,t)u^*(0,x,t_n,t)dt$$

$$= -2.1588386 \times 10^{-9}$$

$$\text{Singular integral} = \int_{x(t_{42})}^{x(t_{43})} u(x,t)u^*(0,x(t),t_n,t)dt =$$

$$= e^{\frac{4A}{Q}}\left[erf\left(\frac{D+A}{\sqrt{QD}}\right) - erf\left(\frac{G+A}{\sqrt{QG}}\right)\right] + \left[erf\left(\frac{D-A}{\sqrt{QD}}\right) - erf\left(\frac{G-A}{\sqrt{QG}}\right)\right]$$

In this case, the parameters will be:

$$A = -0.999$$

$$D = 0$$

$$Q = \frac{4}{S_{42}} = 80$$

Therefore

$$erf\left(\frac{D+A}{\sqrt{QD}}\right) = -1$$

$$erf\left(\frac{G+A}{\sqrt{QG}}\right) = -0.9999993683$$

$$erf\left(\frac{D-A}{\sqrt{QD}}\right) = 1$$

$$erf\left(\frac{F-A}{\sqrt{QF}}\right) = 0.9999993996$$

Hence

$$\text{Singular integral} = \int_{x(t_{42})}^{x(t_{43})} u(x,t)u^*(0,x(t),t_n,t)dt = 0$$

INTEGRALS FROM t_m TO t_n, WITH SOURCE AT $\xi = x(t)$

Let us start with the first integral:

$$\text{COLYY2}(2) = \int_{t_{41}}^{t_{43}} q(x(t),t)u^*(x(t),x(t),t_{43},t)dt$$

$$= \underbrace{\int_{t_{41}}^{t_{42}} q(x(t),t)u^*(x(t),x(t),t_{43},t)dt}_{\text{Non-singular integral}} + \underbrace{\int_{t_{42}}^{t_{43}} q(x(t),t)u^*(x(t),x(t),t_{43},t)dt}_{\text{Singular integral}}$$

Case (1): Non-Singular Time Integral

$$I = \int_{t_{41}}^{t_{42}} q(x(t),t)u^*(x(t),x(t),t_{43},t)dt =$$

$$= \frac{q(x(t),t)}{S_i}\left\{ \begin{aligned} & e^{\frac{4A}{Q}}\left[erf\left(\frac{G+A}{\sqrt{QG}}\right) - erf\left(\frac{F+A}{\sqrt{QF}}\right)\right] \\ & + \left[erf\left(\frac{G-A}{\sqrt{QG}}\right) - erf\left(\frac{F-A}{\sqrt{QF}}\right)\right] \end{aligned} \right\}$$

Where,

$$A=(t_n-t_{i-1})S_i+(\xi-x_{i-1})$$
$$F=(t_n-t_i)S_i$$
$$G=(t_n-t_{i-1})S_{i-1}$$
$$Q=\frac{4}{S_i}$$
$$i=m+1,m+2,\ldots\ldots$$

Numerical computations,

$$A=(t_{43}-t_{41})S_{42}+(x_{43}-x_{41})=(0.43-0.41)\times0.05+(0.999-1.0)=0$$
$$G=(t_{43}-t_{41})S_{42}=(0.43-0.41)\times0.05=0.001$$
$$F=(t_{43}-t_{42})S_{42}=(0.43-0.42)\times0.05=5\times10^{-4}$$
$$Q=\frac{4}{S_{42}}=\frac{4}{0.05}=80$$

Now,

$$\frac{G+A}{\sqrt{QG}}=\sqrt{\frac{G}{Q}}=3.535533906\Rightarrow erf\left(\sqrt{\frac{G}{Q}}\right)=3.989372566\times10^{-3}$$

$$\frac{G-A}{\sqrt{QG}}=\sqrt{\frac{G}{Q}}=3.535533906\Rightarrow erf\left(\sqrt{\frac{G}{Q}}\right)=3.989372566\times10^{-3}$$

$$\frac{F+A}{\sqrt{QF}}=\sqrt{\frac{F}{Q}}=2.5\times10^{-3}\Rightarrow erf\left(\sqrt{\frac{F}{Q}}\right)=2.820917985\times10^{-3}$$

$$\frac{F-A}{\sqrt{QF}}=\sqrt{\frac{F}{Q}}=2.5\times10^{-3}\Rightarrow erf\left(\sqrt{\frac{F}{Q}}\right)=2.820917985\times10^{-3}$$

Then

$$I = \int_{t_{41}}^{t_{42}} q(x(t),t)u^*(x(t),x(t),t_{43},t)dt = 0.2555503646 = \text{GG5}$$

Case (2): Singular time integral

$$I = \int_{t_{42}}^{t_{43}} q(x(t),t)u^*(x(t),x(t),t_{43},t)dt =$$

$$= \left(\frac{\partial u}{\partial x}\right)_{43}^{(1)} \left(\frac{1}{2S_i}\right) \left\{ \begin{array}{l} e^{\frac{4A}{Q}} \left[erf\left(\frac{G+A}{\sqrt{QG}}\right) - erf\left(\frac{F+A}{\sqrt{QF}}\right) \right] \\ + \left[erf\left(\frac{G-A}{\sqrt{QG}}\right) - erf\left(\frac{F-A}{\sqrt{QF}}\right) \right] \end{array} \right\}$$

Where,

$$A = (t_n - t_{i-1})S_i + (\xi - x_{i-1}) = \left(x_{43}^{(1)} - x_{42}\right) = 0$$

$$G = (t_n - t_{i-1})S_i = (t_{43} - t_{42})S_{43} = 5\times10^{-4}$$

$$F = 0$$

$$Q = 80$$

Therefore,

$$\frac{G+A}{\sqrt{QG}} = \sqrt{\frac{G}{Q}} = 2.5\times10^{-3} \Rightarrow erf\left(\sqrt{\frac{G}{Q}}\right) = 2.820917985\times10^{-3}$$

$$\frac{G-A}{\sqrt{QG}} = \sqrt{\frac{G}{Q}} = 2.5\times10^{-3} \Rightarrow erf\left(\sqrt{\frac{G}{Q}}\right) = 2.820917985\times10^{-3}$$

$$\frac{F+A}{\sqrt{QF}} = 0 \Rightarrow erf\left(\frac{F+A}{\sqrt{QF}}\right) = 0$$

$$\frac{F-A}{\sqrt{QF}}=0 \Rightarrow erf\left(\frac{F-A}{\sqrt{QF}}\right)=0$$

$$I=\int_{t_{42}}^{t_{43}} q(x(t),t)u^{*}(x(t),x(t),t_{43},t)dt=0.61696$$

Evaluation Flux Integral

$$\text{COLVV1}(2)=-\int_{t_{41}}^{t_{43}} u(x(t),t)q^{*}(x(t),x(t),t_{43},t)dt$$

$$=-\int_{t_{41}}^{t_{42}} u(x(t),t)q^{*}(x(t),x(t),t_{43},t)dt-\int_{t_{42}}^{t_{43}} u(x(t),t)q^{*}(x(t),x(t),t_{43},t)dt$$

Case (1): Non-singular time integral

$$I_1=-\int_{t_{41}}^{t_{42}} u(x(t),t)q^{*}(x(t),x(t),t_{43},t)dt=$$

$$=-u(x(t),t)\left\{e^{\frac{4A}{Q}}\left[erf\left(\frac{F+A}{\sqrt{QF}}\right)-erf\left(\frac{G+A}{\sqrt{QG}}\right)\right]\right\}$$

$$A=0$$
$$G=0.001$$
$$F=5\times10^{-4}$$
$$Q=80$$

$$erf\left(\frac{F+A}{\sqrt{QF}}\right) = -1$$

$$erf\left(\frac{G+A}{\sqrt{QG}}\right) = -0.9999993683$$

Therefore

$$\text{COLVV1}(2) = -\int_{t_{41}}^{t_{42}} u(x(t),t)q^*(x(t),x(t),t_{43},t)dt = 4.83542\times 10^{-3}$$

Case (2): Singular time integral

$$I_2 = -\int_{t_{42}}^{t_{43}} u(x(t),t)q^*(x(t),x(t),t_{43},t)dt = -u(x(t),t)\int_{t_{42}}^{t_{43}} q^*(0,x(t),t_{43},t)dt$$

$$= -u(x(t),t)\left\{e^{\frac{4A}{Q}}\left[erf\left(\frac{F+A}{\sqrt{QF}}\right) - erf\left(\frac{G+A}{\sqrt{QG}}\right)\right]\right\}$$

Note that the temperature $u(x(t),t)$ is taken out of the integral because it is unknown now the singular integral will be:

$A = 0$

$G = 5\times 10^{-4}$

$F = 0$

$Q = 80$

$$I_2 = -\int_{t_{42}}^{t_{43}} q^*(x(t),x(t),t_{43},t)dt = \left\{e^{\frac{4A}{Q}}\left[erf\left(\frac{F+A}{\sqrt{QF}}\right) - erf\left(\frac{G+A}{\sqrt{QG}}\right)\right]\right\}$$

$$= -2.820917985\times 10^{-3}$$

EVALUATION SINGULAR AND NON-SINGULAR DOMAIN INTEGRALS

In the next, we will evaluate the singular and non-singular domain integrals. Let us start the evaluation by the following integral in which the source point will be at $\xi = x(t)$:

$$\text{COLVV2}(2) = \int_{\ell}^{x(t)} u(x,t)u^*(x(t),x,t_n,t)dt =$$

$$= \underbrace{\int_{\ell}^{x(t_{42})} u(x,t)u^*(x(t),x(t),t_n,t)dt}_{\text{Non–singular integral}} + \underbrace{\int_{x(t_{42})}^{x(t_{43})} u(x,t)u^*(x(t),x(t),t_n,t)dt}_{\text{Singular integral}}$$

The next step is to evaluate the non-singular and singular integrals. Let us start by evaluating the non-singular integral as follows:

The Non-Singular Domain Integral

$$\text{Non-singular integral} = \int_{\ell}^{x(t_{42})} u(x,t)u^*(x(t),x,t_n,t)dt =$$

$$= \frac{u_{\text{melt}}}{2}\left\{\begin{array}{l} e^{\frac{4A}{Q}}\left[erf\left(\dfrac{D+A}{\sqrt{QD}}\right) - erf\left(\dfrac{G+A}{\sqrt{QG}}\right)\right] \\ +\left[erf\left(\dfrac{D-A}{\sqrt{QD}}\right) - erf\left(\dfrac{G-A}{\sqrt{QG}}\right)\right] \end{array}\right\}$$

In this case, the different parameters are as follow:

$$A = (t_n - t_{i-1})S_i + (\xi - x_{i-1}) = (0.43 - 0.41) \times 0.05 + (0.999 - 1) = 0$$
$$G = (t_n - t_{i-1})S_i = (0.43 - 0.41) \times 0.05 = 1 \times 10^{-3}$$
$$D = (t_n - t_{i-1})S_i + (x_i - x_{i-1}) = (0.43 - 0.41) \times 0.05 + (0.9995 - 1) = 5 \times 10^{-4}$$
$$Q = 80$$

Therefore,

$$\frac{D+A}{\sqrt{QD}} = \sqrt{\frac{D}{Q}} = 2.5 \times 10^{-3}$$

$\Rightarrow$

$$erf\sqrt{\frac{D}{Q}} = 2.820917985 \times 10^{-3}$$

$$\frac{D-A}{\sqrt{QD}} = \sqrt{\frac{D}{Q}} = 2.5 \times 10^{-3}$$

$\Rightarrow$

$$erf\sqrt{\frac{D}{Q}} = 2.820917985 \times 10^{-3}$$

$$\frac{G+A}{\sqrt{QG}} = \sqrt{\frac{G}{Q}} = 3.535533906 \times 10^{-3}$$

$\Rightarrow$

$$erf\sqrt{\frac{G}{Q}} = 3.989372566 \times 10^{-3}$$

$$\frac{F-A}{\sqrt{QF}} = 3.535533906 \times 10^{-3}$$

$\Rightarrow$

$$erf\sqrt{\frac{G}{Q}} = 3.989372566 \times 10^{-3}$$

Hence,

$$\text{Non-singular integral} = \int_{x(t_{41})=\ell}^{x(t_{42})} u(x,t)u^*(x(t),x,t_n,t)dt$$

$$= -9.67083 \times 10^{-3}$$

The singular domain integral

$$\text{Singular integral} = \int_{x(t_{42})}^{x(t_{43})} u(x,t)u^*(x(t),x(t),t_n,t)dt =$$

$$= e^{\frac{4A}{Q}}\left[erf\left(\frac{D+A}{\sqrt{QD}}\right) - erf\left(\frac{G+A}{\sqrt{QG}}\right)\right] + \left[erf\left(\frac{D-A}{\sqrt{QD}}\right) - erf\left(\frac{G-A}{\sqrt{QG}}\right)\right]$$

In this case, the parameters will be:

$$A = (\xi - x_i) = 0$$
$$D = (t_n - t_{i-1})S_{43}^{(1)} + (x_i - x_{i-1}) = (0.43 - 0.42) \times 0.05 + (0.999 - 0.9995)$$
$$= 0$$
$$F = (t_n - t_i)S_{43}^{(1)} = (0.43 - 0.43) \times 0.05 = 0$$
$$G = (t_n - t_{i-1})S_{43}^{(1)} = (0.43 - 0.42) \times 0.05 = 5 \times 10^{-4}$$
$$Q = 80$$

Therefore

$$erf\left(\frac{D+A}{\sqrt{QD}}\right) = 0$$

$$erf\left(\frac{D-A}{\sqrt{QD}}\right)=0$$

$$erf\left(\frac{G+A}{\sqrt{QG}}\right)=\sqrt{\frac{G}{Q}}=2.5\times10^{-3}\Rightarrow erf\left(\sqrt{\frac{G}{Q}}\right)=2.820917985\times10^{-3}$$

$$erf\left(\frac{G-A}{\sqrt{QG}}\right)=2.5\times10^{-3}$$

Hence

$$\text{Singular integral}=\int_{x(t_{42})}^{x(t_{43})}u(x,t)u^{*}(x(t),x(t),t_n,t)dt=-5.64184\times10^{-3}$$

Now we have the following system of equations:

$$\begin{pmatrix}a_{11} & a_{12}\\ a_{21} & a_{22}\end{pmatrix}\begin{pmatrix}u(0,t)\\ u_{\text{mov}}\end{pmatrix}=\begin{pmatrix}b_1\\ b_2\end{pmatrix}$$

Where

$$a_{11}=1-\text{COLYY3}(1)=1$$

$$a_{12}=\text{COLVV1}(1)-\text{COLVV2}(1)=0-0=0$$

$$a_{21}=-\text{COLYY3}(2)=0$$

$$a_{22}=1+\text{COLVV1}(2)-\text{COLVV2}(2)$$
$$=1+\left(-2.820917985\times10^{-3}\right)+5.564184\times10^{-3}$$
$$=1.002743266$$

$$b_1 = \text{SOLX1}(1) + \text{COLYY2}(2) + \text{COLYY2}(1) + \sum_{t_m}^{t_{n-1}} \text{COLVV2}(1)$$

$$+ \sum_{t_m}^{t_{n-1}} (\text{COLYY1}, \text{COLYY2}, \text{COLYY3}, \text{COLVV1})_{(1)} =$$

$$= 1.4823 + 5.704700892 \times 10^{-8} + 0 - 2.1588386 \times 10^{-4}$$

$$+ 0 + 0 + 2.486794159 \times 10^{-6} + 0 = 1.482302542$$

$$b_2 = \text{SOLX1}(2) + \text{COLYY2}(2) + \text{COLYY1}(2) + \sum_{t_m}^{t_{n-1}} \text{COLVV2}(2)$$

$$+ \sum_{t_m}^{t_{n-1}} (\text{COLYY2}, \text{COLYY1}, \text{COLVV1}, \text{COLYY3})_{(2)} =$$

$$= 3.3340 + 0.25555 + 0.61696 - +0 + 4.83542 \times 10^{-3}$$

$$-9.67083 \times 10^{-3} = 4.20167459$$

Now by solving the system, one can get:

$$u(0,t) = 1.4823$$

$$u_{\text{mov}} = 4.190198$$

SEMI ANALYTICAL FORMULA FOR STARTING AND COMPLETE MELTING TIMES FOR FINITE DIFFUSION PROBLEM

The following derivation is used for the determination of both starting and complete melting times for finite diffusion problems based on Citron 1962.

Determination of the starting time of melting

This time comes from solving the following integral:

$$t^* = \frac{1}{Q_0} \int_0^{\ell} \rho C u_0 \left(\xi, t^* \right) d\xi \qquad \textbf{(SA-1)}$$

Where,

Q_0: Constant heat input.

By using the following transformation:

$$u_0\left(\xi, t^*\right) = u_m \times \left[1 + \frac{z^2 - 2z}{2r}\right] \quad \textbf{(SA-2)}$$

In equation (SA-2) u_m is the melting temperature.

In the present case:

$$z = x$$
$$r = \frac{u_m}{Q_0} \quad \textbf{(SA-3)}$$
$$\ell = 1$$

By making use of equations (SA-2) and (SA-3) into equation (SA-1), one can get:

$$t^* = \frac{1}{Q_0}\int_0^1 u_m \times \left[1 + \frac{x^2 - 2x}{2\left(\frac{u_m}{Q}\right)}\right] dx = \left(\frac{u_m}{Q_0} - \frac{1}{3}\right)$$

Determination of the complete time of melting

The complete time of melting can be found from the following formula:

$$t_{\text{Complete}} = \left\{t_m + \left(\frac{r\sqrt{\pi}}{2M}\right) + \frac{1}{3}\right\}$$

The following parameters play an important play in determing the input heat flux Q and the melting melting temperature u_m.

$$m = \frac{\sqrt{\pi}}{2}\left\{\frac{c(u_m - u_0)}{L}\right\} \; \& \; r = \left\{\frac{k(u_m - u_0)}{\ell Q}\right\}$$

Note that:

m: Ratio between the sensible c and the latent heat L per unit volume required to raise the solid to melting temperature.

r: A parameter is required to determine the heat input.

The following formulae are widely used in the computations:

$$Q = \left(\frac{2}{\sqrt{\pi}} \times \frac{M}{r}\right) \qquad u_m = (r \times Q)$$

$$u_\alpha = (u_m + Q) \qquad t_m = \left(\frac{u_m}{Q} - \frac{1}{3}\right)$$

$$t_{\text{Complete}} = \left\{t_m + \left(\frac{r\sqrt{\pi}}{2M}\right) + \frac{1}{3}\right\}$$

The following numerical results are in the next table:

$\frac{M}{r}$	r	M	t_m	t_{Complete}	Q	u_m	u_a	Tol
2	0.75	1.5	0.4166	1.1930	2.2568	1.6926	3.9494	0.001
4	0.75		0.4166	0.9715	4.5135	3.3851	7.8986	0.0001
4.89	0.75		0.4166	0.9312	5.5178	4.1383	9.656	0.0001
10	1	10	0.4167	1.0887	11.284	11.284	22.568	0.001
20	1	20	0.4167	1.0443	22.568	22.568	45.136	0.0001
30	1	30	0.4167	1.0296	33.851	33.851	67.702	0.01

REFERENCES

[1] A. Bejan, and P.A. Tyvand, "The pressure melting of ice to an embedded cylinder", *J. Heat Transfer,* vol. 114, no. 2, pp. 532-535, 1992. [http://dx.doi.org/10.1115/1.2911311]

[2] M.K. Moallemi, B.W. Webb, and R. Viskanta, "An experimental and analytical study of close-contact melting", *J. Heat Transfer,* vol. 108, no. 4, pp. 894-899, 1986. [http://dx.doi.org/10.1115/1.3247030]

[3] A. Mehdi, and C.K. Hsieh, "Solution of ablation and combination of ablation and Stefan problems by a source and sink method", *Numer. Heat Transf. A,* vol. 26, no. 1, pp. 67-86, 1994. [http://dx.doi.org/10.1080/10407789408955981]

[4] S.G. Ahmed, "An approximate method for evaporation problem with mushy zone", *Eng. Anal. Bound. Elem.,* vol. 29, no. 2, pp. 161-165, 2005. [http://dx.doi.org/10.1016/j.enganabound.2004.10.002]

[5] H.G. Landau, "Heat Conduction in a Melting Solid", *Q. Appl. Math.,* vol. 8, no. 1, pp. 81-94, 1950. [http://dx.doi.org/10.1090/qam/33441]

[6] T.R. Goodman, "Application of Integral Methods to Transient Nonlinear Heat Transfer", *Adv. Heat Transf.,* vol. 1, pp. 55-122, 1964. [http://dx.doi.org/10.1016/S0065-2717(08)70097-2]

[7] E. Vallerani, "Integral Technique Solution to a Class of Simple Ablation Problems", *Meccanica,* vol. 9, no. 2, pp. 94-101, 1974. [http://dx.doi.org/10.1007/BF02209512]

[8] B.F. Blackwell, "Numerical prediction of one dimensional ablation using a finite control volume procedure with exponential differencing", *Numer. Heat Transf.,* vol. 14, no. 1, pp. 17-34, 1988. [http://dx.doi.org/10.1080/10407788808913631]

[9] M. Hogge, and P. Gerrekens, "Two dimensional finite element methods for surface ablation", *AIAA J.,* vol. 23, no. 3, pp. 465-472, 1985. [http://dx.doi.org/10.2514/3.8936]

[10] H. Abakians, and M.F. Modest, "Evaporative cutting of a semi-transparent body with a moving CW laser", *J. Heat Transfer,* vol. 110, no. 4a, pp. 924-930, 1988. [http://dx.doi.org/10.1115/1.3250594]

[11] L.C. Delima-Silva Jnr And Wrobel, "BEM analysis of melting problems", In: *Boundary Element Technology IX,* C.A. Brebbia, A.J.L Kassab, Eds., Computational Mechanics Publications: Southampton, 1994.

[12] R.N. Hills, D.E. Loper, and P.H. Roberts, "A thermodynamically consistent model of a mushy zone", *Q. J. Mech. Appl. Math.,* vol. 36, no. 4, pp. 505-539, 1983. [http://dx.doi.org/10.1093/qjmam/36.4.505]

[13] H.E. Huppert, and M.G. Worster, "Dynamic solidification of a binary melt", *Nature,* vol. 314, no. 6013, pp. 703-707, 1986. [http://dx.doi.org/10.1038/314703a0]

[14] AS Burns, LA Stickler, and Jr Stewart, "Solidification of an aqueous salt solution in a circular cylinder", vol. 114, 1992.

[15] D.A. Tarzia, "On the determination of the unknown coefficients through phase change process", In. *Third Int. Conference of Free and Moving Boundary Problems.* Bled: Slovenia, 1995.

[16] P.M. Pavel, ph.D. thesis, Bucharest, 1999.

[17] S.G. Ahmed, "Boundary integral formulation and mushy zone model for phase change problem", *Int. J. Numer. Methods Heat Fluid Flow,* vol. 11, no. 4, pp. 371-379, 2001. [http://dx.doi.org/10.1108/09615530110389207]

[18] P.M. Pavel, and N. Ene, "Planar solidification solution with mushy zone", *Energy Convers. Manage.,* vol. 44, no. 18, pp. 2977-2989, 2003. [http://dx.doi.org/10.1016/S0196-8904(03)00052-9]

[19] M. Mekey, "L M. and S. G. Ahmed, Semi-analytical method for combined Stefan and ablation problems with mushy zones, Ain Shams University, Scientific Bulletin, Faculty of Engineering", *Part III,* vol. 42, no. 3, pp. 1241-1256, 2007.

[20] I.V. Alexandrova, D.V. Alexandrov, D.L. Aseev, and S.V. Bulitcheva, "Mushy layer formation during solidification of binary alloys from a cooled wall: The role of boundary conditions", *Acta Phys. Pol. A,* vol. 115, no. 4, pp. 791-794, 2009. [http://dx.doi.org/10.12693/APhysPolA.115.791]

[21] S. Nouri, M. Benzeghiba, and A. Benzeoui, "Numerical computation of magnetic field effect of an alloy with mushy zone", *ISESCO science and technology,* vol. 6, 2010.

[22] S.G. Ahmed, "Numerical solution of ablation problem using modified constraint integral method", *J. Eng. Appl. Sci. (Asian Res. Publ. Netw.),* vol. 47, no. 3, 2000.

[23] J. Mennig, and M.N. Ozisik, "Coupled Integral Equation Approach for solving Melting or Solidification", *Int. J. Heat Mass Transf.,* vol. 28, no. 8, pp. 1481-1485, 1985. [http://dx.doi.org/10.1016/0017-9310(85)90250-9]

[24] J.S. Langer, "Instabilities and pattern formation in crystal growth", *Rev. Mod. Phys.,* vol. 52, no. 1, pp. 1-28, 1980. [http://dx.doi.org/10.1103/RevModPhys.52.1]

[25] Ungar Lyle H, and A. Brown Robert, "Cellular interface morphologies in directional solidification. IV. The formation of deep cells", In: *Phys. Rev* vol. vB 31. , 1985, pp. 5931-5940.

[26] S.H. Davis, *Theory of Solidification.* Cambridge University Press: New York, 2001. [http://dx.doi.org/10.1017/CBO9780511546747]

[27] A. Saito, H.K. Hong, and O. Hirokane, "Heat transfer enhancement in the direct contact melting process", *Int. J. Heat Mass Transf.,* vol. 35, no. 295, pp. 295-305, 1992. [http://dx.doi.org/10.1016/0017-9310(92)90268-W]

[28] Y. Zhang, and A. Faghri, "Analysis of Freezing in an Eccentric Annulus", *J. Sol. Energy Eng.,* vol. 119, no. 3, pp. 237-241, 1997. [http://dx.doi.org/10.1115/1.2888025]

[29] E. Ben-Jacob, and J.S. Nigel Goldenfeld, "Langer, and Gerd Schön, Dynamics of Interfacial Pattern Formation", *Phys. Rev. Lett.,* vol. 51, no. 21, pp. 1930-1932, 1983. [http://dx.doi.org/10.1103/PhysRevLett.51.1930]

[30] E. Obryk, Karl Gabriel, and L.A. Copley, "Electroproduction from Nucleons in a Relativistic Quark Model", In: *Phys. Rev* vol. D 4. , 1971, pp. 2844-2848.

[31] C. Ross, "Kerr, Andrew W. Woods, M. Grae Worster, and Herbert E. Huppert, Solidification of an alloy cooled from above, Part 3. Compositional stratification within the solid", *J. Fluid Mech.,* vol. 218, no. -1, pp. 337-354, 1990. [http://dx.doi.org/10.1017/S0022112090001021]

[32] The Development of Crystal Growth Technology, *HANS J. SCHEEL CONSULTING, CH-8808 Pfaeffikon SZ.* John Wiley & Sons, Ltd.: Switzerland, 2003.

[33] L.C. Wrobel, "A boundary element solution to Stefan's problem", In: *Boundary Elements V.* Springer-

Verlag: Berlin, 1983.

[34] R. Bonnerot, and P. Jamet, "A Third order accurate discontinuous finite element method for one-dimensional Stefan problem", *J. Comput. Phys.,* vol. 32, no. 2, pp. 109-128, 1979. [http://dx.doi.org/10.1016/0021-9991(79)90127-X]

[35] W.W. Mullins, and R.F. Sekerka, "The Stability of a Planar Interface During Solidification of a Dilute Binary Alloy", *J. Appl. Phys.,* vol. 35, no. 2, pp. 444-451, 1964. [http://dx.doi.org/10.1063/1.1713333]

[36] S.L. Braga, and R. Viskanta, "Solidification of a Binary Solution on a Cold Isothermal Surface", *Int. J. Heat Mass Transf.,* vol. 33, no. 4, pp. 745-754, 1990. [http://dx.doi.org/10.1016/0017-9310(90)90172-Q]

[37] W.Z. Cao, and D. Poulikakos, "Freezing of a Binary Alloy Saturating a Packed Bed of Spheres", *J. Thermophys. Heat Trans.,* vol. 5, no. 1, pp. 46-53, 1991. [http://dx.doi.org/10.2514/3.225]

[38] A. Faghri, Y. Zhang, and J.R. Howell, *Advanced Heat and Mass Transfer.* Global Digital Press: Columbia, MO, 2010.

[39] D. Juric, and G. Tryggvason, *A Front-Tracking Method for Liquid-Vapor Phase Change," in Advances in Numerical Modeling of Free Surface and Interface Fluid Dynamics, FED.* vol. Vol. 234. ASME, 1995, pp. 141-148.

[40] G. Segal, C. Vuik, and F. Vermolen, "A conserving discretization for the free boundary in a two-dimensional Stefan problem", *J. Comput. Phys.,* vol. 141, no. 1, pp. 1-21, 1998. [http://dx.doi.org/10.1006/jcph.1998.5900]

[41] W.D. Murray, and F. Landis, "Numerical and machine solutions of transient heat-conduction problems involving melting or freezing, Trans. ASME (C)", *J. Heat Transfer,* vol. 81, pp. 106-112, 1959. [http://dx.doi.org/10.1115/1.4008149]

[42] V.R. Voller, "An implicit enthalpy solution for phase change problems: with application to a binary alloy solidification", *Appl. Math. Model.,* vol. 11, no. 2, pp. 110-116, 1987. [http://dx.doi.org/10.1016/0307-904X(87)90154-5]

[43] S. Chen, B. Merriman, S. Osher, and P. Smereka, "A simple level set method for solving Stefan problems", *J. Comput. Phys.,* vol. 135, no. 1, pp. 8-29, 1997. [http://dx.doi.org/10.1006/jcph.1997.5721]

[44] S. Osher, and R. Fedkiw, *Level Set Methods and Dynamic Implicit Surfaces.* Springer: NewYork, 2003. [http://dx.doi.org/10.1007/b98879]

[45] M. Sussman, P. Smereka, and S. Osher, "A level set approach for computing solutions to incompressible two-phase flow", *J. Comput. Phys.,* vol. 114, no. 1, pp. 146-159, 1994. [http://dx.doi.org/10.1006/jcph.1994.1155]

[46] C.A. Brebbia, J.C.F. Telles, and L.C. Wrobel, *Boundary Element Techniques Theory and Applications in Engineering* Spring-Verlag: Berlin and New York, 1984.

[47] I. Fredholm, "Sur Une Classe d'equations Fonctionelles", *Acta Math.,* vol. 27, no. 0, pp. 365-390, 1903. [http://dx.doi.org/10.1007/BF02421317]

[48] M.A. Jaswon, "Integral Equation Methods in Potential Theory, I", *Proc. Roy. Soc. Ser. A,* vol. 275, pp. 23-32, 1963.

[49] G.T. Symm, "Integral Equation Methods in Potential Theory, II", *Proc. Roy. Soc. Ser. A,* vol. 275, pp. 33-46, 1963.

[50] M.A. Jaswon, and A.R. Ponter, "An Integral Equation Solution of the Torsion Problem", *Proc. Roy. Soc. Ser. A,* vol. 273, pp. 237-246, 1963.

[51] J.L. Hess, and A.M.O. Smith, *Calculation of Potential Flow about Arbitrary Bodies, Progress in Aeronautical Science.*, D. Kuchemann, Ed., vol. Vol. 8. Pergamon Press: London, 1967.

[52] Y.K. Chuang, and J. Szekely, "On the use of Green's functions for solving melting or solidification problems", *Int. J. Heat Mass Transfer,* vol. 14, 1972.

[53] Y.K. Chuang, and J. Szekely, "The use of Green's functions for solving melting or solidification problems in a cylindrical coordinate system", *Int. J. Heat Mass Transfer,* vol. 15, 1972.

[54] R.P. Shaw, "A boundary integral equation approach to the one dimensional ablation problem", In: *Boundary Element Methods in Engineering.* Springer-Verlag: Berlin, 1982. [http://dx.doi.org/10.1007/978-3-662-11273-1_10]

[55] C.J. Coleman, "a boundary integral formulation of the Stefan problem, appl", *Math. Model.,* vol. 10, 1986.

[56] C.J. Coleman, "a boundary integral approach to the solidification of dilute alloys", *Int. J. Heat Mass Transf.,* vol. 30, no. 8, pp. 1727-1732, 1987. [http://dx.doi.org/10.1016/0017-9310(87)90319-X]

[57] K. O'Neill, "Boundary integral equation solution of moving boundary phase change problems", *Int. J. Num. Meth. Eng,* vol. 19, 1983.

[58] *Heinlein, M., Mukherjee, S., and Richmond, A boundary element method analysis of temperature fields and stress during solidification.* vol. Vol. 59. Acta Mechanica, 1986.

[59] N. Zabaras, and S. Mukherjee, *An analysis of the solidification problems by the boundary element method, Int.* vol. Vol. 24. J. Num. Methods, 1987.

[60] D-S. Jnr, and L.C. Wrobel, "Wrobel, boundary element formulation for multi-dimensional ablation problems", In: *Boundary Elements XIV* vol. 1. Computational Mechanics Publications, 1992.

[61] S.G. Ahmed, and L.C. Wrobel, "Numerical solution of solidification problems using integral equations", In: *Third Int. Conference of free and moving boundary problems* Bled: Slovenia, 1995.

[62] H.S. Carslaw, and J.C. Jaejer, *Conduction of Heat in Solids.* 2nd ed. Clarendon Press: Oxford, 1959.

[63] J.S. Wettlaufer, M.G. Worster, and H.E. Huppert, "Solidification of leads: Theory, experiment and field observations", *J. Geophys. Res.,* vol. 105, no. C1, pp. 1123-1134, 2000. [http://dx.doi.org/10.1029/1999JC900269]

[64] W.M. Abd-El Fatah, M.M. Helal, S.G. Ahmed, and M.I.M. Hessein, "Heat balance integral method applied to ablation problems with mushy zones", *Ninth International Congress of Fluid Dynamics & Propulsion* Alexandria, Egypt.

[65] W.M. Abd-el Fatah, M.M. Helal, M.I.M. Hessein, and S.G. Ahmed, "Integral solution of unidirectional solidification of binary melts from a cooled wall", In: *imbic*culcutta, India, 2010.

[66] B.F. Zalewski, R.L. Mullen, and R.L. Muhanna, *Boundary Element Analysis of Systems Using Interval Methods, authors.*U. S. A, 2006.

[67] W.M. Abd-El Fatah, and S.G. Ahmed, "Boundary Integral Formulation for Binary Alloys from a Cooling Solid Wall", *International Journal of Computational and Applied Mathematics,* vol. 5, no. 5, pp. 687-696, 2010.

[68] S.G. Ahmed, *On Diffusion and Hydrodynamics with Moving Boundaries, Ph. D., Chanel system between Portsmouth University, Uk.* And Zagazig University, 1996.

CHAPTER 6

New Approach of BEM

Abstract: In the present chapter, the author suggested a new approach for applying the direct boundary element methods; however, this approach is not published yet. It is a new idea based on Artificial quadratic linear approximation in the boundary element method.

Keywords: Quadratic linear approximation, theory of approximation, direct approach boundary element method.

INTRODUCTION

In the late 1970s and early 1980s, boundary elements began to make strides in the domain of numerical analysis, demonstrating their efficacy in solving a variety of engineering problems with high accuracy. When dealing with boundary elements that appear full range, unique skills are required to get valuable results.The present method evaluates these integrals in an analytical manner, using artificial quadratic approximation (AQA) by assuming the linear element contains three points, the end points, and the midway point. Analytical results for the required integrals can be obtained by expressing the coordinates, potential, and potential derivatives in terms of a quadratic shape function in a special manner [1-19].

Boundary Element Formulation for Potential Problems

<u>Stage (1)</u>

Assume all related parameters in terms of the new shape function (Fig. **1**) as follow:

$$x(\zeta) = I_1 x_n + I_2 x_{n+\frac{1}{2}} + I_3 x_{n+1} \qquad \textbf{(6.1)}$$

S. G. Ahmed

$$y(\zeta) = I_1 y_n + I_2 y_{n+\frac{1}{2}} + I_3 y_{n+1} \tag{6.2}$$

$$u(\zeta) = I_1 u_n + I_2 u_{n+\frac{1}{2}} + I_3 u_{n+1} \tag{6.3}$$

$$q(\zeta) = I_1 q_n + I_2 q_{n+\frac{1}{2}} + I_3 q_{n+1} \tag{6.4}$$

In these equations,

$$I_1 = \frac{1}{2}\zeta(\zeta - 1) \tag{6.5}$$

$$I_2 = (1-\zeta)(1+\zeta) \tag{6.6}$$

$$I_3 = \frac{1}{2}\zeta(1+\zeta) \tag{6.7}$$

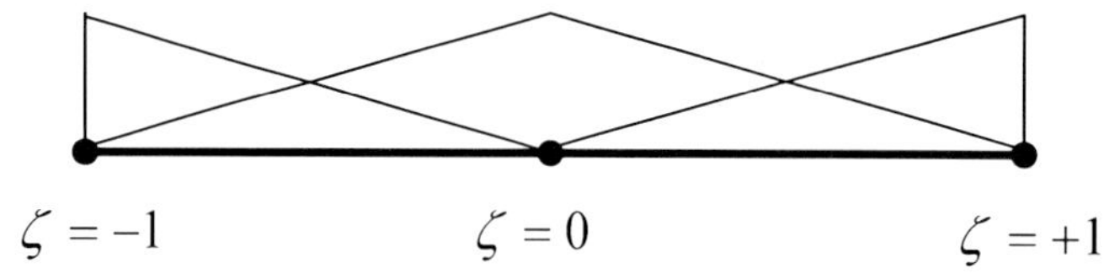

Fig. (1). Shape function distribution.

Stage (2)

Let us start with the boundary integral equation for the Laplace equation as a case study:

$$u(x, y) + \int_\Gamma Hu\, ds = \int_\Gamma Gq\, ds \tag{6.8}$$

After discretization, equation **(6.8)** can take the following form:

$$u(x,y)+\sum_{n=1}^{N}\left[H_n^1 u_n+\left(H_n^{0,\frac{1}{2}}+H_n^{\frac{1}{2},1}\right)u_{n+\frac{1}{2}}+H_n^2 u_{n+1}\right]$$
$$=\sum_{n=1}^{N}\left[G_n^1 q_n+\left(G_n^{0,\frac{1}{2}}+G_n^{\frac{1}{2},1}\right)q_{n+\frac{1}{2}}+G_n^2 q_{n+1}\right] \quad \textbf{(6.9)}$$

Now, we have 8-integrals, they are:

$$H_n^1=\frac{-1}{2\pi}\int_{-1}^{0} I_1 \frac{\chi_i n_i}{r^2}\,ds \quad \textbf{(6.10)}$$

$$H_n^2=\frac{-1}{2\pi}\int_{0}^{1} I_3 \frac{\chi_i n_i}{r^2}\,ds \quad \textbf{(6.11)}$$

$$H_n^{0,\frac{1}{2}}=\frac{-1}{2\pi}\int_{-1}^{0} I_2 \frac{\chi_i n_i}{r^2}\,ds \quad \textbf{(6.12)}$$

$$H_n^{\frac{1}{2},1}=\frac{-1}{2\pi}\int_{0}^{1} I_2 \frac{\chi_i n_i}{r^2}\,ds \quad \textbf{(6.13)}$$

$$G_n^1=\frac{-1}{2\pi}\int_{-1}^{0} I_1 \ln r\,ds \quad \textbf{(6.14)}$$

$$G_n^2=\frac{-1}{2\pi}\int_{0}^{1} I_3 \ln r\,ds \quad \textbf{(6.15)}$$

$$G_n^{0,\frac{1}{2}} = \frac{-1}{2\pi}\int_{-1}^{0} I_2 \ln r\, ds \tag{6.16}$$

$$G_n^{\frac{1}{2},1} = \frac{-1}{2\pi}\int_{0}^{1} I_2 \ln r\, ds \tag{6.17}$$

Note that both shape function I and local coordinates ζ vary according to the following table:

	I_1	I_2	I_3
-1	1	0	0
0	0	1	0
1	0	0	1

Stage (**3**): Evaluation of the integrals

According to [#] the following expressions are used:

$$r^2(\zeta) = a\zeta^2 + b\zeta + c \tag{6.18}$$

$$ds = \frac{\ell_n}{2} d\zeta \tag{6.19}$$

The length ℓ_n will vary according to the distance between two successive nodes over the same element under consideration.

(1) Integral " G_n^1

In this case, the following expressions will be used, see Fig. (**2**):

Fig. (2). Element for integral G_n^1.

The G_n^1 takes the form:

$$G_n^1 = \frac{-\ell_n}{8\beta}\int_{-1}^{0}\zeta(\zeta-1)\ln\left(a\zeta^2+b\zeta+c\right)d\zeta \tag{6.20}$$

$$\ell_n = \sqrt{\left(x_{n+\frac{1}{2}}-x_n\right)^2+\left(y_{n+\frac{1}{2}}-y_n\right)^2} \tag{6.21}$$

$$a = \left(\frac{\ell_n}{2}\right)^2 \tag{6.22}$$

$$b = \left(\frac{x_{n+\frac{1}{2}}+x_n}{2}-\varsigma\right)\left(x_{n+\frac{1}{2}}-x_n\right)+\left(\frac{y_{n+\frac{1}{2}}+y_n}{2}-\mu\right)\left(y_{n+\frac{1}{2}}-y_n\right) \tag{6.23}$$

$$c = \left(\frac{x_{n+\frac{1}{2}}+x_n}{2}-\varsigma\right)^2+\left(\frac{y_{n+\frac{1}{2}}+y_n}{2}-\mu\right)^2 \tag{6.24}$$

$$d = 4ac-b^2 \tag{6.25}$$

$$G_n^1 = -\left(\frac{\ell}{288\beta a^3\sqrt{\left(4ac-b^2\right)}}\right)$$

$$\times\left[\begin{array}{l}\sqrt{\left(4ac-b^2\right)}\left(-26a^3-24a^2b+24a^2c-12ab^2\right)\\ \ln c\sqrt{\left(4ac-b^2\right)}\left(-18abc+9ab^2+6b^3-18a^2c\right)\\ +\ln\left(a-b+c\right)\sqrt{\left(4ac-b^2\right)}\left(30a^3+18a^2c-9ab^2+18abc-6b^3\right)\\ +\tan^{-1}\left(\frac{b}{\sqrt{\left(4ac-b^2\right)}}\right)\left(-48a^2c^2-18ab^3+60ab^2c+72a^2bc-12b^4\right)\\ +\tan^{-1}\left(\frac{2a-b}{\sqrt{\left(4ac-b^2\right)}}\right)\left(72a^2bc-18ab^3-48a^2c^2+60ab^2c-12b^4\right)\end{array}\right] \quad \textbf{(6.26)}$$

(2) Integral " G_n^2 "

In this case, the following expressions will be used, see Fig. **(3)**:

Fig. (3). Element for integral G_n^2.

The G_n^2 takes the form:

$$G_n^2 = \frac{-\ell_n}{8\beta}\int_0^1 \zeta(\zeta+1)\ln\left(a\zeta^2+b\zeta+c\right)d\zeta \quad \textbf{(6.27)}$$

$$G_n^2 = -\left(\frac{\ell}{288\beta a^3\sqrt{\left(4ac-b^2\right)}}\right)\times$$

$$\left[\begin{array}{l} \sqrt{\left(4ac-b^2\right)}\left(-26a^3+24a^2b+24a^2c-12ab^2\right) \\ \ln c\sqrt{\left(4ac-b^2\right)}\left(18abc+9ab^2-6b^3-18a^2c\right) \\ +\ln(a+b+c)\sqrt{\left(4ac-b^2\right)}\left(30a^3+18a^2c-9ab^2-18abc+6b^3\right) \\ +\tan^{-1}\left(\frac{b}{\sqrt{\left(4ac-b^2\right)}}\right)\left(48a^2c^2-18ab^3-60ab^2c+72a^2bc+12b^4\right) \\ +\tan^{-1}\left(\frac{2a+b}{\sqrt{\left(4ac-b^2\right)}}\right)\left(-72a^2bc+18ab^3-48a^2c^2+60ab^2c-12b^4\right) \end{array}\right] \quad \textbf{(6.28)}$$

$$\ell_n = \sqrt{\left(x_{n+1}-x_{n+\frac{1}{2}}\right)^2+\left(y_{n+1}-y_{n+\frac{1}{2}}\right)^2} \quad \textbf{(6.29)}$$

$$a = \left(\frac{\ell_n}{2}\right)^2 \quad \textbf{(6.30)}$$

$$b = \left(\frac{x_{n+1}+x_{n+\frac{1}{2}}}{2}-\varsigma\right)\left(x_{n+1}-x_{n+\frac{1}{2}}\right)+\left(\frac{y_{n+1}+y_{n+\frac{1}{2}}}{2}-\mu\right)\left(y_{n+1}-y_{n+\frac{1}{2}}\right) \quad \textbf{(6.31)}$$

$$c = \left(\frac{x_{n+1}+x_{n+\frac{1}{2}}}{2}-\varsigma\right)^2+\left(\frac{y_{n+1}+y_{n+\frac{1}{2}}}{2}-\mu\right)^2 \quad \textbf{(6.32)}$$

(3) Integral " $G_n^{0,\frac{1}{2}}$ "

In this case, the following expressions will be used, see Fig. **(4)**:

Fig. (4). Element for integral $G_n^{0,\frac{1}{2}}$.

The $G_n^{0,\frac{1}{2}}$ takes the form:

$$G_n^{0,\frac{1}{2}} = \frac{-\ell_n}{8\beta}\int_{-1}^{0}(1-\zeta)(\zeta+1)\ln\left(a\zeta^2+b\zeta+c\right)d\zeta \tag{6.33}$$

$$G_n^{0,\frac{1}{2}} = -\left(\frac{\ell}{144\beta a^3\sqrt{(4ac-b^2)}}\right)\times$$

$$\left[\begin{array}{l} \sqrt{(4ac-b^2)}(3a^2b-32a^3-12a^2c+6ab^2) \\ \ln c\sqrt{(4ac-b^2)}(9abc+9a^2b-3b^3) \\ +\ln(a-b+c)\sqrt{(4ac-b^2)}(12a^3-9a^2b-9abc+3b^3) \\ +\tan^{-1}\left(\frac{b}{\sqrt{(4ac-b^2)}}\right)(24a^2c^2+72a^3c-18a^2b^2-30ab^2c+6b^4) \\ +\tan^{-1}\left(\frac{2a-b}{\sqrt{(4ac-b^2)}}\right)(72a^3c-18a^2b^2+24a^2c^2-30ab^2c+6b^4) \end{array}\right] \tag{6.34}$$

$$\ell_n = \sqrt{\left(x_{n+\frac{1}{2}} - x_n\right)^2 + \left(y_{n+\frac{1}{2}} - y_n\right)^2} \tag{6.35}$$

$$a = \left(\frac{\ell_n}{2}\right)^2 \tag{6.36}$$

$$b = \left(\frac{x_{n+\frac{1}{2}} + x_n}{2} - \varsigma\right)\left(x_{n+\frac{1}{2}} - x_n\right) + \left(\frac{y_{n+\frac{1}{2}} + y_n}{2} - \mu\right)\left(y_{n+\frac{1}{2}} - y_n\right) \tag{6.37}$$

$$c = \left(\frac{x_{n+\frac{1}{2}} + x_n}{2} - \varsigma\right)^2 + \left(\frac{y_{n+\frac{1}{2}} + y_n}{2} - \mu\right)^2 \tag{6.38}$$

(4) Integral " $G_n^{\frac{1}{2},1}$ "

In this case, the following expressions will be used, see Fig. **(5)**:

Fig. (5). Element for integral $G_n^{\frac{1}{2},1}$.

The $G_n^{\frac{1}{2},1}$ takes the form:

$$G_n^{\frac{1}{2},1} = \frac{-\ell_n}{8\pi}\int_0^1 (1-\zeta)(\zeta+1)\ln\left(a\zeta^2+b\zeta+c\right)d\zeta \quad \textbf{(6.39)}$$

$$G_n^{\frac{1}{2},1} = -\left(\frac{\ell}{144\beta a^3\sqrt{\left(4ac-b^2\right)}}\right)\times$$

$$\left[\begin{array}{l} \sqrt{\left(4ac-b^2\right)}\left(3a^2b+32a^3+12a^2c-6ab^2\right) \\ \ln c\sqrt{\left(4ac-b^2\right)}\left(9abc+9a^2b-3b^3\right) \\ +\ln(a+b+c)\sqrt{\left(4ac-b^2\right)}\left(-12a^3-9a^2b-9abc+3b^3\right) \\ +\tan^{-1}\left(\dfrac{b}{\sqrt{\left(4ac-b^2\right)}}\right)\left(24a^2c^2+72a^3c-18a^2b^2-30ab^2c+6b^4\right) \\ +\tan^{-1}\left(\dfrac{2a+b}{\sqrt{\left(4ac-b^2\right)}}\right)\left(-72a^3c+18a^2b^2-24a^2c^2+30ab^2c-6b^4\right) \end{array}\right] \quad \textbf{(6.40)}$$

$$\ell_n = \sqrt{\left(x_{n+1} - x_{n+\frac{1}{2}}\right)^2 + \left(y_{n+1} - y_{n+\frac{1}{2}}\right)^2} \quad \textbf{(6.41)}$$

$$a = \left(\frac{\ell_n}{2}\right)^2 \quad \textbf{(6.42)}$$

$$b = \left(\frac{x_{n+1} + x_{n+\frac{1}{2}}}{2} - \varsigma\right)\left(x_{n+1} - x_{n+\frac{1}{2}}\right) + \left(\frac{y_{n+1} + y_{n+\frac{1}{2}}}{2} - \mu\right)\left(y_{n+1} - y_{n+\frac{1}{2}}\right) \quad \textbf{(6.43)}$$

$$c = \left(\frac{x_{n+1} + x_{n+\frac{1}{2}}}{2} - \varsigma\right)^2 + \left(\frac{y_{n+1} + y_{n+\frac{1}{2}}}{2} - \mu\right)^2 \quad \textbf{(6.44)}$$

(5) Integral " H_n^1 "

In this case, the following expressions will be used, see Fig. **(6)**:

Fig. (6). Element for integral H_n^1.

The H_n^1 takes the form:

$$H_n^1 = \frac{-\ell_n}{8\beta}\int_{-1}^{0}\zeta(\zeta-1)\left(\frac{\left(\dfrac{x_{n+\frac{1}{2}}+x_n}{2}-\varsigma\right)n_x+\left(\dfrac{y_{n+\frac{1}{2}}+y_n}{2}-\mu\right)n_y}{a\zeta^2+b\zeta+c}\right)d\zeta \tag{6.45}$$

It can be written as:

$$H_n^1 = \frac{-\ell_n}{8\beta}\int_{-1}^{0}\zeta(\zeta-1)\left(\frac{D_1 n_x + D_2 n_y}{a\zeta^2+b\zeta+c}\right)d\zeta \tag{6.46}$$

$$H_n^1 = -\left(\frac{\ell}{16\beta a^2\sqrt{\left(4ac-b^2\right)}}\right)\times$$
$$\left[\begin{array}{l}\sqrt{\left(4ac-b^2\right)}\left(2D_1n_xa+2D_2n_ya\right)\\ \ln c\sqrt{\left(4ac-b^2\right)}\left(-D_2n_ya-D_2n_yb-D_1n_xa-D_1n_xb\right)\\ +\ln\left(a-b+c\right)\sqrt{\left(4ac-b^2\right)}\left(D_2n_ya+D_2n_yb+D_1n_xa+D_1n_xb\right)\\ +\tan^{-1}\left(\dfrac{b}{\sqrt{\left(4ac-b^2\right)}}\right)\left(-4D_2n_yac+2D_2n_yab+2D_2n_yb^2-4D_1n_xac+2D_1n_xab+2D_1n_xb^2\right)\\ +\tan^{-1}\left(\dfrac{2a-b}{\sqrt{\left(4ac-b^2\right)}}\right)\left(-4D_2n_yac+2D_2n_yab+2D_2n_yb^2-4D_1n_xac+2D_1n_xab+2D_1n_xb^2\right)\end{array}\right] \tag{6.47}$$

(6) Integral " H_n^2 "

In this case, the following expressions will be used, see Fig. **(7)**:

Fig. (7). Element for integral ^2H_n.

The H_n^2 takes the form:

$$H_n^1 = \frac{-\ell_n}{8\beta}\int_0^1 \varsigma(\zeta+1)\left(\frac{\left(\frac{x_{n+1}+x_{n+\frac{1}{2}}}{2}-\varsigma\right)n_x+\left(\frac{y_{n+1}+y_{n+\frac{1}{2}}}{2}-\mu\right)n_y}{a\zeta^2+b\zeta+c}\right)d\zeta \tag{6.48}$$

$$H_n^1 = \frac{-\ell_n}{8\beta}\int_0^1 (\zeta+1)\left(\frac{D_1n_x+D_2n_y}{a\zeta^2+b\zeta+c}\right) \tag{6.49}$$

$$H_n^2 = -\left(\frac{\ell}{16\beta a^2\sqrt{(4ac-b^2)}}\right)\times$$

$$\left[\begin{array}{l}\sqrt{(4ac-b^2)}(2D_1n_xa+2D_2n_ya)\\ \ln c\sqrt{(4ac-b^2)}(D_2n_yb-D_2n_ya+D_1n_xb-D_1n_xa)\\ +\ln(a+b+c)\sqrt{(4ac-b^2)}(D_2n_ya-D_2n_yb+D_1n_xa-D_1n_xb)\\ +\tan^{-1}\left(\frac{b}{\sqrt{(4ac-b^2)}}\right)(4D_2n_yac+2D_2n_yab-2D_2n_yb^2+4D_1n_xac+2D_1n_xab-2D_1n_xb^2)\\ +\tan^{-1}\left(\frac{2a+b}{\sqrt{(4ac-b^2)}}\right)(-4D_2n_yac-2D_2n_yab+2D_2n_yb^2-4D_1n_xac-2D_1n_xab+2D_1n_xb^2)\end{array}\right] \tag{6.50}$$

(7) Integral " $H_n^{0,\frac{1}{2}}$ "

In this case, the following expressions will be used, see Fig. **(8)**:

Fig. (8): Element for integral $H_n^{0,\frac{1}{2}}$.

The $H_n^{0,\frac{1}{2}}$ takes the form:

$$H_n^{0,\frac{1}{2}} = \frac{-\ell_n}{8\pi}\int_{-1}^{0}(1-\zeta)(\zeta+1)\left(\frac{\left(\dfrac{x_{n+\frac{1}{2}}+x_n}{2}-\varsigma\right)n_x+\left(\dfrac{y_{n+\frac{1}{2}}+y_n}{2}-\mu\right)n_y}{a\zeta^2+b\zeta+c}\right)d\zeta \quad \textbf{(6.51)}$$

$$H_n^{0,\frac{1}{2}} = \frac{-\ell_n}{8\pi}\int_{-1}^{0}(1-\zeta)(\zeta+1)\left(\frac{D_1 n_x + D_2 n_y}{a\zeta^2+b\zeta+c}\right)d\zeta \quad \textbf{(6.52)}$$

$$H_n^{0,\frac{1}{2}} = -\left(\frac{\ell}{16\beta a^2\sqrt{(4ac-b^2)}}\right)\times$$

$$\left[\begin{array}{l}\sqrt{(4ac-b^2)}(-2D_1n_xa-2D_2n_ya)\\ \ln c\sqrt{(4ac-b^2)}(D_1n_xb+D_2n_yb)\\ +\ln(a-b+c)\sqrt{(4ac-b^2)}(-D_1n_xb-D_2n_yb)\\ +\tan^{-1}\left(\frac{b}{\sqrt{(4ac-b^2)}}\right)(4D_2n_ya^2+4D_1n_xa^2-2D_2n_yb^2+4D_1n_xac+4D_2n_yac-2D_1n_xb^2)\\ +\tan^{-1}\left(\frac{2a-b}{\sqrt{(4ac-b^2)}}\right)(4D_2n_ya^2+4D_2n_yac+4D_1n_xa^2-2D_2n_yb^2+4D_1n_xac-2D_1n_xb^2)\end{array}\right] \quad (6.53)$$

(8) Integral "$H_n^{\frac{1}{2},1}$"

In this case, the following expressions will be used, see Fig. **(9)**:

Fig. (9): Element for integral $H_n^{\frac{1}{2},1}$.

The $H_n^{\frac{1}{2},1}$ takes the form:

$$H_n^{\frac{1}{2},1} = \frac{-\ell_n}{2\pi}\int_0^1 (1-\zeta)(\zeta+1)\left(\frac{\left(\frac{x_{n+\frac{1}{2}}+x_{n+1}}{2}-\varsigma\right)n_x+\left(\frac{y_{n+\frac{1}{2}}+y_{n+1}}{2}-\mu\right)n_y}{a\zeta^2+b\zeta+c}\right)d\zeta \quad \textbf{(6.54)}$$

$$H_n^{\frac{1}{2},1} = \frac{-\ell_n}{2\pi}\int_0^1 (1-\zeta)(\zeta+1)\left(\frac{D_1 n_x + D_2 n_y}{a\zeta^2+b\zeta+c}\right)d\zeta \quad \textbf{(6.55)}$$

$$H_n^{\frac{1}{2},1} = \left(\frac{\ell}{16\beta a^2\sqrt{(4ac-b^2)}}\right)\times$$

$$\left[\begin{array}{l} \sqrt{(4ac-b^2)}(2D_1 n_x a + 2D_2 n_y a) \\ \ln c\sqrt{(4ac-b^2)}(D_1 n_x b + D_2 n_y b) \\ +\ln(a+b+c)\sqrt{(4ac-b^2)}(-D_1 n_x b - D_2 n_y b) \\ +\tan^{-1}\left(\frac{b}{\sqrt{(4ac-b^2)}}\right)(4D_2 n_y a^2 + 4D_1 n_x a^2 - 2D_2 n_y b^2 + 4D_1 n_x ac + 4D_2 n_y ac - 2D_1 n_x b^2) \\ +\tan^{-1}\left(\frac{2a+b}{\sqrt{(4ac-b^2)}}\right)(-4D_2 n_y a^2 - 4D_2 n_y ac - 4D_1 n_x a^2 + 2D_2 n_y b^2 - 4D_1 n_x ac + 2D_1 n_x b^2) \end{array}\right] \quad \textbf{(6.56)}$$

REFERENCES

[1] S.G. Ahmed, "An approximate method for oxygen diffusion in a sphere with simultaneous absorption", *Int. J. Numer. Methods Heat Fluid Flow,* vol. 9, no. 6, pp. 631-643, 1999. [http://dx.doi.org/10.1108/09615539910276115]

[2] S.G. Ahmed, "Boundary integral formulation and mushy zone model for phase change problem", *Int. J. Numer. Methods Heat Fluid Flow,* vol. 11, no. 4, pp. 371-380, 2001. [http://dx.doi.org/10.1108/09615530110389207]

[3] S.G. Ahmed, "'An approximate method for evaporation problem with mushy zone', Int", *Eng. Anal. Bound. Elem.,* vol. 29, no. 2, pp. 161-165, 2005. [http://dx.doi.org/10.1016/j.enganabound.2004.10.002]

[4] S.G. Ahmed, "An integral method for solving oxygen diffusion on unbounded domain with moving boundary", *J. Appl. Sci. (Faisalabad),* vol. 5, no. 7, pp. 1244-1249, 2005. [http://dx.doi.org/10.3923/jas.2005.1244.1249]

[5] S. G. Ahmed, *A new algorithm for moving boundary problems subject to periodic boundary conditions,* vol. 16, Int. J. Numerical Methods for Heat and Fluid Flow: U.K, no. 1, 2005.

[6] S.G. Ahmed, *A numerical method for oxygen diffusion and absorption in sike cell* vol. 173/1. Applied Mathematics and Computations: Netherlands, 2005, pp. 668-682.

[7] S.G. Ahmed, "'A collocation method using new combined radial basis function of thin plate and multi-quadric types', Int", *Eng. Anal. Bound. Elem.,* vol. 30, no. 8, pp. 697-701, 2006. [http://dx.doi.org/10.1016/j.enganabound.2006.03.001]

[8] S. G. Ahmed, "A semi-analytical method for phase change transformation in binary alloys", *J. Comput. Appl. Math.,* Netherlands, Available online, August, 2006.

[9] S.G. Ahmed, and S.A. Meshrif, "A New Numerical Algorithm for 2-D Moving Boundary Problems using Boundary Element Method", *Comput. Math. Appl.,* vol. 58, no. 7, pp. 1302-1308, 2009. [http://dx.doi.org/10.1016/j.camwa.2009.03.115]

[10] S. G. Ahmed, and M. L. Mekey, "A collocation and Cartesian grid methods using new radial basis function to solve class of partial differential equations", *Int. J. Comput. Math,* pp. 1-14, 2009.

[11] S.G. Ahmed, "Boundary Integral Formulation for Binary Alloys from a Cooling Solid Wall", *J. Comput. Appl. Math.,* vol. 5, no. 5, pp. 687-696, 2010.

[12] S. G. Ahmed, "A collocation mesh-free method based on multiple basis functions", *Int. J. of Engineering Analysis with Boundary Elements, Sept,* 2011.

[13] S.G. Ahmed, "Numerical solution for advection-diffusion equation using collocation based new radial basis functions", *Indian J. Pure Appl. Math.,* vol. II, no. II, 2015.

[14] S. G. Ahmed, "An integral method to solve phase-change problems with/without mushy zones", *Am. J. Appl. Math,* vol. II, no. I, 2015.

[15] S. G. Ahmed, "Mesh-Free Collocation Based for Phase Transformations in Alloys", *Journal of Applied Mathematics and Mechanic,* vol. II, no. I, 2015.

[16] S.G. Ahmed, "A numerical method based boundary integral equation to solve multiphase moving boundary problems", *British Journal of Mathematics & Computer Science,* vol. 15, no. 1, pp. 1-10, 2016.

[17] S. G. Ahmed, "Numerical solution of moving boundary problems using a new hybrid grid and meshless method", *Int. J. of Engineering Analysis with Boundary Elements,* 2019.

[18] M. Lellis Thivagar, M. Lellis Thivagar, H. Abdulsattar Abdullah, and S.G. Ahmed, "Conforming dynamics in the metric spaces", *Journal of Information Science and Engineering,* 2019.

[19] DATJA Roberto, "Mathematical modeling and numerical algorithm for diffusion problem over moving domain based direct approach of BEM", *Sci. J.,* Suleyman Demiral, University Bulletin: Kazakh, 2019.

CHAPTER 7

Numerical Results

Abstract: The present chapter is designated to cover overall discussion of different numerical examples, and as we will see, the results are divided into four types.

Keywords: Potential problems, Diffusion problems, Factors affecting accuracy of the boundary element method.

INTRODUCTION

The present chapter consists of three main parts:

1- the ***first part*** focuses on the results of potential problems in the fixed domain. The main object of these results is to show how the boundary element method can be applied and how to run the computer code. Another main objective is to study the main factors that affect the accuracy of the results using the boundary element method [1-10].

2- The ***second part*** deals with the results of diffusion problems in a fixed domain. In this part, we want to focus on the time marching scheme when dealing with time [11-25].

3- The ***third part*** shows the results due to different factors affecting the accuracy of the boundary element method [26-40].

4- The ***fourth part*** focuses on the diffusion problem with moving boundaries which is the main core of the present thesis. In this part, we focus on singularities that appear in space and time and show how to overcome these difficulties [41-59].

We will present the results for potential problems in the fixed domain in detail. In this context, we will solve the Laplace equation over a square domain of side length 1 unit and subject it to the boundary conditions shown in Fig.**(1)**.

S. G. Ahmed

Part (I)

Potential Problems with Fixed Boundaries

Test problem (1)

Solve $\frac{\partial^2 u}{\partial x^2}+\frac{\partial^2 u}{\partial y^2}=0$ over the square shown in Fig. (**2**) with the boundary conditions.

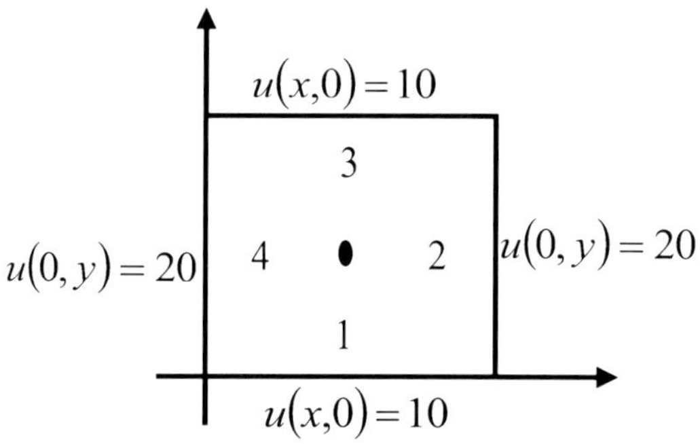

Fig. (7.1). Problem configuration of test problem (1).

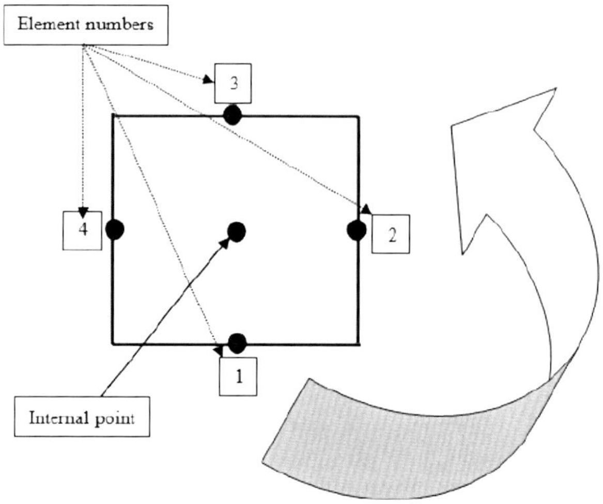

Fig. (7.2). Numbering of elements and their direction.

The Steps of Using the Boundary Elements Code are as Follow

1- Choose the element type (constant, linear, or quadratic),

2- Divide the boundary of the domain into a suitable number of elements, taking into consideration the criteria of each type,

3- Numbering the elements in an anticlockwise direction,

4- Determine the coordinates of the extreme points of each element,

5- Determine the coordinate(s) of the internal points according to the necessity required,

6- Run the code.

The input and output screens from the code are shown below.

Input Data in the Code

Data

Number of Boundary Elements = 4

Number of Internal Points Where the Function is Calculated = 1

Coordinates of the Extreme Points of the Boundary Elements

Point	X	Y
1	0.00000E+00	0.00000E+00
2	0.10000E+01	0.00000E+00
3	0.10000E+01	0.10000E+01
4	0.00000E+00	0.10000E+01

Boundary Conditions

Node	Code	Prescribed Value
1	0	0.10000E+02
2	1	0.20000E+02

3	0	0.10000E+02
4	1	0.20000E+02

Results

Boundary Nodes

X	Y	POTENTIAL	POTENTIAL DERIVATIVE
0.50000E+00	0.00000E+00	0.10000E+02	-0.19990E+02
0.10000E+01	0.50000E+00	0.18893E+02	0.20000E+02
0.50000E+00	0.10000E+01	0.10000E+02	-0.19990E+02
0.00000E+00	0.50000E+00	0.18893E+02	0.20000E+02

Internal Points

X	Y	POTENTIAL
0.50000E+00	0.50000E+00	0.14428E+02

Part (II)

Factors Affecting The B.E.M.

Factors Affecting the Accuracy of B.E.M.

The results due to the boundary element method are affected by four major factors. These factors can be summarized as shown in the following Fig. **(7.3)**. We investigate the direct effect of these factors by solving the test example below.

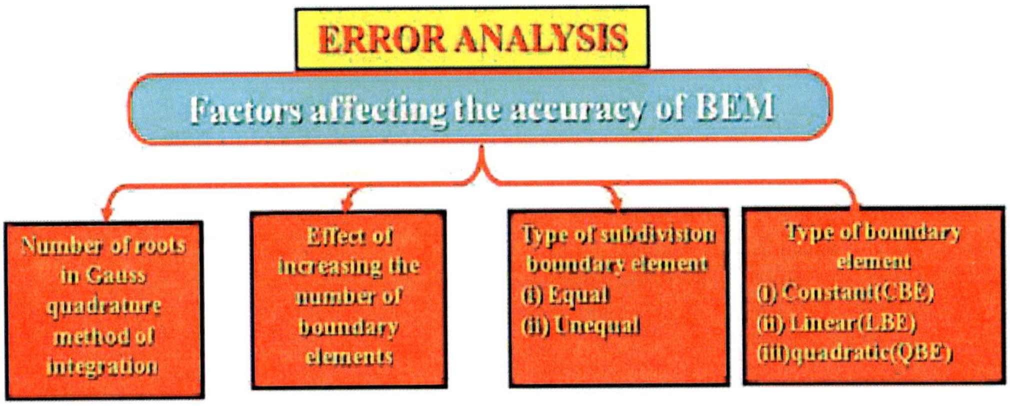

Fig. (7.3). Factors affecting the results of B.E.M.

Test Problem (2)

The parameters affecting the accuracy of the boundary element method are tested on a rectangle of dimensions (1×2) regulated by the Laplace equation Fig. (**7.4**).

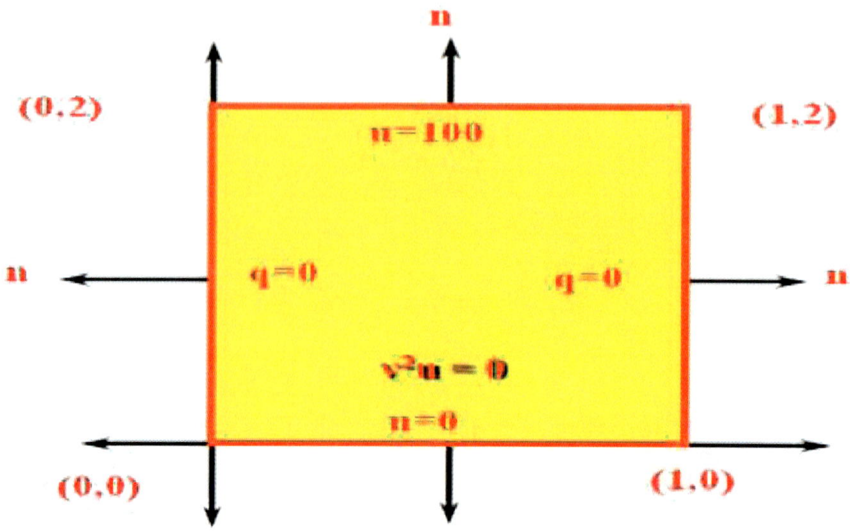

Fig. (7.4). Problem configuration of test problem (2).

In the following, we will show each factor table and figure. The table id for the computed results compared with other solutions and the corresponding errors. The figure summarizes these results in one graph for each factor.

Effect of Gauss Quadrature points

The first and main factor affecting the accuracy of the boundary element method is the number of Gauss quadrature points. The main idea is to carry the integration numerically by dividing the element containing the field point into a suitable number of Gauss quadrature points. Herein, three types of Gauss quadrature points are used to study their direct effect on the accuracy of the results. The following Table (**1**) shows the results of the present example due to the direct boundary element method. These results are compared with the exact analytical solution for different Gaussian quadrature points and the corresponding error at different domain points (Fig. **7.5**).

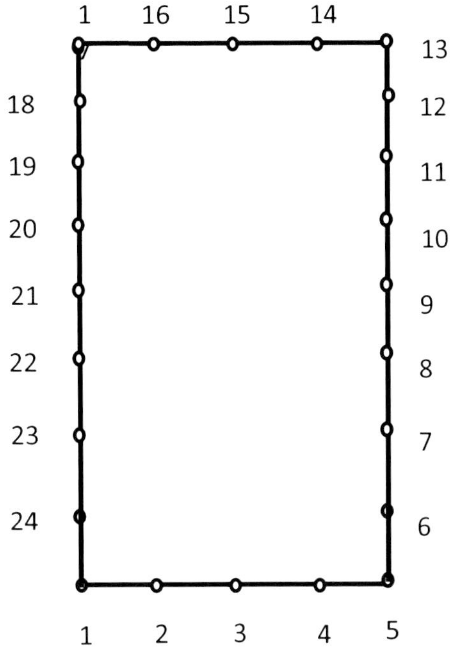

Fig. (7.5). Domain discretization of test problem (2).

Table 1. Effect of Gauss quadrature points

Coordinates (x,y)	**Exact u**	**Using G.I. = 4**	**Relative Error**	**U using G.I.=8**	**Relative Error**	**U using G.I.=12**	**Relative Error.**
1.00 , 0.25	12.5	11.832	0.053	11.838	0.052	11.838	0.052
1 , 0.75	37.5	37.275	0.0059	37.288	0.003	37.288	0.003
1.00,1.25	62.5	62.705	.0032	62.702	0.003	62.702	0.003
1.00,1.75	87.5	88.149	0.0074	88.126	0.0071	88.126	0.0071
0.00,1.75	87.5	88.149	0.0074	88.126	0.0071	88.126	0.0071

0.00,1.25	62.5	62.705	.0032	62.702	0.003	62.702	0.003
0 , 0.75	37.5	37.275	0.0059	37.288	0.003	37.288	0.003
0.00 , 0.25	12.5	11.832	0.053	11.838	0.052	11.838	0.052
0.25 , 0.5	25	24.712	0.011	24.722	0.011	24.722	0.011
0.75 , 0.5	25	24.712	0.011	24.722	0.011	24.722	0.011
0.25 , 1.00	50	49.986	0.0002	50	0	50	0
0.75 , 1.00	50	49.986	0.0002	50	0	50	0
0.25 , 1.5	75	75.264	0.0035	75.218	0.002	75.218	0.002
0.75 , 1.5	75	75.264	0.0035	75.218	0.002	75.218	0.002

Effect of Number of Boundary Elements

The second factor that directly affects the accuracy of the boundary elements methods is the number of boundary elements used. Table (**2**) shows the results of the present example due to the direct boundary element method using different numbers of constant boundary elements. These results are compared with the exact analytical solution at different domain points.

Table 2. Effect of number of boundary elements

(x,y)	Exact potential	B.E.=12	**R. E.**	B.E.=2 4	**R. E.**	B.E.= 30	**R. E.**
1.00,	12.5	12.553	**4.240e-**	12.524	**1.920e-3**	12.50	**2.400e-4**
1.00,	37.5	37.546	**1.227e-**	37.534	**9.067e-4**	37.50	**2.667e-5**
1.00,1.2	62.5	62.562	**9.920e-**	62.541	**6.560e-4**	62.50	**8.000e-5**
1.00,1.7	87.5	87.535	**4.000e-**	87.521	**2.400e-4**	87.50	**1.142e-3**

0.75,	25	24.981	**7.600e-**	25.012	**4.800e-4**	25.00	**4.000e-5**

The first column contains some selected points over the domain of interest, while the second column contains the exact potential from the analytical solution. Column number four contains the relative error between exact and the present solution corresponding to using twelve constant elements. It is clear from columns four, six, and eight that the relative error decreases by increasing the number of boundary elements. On the other hand, no one can make use of this advantage without first considering it. In some engineering applications, the increasing process of boundary elements has the opposite effect on the stability and convergence of the computed results. Therefore, care should be taken when dealing with the number of boundary elements as a tool for improving the required results (Figs. **7.6-7.9**).

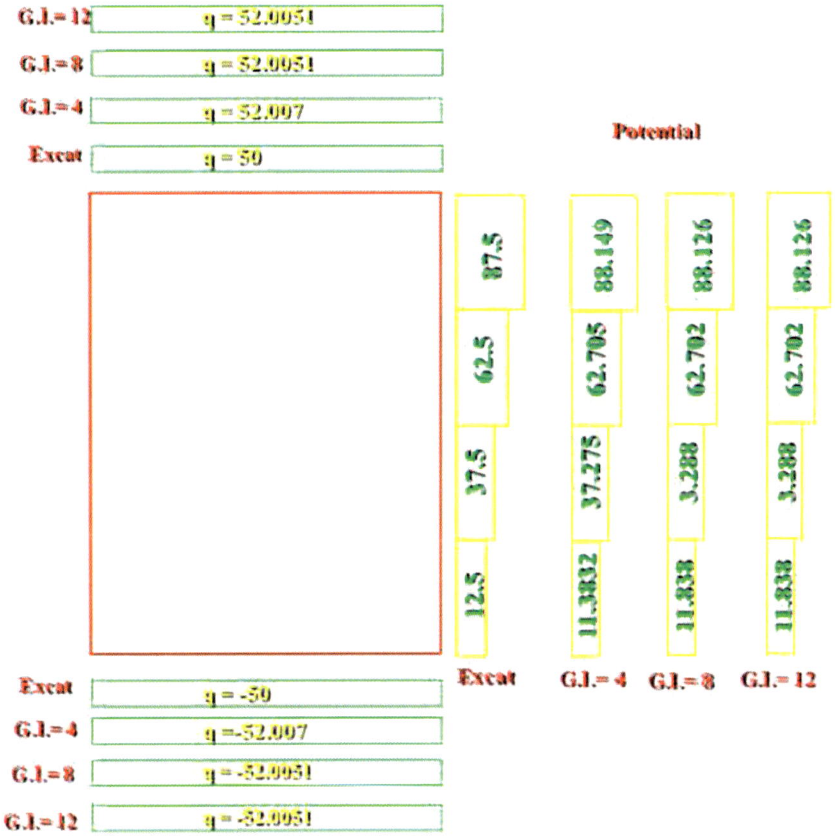

Fig. (7.6). Effect of Gauss Quadrature points.

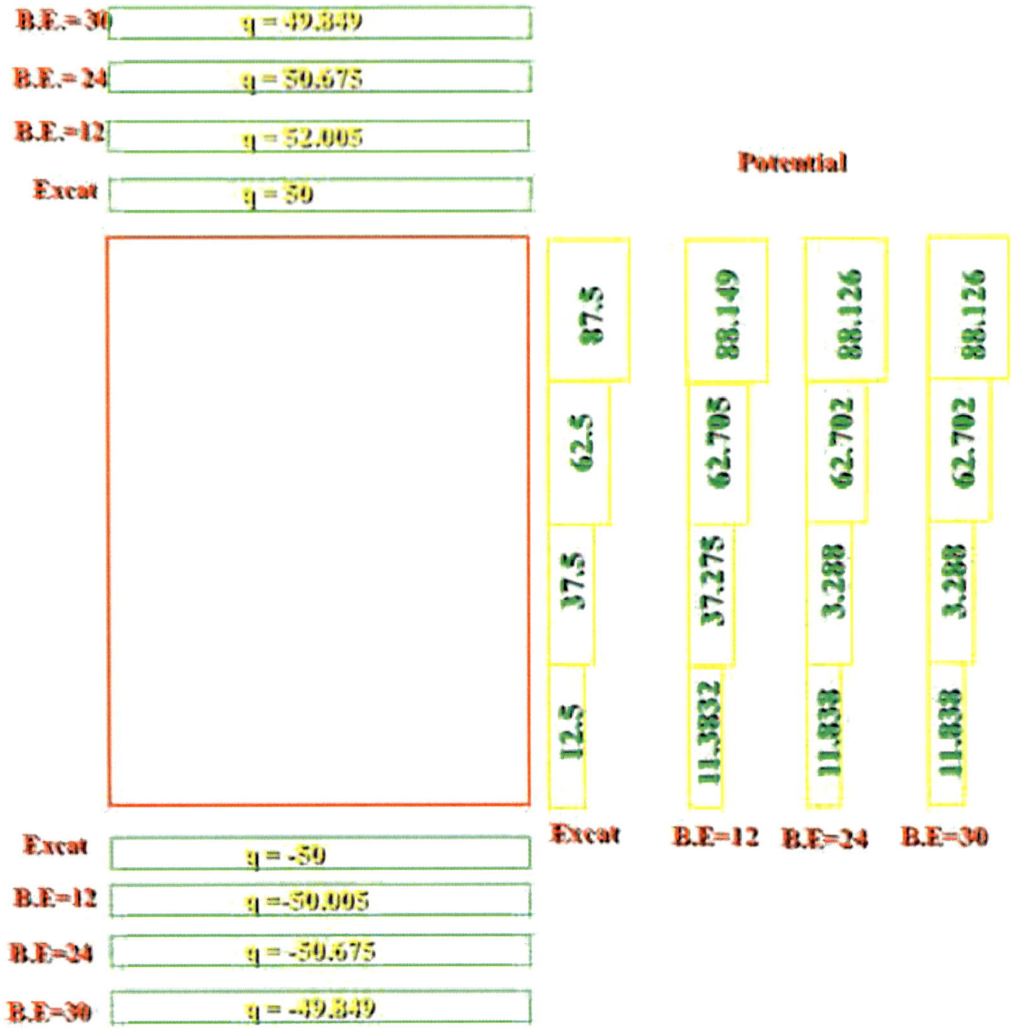

Fig. (7.7). Effect of number of boundary element.

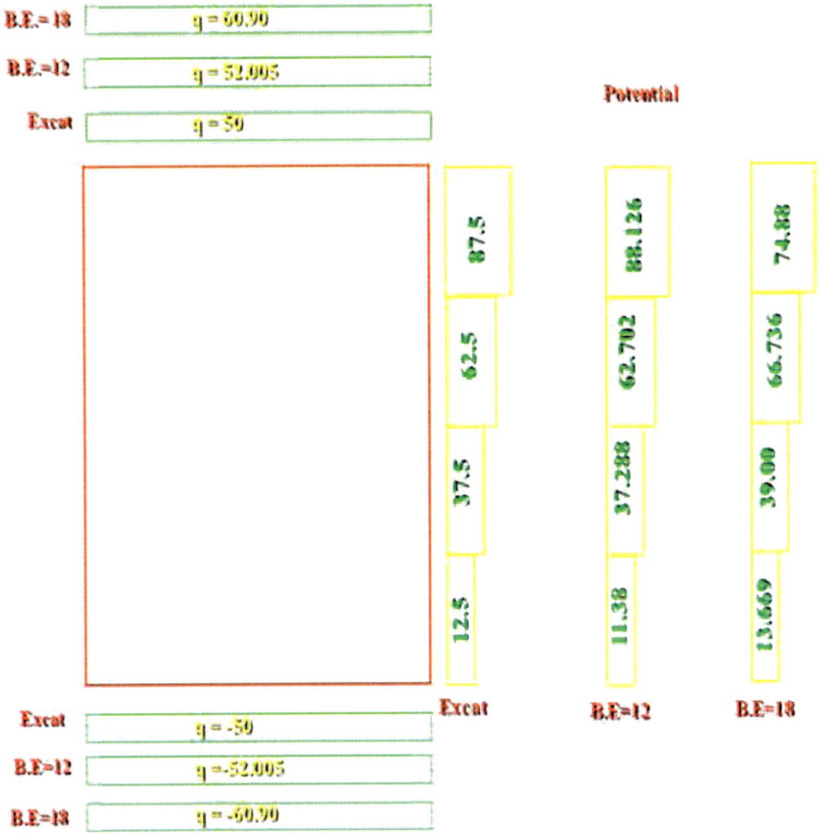

Fig. (7.8). Effect of uniformity of the grid.

Effect of Uniformity of the Grid

Coordinates	U	B.E.=12	Relative	B.E.=18	Relative
1.00 , 0.25	12.5	11.838	0.052	13.669	0.093
1 , 0.75	37.5	37.288	0.003	39.00	0.04
1.00,1.25	62.5	62.702	0.003	66.736	0.11
1.00,1.75	87.5	88.126	0.0071	74.88	0.144

0.00,1.75	87.5	88.126	0.0071	74.88	0.144
0.00,1.25	62.5	62.702	0.003	66.736	0.11
0 , 0.75	37.5	37.288	0.003	39.00	0.04
0.00 , 0.25	12.5	11.838	0.052	13.669	0.093
0.25 , 0.5	25	24.722	0.011	26.846	0.073
0.75 , 0.5	25	24.722	0.011	25.905	0.073
0.25 , 1.00	50	50	0	45.158	0.096
0.75 , 1.00	50	50	0	50.270	0.005
0.25 , 1.5	75	75.218	0.002	65.212	0.13
0.75 , 1.5	75	75.218	0.002	77.059	0.027

Effect of Type of Boundary Elements

Coordinates (x,y)	U Exact	U using CBEM	Relative Error	U using LBEM	Relative Error	U using QBEM	Relative Error
1.00 , 0.25	12.5	12.296	.016	12.5	0	12.5	0
1 , 0.75	37.5	37.445	0.001	37.5	0	37.5	0
1.00,1.25	62.5	62.555	0.0008	62.5	0	62.5	0
1.00,1.75	87.5	87.704	0.002	87.5	0	87.5	0
0.00,1.75	87.5	87.704	0.002	87.5	0	87.5	0
0.00,1.25	62.5	62.555	0.0008	62.5	0	62.5	0

0 , 0.75	37.5	37.445	0.001	37.5	0	37.5	0
0.00 , 0.25	12.5	12.296	.016	12.5	0	12.5	0
0.25 , 0.5	25	24.922	0.003	25	0	25	0
0.75 , 0.5	25	24.922	0.003	25	0	25	0
0.25 , 1.00	50	50	0.00	50	0	50	0
0.75 , 1.00	50	50	0.00	50	0	50	0
0.25 , 1.5	75	75.078	0.001	75	0	75	0
0.75 , 1.5	75	75.078	0.001	75	0	75	0

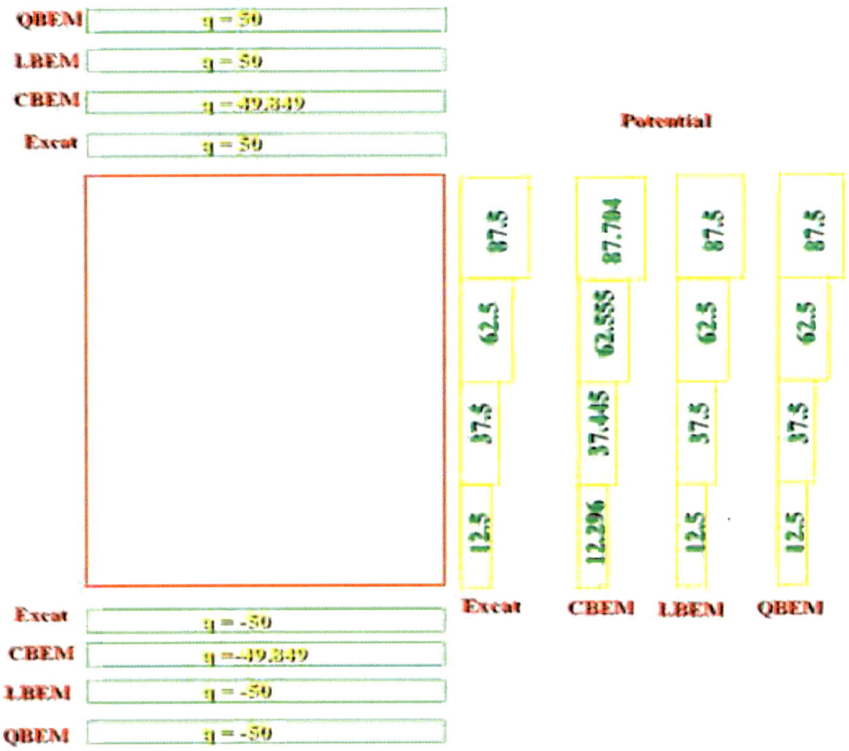

Fig. (7.9). Effect of type of boundary elements.

Part (III)

Diffusion Problems with Fixed Boundaries

ELEMENTARY EXAMPLE (1)

ONE-DIMENSIONAL, ONE-PHASE DIFFUSION EQUATION

In this section, an elementary example will be introduced in detail, these details are the preliminary stage of designing code for these types of problems. Consider a bar of length ℓ subjected to the uniform initial temperature $u_0 = 1.0$. At time $t > 0$, the ends $x = 0$ and $x = \ell$ are kept at zero temperature. It is required to find flux at both ends using the boundary integral equation.

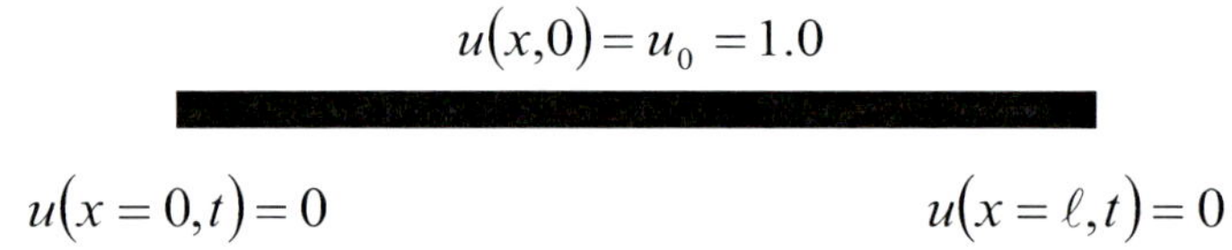

Fig. (7.10). One-dimensional diffusion problem domain.

Let us start the solution by writing the general boundary integral equation as follows:

$$u(\xi,t_n) = \frac{u(x,0)}{2}\left[erf\left(\frac{\xi}{2\sqrt{kt_n}}\right) - erf\left(\frac{\xi - \ell}{2\sqrt{kt_n}}\right)\right]$$

$$+\sum_{i=1}^{n}\left[q_{x=\ell}\int_{t_{i-1}}^{t_i} u^* dt - u_{x=\ell}\int_{t_{i-1}}^{t_i} q^* dt \right]_{x=\ell} \qquad \textbf{(PART III-1)}$$

$$-\sum_{i=1}^{n}\left[q_{x=0}\int_{t_{i-1}}^{t_i} u^* dt - u_{x=0}\int_{t_{i-1}}^{t_i} q^* dt \right]_{x=0}$$

The second step is to substitute by the numerical values of the associated boundary conditions; therefore, equation (PART III-1) takes the following simplified form:

$$u(\xi,t_n)=\frac{u(x,0)}{2}\left[erf\left(\frac{\xi}{2\sqrt{kt_n}}\right)-erf\left(\frac{\xi-\ell}{2\sqrt{kt_n}}\right)\right]$$
$$+\sum_{i=1}^{n-1}\left[q_{x=\ell}\int_{t_{i-1}}^{t_i}u^*\,dt-q_{x=0}\int_{t_{i-1}}^{t_i}u^*\,dt\right] \qquad \textbf{(PART III-2)}$$

Before introducing the details of the computation procedure, let us make a general overview of the equation (PART III-2) at three successive time steps to build a concept of computation for the readers.

At the first time step

$$u(\xi,t_1)=\frac{u_0}{2}\left[erf\left(\frac{\xi}{2\sqrt{kt_1}}\right)-erf\left(\frac{\xi-\ell}{2\sqrt{kt_1}}\right)\right]$$
$$+\left[q_{x=\ell}\int_{t_0}^{t_1}u^*\,dt-q_{x=0}\int_{t_0}^{t_1}u^*\,dt\right] \qquad \textbf{(PART III-3-1)}$$

At the second time step

$$u(\xi,t_2)=\frac{u_0}{2}\left[erf\left(\frac{\xi}{2\sqrt{kt_2}}\right)-erf\left(\frac{\xi-\ell}{2\sqrt{kt_2}}\right)\right]$$
$$+\left[q_{x=\ell}\int_{t_0}^{t_2}u^*\,dt-q_{x=0}\int_{t_0}^{t_2}u^*\,dt\right] \qquad \textbf{(PART III-3-2)}$$

(5) The integral $\int_{t_0}^{t_2}u^*\,dt=\int_{t_0}^{t_1}u^*\,dt+\int_{t_1}^{t_2}u^*\,dt$

(6) Limits of integrations as follow:

$$t_0 \Leftrightarrow a_0 = \frac{r^2}{4(t_2 - t_0)} \quad \& \quad t_1 \Leftrightarrow a_1 = \frac{r^2}{4(t_2 - t_1)}$$

$$t_2 \Leftrightarrow a_2 = \frac{r^2}{4(t_2 - t_2)} \Leftrightarrow \text{Time Singularit y}$$

At the third time step

$$u(\xi, t_3) = \frac{u_0}{2}\left[erf\left(\frac{\xi}{2\sqrt{kt_3}}\right) - erf\left(\frac{\xi - \ell}{2\sqrt{kt_3}}\right)\right]$$

$$+\left[q_{x=\ell} \int_{t_0}^{t_3} u^* dt - q_{x=0} \int_{t_0}^{t_3} u^* dt \right]$$

(PART III-3-3)

(7) The integral $\int_{t_0}^{t_3} u^* dt = \int_{t_0}^{t_1} u^* dt + \int_{t_1}^{t_2} u^* dt + \int_{t_2}^{t_3} u^* dt$

(8) Limits of integrations as follow:

$$t_0 \Leftrightarrow a_0 = \frac{r^2}{4(t_3 - t_0)} \quad \& \quad t_1 \Leftrightarrow a_1 = \frac{r^2}{4(t_3 - t_1)}$$

$$t_2 \Leftrightarrow a_2 = \frac{r^2}{4(t_3 - t_2)} \quad \& \quad t_3 \Leftrightarrow a_3 = \frac{r^2}{4(t_3 - t_3)} \Leftrightarrow \text{Time Singularit y}$$

First Stage: Evaluation for the first time step

In this section, we will introduce a detailed numerical computation for the elementary example defined before. We will start the computation by evaluating the time integrals. To do so, let us start the computation by re-writing the boundary integral equation as follows:

$$u(\xi,t_n)=\frac{u(x,0)}{2}\left[erf\left(\frac{\xi}{2\sqrt{t_n}}\right)-erf\left(\frac{\xi-\ell}{2\sqrt{t_n}}\right)\right]$$
$$+\sum_{i=1}^{n-1}\left[q_{x=\ell}\int_{t_{i-1}}^{t_i}u^{*}dt-q_{x=0}\int_{t_{i-1}}^{t_i}u^{*}dt\right] \quad \textbf{(FS-E1STTS-1)}$$

At the first time step $t_n = t_1$

The corresponding boundary integral equation takes the form:

$$u(\xi,t_1)=\frac{u_0}{2}\left[erf\left(\frac{\xi}{2\sqrt{t_1}}\right)-erf\left(\frac{\xi-\ell}{2\sqrt{t_1}}\right)\right]$$
$$+\left[q_{x=\ell}\int_{t_0}^{t_1}u^{*}dt-q_{x=0}\int_{t_0}^{t_1}u^{*}dt\right] \quad \textbf{(FS-E1STTS-2)}$$

The limits of integration take the following form:

$$a_0=\frac{r^2}{4(t_1-t_0)}=\frac{r^2}{4\Delta t}\ \&\ a_1=\frac{r^2}{4(t_1-t_i)}=\infty \ \text{Time singularit y} \quad \textbf{(FS-E1STTS-3)}$$

The next step is to apply the boundary integral equation (FS-E1STTS-1) at the two end points of the domain, and each point will be considered as a source point ξ.

Source point at $x=0$ **this corresponds to** $\xi=0$

The boundary integral equation (1) takes the form:

$$u(0,t_1)=\frac{u_0}{2}\left[erf(0)-erf\left(\frac{-\ell}{2\sqrt{t_1}}\right)\right]+\left[q_{x=\ell}\int_{t_0}^{t_1}u^{*}dt-q_{x=0}\int_{t_0}^{t_1}u^{*}dt\right] \quad \textbf{(FS-E1STTS-4)}$$

Note that $u(0,t_1)=0$ is already a boundary condition in our elementary example; therefore equation (FS-E1STTS-4) becomes:

$$-\frac{u_0}{2}erf\left(\frac{-\ell}{2\sqrt{t_1}}\right)=q_{x=\ell}\left(\int_{t_0}^{t_1}u^*dt\right)_{x=\ell}-q_{x=0}\left(\int_{t_0}^{t_1}u^*dt\right)_{x=0} \quad \textbf{(FS-E1STTS-5)}$$

Source point at $x=\ell$ **this corresponds to** $\xi=\ell$

The boundary integral equation (FS-E1STTS-1) takes the form:

$$u(\ell,t_1)=\frac{u_0}{2}\left[erf\left(\frac{\ell}{2\sqrt{t_1}}\right)-erf(0)\right]+\left[q_{x=\ell}\int_{t_0}^{t_1}u^*dt-q_{x=0}\int_{t_0}^{t_1}u^*dt\right] \quad \textbf{(FS-E1STTS-6)}$$

Note that $u(\ell,t_1)=0$ is already boundary condition in our elementary example, therefore equation (6) becomes:

$$-\frac{u_0}{2}erf\left(\frac{\ell}{2\sqrt{t_1}}\right)=q_{x=\ell}\left(\int_{t_0}^{t_1}u^*dt\right)_{x=\ell}-q_{x=0}\left(\int_{t_0}^{t_1}u^*dt\right)_{x=0} \quad \textbf{(FS-E1STTS-7)}$$

Now equations (FS-E1STTS-5) and (FS-E1STTS-7) can be written in a matrix form as follows:

$$\begin{bmatrix}-u_{11} & u_{12}\\ -u_{21} & u_{22}\end{bmatrix}\begin{bmatrix}q_0\\ q_1\end{bmatrix}=\begin{bmatrix}b_1\\ b_2\end{bmatrix} \quad \textbf{(FS-E1STTS-8)}$$

Note:

The negative sign that appears in the first column in the u_{ij} matrix comes from the sign of q_0 in equations (FS-E1STTS-5) and (FS-E1STTS-7). The next step is the numerical calculation of the elements u_{ij}. Let us start the calculation by the element u_{11}.

$u_{11}\Leftrightarrow$ Source $\xi=0$ and Field $x=0$

In this case:

$$r = 0 \text{ because } r = \xi - x = 0$$

$$a_0 = \frac{r^2}{4(t_1 - t_0)} = 0 \text{ because r equals zero } \& \ a_1 = \frac{r^2}{4(t_1 - t_1)} = \frac{0}{0} \qquad \textbf{(FS-E1STTS-9)}$$

$$u_{11} = \int_{t_0}^{t_1} u^*(\xi = 0, x = 0; t_1, t)dt = \frac{r}{2\sqrt{\pi}} \left\{ \left(\frac{e^{-a_0}}{\sqrt{a_0}} - \frac{e^{-a_1}}{\sqrt{a_1}} \right) + \sqrt{\pi}\left(erf\left(\sqrt{a_0}\right) - erf\left(\sqrt{a_1}\right)\right) \right\}$$

(FS-E1STTS-10)

Substituting from equation (FS-E1STTS-9) into equation (FS-E1STTS-10), we get:

$$u_{11} = \int_{t_0}^{t_1} u^*(\xi = 0, x = 0; t_1, t)dt = \frac{r}{2\sqrt{\pi}} \left\{ \left(\frac{e^{-a_0}}{\sqrt{\frac{r^2}{4(t_1 - t_0)}}} - \frac{e^{-a_1}}{\sqrt{\frac{r^2}{4(t_1 - t_1)}}} \right) \right\}$$

$$+ \frac{r}{2\sqrt{\pi}} \left\{ \sqrt{\pi} \left(erf\left(\sqrt{\frac{r^2}{4(t_1 - t_0)}} \right) - erf\left(\sqrt{\frac{r^2}{4(t_1 - t_1)}} \right) \right) \right\}$$

(FS-E1STTS-11)

Simplify equation (FS-E1STTS-11), leads to:

$$u_{11} = \int_{t_0}^{t_1} u^*(\xi = 0, x = 0; t_1, t)dt =$$

$$= \frac{r}{2\sqrt{\pi}} \left\{ \left(\frac{2e^{-a_0}\sqrt{(t_1 - t_0)}}{r} - \frac{2e^{-a_1}\sqrt{(t_1 - t_1)}}{r} \right) \right\} + \frac{r}{2} erf\left(\sqrt{a_1}\right) \qquad \textbf{(FS-E1STTS-12)}$$

Simplify equation (FS-E1STTS-12) once more, it becomes:

$$u_{11} = \int_{t_0}^{t_1} u^*(\xi = 0, x = 0; t_1, t)dt = \frac{1}{\sqrt{\pi}}\left\{\sqrt{(t_1 - t_0)}\right\} \qquad \textbf{(FS-E1STTS-13)}$$

$u_{12} \Leftrightarrow$ Source $\xi = 0$ and Field $x = \ell$

In this case:

$$r \neq 0 \;\&\; a_0 = \frac{r^2}{4(t_1 - t_0)} \;\&\; a_1 = \frac{r^2}{4(t_1 - t_1)} \qquad \textbf{(FS-E1STTS-14)}$$

$$u_{12} = \int_{t_0}^{t_1} u^*(\xi = 0, x = \ell; t_1, t)dt =$$

$$= \frac{r}{2\sqrt{\pi}}\left\{\left(\frac{e^{-a_0}}{\sqrt{a_0}} - \frac{e^{-a_1}}{\sqrt{a_1}}\right) + \sqrt{\pi}\left(erf\left(\sqrt{a_0}\right) - erf\left(\sqrt{a_1}\right)\right)\right\} \qquad \textbf{(FS-E1STTS-15)}$$

Substituting from equation (FS-E1STTS-14) into equation (FS-E1STTS-15), we get:

$$u_{12} = \int_{t_0}^{t_1} u^*(\xi = 0, x = \ell; t_1, t)dt = \frac{r}{2\sqrt{\pi}}\left\{\left(\frac{e^{-a_0}}{\sqrt{\frac{r^2}{4(t_1 - t_0)}}} - \frac{e^{-a_1}}{\sqrt{\frac{r^2}{4(t_1 - t_1)}}}\right)\right\} \qquad \textbf{(FS-E1STTS-16)}$$

$$+ \frac{r}{2\sqrt{\pi}}\left\{\sqrt{\pi}\left(erf\left(\sqrt{\frac{r^2}{4(t_1 - t_0)}}\right) - erf\left(\sqrt{\frac{r^2}{4(t_1 - t_1)}}\right)\right)\right\}$$

Simplify equation (FS-E1STTS-16), leads to:

$$u_{12} = \int_{t_0}^{t_1} u^*(\xi = 0, x = \ell; t_1, t)dt = \frac{1}{\sqrt{\pi}}\left\{e^{-a_0}\sqrt{(t_1 - t_0)}\right\} + \frac{r}{2}\left\{erf\left(\sqrt{a_0}\right) - erf\left(\sqrt{\infty}\right)\right\}$$

(FS-E1STTS-17)

Simplify equation (FS-E1STTS-17) once more, it becomes:

$$u_{12} = \int_{t_0}^{t_1} u^*(\xi = 0, x = \ell; t_1, t)dt = \frac{1}{\sqrt{\pi}}\left\{e^{-a_0}\sqrt{(t_1 - t_0)}\right\} + \frac{r}{2}\left\{erf\left(\sqrt{a_0}\right) - 1\right\}$$

(FS-E1STTS-18)

$u_{21} \Leftrightarrow$ Source $\xi = \ell$ and Field $x = 0$

In this case:

$$r \neq 0 \,\&\, a_0 = \frac{r^2}{4(t_1 - t_0)} \,\&\, a_1 = \frac{r^2}{4(t_1 - t_1)}$$ **(FS-E1STTS-19)**

$$u_{21} = \int_{t_0}^{t_1} u^*(\xi = \ell, x = 0; t_1, t)dt = \frac{r}{2\sqrt{\pi}}\left\{\left(\frac{e^{-a_0}}{\sqrt{a_0}} - \frac{e^{-a_1}}{\sqrt{a_1}}\right) + \sqrt{\pi}\left(erf\left(\sqrt{a_0}\right) - erf\left(\sqrt{a_1}\right)\right)\right\}$$

(FS-E1STTS-20)

Substituting from equation (FS-E1STTS-19) into equation (FS-E1STTS-20), we get:

$$u_{22} = \int_{t_0}^{t_1} u^*(\xi = \ell, x = 0; t_1, t)dt = \frac{r}{2\sqrt{\pi}}\left\{\left(\frac{e^{-a_0}}{\sqrt{\frac{r^2}{4(t_1 - t_0)}}} - \frac{e^{-a_1}}{\sqrt{\frac{r^2}{4(t_1 - t_1)}}}\right)\right\}$$

$$+ \frac{r}{2\sqrt{\pi}}\left\{\sqrt{\pi}\left(erf\left(\sqrt{\frac{r^2}{4(t_1 - t_0)}}\right) - erf\left(\sqrt{\frac{r^2}{4(t_1 - t_1)}}\right)\right)\right\}$$

(FS-E1STTS-21)

Simplify equation (FS-E1STTS-21), leads to:

$$u_{21} = \int_{t_0}^{t_1} u^*(\xi = \ell, x = 0; t_1, t)dt = \frac{1}{\sqrt{\pi}}\left\{e^{-a_0}\sqrt{(t_1 - t_0)}\right\} + \frac{r}{2}\left\{erf\left(\sqrt{a_0}\right) - erf\left(\sqrt{\infty}\right)\right\}$$

(FS-E1STTS-22)

Simplify equation (FS-E1STTS-22) once more, it becomes:

$$u_{21} = \int_{t_0}^{t_1} u^*(\xi = 0, x = \ell; t_1, t)dt = \frac{1}{\sqrt{\pi}}\left\{e^{-a_0}\sqrt{(t_1 - t_0)}\right\} + \frac{r}{2}\left\{erf\left(\sqrt{a_0}\right) - 1\right\}$$

(FS-E1STTS-23)

$u_{22} \Leftrightarrow$ Source $\xi = \ell$ and Field $x = \ell$

In this case:

$$r = \xi - x = 0 \,\&\, a_0 = \frac{r^2}{4(t_1 - t_0)} = 0 \,\&\, a_1 = \frac{r^2}{4(t_1 - t_1)} = \frac{0}{0} \qquad \textbf{(FS-E1STTS-24)}$$

$$u_{22} = \int_{t_0}^{t_1} u^*(\xi = \ell, x = 0; t_1, t)dt = \frac{r}{2\sqrt{\pi}}\left\{\left(\frac{e^{-a_0}}{\sqrt{a_0}} - \frac{e^{-a_1}}{\sqrt{a_1}}\right) + \sqrt{\pi}\left(erf\left(\sqrt{a_0}\right) - erf\left(\sqrt{a_1}\right)\right)\right\}$$

(FS-E1STTS-25)

Substituting from equation (FS-E1STTS-24) into equation (FS-E1STTS-25), we get:

$$u_{22} = \int_{t_0}^{t_1} u^*(\xi = \ell, x = 0; t_1, t)dt = \frac{r}{2\sqrt{\pi}}\left\{\left(\frac{e^{-a_0}}{\sqrt{\frac{r^2}{4(t_1 - t_0)}}} - \frac{e^{-a_1}}{\sqrt{\frac{r^2}{4(t_1 - t_1)}}}\right)\right\}$$

$$+ \frac{r}{2\sqrt{\pi}}\left\{\sqrt{\pi}\left(erf\left(\sqrt{\frac{r^2}{4(t_1 - t_0)}}\right) - erf\left(\sqrt{\frac{r^2}{4(t_1 - t_1)}}\right)\right)\right\} \qquad \textbf{(FS-E1STTS-26)}$$

Simplify equation (FS-E1STTS-26), leads to:

$$u_{22}=\int_{t_0}^{t_1}u^*\left(\xi=\ell,x=0;t_1,t\right)dt=\frac{1}{\sqrt{\pi}}\left\{e^{-a_0}\sqrt{\left(t_1-t_0\right)}\right\}+\frac{r}{2}\left\{erf\left(\sqrt{a_0}\right)-erf\left(\sqrt{\infty}\right)\right\}$$

(FS-E1STTS-27)

Simplify equation (FS-E1STTS-27) once more, it becomes:

$$u_{22}=\int_{t_0}^{t_1}u^*\left(\xi=0,x=\ell;t_1,t\right)dt=\frac{1}{\sqrt{\pi}}\left\{\sqrt{\left(t_1-t_0\right)}\right\} \quad \textbf{(FS-E1STTS-28)}$$

Now the system of equations for the first time step is defined by equations (FS-E1STTS-13), (FS-E1STTS-18), (FS-E1STTS-23), and (FS-E1STTS-28).

We will take the time step $\Delta t=0.0025$ and the initial time equals zero. Now let us start computing the elements u_{11}, u_{12}, u_{21} and u_{22}.

Evaluation u_{11}

$$u_{11}=\int_{t_0}^{t_1}u^*\left(\xi=0,x=0;t_1,t\right)dt=\frac{1}{\sqrt{\pi}}\left\{\sqrt{\left(t_1-t_0\right)}\right\}$$

$$u_{11}=\frac{1}{\sqrt{\pi}}\sqrt{0.0025-0}=-0.0282094$$

Evaluation u_{12}

$$u_{12}=\int_{t_0}^{t_1}u^*\left(\xi=0,x=\ell;t_1,t\right)dt=\frac{1}{\sqrt{\pi}}\left\{e^{-a_0}\sqrt{\left(t_1-t_0\right)}\right\}+\frac{r}{2}\left\{erf\left(\sqrt{a_0}\right)-1\right\}$$

$$u_{12} = \frac{1}{2\sqrt{\pi}}\left[\frac{e^{-100}}{\sqrt{100}}\right] + \frac{1}{2}\left[erf\left(\sqrt{100}\right) - 1\right] = 1.04941405 \times 10^{-45}$$

Evaluation u_{21}

$$u_{21} = \int_{t_0}^{t_1} u^*(\xi = 0, x = \ell; t_1, t)dt = \frac{1}{\sqrt{\pi}}\left\{e^{-a_0}\sqrt{(t_1 - t_0)}\right\} + \frac{r}{2}\left\{erf\left(\sqrt{a_0}\right) - 1\right\}$$

$$u_{12} = 1.04941405 \times 10^{-45}$$

Evaluation u_{22}

$$u_{22} = \int_{t_0}^{t_1} u^*(\xi = 0, x = \ell; t_1, t)dt = \frac{1}{\sqrt{\pi}}\left\{\sqrt{(t_1 - t_0)}\right\}$$

$$u_{22} = 0.0282094$$

Evaluation b_1

$$b_1 = -\frac{u_0}{2} erf\left(\frac{-\ell}{2\sqrt{t_1}}\right) = -\frac{u_0}{2} erf\left(\frac{-\ell}{2\sqrt{t_0 + \Delta t}}\right) \cong -0.5$$

Evaluation b_2

$$b_1 = -\frac{u_0}{2} erf\left(\frac{-\ell}{2\sqrt{t_1}}\right) = -\frac{u_0}{2} erf\left(\frac{-\ell}{2\sqrt{t_0 + \Delta t}}\right) \cong -0.5$$

Now the system is given by equation (8) takes the following form:

$$\begin{bmatrix} -u_{11} & u_{12} \\ -u_{21} & u_{22} \end{bmatrix} \begin{bmatrix} q_0 \\ q_1 \end{bmatrix} = \begin{bmatrix} b_1 \\ b_2 \end{bmatrix}$$

$\Rightarrow$

$$\begin{bmatrix} -0.0282094 & 1.04941405 \times 10^{-45} \\ -1.04941405 \times 10^{-45} & 0.0282094 \end{bmatrix} \begin{bmatrix} q_0 \\ q_1 \end{bmatrix} = \begin{bmatrix} -0.5 \\ -0.5 \end{bmatrix}$$

Numerical solution leads to:

$$q_0 = +17.72458933$$
$$q_\ell = -17.72458933$$

Second Stage: Evaluation for the second time step

The next stage is to extend the computation procedure to compute the flux for the elementary example at the second time step. Let us start this stage by re-writing the boundary integral equation in its general case for the elementary example:

$$u(\xi,t_n) = \frac{u(x,0)}{2}\left[erf\left(\frac{\xi}{2\sqrt{t_n}}\right) - erf\left(\frac{\xi-\ell}{2\sqrt{t_n}}\right)\right]$$
$$+\sum_{i=1}^{n-1}\left[q_{x=\ell}\left(\int_{t_{i-1}}^{t_i} u^*(\xi,x;t_n,t)dt\right)_{x=\ell} - q_{x=0}\left(\int_{t_{i-1}}^{t_i} u^*(\xi,x;t_n,t)dt\right)_{x=0}\right] \quad \textbf{(SS-E2NTTS-1)}$$

Equation (SS-E2NTTS-1) for the second time step takes the following form:

$$u(\xi,t_2) = \frac{u_0}{2}\left[erf\left(\frac{\xi}{2\sqrt{t_2}}\right) - erf\left(\frac{\xi-\ell}{2\sqrt{t_2}}\right)\right]$$
$$+\sum_{i=1}^{2}\left[q_{x=\ell}\left(\int_{t_0}^{t_2} u^*(\xi,\ell;t_2,t)dt\right) - q_{x=0}\left(\int_{t_0}^{t_2} u^*(\xi,0;t_2,t)dt\right)\right] \quad \textbf{(SS-E2NTTS-2)}$$

Re-write equation (SS-E2NTTS-2) in some more detail:

$$u(\xi,t_2)=\frac{u_0}{2}\left[erf\left(\frac{\xi}{2\sqrt{t_2}}\right)-erf\left(\frac{\xi-\ell}{2\sqrt{t_2}}\right)\right]$$

$$+q_{x=\ell}\left(\int_{t_0}^{t_1}u^*(\xi,\ell;t_2,t)dt\right)+q_{x=\ell}\left(\int_{t_1}^{t_2}u^*(\xi,\ell;t_2,t)dt\right) \qquad \textbf{(SS-E2NTTS-3)}$$

$$-q_{x=0}\left(\int_{t_0}^{t_1}u^*(\xi,0;t_2,t)dt\right)-q_{x=0}\left(\int_{t_1}^{t_2}u^*(\xi,0;t_2,t)dt\right)$$

It is known that ξ is the source point, and in our example, it takes two different values $\xi=0$ & $\xi=\ell$ and in both cases, we have zero boundary condition, therefore, $u(\xi,t)=0$ and so equation (SS-E2NTTS-3) takes the following form:

$$-\frac{u_0}{2}\left[erf\left(\frac{\xi}{2\sqrt{t_2}}\right)-erf\left(\frac{\xi-\ell}{2\sqrt{t_2}}\right)\right]$$

$$=q_{x=\ell}\left(\int_{t_0}^{t_1}u^*(\xi,\ell;t_2,t)dt\right)+q_{x=\ell}\left(\int_{t_1}^{t_2}u^*(\xi,\ell;t_2,t)dt\right) \qquad \textbf{(SS-E2NTTS-4)}$$

$$-q_{x=0}\left(\int_{t_0}^{t_1}u^*(\xi,0;t_2,t)dt\right)-q_{x=0}\left(\int_{t_1}^{t_2}u^*(\xi,0;t_2,t)dt\right)$$

Note that $\int_{t_0}^{t_1}u^*(\xi,\ell;t_2,t)dt$ and $\int_{t_0}^{t_1}u^*(\xi,0;t_2,t)dt$ are two known values from the first time step, and so the other two values are unknown and should be calculated at the second time step. The source point in equation (SS-E2NTTS-4) will be applied for our elementary example at two different points and . By applying equation (SS-E2NTTS-4) at these two source points, one gets the following two boundary integral equations:

$$
\begin{aligned}
&-\frac{u_0}{2}\left[-erf\left(\frac{-\ell}{2\sqrt{t_2}}\right)\right] = q_{x=\ell}\left(\int_{t_0}^{t_1} u^*(\xi=0,x=\ell;t_2,t)dt\right) \\
&+q_{x=\ell}\left(\int_{t_1}^{t_2} u^*(\xi=0,x=\ell;t_2,t)dt\right) - q_{x=0}\left(\int_{t_0}^{t_1} u^*(\xi=0,x=0;t_2,t)dt\right) \\
&-q_{x=0}\left(\int_{t_1}^{t_2} u^*(\xi=0,x=0;t_2,t)dt\right)
\end{aligned}
$$

(SS-E2NTTS-5)

And

$$
\begin{aligned}
&-\frac{u_0}{2}\left[erf\left(\frac{\ell}{2\sqrt{t_2}}\right)\right] = q_{x=\ell}\left(\int_{t_0}^{t_1} u^*(\xi=\ell,x=\ell;t_2,t)dt\right) \\
&+q_{x=\ell}\left(\int_{t_1}^{t_2} u^*(\xi=\ell,x=\ell;t_2,t)dt\right) - q_{x=0}\left(\int_{t_0}^{t_1} u^*(\xi=\ell,x=0;t_2,t)dt\right) \\
&-q_{x=0}\left(\int_{t_1}^{t_2} u^*(\xi=\ell,x=0;t_2,t)dt\right)
\end{aligned}
$$

(SS-E2NTTS-6)

Evaluation of the time integrals

At ***first*** time step with a source at $\xi = 0$ and field at $x = 0$

$$(u_{11})_{t=t_1} = \int_{t_0}^{t_1} u^*(\xi=0,x=0;t_2,t)dt$$

$$= \frac{r}{2\sqrt{\pi}}\left\{\left(\frac{e^{-a_0}}{\sqrt{a_0}} - \frac{e^{-a_1}}{\sqrt{a_1}}\right) + \sqrt{\pi}\ erf\left(\sqrt{a_0}\right) - erf\left(\sqrt{a_1}\right)\right\} \qquad \textbf{(SS-E2NTTS-7)}$$

In this case:

$$r = \xi - x = 0 \,\&\, a_0 = \frac{r^2}{4(t_2 - t_0)} \,\&\, a_1 = \frac{r^2}{4(t_2 - t_1)} \tag{SS-E2NTTS-8}$$

Using equation (SS-E2NTTS-8) into equation (SS-E2NTTS-7) and simplify, we get:

$$(u_{11})_{t=t_1} = \int_{t_0}^{t_1} u^*(\xi = 0, x = 0; t_2, t)dt = \frac{1}{\sqrt{\pi}}\left\{\sqrt{(t_2 - t_0)} + \sqrt{(t_2 - t_1)}\right\}$$

(SS-E2NTTS-9)

$$r = \xi - x = 0 \,\&\, a_0 = \frac{r^2}{4(t_2 - t_0)} = 0 \,\&\, a_1 = \frac{r^2}{4(t_2 - t_1)} = 0$$

$$(u_{11})_{t=t_1} = \frac{1}{\sqrt{\pi}}\left\{\sqrt{(t_2 - t_0)} + \sqrt{(t_2 - t_1)}\right\}$$

$$= \frac{1}{\sqrt{\pi}}\left\{\sqrt{(0.005 - 0)} + \sqrt{(0.005 - 0.0025)}\right\} = 0.01168475$$

At ***second*** time step with a source at and field at

$$(u_{11})_{t=t_2} = \int_{t_1}^{t_2} u^*(\xi = 0, x = 0; t_2, t)dt =$$

$$= \frac{r}{2\sqrt{\pi}}\left\{\left(\frac{e^{-a_1}}{\sqrt{a_1}} - \frac{e^{-a_2}}{\sqrt{a_2}}\right) + \sqrt{\pi}\left(erf\left(\sqrt{a_1}\right) - erf\left(\sqrt{a_2}\right)\right)\right\} \tag{SS-E2NTTS-10}$$

In this case:

$$r = \xi - x = 0 \,\&\, a_1 = \frac{r^2}{4(t_2 - t_1)} = 0 \,\&\, a_2 = \frac{r^2}{4(t_2 - t_2)} = \frac{0}{0} \text{.............Singularit y}$$

(SS-E2NTTS-11)

Using equation (SS-E2NTTS-11) into equation (SS-E2NTTS-10) and simplify, we get:

$$(u_{11})_{t=t_2} = \int_{t_1}^{t_2} u^*(\xi = 0, x = 0; t_2, t)dt = \frac{1}{\sqrt{\pi}}\left\{\sqrt{(t_2 - t_1)}\right\} \quad \textbf{(SS-E2NTTS-12)}$$

$$(u_{11})_{t=t_2} = \frac{1}{\sqrt{\pi}}\left\{\sqrt{(t_2 - t_1)}\right\} = \frac{1}{\sqrt{\pi}}\left(\sqrt{0.005 - 0.0025}\right) = 0.02820947$$

At ***first*** time step with a source at and field at $x = \ell$

$$(u_{12})_{t=t_1} = \int_{t_0}^{t_1} u^*(\xi = 0, x = \ell; t_2, t)dt =$$

$$= \frac{r}{2\sqrt{\pi}}\left\{\left(\frac{e^{-a_0}}{\sqrt{a_0}} - \frac{e^{-a_1}}{\sqrt{a_1}}\right) + \sqrt{\pi}\left(erf\left(\sqrt{a_0}\right) - erf\left(\sqrt{a_1}\right)\right)\right\} \quad \textbf{(SS-E2NTTS-13)}$$

In this case:

$$r = \xi - x = -\ell \ \& \ a_0 = \frac{r^2}{4(t_2 - t_0)} \ \& \ a_1 = \frac{r^2}{4(t_2 - t_1)} \quad \textbf{(SS-E2NTTS-14)}$$

$$r = \xi - x = -\ell$$

$$a_0 = \frac{r^2}{4(t_2 - t_0)} = \frac{1}{4(0.005 - 0.000)} = 50$$

$$a_1 = \frac{r^2}{4(t_2 - t_1)} = \frac{1}{4(0.005 - 0.0025)} = 100$$

$$(u_{12})_{t=t_1} = \int_{t_0}^{t_1} u^*(\xi = 0, x = \ell; t_2, t)dt =$$

$$= \frac{1}{2\sqrt{\pi}}\left\{\left(\frac{e^{-50}}{\sqrt{50}} - \frac{e^{-100}}{\sqrt{100}}\right) + \sqrt{\pi}\left(erf\left(\sqrt{50}\right) - erf\left(\sqrt{100}\right)\right)\right\}$$

$$= 7.694598628 \times 10^{-24}$$

At ***second*** time step with a source at and field at $x = \ell$

$$(u_{12})_{t=t_2} = \int_{t_1}^{t_2} u^*(\xi = 0, x = \ell; t_2, t)dt$$

$$= \frac{r}{2\sqrt{\pi}}\left\{\left(\frac{e^{-a_1}}{\sqrt{a_1}} - \frac{e^{-a_2}}{\sqrt{a_2}}\right) + \sqrt{\pi}\left(erf\left(\sqrt{a_1}\right) - erf\left(\sqrt{a_2}\right)\right)\right\}$$

(SS-E2NTTS-15)

In this case:

$$r = \xi - x = -\ell \; \& \; a_1 = \frac{r^2}{4(t_2 - t_1)} \; \& \; a_2 = \frac{r^2}{4(t_2 - t_2)} \text{.............Singularit y}$$

(SS-E2NTTS-16)

Using equation (SS-E2NTTS-16) into equation (SS-E2NTTS-15) and simplify, we get:

$$(u_{12})_{t=t_2} = \int_{t_1}^{t_2} u^*(\xi = 0, x = \ell; t_2, t)dt = \frac{r}{2\sqrt{\pi}}\frac{e^{-a_1}}{\sqrt{a_1}} + \frac{r}{2}\left\{erf\left(\sqrt{a_1}\right) - 1\right\}$$

(SS-E2NTTS-17)

$$(u_{12})_{t=t_2} = \int_{t_1}^{t_2} u^*(\xi = 0, x = \ell; t_2, t)dt =$$

$$= \frac{1}{2\sqrt{\pi}} \frac{e^{-100}}{\sqrt{100}} + \frac{1}{2}\left\{erf\left(\sqrt{100}\right) - 1\right\} = 1.04941405 \times 10^{-45}$$

At ***first*** time step with a source at $\xi = \ell$ and field at $x = 0$

$$(u_{21})_{t=t_1} = \int_{t_0}^{t_1} u^*(\xi = \ell, x = 0; t_2, t)dt =$$

$$= \frac{r}{2\sqrt{\pi}}\left\{\left(\frac{e^{-a_0}}{\sqrt{a_0}} - \frac{e^{-a_1}}{\sqrt{a_1}}\right) + \sqrt{\pi}\left(erf\left(\sqrt{a_0}\right) - erf\left(\sqrt{a_1}\right)\right)\right\} \quad \textbf{(SS-E2NTTS-18)}$$

In this case:

$$r = \xi - x = \ell \ \& \ a_0 = \frac{r^2}{4(t_2 - t_0)} \ \& \ a_1 = \frac{r^2}{4(t_2 - t_1)} \quad \textbf{(SS-E2NTTS-19)}$$

$$r = \xi - x = \ell = 1$$

$$a_0 = \frac{r^2}{4(t_2 - t_0)} = \frac{1}{4(0.005 - 0.000)} = 50$$

$$a_1 = \frac{r^2}{4(t_2 - t_1)} = \frac{1}{4(0.005 - 0.0025)} = 100$$

$$(u_{21})_{t=t_1} = \int_{t_0}^{t_1} u^*(\xi = \ell, x = 0; t_2, t)dt =$$

$$= \frac{1}{2\sqrt{\pi}}\left\{\left(\frac{e^{-50}}{\sqrt{50}} - \frac{e^{-100}}{\sqrt{100}}\right) + \sqrt{\pi}\left(erf\left(\sqrt{50}\right) - erf\left(\sqrt{100}\right)\right)\right\} = 7.694598628 \times 10^{-24}$$

At ***second*** time step with a source at $\xi = \ell$ and field at $x = 0$

$$(u_{21})_{t=t_2} = \int_{t_1}^{t_2} u^*(\xi = \ell, x = 0; t_2, t)dt =$$

$$= \frac{r}{2\sqrt{\pi}}\left\{\left(\frac{e^{-a_1}}{\sqrt{a_1}} - \frac{e^{-a_2}}{\sqrt{a_2}}\right) + \sqrt{\pi}\left(erf\left(\sqrt{a_1}\right) - erf\left(\sqrt{a_2}\right)\right)\right\}$$

(SS-E2NTTS-20)

In this case:

$$r = \xi - x = \ell \;\&\; a_1 = \frac{r^2}{4(t_2 - t_1)} \;\&\; a_2 = \frac{r^2}{4(t_2 - t_2)} \text{.............Singularit y}$$

(SS-E2NTTS-21)

Using equation (SS-E2NTTS-21) into equation (SS-E2NTTS-20) and simplify, we get:

$$(u_{21})_{t=t_2} = \int_{t_1}^{t_2} u^*(\xi = \ell, x = 0; t_2, t)dt = \frac{r}{2\sqrt{\pi}}\frac{e^{-a_1}}{\sqrt{a_1}} + \frac{r}{2}\left\{erf\left(\sqrt{a_1}\right) - 1\right\}$$

(SS-E2NTTS-22)

$$r = \xi - x = \ell = 1$$

$$a_1 = \frac{r^2}{4(t_2 - t_1)} = \frac{1}{4(0.005 - 0.025)} = 100$$

$$a_2 = \frac{r^2}{4(t_2 - t_2)} = \frac{1}{4(0.005 - 0.005)} = \infty$$

$$(u_{21})_{t=t_2} = \int_{t_1}^{t_2} u^*(\xi = \ell, x = 0; t_2, t)dt =$$

$$= \frac{1}{2\sqrt{\pi}}\frac{e^{-100}}{\sqrt{100}} + \frac{r}{2}\left\{erf\left(\sqrt{100}\right) - 1\right\} = 1.04941405 \times 10^{-45}$$

At ***first*** time step with a source at $\xi = \ell$ and field at $x = \ell$

$$(u_{22})_{t=t_1} = \int_{t_0}^{t_1} u^*(\xi = \ell, x = \ell; t_2, t)dt = $$

$$= \frac{r}{2\sqrt{\pi}}\left\{\left(\frac{e^{-a_0}}{\sqrt{a_0}} - \frac{e^{-a_1}}{\sqrt{a_1}}\right) + \sqrt{\pi}\left(erf\left(\sqrt{a_0}\right) - erf\left(\sqrt{a_1}\right)\right)\right\} \quad \textbf{(SS-E2NTTS-23)}$$

In this case:

$$r = \xi - x = 0 \;\&\; a_0 = \frac{r^2}{4(t_2 - t_0)} \;\&\; a_1 = \frac{r^2}{4(t_2 - t_1)} \quad \textbf{(SS-E2NTTS-24)}$$

Using equation (SS-E2NTTS-24) into equation (SS-E2NTTS-23) and simplify, we get:

$$(u_{22})_{t=t_1} = \int_{t_1}^{t_2} u^*(\xi = \ell, x = \ell; t_2, t)dt = \frac{1}{\sqrt{\pi}}\left\{\sqrt{t_2 - t_0} - \sqrt{t_2 - t_1}\right\} \quad \textbf{(SS-E2NTTS-25)}$$

$$r = \xi - x = 0 \;\&\; a_0 = \frac{r^2}{4(t_2 - t_0)} = 0 \;\&\; a_1 = \frac{r^2}{4(t_2 - t_1)} = 0$$

$$(u_{22})_{t=t_1} = \int_{t_1}^{t_2} u^*(\xi = \ell, x = \ell; t_2, t)dt = $$

$$= \frac{1}{\sqrt{\pi}}\left\{\sqrt{0.005 - 0.000} - \sqrt{0.005 - 0.0025}\right\} = 0.01168475$$

At ***second*** time step with a source at $\xi = \ell$ and field at $x = \ell$

$$(u_{22})_{t=t_2} = \int_{t_1}^{t_2} u^*(\xi = \ell, x = \ell; t_2, t)dt =$$

$$= \frac{r}{2\sqrt{\pi}}\left\{\left(\frac{e^{-a_1}}{\sqrt{a_1}} - \frac{e^{-a_2}}{\sqrt{a_2}}\right) + \sqrt{\pi}\left(erf\left(\sqrt{a_1}\right) - erf\left(\sqrt{a_2}\right)\right)\right\}$$

(SS-E2NTTS-26)

In this case:

$$r = \xi - x = 0 \;\&\; a_1 = \frac{r^2}{4(t_2 - t_1)} \;\&\; a_2 = \frac{r^2}{4(t_2 - t_2)} \text{.............Singularit y}$$

(SS-E2NTTS-27)

Using equation (SS-E2NTTS-27) into equation (SS-E2NTTS-26) and simplify, we get:

$$(u_{22})_{t=t_2} = \int_{t_1}^{t_2} u^*(\xi = \ell, x = \ell; t_2, t)dt = \frac{1}{\sqrt{\pi}}\left(\sqrt{t_2 - t_1}\right)$$

(SS-E2NTTS-28)

$$r = \xi - x = 0 \;\&\; a_1 = \frac{r^2}{4(t_2 - t_1)} = 0 \;\&\; a_2 = \frac{r^2}{4(t_2 - t_2)} = \frac{0}{0}$$

$$(u_{22})_{t=t_2} = \int_{t_1}^{t_2} u^*(\xi = \ell, x = \ell; t_2, t)dt$$

$$= \frac{1}{\sqrt{\pi}}\left(\sqrt{0.005 - 0.0025}\right) = 0.02820947$$

Summary of the numerical results

$$q_0 = +17.72458933$$

$$q_\ell = -17.72458933$$

$$(u_{11})_{t=t_1} = \frac{1}{\sqrt{\pi}}\left\{\sqrt{(t_2-t_0)}+\sqrt{(t_2-t_1)}\right\}=$$

$$=\frac{1}{\sqrt{\pi}}\left\{\sqrt{(0.005-0)}+\sqrt{(0.005-0.0025)}\right\}=0.01168475$$

$$(u_{11})_{t=t_2} = \frac{1}{\sqrt{\pi}}\left\{\sqrt{(t_2-t_1)}\right\}=\frac{1}{\sqrt{\pi}}\left(\sqrt{0.005-0.0025}\right)=0.02820947$$

$$(u_{12})_{t=t_1} = \int_{t_0}^{t_1} u^*(\xi=0,x=\ell;t_2,t)dt=$$

$$=\frac{1}{2\sqrt{\pi}}\left\{\left(\frac{e^{-50}}{\sqrt{50}}-\frac{e^{-100}}{\sqrt{100}}\right)+\sqrt{\pi}\left(erf\left(\sqrt{50}\right)-erf\left(\sqrt{100}\right)\right)\right\}=7.694598628\times10^{-24}$$

$$(u_{12})_{t=t_2} = \int_{t_1}^{t_2} u^*(\xi=0,x=\ell;t_2,t)dt=$$

$$=\frac{1}{2\sqrt{\pi}}\frac{e^{-100}}{\sqrt{100}}+\frac{1}{2}\left\{erf\left(\sqrt{100}\right)-1\right\}=1.04941405\times10^{-45}$$

$$(u_{21})_{t=t_1} = \int_{t_0}^{t_1} u^*(\xi=\ell,x=0;t_2,t)dt=$$

$$=\frac{1}{2\sqrt{\pi}}\left\{\left(\frac{e^{-50}}{\sqrt{50}}-\frac{e^{-100}}{\sqrt{100}}\right)+\sqrt{\pi}\left(erf\left(\sqrt{50}\right)-erf\left(\sqrt{100}\right)\right)\right\}=7.694598628\times10^{-24}$$

$$(u_{21})_{t=t_2} = \int_{t_1}^{t_2} u^*(\xi=\ell,x=0;t_2,t)dt=$$

$$=\frac{1}{2\sqrt{\pi}}\frac{e^{-100}}{\sqrt{100}}+\frac{r}{2}\left\{erf\left(\sqrt{100}\right)-1\right\}=1.04941405\times10^{-45}$$

$$(u_{22})_{t=t_1} = \int_{t_1}^{t_2} u^*(\xi = \ell, x = \ell; t_2, t)dt =$$

$$= \frac{1}{\sqrt{\pi}} \left\{\sqrt{0.005 - 0.000} - \sqrt{0.005 - 0.0025}\right\} = 0.01168475$$

$$(u_{22})_{t=t_2} = \int_{t_1}^{t_2} u^*(\xi = \ell, x = \ell; t_2, t)dt = \frac{1}{\sqrt{\pi}}\left(\sqrt{0.005 - 0.0025}\right) = 0.02820947$$

The next step is to re-write equations (SS-E2NTTS-6) and (SS-E2NTTS-7) in a compact form as follows:

$$-\frac{u_0}{2}\left[-erf\left(\frac{-\ell}{2\sqrt{t_2}}\right)\right] =$$
$$= \{[q_{x=\ell}][u_{12}]\}_{t_1} + \{[q_{x=\ell}][u_{12}]\}_{t_2} - \{[q_{x=0}][u_{11}]\}_{t_1} - \{[q_{x=0}][u_{11}]\}_{t_2}$$

(SS-E2NTTS-29)

Equation (SS-E2NTTS-29) can be re-arranged so that all known terms moved to the left-hand side; therefore equation (SS-E2NTTS-29) becomes:

$$-\frac{u_0}{2}\left[-erf\left(\frac{-\ell}{2\sqrt{t_2}}\right)\right] - \{[q_{x=\ell}][u_{12}]\}_{t_1} + \{[q_{x=0}][u_{11}]\}_{t_1} =$$
$$= \{[q_{x=\ell}][u_{12}]\}_{t_2} - \{[q_{x=0}][u_{11}]\}_{t_2}$$

(SS-E2NTTS-30)

And

$$-\frac{u_0}{2}\left[erf\left(\frac{\ell}{2\sqrt{t_2}}\right)\right] = \{[q_{x=\ell}][u_{22}]\}_{t_1} + \{[q_{x=\ell}][u_{22}]\}_{t_2} - \{[q_{x=0}][u_{21}]\}_{t_1} - \{[q_{x=0}][u_{21}]\}_{t_2}$$

(SS-E2NTTS-31)

Similar procedure can be carried out on equation (SS-E2NTTS-31), therefore it becomes:

$$-\frac{u_0}{2}\left[erf\left(\frac{\ell}{2\sqrt{t_2}}\right)\right]-\{[q_{x=\ell}][u_{22}]\}_{t_1}+\{[q_{x=0}][u_{21}]\}_{t_1}==\{[q_{x=\ell}][u_{22}]\}_{t_2}-\{[q_{x=0}][u_{21}]\}_{t_2}$$

(SS-E2NTTS-32)

Substituting by the numerical results into equation (SS-E2NTTS-30), we get:

$$\begin{aligned}&-0.5-\{[-17.72458933][7.694598628\times 10^{-24}]\}_{t_1}\\&+\{[+17.72458933][0.01168475]\}_{t_1}\\&=\{[q_{x=\ell}][1.04941405\times 10^{-45}]\}_{t_2}-\{[q_{x=0}][0.02820947]\}_{t_2}\end{aligned}$$

(SS-E2NTTS-33)

Substituting by the numerical results into equation (31), we get:

$$\begin{aligned}&-0.5-\{[-17.72458933][0.01168475]\}_{t_1}\\&+\{[+17.72458933][7.694598628\times 10^{-24}]\}_{t_1}\\&=\{[q_{x=\ell}][0.02820947]\}_{t_2}-\{[q_{x=0}][1.04941405\times 10^{-45}]\}_{t_2}\end{aligned}$$

(SS-E2NTTS-34)

Solution of equations (SS-E2NTTS-33) and (SS-E2NTTS-34) leads to the values of both $q_{x=0}$ and $q_{x=\ell}$ at the next time step, these values are:

$q_{x=0}=+10.382775$

&

$q_{x=\ell}=-10.382775$

Third Stage: Evaluation for the Third Time Step

This stage extends the computation procedure to compute the flux for the elementary example at the third step.

General Overview Notes For This Stage

(1)

$$t_n = t_3 = \underbrace{\underbrace{\underbrace{t_0 + \Delta t}_{t_1} + 2\Delta t}_{t_2} + 2\Delta t}_{t_3}$$

(3) The limits of integrations corresponding to this time step will be:

$$a_0 = \frac{r^2}{4(t_3 - t_0)} \qquad a_1 = \frac{r^2}{4(t_3 - t_1)} \qquad a_2 = \frac{r^2}{4(t_3 - t_2)}$$

$$a_3 = \frac{r^2}{4(t_3 - t_3)} \text{.......Singularit y}$$

The boundary integral equation corresponding to this time step takes the form:

$$u(\xi, t_3) = \frac{u_0}{2}\left[erf\left(\frac{\xi}{2\sqrt{t_3}}\right) - erf\left(\frac{\xi - \ell}{2\sqrt{t_3}}\right)\right] +$$

$$\sum_{i=1}^{3}\left[q_{x=\ell}\left(\int_{t_0}^{t_3} u^*(\xi, x; t_3, t)dt\right)_{x=\ell} - q_{x=0}\left(\int_{t_0}^{t_3} u^*(\xi, x; t_3, t)dt\right)_{x=0}\right]$$

(RS-E3DTTS-1)

Equation (RS-E3DTTS-1) can be written in some details as follows:

$$u(\xi, t_3) = \frac{u_0}{2}\left[erf\left(\frac{\xi}{2\sqrt{t_3}}\right) - erf\left(\frac{\xi - \ell}{2\sqrt{t_3}}\right)\right] +$$

$$+ q_{x=\ell}\left(\int_{t_0}^{t_1} u^*(\xi, x; t_3, t)dt + \int_{t_1}^{t_2} u^*(\xi, x; t_3, t)dt + \int_{t_2}^{t_3} u^*(\xi, x; t_3, t)dt\right)_{x=\ell}$$

$$- q_{x=0}\left(\int_{t_0}^{t_1} u^*(\xi, x; t_3, t)dt + \int_{t_1}^{t_2} u^*(\xi, x; t_3, t)dt + \int_{t_2}^{t_3} u^*(\xi, x; t_3, t)dt\right)_{x=0}$$

(RS-E3DTTS-2)

Equation (RS-E3DTTS-2) still has a detailed form as follows:

$$u(\xi,t_3)=\frac{u_0}{2}\left[erf\left(\frac{\xi}{2\sqrt{t_3}}\right)-erf\left(\frac{\xi-\ell}{2\sqrt{t_3}}\right)\right]++[q_{x=\ell}]_{t_1}\int_{t_0}^{t_1}u^*(\xi,x=\ell;t_3,t)dt$$

$$+[q_{x=\ell}]_{t_2}\int_{t_1}^{t_2}u^*(\xi,x=\ell;t_3,t)dt+[q_{x=\ell}]_{t_3}\int_{t_2}^{t_3}u^*(\xi,x=\ell;t_3,t)dt$$

$$-[q_{x=0}]_{t_1}\int_{t_0}^{t_1}u^*(\xi,x=0;t_3,t)dt-[q_{x=0}]_{t_2}\int_{t_1}^{t_2}u^*(\xi,x=0;t_3,t)dt$$

$$-[q_{x=0}]_{t_3}\int_{t_2}^{t_3}u^*(\xi,x=0;t_3,t)dt$$

(RS-E3DTTS-3)

Important Notes

(4) In equation (3), both $[q_{x=0}]_{t_3}$ and $[q_{x=\ell}]_{t_3}$ are unknowns.

(5) All integrals are known except $\int_{t_2}^{t_3}u^*(\xi,x=\ell;t_3,t)dt$ and $\int_{t_2}^{t_3}u^*(\xi,x=0;t_3,t)dt$

(6) It is required to evaluate the integrals in note (2) and then form the system as done before for the first and second-time steps to evaluate the unknown fluex at the end points. The next step is to apply equation (RS-E3DTTS-3) at two different source points; they are $\xi=0$ and $\xi=\ell$. Let us start by taking source point at $\xi=0$, leads to:

$$u(\xi=0,t_3)=\frac{u_0}{2}\left[erf\left(\frac{\ell}{2\sqrt{t_3}}\right)\right]++\left[q_{x=\ell}\right]_{t_1}\int_{t_0}^{t_1}u^*(\xi=0,x=\ell;t_3,t)dt$$
$$+\left[q_{x=\ell}\right]_{t_2}\int_{t_1}^{t_2}u^*(\xi=0,x=\ell;t_3,t)dt+\left[q_{x=\ell}\right]_{t_3}\int_{t_2}^{t_3}u^*(\xi=0,x=\ell;t_3,t)dt$$
$$-\left[q_{x=0}\right]_{t_1}\int_{t_0}^{t_1}u^*(\xi=0,x=0;t_3,t)dt-\left[q_{x=0}\right]_{t_2}\int_{t_1}^{t_2}u^*(\xi=0,x=0;t_3,t)dt$$
$$-\left[q_{x=0}\right]_{t_3}\int_{t_2}^{t_3}u^*(\xi=0,x=0;t_3,t)dt$$

(RS-E3DTTS-4)

$$u_{11}=\int_{t_1}^{t_2}u^*(\xi=0,x=0;t_3,t)dt\ \&\ u_{12}=\int_{t_1}^{t_2}u^*(\xi=0,x=\ell;t_3,t)dt$$

(RS-E3DTTS-5)

Now let us re-write the equation (RS-E3DTTS-4) by making use of equation (RS-E3DTTS-5), we get:

$$\left[u(\xi=0,t_3)=0\right]=$$
$$=\frac{u_0}{2}\left[erf\left(\frac{\ell}{2\sqrt{t_3}}\right)\right]+\left[q_{x=\ell}\right]_{t_1}\left[u_{12}\right]_{t_1}+\left[q_{x=\ell}\right]_{t_2}\left[u_{12}\right]_{t_2}+\left[q_{x=\ell}\right]_{t_3}\left[u_{12}\right]_{t_3}$$
$$-\left[q_{x=0}\right]_{t_1}\left[u_{11}\right]_{t_1}-\left[q_{x=0}\right]_{t_2}\left[u_{11}\right]_{t_2}-\left[q_{x=0}\right]_{t_3}\left[u_{11}\right]_{t_3}$$

(RS-E3DTTS-6)

$$-\frac{u_0}{2}\left[erf\left(\frac{\ell}{2\sqrt{t_3}}\right)\right]=\left[q_{x=\ell}\right]_{t_1}\left[u_{12}\right]_{t_1}+\left[q_{x=\ell}\right]_{t_2}\left[u_{12}\right]_{t_2}+\left[q_{x=\ell}\right]_{t_3}\left[u_{12}\right]_{t_3}$$
$$-\left[q_{x=0}\right]_{t_1}\left[u_{11}\right]_{t_1}-\left[q_{x=0}\right]_{t_2}\left[u_{11}\right]_{t_2}-\left[q_{x=0}\right]_{t_3}\left[u_{11}\right]_{t_3}$$

(RS-E3DTTS-7)

(1) $\left[q_{x=\ell}\right]_{t_1}$ & $\left[q_{x=0}\right]_{t_1}$ Fluxes found from first time step

(2) $\left[u_{12}\right]_{t_1}$ & $\left[u_{11}\right]_{t_1}$ Evaluated from first time step

(3) $\left[q_{x=\ell}\right]_{t_2}$ & $\left[q_{x=0}\right]_{t_2}$ Fluxes found from second time step

(4) $\left[u_{12}\right]_{t_2}$ & $\left[u_{11}\right]_{t_2}$ Evaluated from first time step

(5) $\left[q_{x=\ell}\right]_{t_3}$ & $\left[q_{x=0}\right]_{t_3}$ Unknown to be found from third time step

(6) $\left[u_{12}\right]_{t_3}$ & $\left[u_{11}\right]_{t_3}$ Integral to be calculated in third time step

Now, re-arrange the equation (RS-E3DTTS-7) so that all known values will be on one side and the unknowns on the other side, leads to:

$$-\frac{u_0}{2}\left[erf\left(\frac{\ell}{2\sqrt{t_3}}\right)\right]+\left[q_{x=0}\right]_{t_1}\left[u_{11}\right]_{t_1}+\left[q_{x=0}\right]_{t_2}\left[u_{11}\right]_{t_2}$$
$$-\left[q_{x=\ell}\right]_{t_1}\left[u_{12}\right]_{t_1}-\left[q_{x=\ell}\right]_{t_2}\left[u_{12}\right]_{t_2}=\left[q_{x=\ell}\right]_{t_3}\left[u_{12}\right]_{t_3}-\left[q_{x=0}\right]_{t_3}\left[u_{11}\right]_{t_3}$$

(RS-E3DTTS-8)

Now, let us take source point at $\xi=\ell$, leads to:

$$u(\xi=\ell,t_3)=\frac{u_0}{2}\left[erf\left(\frac{\ell}{2\sqrt{t_3}}\right)\right]+[q_{x=\ell}]_{t_1}\int_{t_0}^{t_1}u^*(\xi=\ell,x=\ell;t_3,t)dt$$
$$+[q_{x=\ell}]_{t_2}\int_{t_1}^{t_2}u^*(\xi=\ell,x=\ell;t_3,t)dt+[q_{x=\ell}]_{t_3}\int_{t_2}^{t_3}u^*(\xi=\ell,x=\ell;t_3,t)dt$$
$$-[q_{x=0}]_{t_1}\int_{t_0}^{t_1}u^*(\xi=\ell,x=0;t_3,t)dt-[q_{x=0}]_{t_2}\int_{t_1}^{t_2}u^*(\xi=\ell,x=0;t_3,t)dt$$
$$-[q_{x=0}]_{t_3}\int_{t_2}^{t_3}u^*(\xi=\ell,x=0;t_3,t)dt$$

(RS-E3DTTS-9)

$$u_{21}=\int_{t_1}^{t_2}u^*(\xi=\ell,x=0;t_3,t)dt \;\&\; u_{22}=\int_{t_1}^{t_2}u^*(\xi=\ell,x=\ell;t_3,t)dt$$

(RS-E3DTTS-10)

Now let us re-write the equation (RS-E3DTTS-9) by making use of equation (10), we get:

$$[u(\xi=\ell,t_3)=0]=\frac{u_0}{2}\left[erf\left(\frac{\ell}{2\sqrt{t_3}}\right)\right]+$$
$$+[q_{x=\ell}]_{t_1}[u_{22}]_{t_1}+[q_{x=\ell}]_{t_2}[u_{22}]_{t_2}+[q_{x=\ell}]_{t_3}[u_{22}]_{t_3}$$
$$-[q_{x=0}]_{t_1}[u_{21}]_{t_1}-[q_{x=0}]_{t_2}[u_{21}]_{t_2}-[q_{x=0}]_{t_3}[u_{21}]_{t_3}$$

(RS-E3DTTS-11)

$$-\frac{u_0}{2}\left[erf\left(\frac{\ell}{2\sqrt{t_3}}\right)\right]=[q_{x=\ell}]_{t_1}[u_{22}]_{t_1}+[q_{x=\ell}]_{t_2}[u_{22}]_{t_2}+[q_{x=\ell}]_{t_3}[u_{22}]_{t_3}$$
$$-[q_{x=0}]_{t_1}[u_{21}]_{t_1}-[q_{x=0}]_{t_2}[u_{21}]_{t_2}-[q_{x=0}]_{t_3}[u_{21}]_{t_3}$$

(RS-E3DTTS-12)

(1) $\left[q_{x=\ell}\right]_{t_1}$ & $\left[q_{x=0}\right]_{t_1}$ Fluxes found from first time step

(2) $\left[u_{22}\right]_{t_1}$ & $\left[u_{21}\right]_{t_1}$ Evaluated from first time step

(3) $\left[q_{x=\ell}\right]_{t_2}$ & $\left[q_{x=0}\right]_{t_2}$ Fluxes found from second time step

(4) $\left[u_{22}\right]_{t_2}$ & $\left[u_{21}\right]_{t_2}$ Evaluated from first time step

(5) $\left[q_{x=\ell}\right]_{t_3}$ & $\left[q_{x=0}\right]_{t_3}$ Unknown to be found from third time step

(6) $\left[u_{22}\right]_{t_3}$ & $\left[u_{21}\right]_{t_3}$ Integral to be calculated in third time step

Now, re-arrange the equation (RS-E3DTTS-12) so that all known values will be on one side and the unknowns on the other side, leads to:

$$-\frac{u_0}{2}\left[erf\left(\frac{\ell}{2\sqrt{t_3}}\right)\right]+\left[q_{x=0}\right]_{t_1}\left[u_{21}\right]_{t_1}+\left[q_{x=0}\right]_{t_2}\left[u_{21}\right]_{t_2}-\left[q_{x=\ell}\right]_{t_1}\left[u_{22}\right]_{t_1}-\left[q_{x=\ell}\right]_{t_2}\left[u_{22}\right]_{t_2}$$
$$=\left[q_{x=\ell}\right]_{t_3}\left[u_{22}\right]_{t_3}-\left[q_{x=0}\right]_{t_3}\left[u_{21}\right]_{t_3}$$

(RS-E3DTTS-13)

Now, equations (RS-E3DTTS-8) and (RS-E3DTTS-13) can be re-written in a matrix form as follow:

$$AX = b$$ **(RS-E3DTTS-14)**

Where,

$$A=\begin{bmatrix} -u_{11} & u_{12} \\ u_{21} & u_{22} \end{bmatrix}$$ **(RS-E3DTTS-15)**

$$X = \begin{bmatrix} q_0 \\ q_1 \end{bmatrix} \qquad \textbf{(RS-E3DTTS-16)}$$

$$b = \begin{bmatrix} -\dfrac{u_0}{2}\left[erf\left(\dfrac{\ell}{2\sqrt{t_3}}\right)\right] + [q_{x=0}]_{t_1}[u_{11}]_{t_1} \\ +[q_{x=0}]_{t_2}[u_{11}]_{t_2} - [q_{x=\ell}]_{t_1}[u_{12}]_{t_1} - [q_{x=\ell}]_{t_2}[u_{12}]_{t_2} \\ -\dfrac{u_0}{2}\left[erf\left(\dfrac{\ell}{2\sqrt{t_3}}\right)\right] + [q_{x=0}]_{t_1}[u_{21}]_{t_1} \\ +[q_{x=0}]_{t_2}[u_{21}]_{t_2} - [q_{x=\ell}]_{t_1}[u_{22}]_{t_1} - [q_{x=\ell}]_{t_2}[u_{22}]_{t_2} \end{bmatrix} \qquad \textbf{(RS-E3DTTS-17)}$$

The next step is to evaluate the time integrals in equation (RS-E3DTTS-15); these integrals can be summarized as follow:

$$u_{11} = \int_{t_1}^{t_2} u^*(\xi = 0, x = 0; t_3, t)dt$$

$$u_{12} = \int_{t_1}^{t_2} u^*(\xi = 0, x = \ell; t_3, t)dt$$

$$u_{21} = \int_{t_1}^{t_2} u^*(\xi = \ell, x = 0; t_3, t)dt \qquad \textbf{(RS-E3DTTS-18)}$$

$$u_{22} = \int_{t_1}^{t_2} u^*(\xi = \ell, x = \ell; t_3, t)dt$$

Let us now illustrate the number of integrals associated within the time steps up to the third one. To do that, refer to the following diagrammatic equation and have a look at the number of integrals associated with each time step.

$$\int_{t_0}^{t_3} = \underbrace{\int_{t_0}^{t_1}}_{\substack{\text{Four integrals}\\ u_{11},u_{12}\\ u_{21},u_{22}}} + \underbrace{\int_{t_1}^{t_2}}_{\substack{\text{Four integrals}\\ u_{11},u_{12}\\ u_{21},u_{22}}} + \underbrace{\int_{t_2}^{t_3}}_{\substack{\text{Four integrals}\\ u_{11},u_{12}\\ u_{21},u_{22}}}$$

Fig. (7.10): Evaluation the time integrals in details.

It is obvious from the Fig. (**7.10**) that we have 12 integrals that should be evaluated within this stage. What we are doing now is to evaluate these integrals one by one simultaneously.

For the first time step

At ***first*** time step with a source at $\xi = 0$ and field at $x = 0$

$$\{u_{11}\}_{t_1} = \int_{t_0}^{t_1} u^*(\xi = 0, x = 0; t_3, t)dt = \tag{1STS-19}$$

$$= \frac{r}{2\sqrt{\pi}}\left\{\left(\frac{e^{-a_0}}{\sqrt{a_0}} - \frac{e^{-a_1}}{\sqrt{a_1}}\right) + \sqrt{\pi}\left\{erf\sqrt{a_0} - erf\sqrt{a_1}\right\}\right\}$$

In equation (19):

$$r = \xi - x = 0 \,\&\, a_0 = \frac{r^2}{4(t_3 - t_0)} \,\&\, a_1 = \frac{r^2}{4(t_3 - t_1)} \tag{1STS-20}$$

Therefore, equation (1STS-19) takes the following final form:

$$\{u_{11}\}_{t_1} = \frac{1}{\sqrt{\pi}}\left(\sqrt{t_3 - t_0} - \sqrt{t_3 - t_1}\right) \tag{1STS-21}$$

At ***first*** time step with source at $\xi - 0$ and field at $x = \ell$

$$\{u_{12}\}_{t_1} = \int_{t_0}^{t_1} u^*(\xi = 0, x = \ell; t_3, t)dt =$$

$$= \frac{r}{2\sqrt{\pi}}\left\{\left(\frac{e^{-a_0}}{\sqrt{a_0}} - \frac{e^{-a_1}}{\sqrt{a_1}}\right) + \sqrt{\pi}\left\{erf\sqrt{a_0} - erf\sqrt{a_1}\right\}\right\} \tag{1STS-22}$$

In equation (1STS-22):

$$r = \xi - x = -\ell \;\&\; a_0 = \frac{r^2}{4(t_3 - t_0)} \;\&\; a_1 = \frac{r^2}{4(t_3 - t_1)} \tag{1STS-23}$$

At ***first*** time step with a source at $\xi = \ell$ and field at $x = 0$

$$\{u_{21}\}_{t_1} = \int_{t_0}^{t_1} u^*(\xi = \ell, x = 0; t_3, t)dt =$$

$$= \frac{r}{2\sqrt{\pi}}\left\{\left(\frac{e^{-a_0}}{\sqrt{a_0}} - \frac{e^{-a_1}}{\sqrt{a_1}}\right) + \sqrt{\pi}\left\{erf\sqrt{a_0} - erf\sqrt{a_1}\right\}\right\} \tag{1STS-24}$$

In equation (1STS-24):

$$r = \xi - x = \ell \;\&\; a_0 = \frac{r^2}{4(t_3 - t_0)} \;\&\; a_1 = \frac{r^2}{4(t_3 - t_1)} \tag{1STS-25}$$

At ***first*** time step with a source at $\xi = \ell$ and field at $x = \ell$

$$\{u_{22}\}_{t_1} = \int_{t_0}^{t_1} u^*(\xi = \ell, x = \ell; t_3, t)dt =$$

$$= \frac{r}{2\sqrt{\pi}}\left\{\left(\frac{e^{-a_0}}{\sqrt{a_0}} - \frac{e^{-a_1}}{\sqrt{a_1}}\right) + \sqrt{\pi}\left\{erf\sqrt{a_0} - erf\sqrt{a_1}\right\}\right\} \tag{1STS-26}$$

In equation (1STS-26):

$$r = \xi - x = 0 \;\&\; a_0 = \frac{r^2}{4(t_3 - t_0)} \;\&\; a_1 = \frac{r^2}{4(t_3 - t_1)} \qquad \textbf{(1STS-27)}$$

Therefore, equation (1STS-26) becomes:

$$\{u_{22}\}_{t_1} = \frac{1}{\sqrt{\pi}}\left(\sqrt{t_3 - t_0} - \sqrt{t_3 - t_1}\right) \qquad \textbf{(1STS-28)}$$

For the second time step

At ***second*** time step with a source at $\xi = 0$ and field at $x = 0$

$$\{u_{11}\}_{t_2} = \int_{t_1}^{t_2} u^*(\xi = 0, x = 0; t_3, t)dt =$$

$$= \frac{r}{2\sqrt{\pi}}\left\{\left(\frac{e^{-a_1}}{\sqrt{a_1}} - \frac{e^{-a_2}}{\sqrt{a_2}}\right) + \sqrt{\pi}\left\{erf\sqrt{a_1} - erf\sqrt{a_2}\right\}\right\} \qquad \textbf{(1STS-29)}$$

In equation (1STS-29):

$$r = \xi - x = 0 \;\&\; a_1 = \frac{r^2}{4(t_3 - t_1)} \;\&\; a_2 = \frac{r^2}{4(t_3 - t_2)} \qquad \textbf{(1STS-30)}$$

Therefore, equation (1STS-29) takes the following final form:

$$\{u_{11}\}_{t_1} = \frac{1}{\sqrt{\pi}}\left(\sqrt{t_3 - t_1} - \sqrt{t_3 - t_2}\right) \qquad \textbf{(1STS-31)}$$

At ***second*** time step with source at $\xi = 0$ and field at $x = \ell$

$$\{u_{12}\}_{t_2} = \int_{t_1}^{t_2} u^*(\xi = 0, x = \ell; t_3, t)dt =$$
$$= \frac{r}{2\sqrt{\pi}}\left\{\left(\frac{e^{-a_1}}{\sqrt{a_1}} - \frac{e^{-a_2}}{\sqrt{a_2}}\right) + \sqrt{\pi}\left\{erf\sqrt{a_1} - erf\sqrt{a_2}\right\}\right\} \tag{1STS-32}$$

In equation (1STS-32):

$$r = \xi - x = -\ell \;\&\; a_1 = \frac{r^2}{4(t_3 - t_1)} \;\&\; a_2 = \frac{r^2}{4(t_3 - t_2)} \tag{1STS-33}$$

At ***second*** time step with source at $\xi = \ell$ and field at $x = 0$

$$\{u_{21}\}_{t_2} = \int_{t_1}^{t_2} u^*(\xi = \ell, x = 0; t_3, t)dt = \frac{r}{2\sqrt{\pi}}\left\{\begin{aligned}&\left(\frac{e^{-a_1}}{\sqrt{a_1}} - \frac{e^{-a_2}}{\sqrt{a_2}}\right)\\ &+ \sqrt{\pi}\left\{erf\sqrt{a_1} - erf\sqrt{a_2}\right\}\end{aligned}\right\} \tag{1STS-34}$$

In equation (1STS-34):

$$r = \xi - x = -\ell$$
$$\&$$
$$a_1 = \frac{r^2}{4(t_3 - t_1)} \tag{1STS-35}$$
$$\&$$
$$a_2 = \frac{r^2}{4(t_3 - t_2)}$$

At ***first*** time step with source at $\xi = \ell$ and field at $x = \ell$

$$\{u_{22}\}_{t_2} = \int_{t_1}^{t_2} u^*(\xi = \ell, x = \ell; t_3, t)dt =$$

$$= \frac{r}{2\sqrt{\pi}}\left\{\left(\frac{e^{-a_1}}{\sqrt{a_1}} - \frac{e^{-a_2}}{\sqrt{a_2}}\right) + \sqrt{\pi}\left\{erf\sqrt{a_1} - erf\sqrt{a_2}\right\}\right\} \quad \textbf{(1STS-36)}$$

In equation (1STS-36):

$$r = \xi - x = 0$$

&

$$a_1 = \frac{r^2}{4(t_3 - t_1)} \quad \textbf{(1STS-37)}$$

&

$$a_2 = \frac{r^2}{4(t_3 - t_2)}$$

Therefore, equation (1STS-36) becomes:

$$\{u_{22}\}_{t_2} = \frac{1}{\sqrt{\pi}}\left(\sqrt{t_3 - t_1} - \sqrt{t_3 - t_2}\right) \quad \textbf{(1STS-38)}$$

For the third time step

At ***third*** time step with a source at $\xi = 0$ and field at $x = 0$

$$\{u_{11}\}_{t_3} = \int_{t_2}^{t_3} u^*(\xi = 0, x = 0; t_3, t)dt =$$

$$= \frac{r}{2\sqrt{\pi}}\left\{\left(\frac{e^{-a_2}}{\sqrt{a_2}} - \frac{e^{-a_3}}{\sqrt{a_3}}\right) + \sqrt{\pi}\left\{erf\sqrt{a_2} - erf\sqrt{a_3}\right\}\right\} \quad \textbf{(1STS-39)}$$

In equation (1STS-39):

$$r = \xi - x = 0 \;\&\; a_2 = \frac{r^2}{4(t_3 - t_2)} \;\&\; a_1 = \frac{r^2}{4(t_3 - t_3)} \text{......Singularity} \qquad \textbf{(1STS-40)}$$

Therefore, equation (1STS-39) takes the following final form:

$$\{u_{11}\}_{t_1} = \frac{1}{\sqrt{\pi}}\left(\sqrt{t_3 - t_2}\right) \qquad \textbf{(1STS-41)}$$

At ***third*** time step with source at $\xi = 0$ and field at $x = \ell$

$$\{u_{12}\}_{t_3} = \int_{t_2}^{t_3} u^*(\xi = 0, x = \ell; t_3, t)dt =$$
$$= \frac{r}{2\sqrt{\pi}}\left\{\left(\frac{e^{-a_2}}{\sqrt{a_2}} - \frac{e^{-a_3}}{\sqrt{a_3}}\right) + \sqrt{\pi}\left\{erf\sqrt{a_2} - erf\sqrt{a_3}\right\}\right\} \qquad \textbf{(1STS-42)}$$

In equation (1STS-42):

$$r = \xi - x = -\ell$$

&

$$a_2 = \frac{r^2}{4(t_3 - t_2)} \qquad \textbf{(1STS-43)}$$

&

$$a_3 = \frac{r^2}{4(t_3 - t_3)} = \infty$$

Therefore, equation (1STS-42) becomes:

$$\{u_{12}\}_{t_3} = \frac{r}{2\sqrt{\pi}}\left\{\left(\frac{e^{-a_2}}{\sqrt{a_2}}\right) + \sqrt{\pi}\left\{erf\sqrt{a_2} - 1\right\}\right\} \qquad \textbf{(1STS-44)}$$

At ***third*** time step with source at $\xi = \ell$ and field at $x = 0$

$$\{u_{21}\}_{t_3} = \int_{t_2}^{t_3} u^*(\xi = \ell, x = 0; t_3, t)dt =$$

$$= \frac{r}{2\sqrt{\pi}}\left\{\left(\frac{e^{-a_2}}{\sqrt{a_2}} - \frac{e^{-a_3}}{\sqrt{a_3}}\right) + \sqrt{\pi}\left\{erf\sqrt{a_2} - erf\sqrt{a_3}\right\}\right\} \quad \textbf{(1STS-45)}$$

In equation (45):

$$r = \xi - x = \ell \;\&\; a_2 = \frac{r^2}{4(t_3 - t_2)} \;\&\; a_3 = \frac{r^2}{4(t_3 - t_3)} \quad \textbf{(1STS-46)}$$

At ***third*** time step with source at $\xi = \ell$ and field at $x = \ell$

$$\{u_{22}\}_{t_3} = \int_{t_2}^{t_3} u^*(\xi = \ell, x = \ell; t_3, t)dt =$$

$$= \frac{r}{2\sqrt{\pi}}\left\{\left(\frac{e^{-a_2}}{\sqrt{a_2}} - \frac{e^{-a_3}}{\sqrt{a_3}}\right) + \sqrt{\pi}\left\{erf\sqrt{a_2} - erf\sqrt{a_3}\right\}\right\} \quad \textbf{(1STS-47)}$$

In equation (1STS-47):

$$r = \xi - x = 0$$

&

$$a_2 = \frac{r^2}{4(t_3 - t_2)} \quad \textbf{(1STS-48)}$$

&

$$a_3 = \frac{r^2}{4(t_3 - t_3)}$$

Therefore, equation (48) becomes:

$$\{u_{22}\}_{t_3} = \frac{1}{\sqrt{\pi}}\left(\sqrt{t_3 - t_2}\right) \qquad \textbf{(1STS-49)}$$

Preliminary Stage For Code Program

In this section, we will prepare the general algorithm for solving the heat diffusion problem over a fixed domain. The first step of this algorithm is to create the time step, and this can be done using the following formula:

$$T(k) = T(k-1) + \Delta T,\ \ k = 1, N$$

Example

For

$$k = 1$$
$$T(1) = T(0) + \Delta T$$
$$k = 2$$
$$T(2) = T(1) + \Delta T$$
$$k = 3$$
$$T(3) = T(2) + \Delta T$$

Notes:

k :Counter
N :Total number of time steps
$T(0)$:Initial time

$kk = 1, k$: Do loop for evaluating time integrals at each time step

$$a(kk-1) = \frac{1}{4(T(k) - T(kk-1))}:$$

Do loop for evaluating limits of integrals at each time step

Check (1)

$$N: \text{Total number of time steps} = 2$$
$$k = 1$$
$$kk = 1,1 \Leftrightarrow a_0 = \frac{1}{4(t_1 - t_0)}$$

$$N: \text{Total number of time steps} = 2$$
$$k = 2$$
$$kk = 1 \Leftrightarrow a_0 = \frac{1}{4(t_2 - t_0)}$$
$$kk = 2 \Leftrightarrow a_1 = \frac{1}{4(t_2 - t_1)}$$

Check (2)

$$N: \text{Total number of time steps} = 5$$
$$k = 1$$
$$kk = 1,1 \Leftrightarrow a_0 = \frac{1}{4(t_1 - t_0)}$$

$$N: \text{Total number of time steps} = 5$$
$$k = 2$$
$$kk = 1 \Leftrightarrow a_0 = \frac{1}{4(t_2 - t_0)}$$
$$kk = 2 \Leftrightarrow a_1 = \frac{1}{4(t_2 - t_1)}$$

N : Total number of time steps = 5

$k = 3$

$$kk = 1 \Leftrightarrow a_0 = \frac{1}{4(t_3 - t_0)}$$

$$kk = 2 \Leftrightarrow a_1 = \frac{1}{4(t_3 - t_1)}$$

$$kk = 3 \Leftrightarrow a_3 = \frac{1}{4(t_3 - t_2)}$$

N : Total number of time steps = 5

$k = 4$

$$kk = 1 \Leftrightarrow a_0 = \frac{1}{4(t_4 - t_0)}$$

$$kk = 2 \Leftrightarrow a_1 = \frac{1}{4(t_4 - t_1)}$$

$$kk = 3 \Leftrightarrow a_3 = \frac{1}{4(t_4 - t_2)}$$

$$kk = 4 \Leftrightarrow a_4 = \frac{1}{4(t_4 - t_3)}$$

N : Total number of time steps = 5

$k = 5$

$$kk = 1 \Leftrightarrow a_0 = \frac{1}{4(t_5 - t_0)}$$

$$kk = 2 \Leftrightarrow a_1 = \frac{1}{4(t_5 - t_1)}$$

$$kk = 3 \Leftrightarrow a_3 = \frac{1}{4(t_5 - t_2)}$$

$$kk = 4 \Leftrightarrow a_4 = \frac{1}{4(t_5 - t_3)}$$

$$kk = 5 \Leftrightarrow a_5 = \frac{1}{4(t_5 - t_4)}$$

General algorithm for the computation of the ***first*** time step

$$t_n = t_1$$
$$t_1 = t_0 + \Delta t$$

$$a_0 = \frac{r^2}{4\left(\underbrace{t_1}_{\text{The index corresponds to "k"}} - \underbrace{t_0}_{\text{The index corresponds to "kk"}}\right)}$$

$$u_{11} = -\frac{1}{\sqrt{\pi}}\left(\sqrt{t_1 - t_0}\right)$$

$$u_{12} = \frac{r}{2\sqrt{\pi}}\left(\frac{e^{-a_0}}{\sqrt{a_0}}\right) + \frac{r}{2}\left\{erf\left(\sqrt{a_0}\right) - 1\right\}$$

$$u_{21} = -u_{12}$$

$$u_{22} = \frac{1}{\sqrt{\pi}}\left(\sqrt{t_1 - t_0}\right)$$

$$b_1 = b_1 = \frac{-u_0}{2}\left(erf\left[\frac{\ell}{2\sqrt{t_1}}\right]\right)$$

$$q_1 = \frac{\left(b_1 \times u_{22} - b_2 \times u_{12}\right)}{\left(u_{11} \times u_{22} - u_{21} \times u_{12}\right)}$$

$$q_2 = \frac{\left(b_1 \times u_{11} - b_2 \times u_{21}\right)}{\left(u_{11} \times u_{22} - u_{21} \times u_{12}\right)}$$

General algorithm for the computation of the <u>*second*</u> time step

The following algorithm will consist of two major parts the first one is for the first time step while the second one is for the second time step.

$$t_n = t_2$$

$$t_1 = t_0 + \Delta t$$

$$t_2 = t_1 + \Delta t = t_0 + 2\Delta t$$

Fisrt part of the algorithm corresponds to the time integral in the first time step

$$a_0 = \frac{r^2}{4(t_2 - t_0)}$$

$$a_1 = \frac{r^2}{4(t_2 - t_1)}$$

$$u_{11} = \frac{1}{\sqrt{\pi}}\left(\sqrt{t_2 - t_0} - \sqrt{t_2 - t_1}\right)$$

$$u_{12} = \frac{r}{2\sqrt{\pi}}\left(\frac{e^{-a_0}}{\sqrt{a_0}} - \frac{e^{-a_1}}{\sqrt{a_1}}\right) + \sqrt{\pi}\left\{erf\left(\sqrt{a_0}\right) - erf\left(\sqrt{a_1}\right)\right\}$$

$$u_{21} = u_{12}$$

$$u_{22} = \frac{1}{\sqrt{\pi}}\left(\sqrt{t_2 - t_0} - \sqrt{t_2 - t_1}\right)$$

$$b_{11} = \frac{-u_0}{2}\left(erf\left[\frac{\ell}{2\sqrt{t_2}}\right]\right) +$$

$$+ [q_1]_{\text{Evaluated at the first time step}}\, u_{11} - [q_2]_{\text{Evaluated at the first time step}}\, u_{12}$$

$$b_{22} = \frac{-u_0}{2}\left(erf\left[\frac{\ell}{2\sqrt{t_2}} \right] \right) +$$

$+[q_1]$ Evaluated at the first time step $u_{21} - [q_2]$ Evaluated at the first time step u_{22}

Second part of the algorithm corresponds to the time integral in the second time step

$$u_{11} = -\frac{1}{\sqrt{\pi}}\left(\sqrt{t_2 - t_1}\right)$$

$$u_{12} = \frac{r}{2\sqrt{\pi}}\left(\frac{e^{-a_1}}{\sqrt{a_1}} \right) + \frac{r}{2}\left\{ erf\left(\sqrt{a_1}\right) - 1 \right\}$$

$$u_{21} = -u_{12}$$

$$u_{22} = \frac{1}{\sqrt{\pi}}\left(\sqrt{t_2 - t_1}\right)$$

$$b_{11} = \frac{-u_0}{2}\left(erf\left[\frac{\ell}{2\sqrt{t_2}} \right] \right) +$$

$+[q_1]$Evaluated at the first time step $u_{11} - [q_2]$Evaluated at the first time step u_{12}

$$q_1 = \frac{(b_{11} \times u_{22} - b_{22} \times u_{12})}{(u_{11} \times u_{22} - u_{21} \times u_{12})}$$

$$q_2 = \frac{(b_{11} \times u_{11} - b_{22} \times u_{21})}{(u_{11} \times u_{22} - u_{21} \times u_{12})}$$

Now let us re-write the summary of the numerical results for the first and second time steps, so as we use these results for the computation for the third time step.

Summary For The First Time Step

$q_0 = +17.72458933$

&

$q_\ell = -17.72458933$

Summary For The Second Time Step

$q_{x=0} = +10.382775$

&

$q_{x=\ell} = -10.382775$

Numerical computation for the third time step

As described before, the computation for the third time step requires computations of the integrals from t_0 to t_1, t_1 to t_2 and t_2 to t_3. Each integral consists of four elements, and now the computations of these elements are as follows:

For the first time step

At ***first*** time step with a source at $\xi = 0$ and field at $x = 0$

$$\{u_{11}\}_{t_1} = \frac{1}{\sqrt{\pi}}\left(\sqrt{t_3 - t_0} - \sqrt{t_3 - t_1}\right) = 8.96602 \times 10^{-3}$$

At ***first*** time step with a source at $\xi = 0$ and field at $x = \ell$

$$\{u_{12}\}_{t_1} = \int_{t_0}^{t_1} u^*(\xi = 0, x = \ell; t_3, t)dt =$$

$$= \frac{r}{2\sqrt{\pi}}\left\{\left(\frac{e^{-a_0}}{\sqrt{a_0}} - \frac{e^{-a_1}}{\sqrt{a_1}}\right) + \sqrt{\pi}\left\{erf\sqrt{a_0} - erf\sqrt{a_1}\right\}\right\}$$

$$r=\xi-x=-\ell \;\&\; a_0=\frac{r^2}{4(t_3-t_0)} \;\&\; a_1=\frac{r^2}{4(t_3-t_1)}$$

$$\{u_{12}\}_{t_1}=1.6316\times10^{-16}$$

At ***first*** time step with a source at $\xi=\ell$ and field at $x=0$

$$\{u_{21}\}_{t_1}=\int_{t_0}^{t_1}u^*(\xi=\ell,x=0;t_3,t)dt=$$

$$=\frac{r}{2\sqrt{\pi}}\left\{\left(\frac{e^{-a_0}}{\sqrt{a_0}}-\frac{e^{-a_1}}{\sqrt{a_1}}\right)+\sqrt{\pi}\left\{erf\sqrt{a_0}-erf\sqrt{a_1}\right\}\right\}$$

$$r=\xi-x=\ell \;\&\; a_0=\frac{r^2}{4(t_3-t_0)} \;\&\; a_1=\frac{r^2}{4(t_3-t_1)}$$

$$\{u_{12}\}_{t_1}=1.6316\times10^{-16}$$

At ***first*** time step with a source at $\xi=\ell$ and field at $x=\ell$

$$\{u_{22}\}_{t_1}=\frac{1}{\sqrt{\pi}}\left(\sqrt{t_3-t_0}-\sqrt{t_3-t_1}\right)=8.96602\times10^{-3}$$

For the second time step

At ***second*** time step with a source at $\xi=0$ and field at $x=0$

$$\{u_{11}\}_{t_1}=\frac{1}{\sqrt{\pi}}\left(\sqrt{t_3-t_1}-\sqrt{t_3-t_2}\right)=0.0116848$$

At ***second*** time step with a source at $\xi=0$ and field at $x=\ell$

$$\{u_{12}\}_{t_2} = \int_{t_1}^{t_2} u^*(\xi = 0, x = \ell; t_3, t)dt = \frac{r}{2\sqrt{\pi}}\left\{\left(\frac{e^{-a_1}}{\sqrt{a_1}} - \frac{e^{-a_2}}{\sqrt{a_2}}\right) + \sqrt{\pi}\left\{erf\sqrt{a_1} - erf\sqrt{a_2}\right\}\right\}$$

$$r = \xi - x = -\ell \;\&\; a_1 = \frac{r^2}{4(t_3 - t_1)} \;\&\; a_2 = \frac{r^2}{4(t_3 - t_2)}$$

$$\{u_{12}\}_{t_2} = 7.6946 \times 10^{-24}$$

At ***second*** time step with source at $\xi = \ell$ and field at $x = 0$

$$\{u_{21}\}_{t_2} = \int_{t_1}^{t_2} u^*(\xi = \ell, x = 0; t_3, t)dt = \frac{r}{2\sqrt{\pi}}\left\{\left(\frac{e^{-a_1}}{\sqrt{a_1}} - \frac{e^{-a_2}}{\sqrt{a_2}}\right) + \sqrt{\pi}\left\{erf\sqrt{a_1} - erf\sqrt{a_2}\right\}\right\}$$

$$r = \xi - x = -\ell \;\&\; a_1 = \frac{r^2}{4(t_3 - t_1)} \;\&\; a_2 = \frac{r^2}{4(t_3 - t_2)}$$

$$\{u_{21}\}_{t_2} = 7.6946 \times 10^{-24}$$

At ***first*** time step with a source at $\xi = \ell$ and field at $x = \ell$

$$\{u_{22}\}_{t_2} = \frac{1}{\sqrt{\pi}}\left(\sqrt{t_3 - t_1} - \sqrt{t_3 - t_2}\right) = 0.0116848$$

For the third time step

At ***third*** time step with a source at $\xi = 0$ and field at $x = 0$

$$\{u_{11}\}_{t_3} = \frac{1}{\sqrt{\pi}}\left(\sqrt{t_3 - t_2}\right) = -.028209$$

At ***third*** time step with a source at $\xi = 0$ and field at $x = \ell$

$$\{u_{12}\}_{t_3} = \int_{t_2}^{t_3} u^*(\xi = 0, x = \ell; t_3, t)dt = \frac{r}{2\sqrt{\pi}}\left\{\left(\frac{e^{-a_2}}{\sqrt{a_2}} - \frac{e^{-a_3}}{\sqrt{a_3}}\right) + \sqrt{\pi}\left\{erf\sqrt{a_2} - erf\sqrt{a_3}\right\}\right\}$$

$$r = \xi - x = -\ell \;\&\; a_2 = \frac{r^2}{4(t_3 - t_2)} \;\&\; a_3 = \frac{r^2}{4(t_3 - t_3)} = \infty$$

$$\{u_{12}\}_{t_3} = \frac{r}{2\sqrt{\pi}}\left\{\left(\frac{e^{-a_2}}{\sqrt{a_2}}\right) + \sqrt{\pi}\left\{erf\sqrt{a_2} - 1\right\}\right\} = 1.049414 \times 10^{-45}$$

At ***third*** time step with a source at $\xi = \ell$ and field at $x = 0$

$$\{u_{21}\}_{t_3} = \int_{t_2}^{t_3} u^*(\xi = \ell, x = 0; t_3, t)dt = \frac{r}{2\sqrt{\pi}}\left\{\left(\frac{e^{-a_2}}{\sqrt{a_2}} - \frac{e^{-a_3}}{\sqrt{a_3}}\right) + \sqrt{\pi}\left\{erf\sqrt{a_2} - erf\sqrt{a_3}\right\}\right\}$$

$$r = \xi - x = \ell \;\&\; a_2 = \frac{r^2}{4(t_3 - t_2)} \;\&\; a_3 = \frac{r^2}{4(t_3 - t_3)}$$

$$\{u_{21}\}_{t_3} = -1.149414 \times 10^{-45}$$

At ***third*** time step with a source at $\xi = \ell$ and field at $x = \ell$

$$\{u_{22}\}_{t_3} = \frac{1}{\sqrt{\pi}}\left(\sqrt{t_3 - t_2}\right) = 0.028209$$

$$\{b_{11}\}_{t_3} = -0.5 - \left[-17.7246 \times 1.6316 \times 10^{-16}\right] + \left[17.7246 \times 8.96602 \times 10^{-3}\right]$$
$$-\left[-10.382775 \times 7.6946 \times 10^{-24}\right] + \left[10.382775 \times 0.0116848\right] = -0.220926$$

$$\{b_{22}\}_{t_3} = -0.5 - \left[-17.7246 \times 8.96602 \times 10^{-3}\right] + \left[17.7246 \times 17.7246 \times 10^{-16}\right]$$
$$-\left[-10.382775 \times 0.0116848\right] + \left[10.382775 \times 7.6946 \times 10^{-24}\right] = -0.220926$$

Now the system of equations for the third time step will be:

$$\begin{bmatrix} -0.028209 & 1.049414\times10^{-45} \\ 1.049414\times10^{-45} & 0.028209 \end{bmatrix}\begin{bmatrix} q_0 \\ q_\ell \end{bmatrix}=\begin{bmatrix} -0.220926 \\ -0.220926 \end{bmatrix}$$

The solution leads to the numerical values for q_0 and q_ℓ at the third time step.

Part (IV) Diffusion Problems with Moving Boundaries

Vaporization test problem

In this section, three test problems are solved to check the validity of the proposed algorithm and the high accuracy expected. The first example is a multi-moving boundary problem **[13]**. The thermo-physical properties are listed below in Table (**1**).

Table 1. Numerical data for a vaporization test problem.

Parameter	**Definition**	**Numerical data**
c_s	Specific heat for solid	4.944
c_ℓ	Specific heat for liquid	4.944
K_s	Thermal conductivity for solid	0.259
K_ℓ	Thermal conductivity for liquid	0.259
u_m	Melting temperature	1454
u_v	Vaporization temperature	3000
L_m	Latent heat for melting	2160
L_v	Latent heat for vaporization	37200
$u(x,0)$	Initial temperature	27
$q(t)$	Input heat flux at the boundary $x=0$	2500
ρ	Density	1

The solid material herein is of low thermal conductivity and so the vaporization occurs before the moving boundary separating liquid and solid reach the adiabatic boundary. The result due to the present algorithm is shown in Fig. (**7.11**). In this figure, both vapor/liquid and liquid/solid moving boundaries are plotted both on the same figure. The liquid/solid moving boundary is on the left side, while the other is on the right. From this figure, one can clearly determine the melting, vapor, and time the process ends. These times are determined and summarized in table (**2**), at two different time steps. The absolute errors between the results due to the present method compared with the finite element method are also presented in the same table. As we see, the absolute errors decrease by decreasing the time step. On the other hand, the number of iterations increases by decreasing the time step. The number of iterations is shown in table (**2**), corresponding to the two-time steps used. It is clear that the increase in the number of iterations is not so much, but the accuracy improved to nearly half.

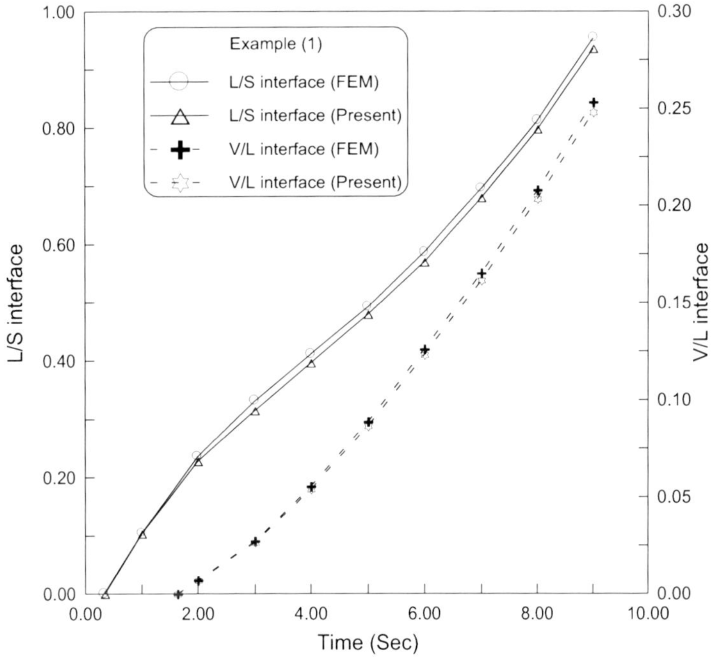

Fig. (7.11). Moving boundaries location for a vaporization test problem

Table 2: Comparison between melting, vaporization, and end time.

Time Step	Type of time	FE	Present	Absolute error	Average number of iterations
$\Delta t = 0.25$	t_m	0.36150	0.36158	8.0×10^{-5}	12-16
	t_v	1.63034	1.63041	7.0×10^{-5}	
	t_e	9.75464	9.75468	3.4×10^{-5}	
$\Delta t = 0.0625$	t_m	0.32767	0.32771	4.0×10^{-5}	20-24
	t_v	1.63446	1.63449	3.25×10^{-5}	
	t_e	9.38719	9.38721	2.0×10^{-5}	

Ablation Test Problem-1

A solid medium initially at uniform temperature, $u_i = 300^o K$, the boundary $x = 0$ exposed to two different cases of input heat flux, constant, $Q(t) = 5\times10^6$ and linear, $Q(t) = 3\times10^4 t$ respectively. The domain in the present problem is still fixed while appearing two moving interfaces solid-liquid and vapor (gas)-liquid (Ablation surface). The results due to the present algorithm are shown in Figs (**7.12-14**), respectively, and the results are compared with the results due to the source and sink method. Fig. (**7.12**) shows the movement of solid-liquid due to constant input heat flux, and the resulting ablation surface due to the same input heat flux is shown in Fig. (**7.3**). The same results due to the linear case are plotted on the same plot as shown in Fig. (**7.14**). From the computations and figures, it is found that the solid-liquid interface has the same behavior in both constant and linear cases of input heat flux that is concave upward. In the case of linear heat flux input, this concavity becomes more apparent than the constant case. On the contrary, the vapor (gas)/liquid interface behaves downward, but in linear case, this concavity increases.

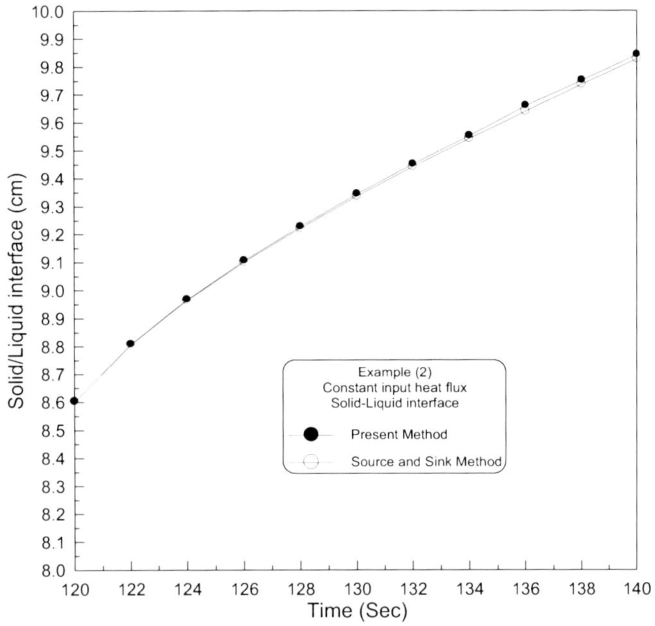

Fig. (7.12). Solid/Liquid due to Constant heat flux-problem-1.

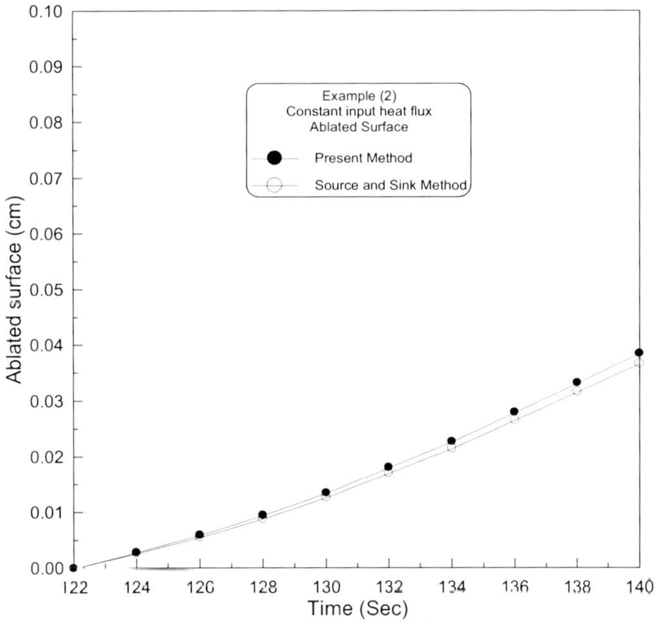

Fig. (7.13). Ablated surface due to constant heat flux-problem-1.

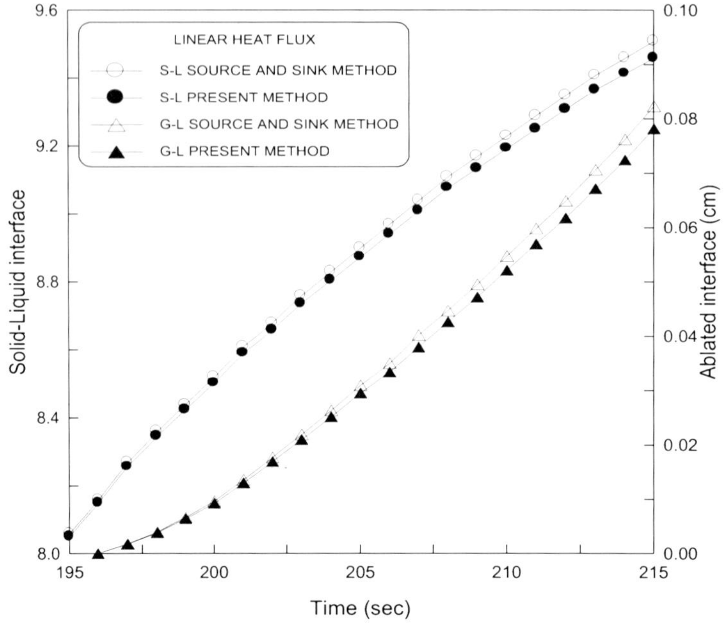

Fig. (7.14). Solid/Liquid and Ablated surface due to linear heat flux-problem-1.

Ablation Test Problem-2

This problem is for a long enough solid mold initially at a uniform temperature.

The surface $x = 0$ exposed to two different high input heat fluxes, constant and linear and the vapor is removed as soon as it forms – ablation problem –. Two specific heat flux boundary conditions are chosen in the present computations, namely, $q(t) = 2.0, 10\ (t / t_c)$ and the following values used in the calculation **[15]**.

Table (3). Thermo-physical parameters.

Symbol	**Physical meaning**	**Numerical value**
α	Thermal diffusivity	$0.1\ ft^2 / \sec$
t_c	Characteristic time	$10 \sec$
L	characteristic length	$1.0\ st$
q_r	Reference heat flux	$10\ Btu / \sec - ft^2$
$T_m - T_0$	Temperature difference	$100^o\ F$

v	Inverse Stefan number	1
T	Reference time	$100\,\text{sec}$

The ablation thickness due to the present method compared with the corresponding from the heat balance integral method is shown in Fig. (**7.15**). It is clear that the ablation thickness is concave downward in both cases of input heat flux and at the same time, the ablation thickness in the case of linear heat flux is higher than that for the constant case. There is a good agreement between the two methods in both cases.

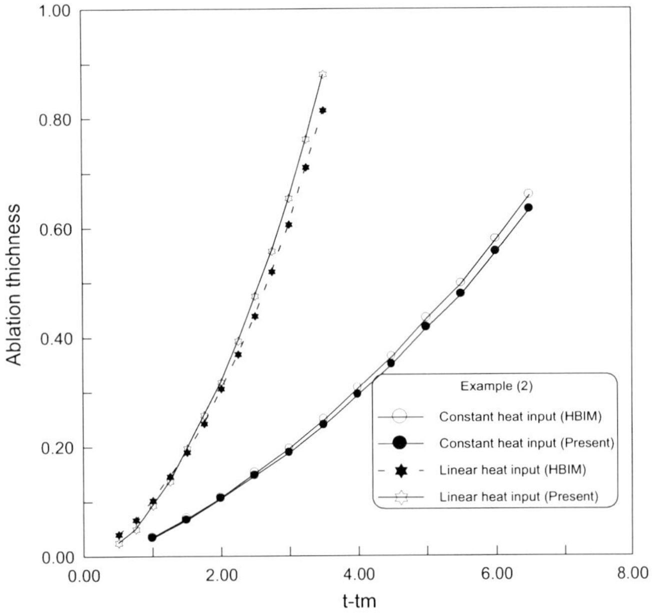

Fig. (7.15). Ablation thickness for ablation test problem-2

Test problem (1)

A solid medium initially at uniform temperature, $U_i = 300^o K$, the boundary $x = 0$ exposed to two different cases of input heat flux, constant, $Q(t) = 5\times10^6$ and linear, $Q(t) = 3\times10^4 t$ respectively. Fig. (**7.16**) shows the movement of solid-liquid due to constant heat flux, and the resulting ablation surface due to the same heat flux is shown in Fig. (**7.17**). The same results due to linear case are plotted on the same plot as shown in Figs (**7.18** and **19**). The above figures show that the solid-liquid interface has the same behavior in both constant and linear cases of heat flux that is concave upward. In the case of linear heat flux input, this concavity becomes more apparent than the constant case. On the contrary, the ablated interface behaves concave downward but in linear case this concavity increases.

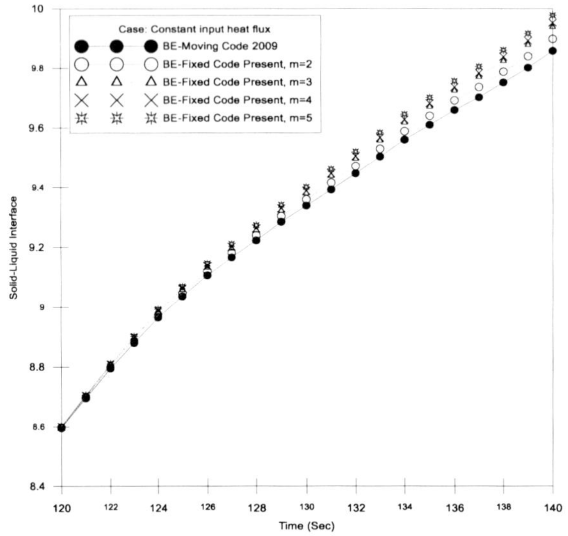

Fig. (7.16). Solid-Liquid due to constant input heat flux.

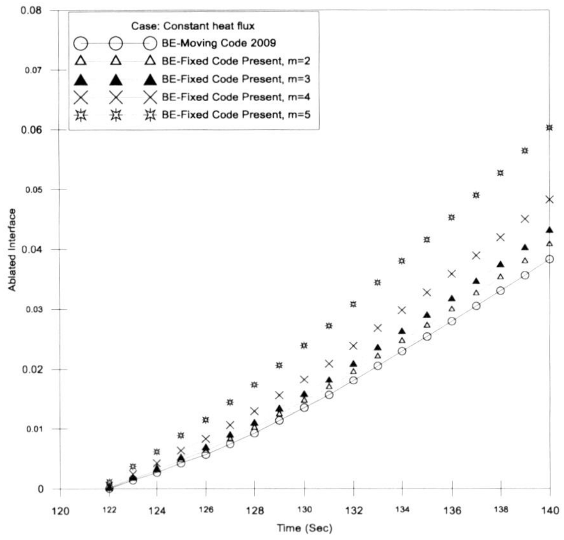

Fig. (7.17). Ablation surface due to constant heat flux.

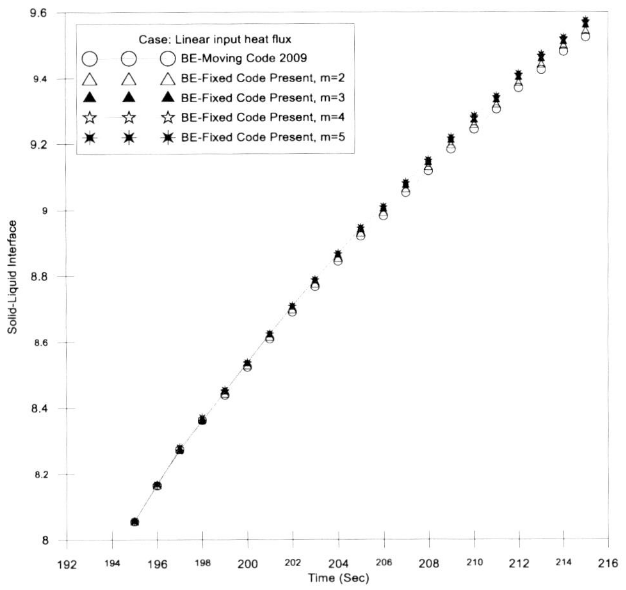

Fig. (7.18). Solid-Liquid due to linear heat flux.

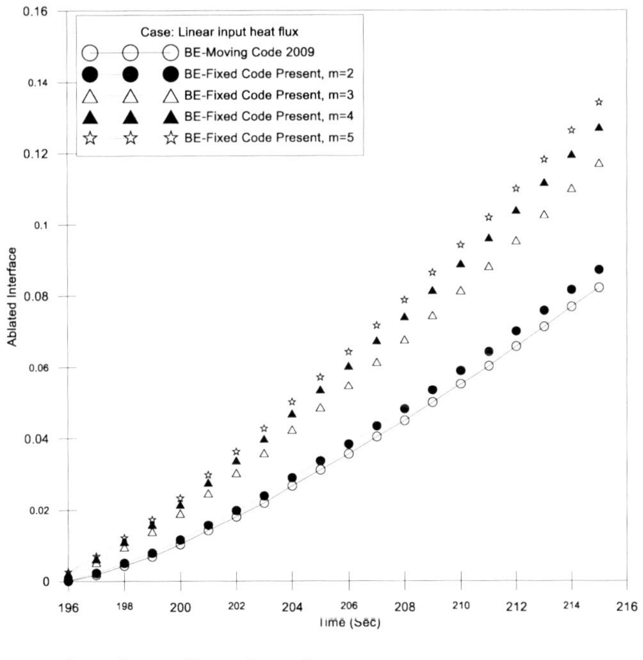

Fig. (7.19). Ablation surface due to linear heat flux.

Also, It is found from the computation that the exponent m has a direct effect in the obtained results where the deviation from the previous moving code results starts increasing by increasing m but still acceptable up to $m = 5$. But this conclusion is not acceptable when tracing the ablation interface.

Test problem (2)

Suppose that a semi-infinite solid be the physical domain. Assume that the free surface of the solid $x = 0$, is exposed to a constant oxygen concentration C_o, and the initial oxygen concentration is zero. Assume that oxygen concentration in oxidized and metallic layers is denoted by C_1 and C_2 respectively. State equations describing this process can be stated as [10]:

$$\frac{\partial C_1(x,t)}{\partial t} = D_1 \frac{\partial^2 C_1(x,t)}{\partial x^2} \qquad 0 \le x \le s(t)$$

$$\frac{\partial C_2(x,t)}{\partial t} = D_2 \frac{\partial^2 C_2(x,t)}{\partial x^2} \qquad s(t) < x < \infty$$

$$s(0) = 0$$

$$C_2(x,0) = 0 \qquad 0 < x < \infty$$

$$C_1(0,t) = C_o \qquad t > 0$$

$$C_1(s(t),t) = C_{cr} \quad t > 0$$

$$C_2(s(t),t) = (C_{cr} - [C]) \qquad t > 0$$

$$D_2 \frac{\partial C_2(s(t),t)}{\partial x} - D_1 \frac{\partial C_1(s(t),t)}{\partial x} = [C]\frac{ds(t)}{dt}$$

D_1 Oxygen diffusivity for oxidized layer

D_2 Oxygen diffusivity for metallic layer

C_{cr} Critical oxygen concentration for the oxide formation

$[C]$ Jump of oxygen concentration across the interface.

The following numerical data are used in the computations:

$D_1 = 0.274\mu m^2$, $D_2 = 0.166\mu m^2$, $C_o = 1.0$, $C_{cr} = 0.65$ **and** $[C] = 0.25$.

Since the exact analytical solution to the planar oxidation problem is available, a comparison between the exact solution and the present method is made, as shown in Fig. (**7.20**). It is clear that there is a good agreement between the two solutions with small acceptable errors. Follow up on the present method results; Figs. (**7.21-25**) show the oxygen concentration in both layers at different times times starts from $t = 0.5h$ up to $t = 2.5h$. It is clear from these figures that the behavior is the same but the distance from the surface increases by growing up the time.

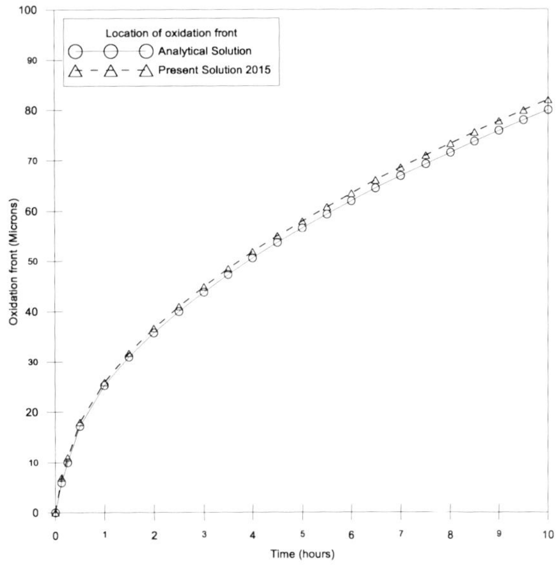

Fig. (7.20). Location of the oxidation front.

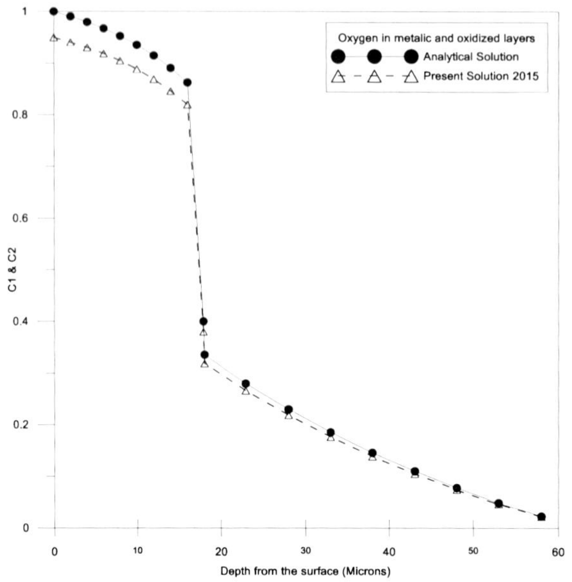

Fig. (7.21). Oxygen concentrations in both layers at time = 0.5 h.

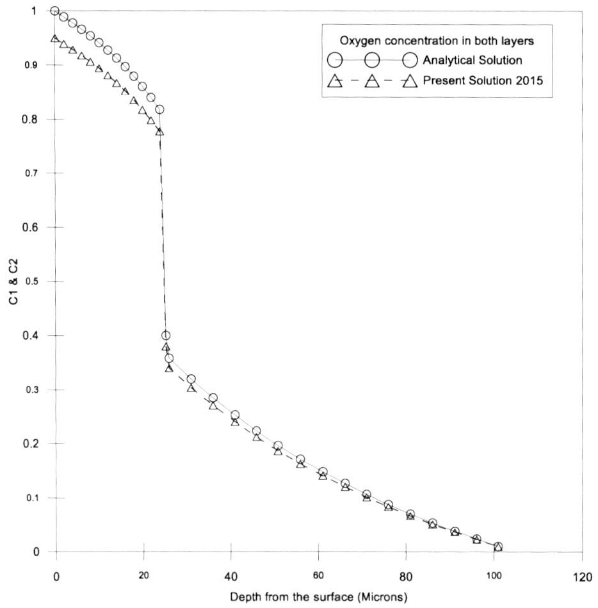

Fig. (7.22). Oxygen concentrations in both layers at time = 1.0 h.

Heat

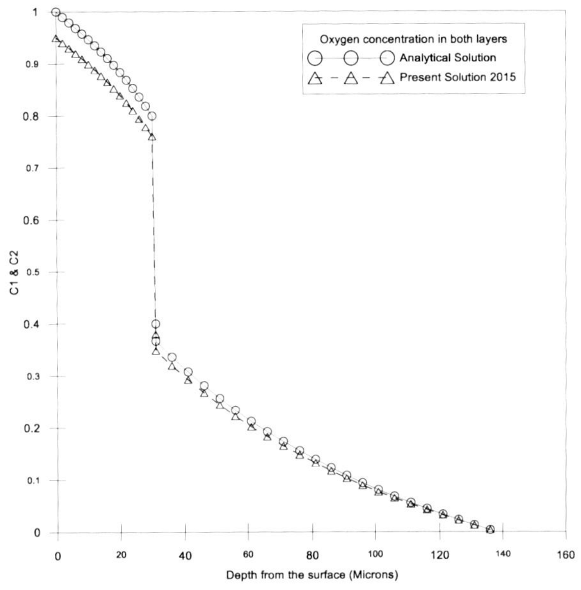

Fig. (7.23). Oxygen concentrations in both layers at time = 1.5h.

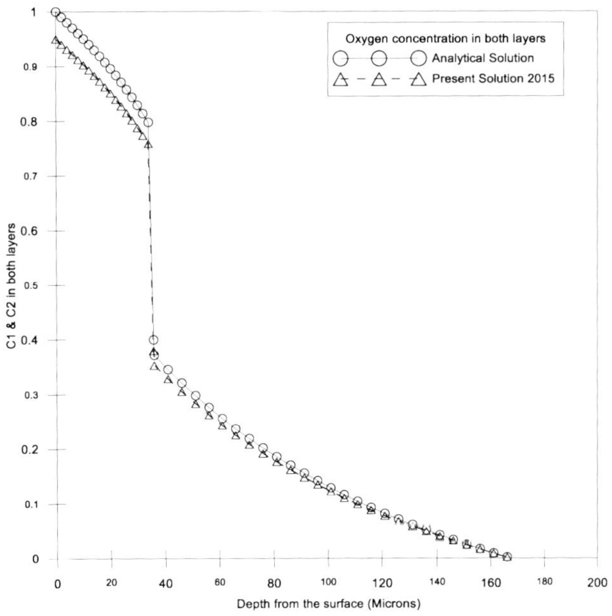

Fig. (7.24). Oxygen concentrations in both layers at time = 2.0h.

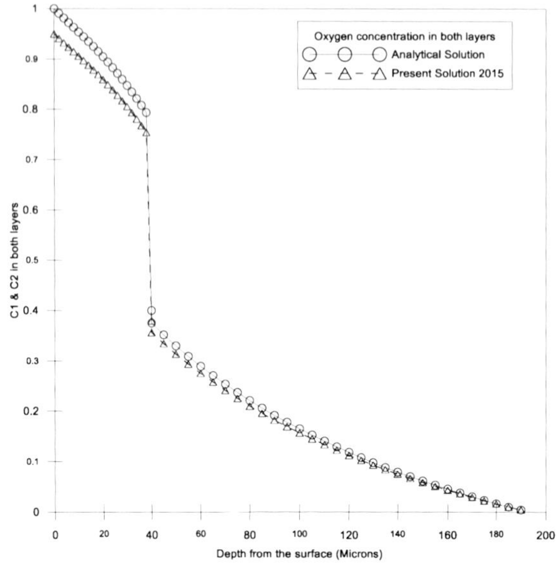

Fig. (7.25). Oxygen concentrations in both layers at time = 2.5h.

REFERENCES

[1] S.G. Ahmed, "An approximate method for oxygen diffusion in a sphere with simultaneous absorption", *Int. J. Numer. Methods Heat Fluid Flow,* vol. 9, no. 6, pp. 631-643, 1999. [http://dx.doi.org/10.1108/09615539910276115]

[2] S.G. Ahmed, "Boundary integral formulation and mushy zone model for phase change problem", *Int. J. Numer. Methods Heat Fluid Flow,* vol. 11, no. 4, pp. 371-380, 2001. [http://dx.doi.org/10.1108/09615530110389207]

[3] S.G. Ahmed, "'An approximate method for evaporation problem with mushy zone', Int", *Eng. Anal. Bound. Elem.,* vol. 29, no. 2, pp. 161-165, 2005. [http://dx.doi.org/10.1016/j.enganabound.2004.10.002]

[4] S.G. Ahmed, "An integral method for solving oxygen diffusion on unbounded domain with moving boundary", *J. Appl. Sci. (Faisalabad),* vol. 5, no. 7, pp. 1244-1249, 2005. [http://dx.doi.org/10.3923/jas.2005.1244.1249]

[5] S. G. Ahmed, "A new algorithm for moving boundary problems subject to periodic boundary conditions", *Int. Journal of Numerical Methods for Heat and Fluidflow, U.K,* vol. 16, no. 1, 2005.

[6] S.G. Ahmed, "A numerical method for oxygen diffusion and absorption in sike cell", *Int. J. Numer. Methods for Heat and Fluid Flow,* vol. 16, U.K., no. 1, 2005.

[7] S.G. Ahmed, "'A Collocation method using new combined radial basis function of thin plate and multi-quadric types', Int", *Eng. Anal. Bound. Elem.,* vol. 30, no. 8, pp. 697-701, 2006. [http://dx.doi.org/10.1016/j.enganabound.2006.03.001]

[8] S. G. Ahmed, "A semi-analytical method for phase change transformation in binary alloys", *J. Comput. Appl. Math,* Netherlands, 2006.

[9] S.G. Ahmed, and S.A. Meshrif, "A New Numerical Algorithm for 2-D Moving Boundary Problems using Boundary Element Method", *Comput. Math. Appl.,* vol. 58, no. 7, pp. 1302-1308, 2009. [http://dx.doi.org/10.1016/j.camwa.2009.03.115]

[10] S. G. Ahmed, and M. L. Mekey, "A collocation and Cartesian grid methods using new radial basis function to solve class of partial differential equations", *J. Comput. Math,* pp. 1-14, 2009.

[11] S.G. Ahmed, "Boundary Integral Formulation for Binary Alloys from a Cooling Solid Wall", *J. Comput. Appl. Math.,* vol. 5, no. 5, pp. 687-696, 2010.

[12] S. G. Ahmed, "A collocation mesh-free method based on multiple basis functions", *Int. J. of Engineering Analysis with Boundary Elements, Sept,* 2011.

[13] S.G. Ahmed, "Numerical solution for advection-diffusion equation using collocation based new radial basis functions", *Indian J. Pure Appl. Math.,* vol. 2, no. 2, 2015.

[14] S. G. Ahmed, "An integral method to solve phase-change problems with/without mushy zones", *American Journal of Applied Mathematics,* vol. 2, no. 1, 2015.

[15] S. G. Ahmed, "Mesh-Free Collocation Based for Phase Transformations in Alloys", *J. Appl. Math. Mech.,* vol. 2, no. 1, 2015.

[16] S.G. Ahmed, "A numerical method based boundary integral equation to solve multiphase moving boundary problems", *Br. J. Math. Comput. Sci,* vol. 15, no. 1, pp. 1-10, 2016.

[17] S. G. Ahmed, "Numerical solution of moving boundary problems using a new hybrid grid and meshless method", *Int. J. of Engineering Analysis with Boundary Elements,* 2019.

[18] M. Lellis Thivagar, "Abdulsattar Abdullah Hamad, and S. G. Ahmed, Conforming dynamics in the

metric spaces, Journal of Information Science and Engineering", *Finally Acceppted, October,* 2019.

[19] DATJA Roberto, "Mathematical modeling and numerical algorithm for diffusion problem over moving domain based direct approach of BEM", *Scientific Journal,* Suleyman Demiral, University Bulletin, Finally Accepted: Kazakh, 2019.

[20] S.G. Ahmed, and L.C. Wrobel, "Numerical solution of solidification problems using integral equations", *Third International conference of Free and Moving Boundary Problems,* 1995 Bled, Slovenia.

[21] S.G. Ahmed, and L.C. Wrobel, "An alternative BEM approach for solving cavity flow problems", *Eleventh International Conference on Boundary Element Technology,* 1996 Hawaii, USA.

[22] S.G. Ahmed, "A new numerical technique developed for polymer processing problems", *Third International Conference on Engineering Mathematics and Physics,* 1997 Cairo, Egypt.

[23] S.G. Ahmed, and M.M. Kassem, "New iteration scheme with integral representation for two phase Stefan problems", *Third International Conference on Engineering Mathematics and Physics,* 1997 Cairo, Egypt.

[24] S.G. Ahmed, *'A combined domain and integral techniques to solve planar oxidation problem', Computational Modeling of Free and Moving Boundary Problems VI.* Lemons: Greece, 2001.

[25] S.G. Ahmed, "A numerical solution for nonlinear heat flow problem using boundary integral method", *Conference of Computer Methods and Experimental Measurements for Surface Treatment,* 2003 Crete, Greece.

[26] S.G. Ahmed, and S.A. Mishref, "A new alternative approach to free-moving boundary problems", *Eight International Conference for Free and Moving Boundary Problems,* 2005 Spain.

[27] S.G. Ahmed, and A.R. Intisar, "New approach of BEM methods to solve solute diffusion problems with moving boundaries", *Eighth International congress of Fluid Dynamics & Propulsion,* 2006 Sharm El-Shiekh, Egypt.

[28] S.G. Ahmed, "A boundary element approach with new algorithm to predict partial cavitation around 2-D hydrofoils,", *Eighth International congress of Fluid Dynamics & Propulsion,* 2006 Sharm El-Shiekh, Egypt.

[29] S.G. Ahmed, "Approximate integral method applied to ablation problem in a finite slab", *International Symposium on Recent Advances in Mathematics and its Applications (ISRAMA 2007),* 15-17 Dec, 2007.

[30] S.G. Ahmed, "A new approach of cartesian grid method to solve inhomogeneous partial differential equations", *International Symposium on Recent Advances in Mathematics and its Applications (ISRAMA 2007),* 15-17 Dec, 2007.

[31] S.G. Ahmed, "Overview of mesh-less methods and its engineering applications", *3rd International Conference of IMBIC,* 21-22, Dec, 2009 Kolkata.

[32] S.G. Ahmed, "Application of boundary integral method to solve combined problems with mushy zone", *3rd International Conference of IMBIC,* 21-22, Dec, 2009 Kolkata.

[33] S.G. Ahmed, *IAENG Conference WCECS,* 2015 San Franscio, USA.

[34] S.G. Ahmed, and A.A. Megahed, "' A New numerical method for solving one-dimensional, multiphase problems'", *J. Eng. Appl. Sci. (Asian Res. Publ. Netw.),* vol. 45, no. 3, 1998.

[35] S.G. Ahmed, "A semi-analytical method for one dimensional multiphase moving boundary problems", *J. Eng. Appl. Sci. (Asian Res. Publ. Netw.),* vol. 46, no. 5, 1999.

[36] S.G. Ahmed, "Numerical solution of ablation problem using constraint integral method", *J. Eng. Appl. Sci. (Asian Res. Publ. Netw.),* vol. 47, no. 3, 2000.

[37] S. G. Ahmed, "A new fixed-moving algorithm using BIM applied to phase change problem", *Journal of Engineering and Applied Science, Cairo, Egypt,* 2002.

[38] S. G. Ahmed, "A coupled moment and finite difference methods for solving combined moving boundary problem", *Journal of Engineering and Applied Science, Cairo, Egypt,* April, 2003.

[39] S. G. Ahmed, "A new algorithm for front tracking of ablation problem in unbounded domain", *Ain Shams University, Egypt,* 2003.

[40] S. G. Ahmed, "A semi-analytical method to track the moving interface in moving boundary problems", In: *Scientific Bulletin, Ain Shams University*, 2006.

[41] S. G. Ahmed, "An approximate method to predict the moving interface in binary alloy problems", *Scientific Bulletin, Ain Shams University,* 2006.

[42] S. G. Ahmed, "Semi-analytical method for combined Stefan and ablation problems with mushy zones", In: *Scientific Bulletin* Ain Shams University, 2007.

[43] "Numerical Analysis for Science", In: *Engineering and Technology* Benetham eBook, 2019.

[44] "Higher Mathematics for Science", In: *Engineering and Technology* Benetham eBook, 2016.

[45] "Basis Mathematics for Law students", In: *Brieme University, UAE* Arabic Edition. University Book House: Ain, UAE, 2016.

[46] Statistics and Probability, *Arabic Edition.* University Book House: Ain, UAE, 2016.

[47] *Complex Analysis in Engineering, Science, and Technology* Bentham eBooks, 2015. eISNB: 978---68108-042-0, ISNB: 978-1-68108-043-7

[48] Elementary Mathematics for Science and Technology, *University Book House.* ISNB: Ain, UAE, 2015.

[49] Encyclopedia in Engineering Mathematics, *(Arabic and English language)* vol. 6. University Book House: Ain, UAE, 2014. Part 1-Calculus, ISNB:978-658909153-X

[50] Encyclopedia in Engineering Mathematics, *(Arabic and English language)* vol. 6. University Book House: Ain, UAE, 2014. Part 2-Calculus, ISNB:978-658909154-7

[51] Encyclopedia in Engineering Mathematics, *(Arabic and English language)* vol. 6. University Book House: Ain, UAE, 2014. Part 3-Calculus, ISNB:978-658909155-4

[52] Encyclopedia in Engineering Mathematics, *(Arabic and English language)* vol. 6. University Book House: Ain, UAE, 2014. Part 4-Calculus, ISNB:978-658909156-1

[53] Encyclopedia in Engineering Mathematics, *(Arabic and English language)* vol. 6. University Book House: Ain, UAE, 2014, 2014. Part 5-Calculus, ISNB:978-658909157-8

[54] Encyclopedia in Engineering Mathematics, *(Arabic and English language)* vol. 6. University Book House: Ain, UAE, 2014, 2014. Part 6-Calculus, ISNB:978-658909158-5

[55] Lectures in Engineering Mathematics "*Analytical Geometry*", 2007 I.S.B.N. 977-17-6294-X

[56] Lectures in Engineering Mathematics "*Algebra*", 2006 I.S.B.N. 977-17-6293-I

[57] Lectures in Engineering Mathematics "*Calculus*", 2005 I.S.B.N. 977-17-6292-3

[58] Algebra for Engineers, 2004 I.S.B.N. 977-5537-8-X.

[59] Numerical Methods for Engineers, I.S.B.N. 977-5537-44-4, 2003

CHAPTER 8

Industrial Applications

Abstract: In this chapter, we will discuss some of the practical and industrial applications that we have developed over the course of our 37-year research career.

Keywords: Phase change problems, moving boundaries, Cavitations, melting, and solidification.

INTRODUCTION

The boundary-value problem is defined as the solution of a differential equation in a given domain while satisfying a set of conditions. The problem is referred to as a moving-boundary problem when one or more of the boundaries are unknown and moving with time. Furthermore, if the governing equation is time independent as well as the boundary condition, then the problem is referred to as free-boundary problem. All these types have practical, industrial, and engineering applications. The phase change problem is a direct example of a moving-boundary problem; meanwhile, the cavitation problem is a direct example of a free-boundary problem. Cavitation is a challenging hydrodynamic phenomenon both in physical science and engineering applications. In physical science, cavitation flows involve many complicated and interdisciplinary physical phenomena in which both physicists and applied mathematicians are interested [1-2]. In engineering applications, cavitation plays a major role in surface sea-going vessel design and operation, as well as in hydraulic equipment. Propellers, hydrofoil ships, hydraulic turbines, and pumps may suffer from its consequences in many ways. All these devices encompass lifting surfaces and a liquid working medium. When cavitation occurs unexpectedly, a severe decrease in performance is observed and in many cases, the machine itself suffers from damage. Systems that operate for long time intervals without maintenance, like a marine propeller, may be completely affected by such damage [3]. In addition, the presence of a cavitation region has a large effect on surface pressure distribution as well as the mass flow capability of internal flow devices operating in this environment. For these reasons, substantial efforts have been undertaken by many researchers to develop capabilities to predict the extent of cavitation for various types of geometries and flow situations [4-6]. The nature of cavitation phenomenon is generally a two-phase, three-component flow. The two phases are most

S. G. Ahmed

importantly involved, *i.e.*, a liquid and its vapor. However, in almost all real cases, at least a trace quantity of non-condensable gas such as air is also involved significantly in bubble collapse and inception, particularly inception. The non-liquid portion, in general, can be either in the form of ***quasi-fixed cavities*** or ***traveling bubbles***. In quasi-fixed cavities (attached cavitation), the vapor region lies near a solid boundary, while traveling bubbles (vortex or traveling cavitation) refer to a bubble field completely surrounded by the parent liquid. The quasi-fixed cavities are usually classified into three main groups to ease the research work. The first group is called the semi-fixed cavities (partially cavitation) usually found in cases of relatively low cavitation numbers. The second group is known as a fixed cavity (super-cavitation) formed for a very much lower cavitation number (σ). For sufficiently high values of σ, cavitation will never happen, known as the third group (see Fig **8.1**). Also, the cavitation photos of the three groups are shown in Fig (**8.2**) [7-15]. The type of cavitation (partial or super) depends on the operating conditions (cavitation number, flow velocity, angle of attack, *etc.*) and the fluid properties as well as the profile of the hydrofoil section.

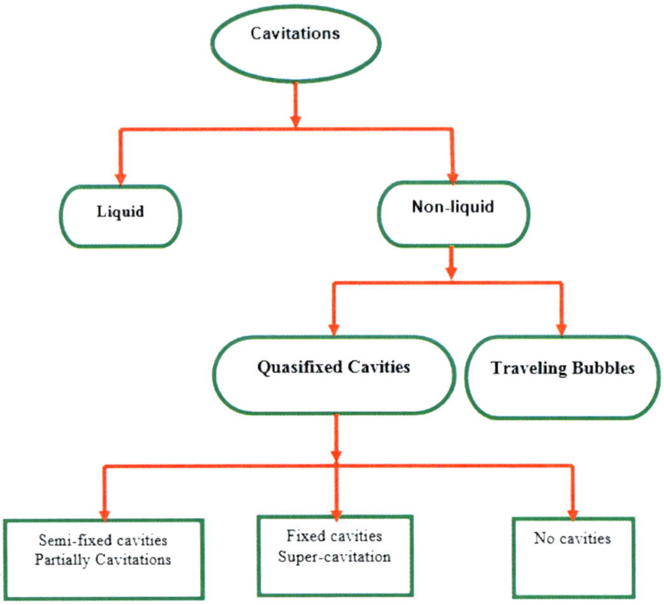

Fig. (8.1). Classification of cavitation.

Fig. (8.2). Cavity photos [14].

Partial cavitations are at the top, moving bubbles are in the middle, and super-cavitation is at the bottom. In extreme cases, the flow regime will become that of super-cavitation, *i.e.*, the cavity termination will be downstream of the cavitating body (hydrofoil, *etc.*). If the cavitation number, σ, is raised from the super-cavitating condition, the cavity length will be reduced so that its termination point will move upstream, and then cavity attachment will occur on the body. A further increase in σ will cause the fixed cavity to disappear completely. However, after the complete disappearance of the fixed cavity, the small cavitation bubbles will remain in the region of minimum pressure. These generally will be entrained in the flow and hence are traveling cavities. Even for those cavitation regimes dominated by a fixed or semi-fixed cavity, it is generally true that traveling bubbles exist in the interface region between the cavity and the main liquid flow. So, the collapse in the region of cavity termination of the traveling bubbles is presumed responsible for the cavitation damage observed in this region. The traveling bubbles begin to collapse under the influence of increasing external pressure. In some cases, and to simplify the problem, the cavity was usually macroscopically treated as a single big bubble of finite length, which encompasses the whole region where cavitation occurs and micro-bubbles dominate. The pressure inside the cavity bubble is usually assumed to be constant.

Cavitation Models

The modeling of cavitating flows has taken two distinctly different approaches. ***The first approach is an interface-tracking scheme,*** which presumes a distinct interface between liquid and vapor and solves governing equations characterized by a cavitated region of nearly fixed dimensions, and is not suitable for traveling cavitation problems. Typically, the interface assumes the cavitated region is at a constant pressure equal to the local vapor pressure. Numerically, it is impossible to impose a constant pressure condition on the entire bubble because of the recovery of pressure at the end of the cavitation region. For this reason, the interface-tracking scheme must employ a wake model to approximate the two-phase behavior in this complex area. Unfortunately, it is difficult to introduce physics into the wake model since the single-phase flow solver cannot address two-phase effects. Most authors agree that the choice of the wake model, within limits, has little influence on the force body of the cavity. Depending on the flow conditions and the extent of the wake, the interface tracking method can predict the overall behavior of cavitating flow fairly well. ***The second approach is a two-phase scheme***, which solves the governing equations in the entire liquid/vapor domain by introducing a pseudo-density, which can vary between liquid and vapor density extremes this two-phase approach is beneficial since it does not require special wake treatment, however, it has a drawback of not determining the interface of gas and liquid , which must be inferred from density values in neighboring cells. Also, it can treat traveling as well as attached cavitation. This technique can also be used in turbulent cavitation flows in which a bubble field (rather than a distinct interface) is present [16-23].

Riabouchinsky Model

When a thin, rigid plane is fixed in a fluid in motion and normal to the direction of flow, it is observed that, near to its edges, the formation of a "free surface", or surface of slip, which remains well defined for a certain distance from the plane. Experience showed that the pressure behind the plane is distributed in an approximately uniform manner over its surface and is less than the pressure in the undisturbed current. Riabouchinsky [15] studied the case in which two planes are disposed of, one behind the other, at right angles to the stream and their edges connected by free surfaces. The velocity at the free surface is greater than the velocity of the stream. The two planes mutually attract each other, but the resultant pressure of the fluid on the system of the two planes is evidently zero, see Fig. (**8.3**).

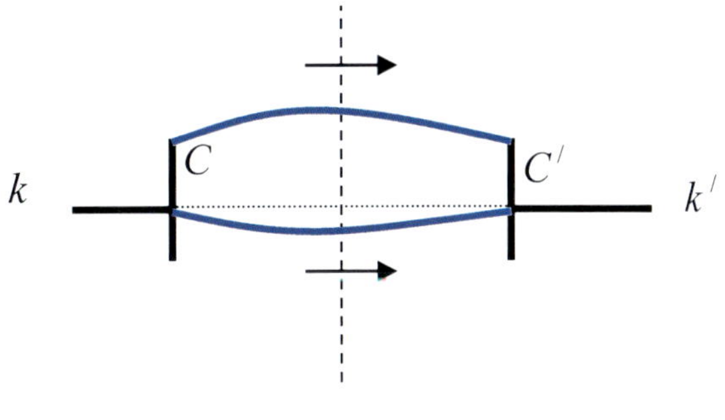

Fig. (8.3). Riabouchinsky model.

Open-wake Model

This model was first introduced by Joukowsky [15] in a fundamental work that was apparently extended by McNown and Yih and Eppler for symmetric bodies in non-lifting flows. Wu [26] modified the model in order to extend it to lifting flows past arbitrary bodies. The wake model for the free-stream line theory is proposed to treat the two-dimensional flow past an obstacle with a wake or cavity formation, as shown in Fig. (**8.4**).

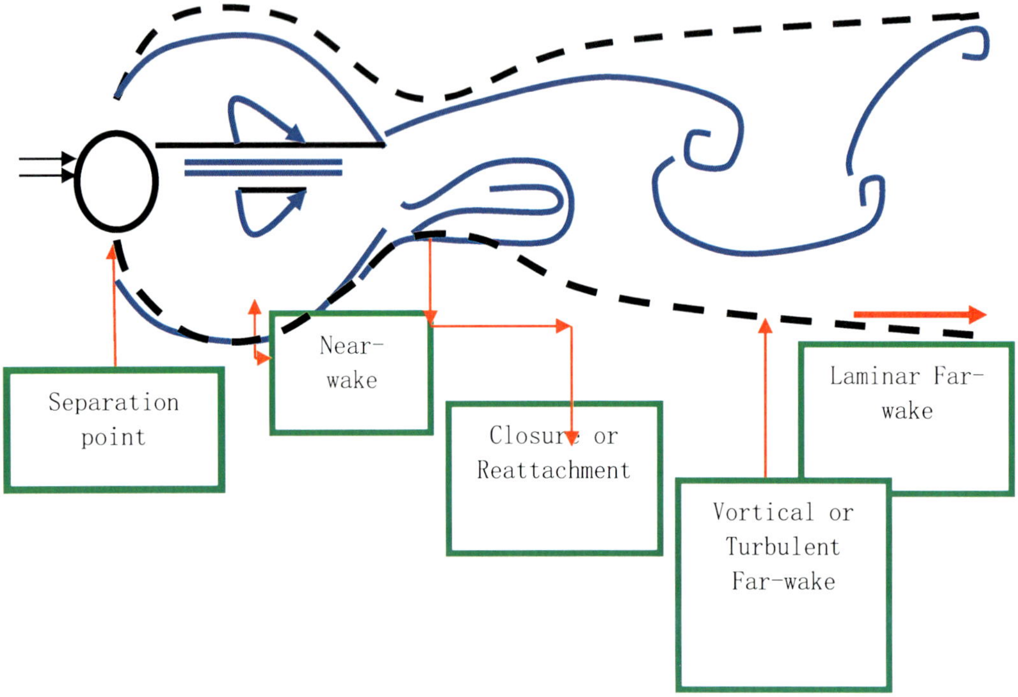

Fig. (8.4). Two-dimensional wake formation.

In this model, the wake flow is approximately described, in the large, by an equivalent potential flow such that, along the wake boundary, the pressure first assumes a prescribed constant under-pressure in a region down stream of the separation points (called the near-wake) and then increases continuously from this under-pressure to the given free-stream value in an infinite wake strip of finite width (called the far-wake). Application of this wake model provides a rather smooth continuous transition of the hydrodynamic forces from the fully developed wake flow to the fully wetted flow as the wake disappears. When applied to the wake flow past an inclined flat plate, this model yields the exact solution in a closed form for the whole range of the wake under-pressure coefficient, see Fig. (**8.5**) [27].

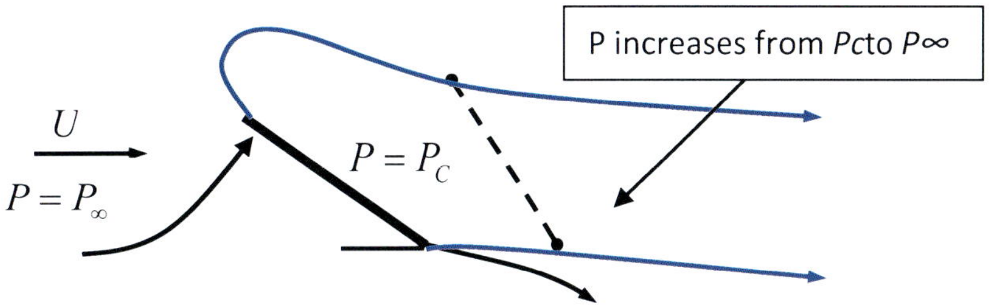

Fig. (8.5). Open-wake model.

Reentrant-jet Model

This model seems to have been first suggested by Prandtl and H. Wagner [27], later formulated by Kreisel [27] and independently by Efros [27]. In this model, the two dividing streamlines reverse direction, carrying a finite part of the main stream into a jet, which flows through the cavity in an undetermined direction for lifting flows and is continued mathematically into a second Riemann sheet, see Fig. (**8.6**).

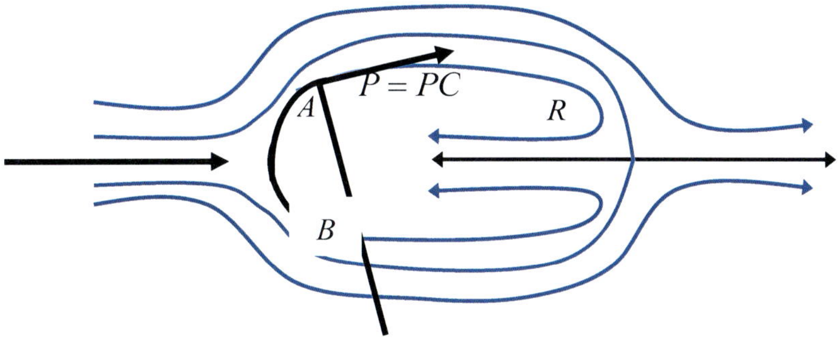

Fig. (8.6). Reentrant- jet model.

Cusped-wake Model

Brillouin [28] appears to have been the first to discuss the finite cavity, which is terminated with a cusp, especially since the minimum pressure no longer falls on the cavity boundary. This model was applied to the circular cylinder by South well and Vaisey [28] and was taken as a realistic model for truncated airfoil by Lighthill, see Fig. (**8.7**) [27].

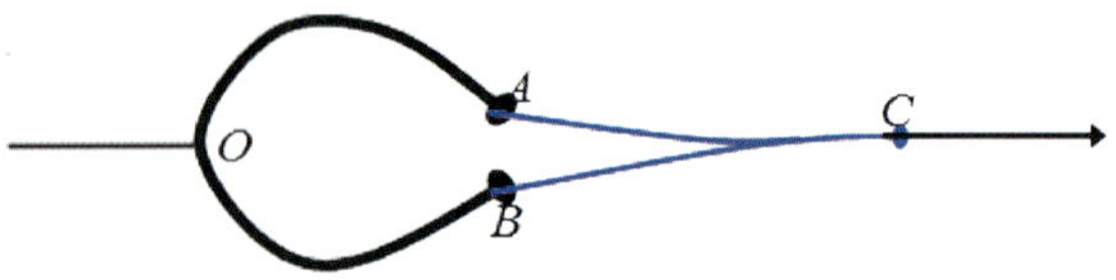

Fig. (8.7). Cusped – wake model.

Batchelor Wake Model

With wakes at high Reynolds numbers in mind, Batchelor [29-30] proposed a closed wake model, which encloses a pair of stationary viscous, eddies of constant vorticity. Of particular concern was the uniform validity of possible wake models as the viscosity. An outstanding property of this model is that the resistance of the fluid on the body vanishes in this limit.

Vortex Models

Systems of stationary vortices are observed in the numerical calculations of wakes at low Reynolds numbers, for example, the early computations Thom [31] and Apelt [32] show a pair of stationary vortices behind a circular cylinder up to a Reynolds number of 40. For high Reynolds numbers, various kinds of twin vortices, such as the single-spiral and double-vortices, have been proposed as possible wake and cavity models by Tulin [33].

Lavrentiev Wake Model

Lavrentiev [34] proposed a closed wake-bubble model, which terminates with a stagnation point at the closure and has a recalculating flow that encloses two separate regions of constant pressure, its potential utility in the determination of viscous near-wake remains to be realized.

Displacement-thickness Model

This model was produced by Lemonnier and Rowe [35]. Their model aims to be flexible, simple and adjustable; flexible to allow it to be used under a variety of conditions, simple so that extrapolations to unsteady or three-dimensional flows can be made and adjustable so that it can be made to fit experimental results. The model is suitable for complex flows such as partial cavitation on a thin foil, cavitation behind a rounded body, consideration of the gravity effect in the case of an axi-symmetrical body and simultaneous calculation of two cavities, see Fig. (**8.8**).

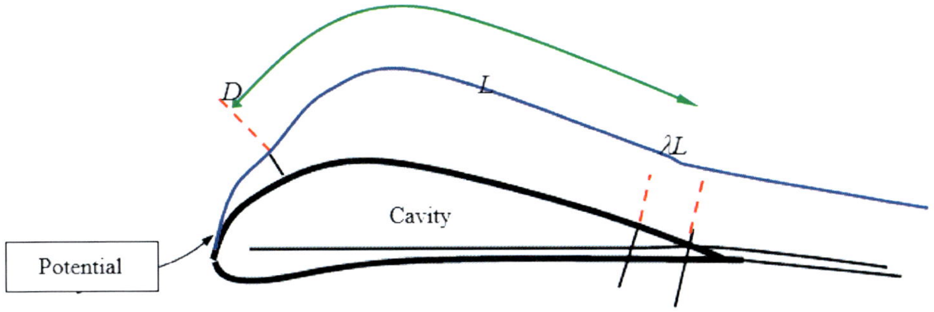

Fig. (8.8). The Displacement – thickness model.

Also, Ingber and Hailey [36] used a displacement thickness model to connect the cavity to the body in the reattachment zone. Their closure model is used not only to approximate the shape of the aft-portion of the cavity but also to determine the point of detachment that point, at which the streamline divided, the angle at which the dividing streamline makes with the attached flow, and the cavity length, see Fig. (**8.9**).

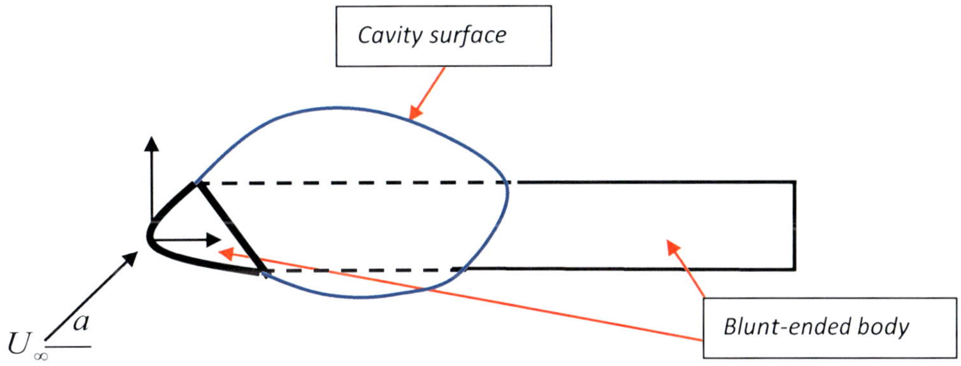

Fig. (8.9). Ingber and Hailey model.

Other Models

A side from the above-mentioned models, Hopkinson considered the flow bounded by two plates and two free streamlines with interior sources and vortices. Artificial plates have been used to intercept free streamlines and control the open wake width (Gadd). The one-side reentrant jet at the edge of a wedge or plate was treated by Cox and Clayden [37-45] and Cumberbatch [46-48]. Also, Wu and Chen [49-50] studied two different closure models; ***the first one*** is the simple closure model. The cavity shape and the tangential speed on the cavity surface are part of the solution to be determined. This is a purely computational model due to its simplicity in numerical implementation. In fact, from the mathematical point of view, a closed finite body with a constant pressure does not exist in the exact potential flow theory. This kind of termination condition is contradictory to the constant pressure condition inside the cavity (or the constant tangential velocity condition on the cavity surface). ***The second model*** is the short plate termination model (or modified Riabouchinsky model). Fig. (**8.10**) shows that the cavity ends at a plate normal to the incoming flow. In addition to the cavity shape and the tangential velocity, the height of the short plate is unknown and needs to be determined as the flow solution is sought.

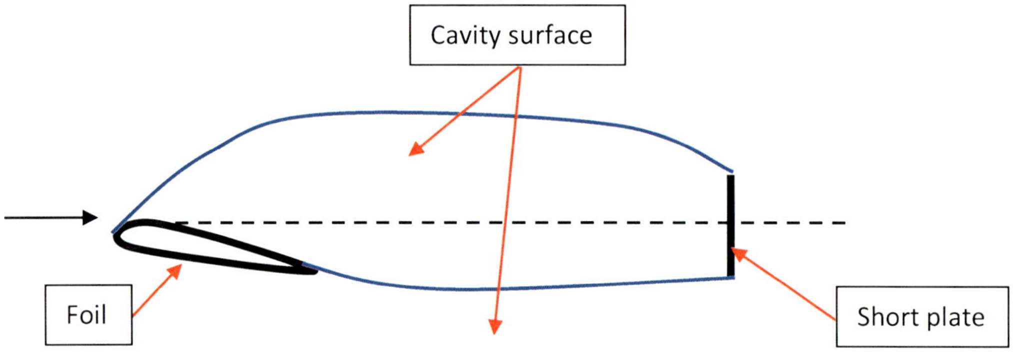

Fig. (8.10). Short plate termination model.

Survey of Some Analytical Methods

The analytical calculation of cavitating flows is based on two techniques: ***the small-perturbation theory*** proposed by Tulin, and ***the non-linear theory*** that was originated from the work by Helmoltz [51-54], Kirchhoff [55-62] and Levi-civita [34]. Following the linear super cavitating flow theory, first introduced by Tulin [33], Acosta [63-166] provided the first partial cavitation solution specifically for a flat plate hydrofoil. He also expressed the cavitation number explicitly in terms of the non-dimensional cavity length and the angle of attack. Of course, all these results are applicable only to flow at a small angle of attack as the assumptions of the linear theory imply. Geurst [167] employed a conformal transformation technique to derive an implicit theoretical relation among the angle of attack, cavitation number, and non-dimensional cavity length for a super-cavitating flow. The relation contains coefficients in integral form, which can be evaluated using a typical numerical Quadrature procedure. Nevertheless, the cavity length can be readily obtained for a given cavitation number, and *vise-versa*, provided that the incident angle of uniform flow is specified. Furthermore, he also studied the special case of a flat plate hydrofoil and obtained a simple analytical expression for these physical quantities. Davies [170-174] formulated the linearized cavitating hydrofoil problem in terms of singular integral equations with respect to unknown vorticity and source distributions, and inverted the resulting integral equations. Unfortunately, the final expressions for the vorticity and source distributions were coupled to each other. Persson [35] focused on a super-cavitating flat plate and inverted the equations to obtain analytical expressions for their distributions. Kinnas [175-176] extended the analytical inversion to super cavitating hydrofoils of arbitrary inversion to super-cavitating hydrofoils of arbitrary shape. He expressed the cavitation number, the vorticity and source distributions in terms of integrals of quantities, which depended only on the foil geometry and the cavity length [175]. The analytical calculation of cavitating flows using conformal transformation is difficult, and extension to three-dimensional flow is impossible outside the framework of perturbation methods. The most comprehensive analytical developments are probably those proposed by Furaya [176], whose non-linear theory dealt successively with the case of a super-cavitating section near a free surface, a super-cavitating foil near a free surface [173] and a partially or fully cavitating cascade [176]. Due to the complexity of the cavitation problem and the rare of analytical solution, many researchers prefer numerical methods.

Survey of Some Numerical Methods

Finite difference method was proposed by Garadedian [177-179]. The solutions are obtained for cavities behind a disk and a sphere. He approached the axisymmetric case by successive corrections to the corresponding planar flow, each correction involving the solution of a linear mixed boundary value problem. In addition, Jeppson [180-182] successfully solved a number of plane and axisymmetric cavity flow problems and avoided the difficulty of dealing with the unknown location of the free surface by considering the velocity potential and stream function as independent variables and the coordinates as dependent variables. Brennen [183-186] also used the finite difference method to obtain the solution for cavities behind a disk and a sphere, the Garadedian case, in different sizes of solid wall tunnels. Later, Mogel and Street [187-188] developed a numerical method for steady state cavity flows that provides a systematic correction of an initial, assumed free streamline position. The method used a finite difference representation with velocities as dependent variables and obtained a solution by successive over-relaxation. ***Boundary integral equation method*** was used in the works of Amromin and Ivanov [188], Gyzevsky [185] and Kojouro [186]. In these works, the vortex surface distribution was used and the resulting system of nonlinear algebraic equations was solved by the different numerical methods such as finite element method and boundary element method [189]. Aitchison [87] described a method for the numerical solution of planar and axisymmetric Riabouchinsky cavity flows. The problem is formulated as the minimization of a function over a variable region, which is solved numerically by the method of variable finite elements. However, several numerical difficulties are associated with domain numerical methods for cavitation problems, which limit their effectiveness. In particular, the cavitation problem is essentially an external flow problem causing discretization problems for either the finite difference method or the finite element method. Not only does the exterior domain need to be truncated, but also, in many cases, a finer mesh is required near the separation point resulting in variable grid spacing. The most popular numerical technique for analyzing cavitation problems has been the ***boundary element method (BEM)*** because of the reduction of the gridding difficulties that the boundary element method affords. Several researchers have successfully analyzed planar and axisymmetric cavitation problems using the boundary element method. Aitchison and Karagearghis [187] presented a boundary element method formulation for axisymmetric Riabouchinsky cavity flow problems, and Lemonnier and Rowe [190] wused the panel technique with minimization of a certain vector characterizing the discretion error which may become important under cavitating condition. Several practical examples were presented: partial cavitation on an isolated foil, cavitation behind a blunt ended body, and the problem of two cavities around an axisymmetric body. Also, Worbel [191] presented a different boundary element method approach to the same problem presented by Aitchison and Karageorghis [192]. It is demonstrated that the determination of the free surface-location is equivalent to the solution of a system of nonlinear equations.

Ingber and Hailey [193] used the boundary element solution for cavitating flows about axisymmetric bodies both at zero and nonzero angles of attack. The nonzero angle of attack case results in a non-axisymmetric cavity were presented. Kinnas and Fine [174] treated the partially cavitating two-dimensional hydrofoil problem using nonlinear theory by employing a low order potential-based boundary element method. Also, they extended their work to treat the partially cavitating three-dimensional hydrofoil problems. Ahmed and Worbel [195] solved the planner cavity flow problem through the boundary integral method and a new algorithm depending on the correction of the free surface location iteratively. Also, Ahmed *et al.* [195] studied deeply cavitations for different and several practical cases of obstacles starting with a vertical flat plate, oblique flat plate with a wide range of inclined angle from 0 л/2 and wedge of the different angle of inclination. Yas'ko [195] presented a new numerical iterative algorithm based on the direct boundary element method, which is proposed for the solution of the steady cavitating. Kinnas *et al.* [196] summarized some applications of the boundary element method to predict sheet or developed tip vortex cavitation on lifting bodies using the nonlinear cavity theory. They applied Dirichlet type boundary condition on cavity surface and Neumann type on wetted body surface for two-dimensional and three-dimensional hydrofoils and submerged marine propellers in non-uniform inflow. Vaz *et al.* [197] developed a new boundary element method in which the elements are located on the foil surface, and the boundary conditions for the cavity surface have been reformulated based upon a Taylor expansion. Kinnas *et al.* [196] studied the steady, fully wetted, cavitating flows over hydrofoil. The non-linear analysis of the inviscid cavitating hydrofoil flow is based on a potential-based boundary element method. Mohamed *et al.* [27] used the boundary element method to solve the cavitation problem around two-dimensional hydrofoil with various NACA sections. A new algorithm is obtained to determine the location of the free surface.

Background of Boundary Element Technique

In recent years, engineers and physical scientists have become very conversant with the numerical technique of analysis. These techniques are based on the approximate solution of an equation or set of equations describing a physical problem. The first widely known approximate method was a finite difference, which approximates the governing equation of the problem by defining a series of nodes at which the discrete version of the differential equation is satisfied. Another approximate method is the finite element method, which has attracted the attention of analysts largely due to its property of dividing the continuum into a series of elements, which can be associated with physical parts. In addition, the differential equation is satisfied in an average sense over a region or element in finite elements. The boundary element method is firmly established as an important alternative technique to the prevailing numerical methods of analysis in continuum mechanics because, unlike the domain discretization techniques such as finite element and finite difference methods, only the boundary needs to be discretized, leading to considerable savings in computer storage and running time. Also, the boundary element method can now be used, especially when the better accuracy is wanted or where the domain extends to infinity. Its important property is that it only needs discretization of the surface rather than the volume. So, it requires less meshing than finite element analysis and thus, it is comparatively faster in generating or refining the mesh. This advantage is particularly important for designing as the process usually involves a series of modifications that are difficult to carry out using finite elements. Moreover, the boundary element method doesn't require complete remeshing when design changes. This technique consists in the transformation of the partial differential equation describing the behavior of the unknown inside and on the boundary of the domain into an integral equation relating only boundary values and then finding out the numerical solution of this equation. If values at internal points are required, they are calculated afterward from the boundary data. Since all numerical approximations take place only at the boundaries, one reduces the dimensionality of the problem and a smaller system of equations obtained compared to those achieved through differential methods. Historically, the application of integral equations to formulate the fundamental boundary-value problems of potential theory dates back to 1903 when Fredholm [198] demonstrated the existence of solutions to such equations. Due to the difficulty of finding an analytical solution, the use of integral equations has, to a great extent, been limited to theoretical investigations of existence and uniqueness to solutions of problems of mathematical physics. However, the advent of high-speed computers made it possible to implement discretization procedures analytically and enabled numerical solutions to be readily achieved. Fredholm integral equations followed from the representation of harmonic potentials by single-layer or double-layer potentials and set up the foundations of the so-called indirect boundary element method. Vector integral equations analogous to the Fredholm integral equations of the potential theory were introduced by Kupradze in the context of the theory of elasticity [199]. Integral equations for linear problems can alternatively be formulated through the application of Green's third identity, which represents a harmonic function as the superposition of a single-layer and a double-layer potential. Taking the field point to the boundary, an integral equation is obtained that relates only boundary values and normal derivatives of the harmonic function. The Green's function method of solving boundary-value problems is most directly applicable to elliptic partial differential equations. In fact, the concept of Green's function grew out of a detailed study of such boundary-value problems, but the method can also be extended to solve parabolic and hyperbolic partial differential equations. Jaswon and Symm [200] presented a numerical technique to solve Fredholm boundary integral equations. The technique consists of discretizing the boundary into a series of small segments (elements). Assuming that the source density remains constant within each segment, they obtained accurate solutions for simple two-dimensional Neumann (all conditions are natural) and Dirichlet (all conditions are essential) problems. They also proposed a more general numerical formulation for solving Cauchy boundary-value problems (also called mixed because the boundary conditions are essential on parts of the boundary and natural on other parts) through the application of Green's third identity, which yields a boundary integral equation where boundary values and normal derivatives of the physical variable play the role of the fictitious source densities. Efforts such as the pioneering work by Hess and Smith [201] remained as special cases rather than being interpreted as a way of generating a whole new method of solutions for general engineering problems. Nevertheless, Hess and Smith [201] developed many powerful programs for the solution of Laplace type boundary-value problems, which were applied to potential flow and arbitrary bodies, using what is now called the indirect boundary element technique. They extended their formulation to analyze the three-dimensional objects and two-dimensional, and their codes are still popular in aerodynamics. The boundary element method is now used widely as an important technique in one of the most important applications such as temperature diffusion, some types of fluid flow motion, flow in porous media, electrostatic and many other applications which their equation can be written as a function of a potential and governed by the Laplace, Poisson or Helmholtz equation in two-dimensional, axisymmetric or fully three-dimensional problems. All these cases are potential problems and they can generally be efficiently and economically analyzed using boundary elements. The boundary element method solves many engineering applications, especially the cavitation problem on the hydrofoils. This problem was treated by many researchers, for example, Widnall treated the case of super-cavitating foil in unsteady flow using the surface-lifting method. Nishiyama and Miyamoto treated the problem of a super-cavitating foil placed under a free surface. Sakir [202] used the thin-foil approximation with linearized free surface condition. Tsen and Guilbaud studied the effect of the plan shape of a super-ventilated foil from both experimental and theoretical standpoints. In (1979), Verron extended Tsen and Guilbaud's method to the case of base-vented foils. In his work the plane shape of the cavity is calculated for nonzero cavitations numbers. Ghassemi [203] dealt with thick-foil methods, which precisely represented the flow near the hydrofoil surface. Yamaguchi and Kato presented methods well suited to the case of foils having a large relative thickness [204]. Kinnas and Fine [205] proposed a nonlinear analysis of the flow around partially or super-cavitating hydrofoils using a potential-based panel method. They stated that their method is promising as it converges to the final cavity shape quicker than the other numerical schemes using surface-vorticity-velocity-based panel methods. Xie and He [206] conducted numerical investigations on submerged cavitating slender axisymmetric bodies with gravity effect. They used the boundary element method to solve the potential flow problem. They also developed a new iteration method named Adaptive Modified Newton iteration to determine the cavity shape. Bal [202] described a potential-based panel method for the hydrodynamic analysis of two-dimensional hydrofoils moving beneath the free-surface with constant speed without considering cavitation. Berntsen *et al.* [207] presented numerical and experimental work to investigate the effect of cavitation on the dynamic features of hydraulic machinery. The numerical calculations were done with the commercial code fluent V. 5. Their results showed that cavitation induces large fluctuations in the lift on hydrofoils. Dular *et al.* [208] made a study of visual and erosion effects of cavitation on simple single hydrofoil configurations in a cavitation tunnel. They used a two-dimensional hydrofoil with a circular leading edge for the experiments. They established a relation between characteristics of cavitation structures and cavitation damage. The models they employed included the constant velocity model and the short plate termination model.

MATHEMATICAL MODELLING AND FORMULATION

Cavitations around hydrofoils belong to a free boundary problem in which a set of partial differential equations with associated boundary conditions must be solved within the domain in which one or more of its boundaries moves freely. This boundary is called free surface; its location is not known in advance and should be found as a major part of the solution of the problem. In the present chapter, different models describing cavitation flow will be presented. In addition to the mathematical formulation, some important models describing the flow should also be introduced. For the steady, irrotational, planar flow of an incompressible, inviscid fluid past an obstacle and infinite domain problems, the following equations apply:

Continuity equation:

$$\frac{\partial u}{\partial x}+\frac{\partial v}{\partial y}=0 \tag{1}$$

Irrotationality equation:

$$\frac{\partial u}{\partial y}-\frac{\partial v}{\partial x}=0 \tag{2}$$

Bernoulli's equation:

$$P+\frac{1}{2}\rho q^{2}+\rho g y=\beta \tag{3}$$

In equation (1), u and v are the horizontal and vertical components of the velocity whose magnitude is given by the following formula:

$$q=\sqrt{u^{2}+v^{2}} \tag{4}$$

Where:

x & y Horizontal and vertical coordinates of a fixed system,

P Pressure,

ρ Fluid density,

g Acceleration due to gravity,

β Bernoulli's constant.

If the effects of gravity are neglected, equation (3) takes the following form:

$$P + \frac{1}{2}\rho q^2 = \beta = const. \tag{5}$$

The pressure is taken to be constant in the cavity. Denoting quantities referenced to the cavity or free-stream line by the subscript c, then:

$$P_\infty + \frac{1}{2}\rho U_\infty^2 = P_c + \frac{1}{2}\rho q_c^2 = const. \tag{6}$$

Where

P_∞ Flow pressure at infinity,

U_∞ Inflow velocity at infinity,

P_c Pressure in the cavity,

q_c Velocity on the cavity free-surface

The Prandtl cavitation number σ is defined as [208]:

$$\sigma = \frac{P_\infty - P_c}{\frac{1}{2}\rho U_\infty^2} \tag{7}$$

From (6) and (7), one can get the following formula:

$$\sigma = \frac{q_c^2}{U_\infty^2} - 1 \tag{8}$$

The velocity potential Φ can be introduced such that:

$$u = \frac{\partial \Phi}{\partial x} \tag{9}$$

$$v = \frac{\partial \Phi}{\partial y} \tag{10}$$

Substituting (9) and (10) into the continuity equation leads to:

$$\frac{\partial^2 \Phi}{\partial x^2} + \frac{\partial^2 \Phi}{\partial y^2} = 0 \tag{11}$$

It can be seen that the potential given by (9) and (10) identically satisfies the Irrotationality equation.

The total potential Φ and perturbation potential are related as follows:

$$\phi(x,y) = \Phi(x,y) - \Phi_{in}(x,y) \tag{12}$$

Where the inflow velocity potential Φ_{in} corresponds to the uniform inflow velocity of magnitude U_∞ at an angle of attack α of the fluid with the obstacle [**75**] as follows:

$$\Phi_{in}(x,y) = U_\infty(x\cos\alpha + y\sin\alpha) \tag{13}$$

In addition, the perturbation potential ϕ will satisfy Laplace's equation in the flow domain outside the cavity and the obstacle:

$$\nabla^2\phi = 0 \tag{14}$$

Boundary Conditions for Infinite Domain Problems

When the obstacle is placed in an infinite length and infinite depth of a channel (an infinite domain), the flow separates the obstacle and makes a cavity flow. Then, the boundary conditions must be known on the different surfaces to solve the cavity problem.

Kinematic Boundary Condition

According to this condition, the flow is required to be tangent to the wetted surface as well as to the cavity surface. The kinematic boundary condition is given by:

$$\frac{\partial \Phi}{\partial n} = 0 \quad \textbf{(15)}$$

Where n is the outward unit normal at any point on the wetted surface or the free-surface.

Substituting from equation (12) into (15), one can get:

$$\frac{\partial \phi}{\partial n} = -\frac{\partial \Phi_{in}}{\partial n} \quad \textbf{(16)}$$

Apply equation (16) on the wetted surfaces and the cavity surface, the right-hand side of equation (16) can be written as:

$$\frac{\partial \Phi_{in}}{\partial n} = U_{\infty}(n_x \cos\alpha + n_y \sin\alpha) \quad \textbf{(17)}$$

Where, n_x and n_y are the components of n in the *x* and *y* direction, respectively, for each element of the free-surface [**75**].

Substituting from equation (16) into (17), then equation (16) takes the following form:

$$\frac{\partial \phi}{\partial n} = -U_{\infty}(n_x \cos\alpha + n_y \sin\alpha) = -\bar{U}_{\infty} \cdot \hat{n} \quad \textbf{(18)}$$

Dynamic Boundary Condition

This condition is applied to the cavity surface. Simply, it means that the pressure is required to be constant at each point of the free-surface and equals to P_c. Also, the magnitude of the total velocity on the cavity surface, q_c, is found to be constant, where:

$$q_c = U_{\infty}\sqrt{1+\sigma} \quad \textbf{(19)}$$

Then, the dynamic condition on the cavity surface is given by:

$$\frac{\partial \Phi}{\partial s} = q_c \tag{20}$$

In equation (20), s denotes the curvilinear length from the plate tip on the cavity free-surface.

Substituting of equation (12) into (20), then:

$$\frac{\partial \phi}{\partial s} + \frac{\partial \Phi_{in}}{\partial s} = q_c \tag{21}$$

Or,

$$\frac{\partial \phi}{\partial s} = q_c - \frac{\partial \Phi_{in}}{\partial s} \tag{22}$$

Integrate both sides with respect to *s,* one can get:

$$\phi\Big|_A^{s_i} = q_c s_i - \Phi_{in}\Big|_A^{s_i} \tag{23}$$

Or, in some details:

$$\phi_{s_i} - \phi_A = q_c s_i - \Phi_{in}(s_i) + \Phi_{in}(A) \tag{24}$$

where, s_i is the curvilinear length from the plate tip (A) to the node of each element. From equation (13), the Φ_{in} over the curvilinear distance s_i is:

$$\Phi_{in}(s_i) = U_\infty(x\big|_{s_i} \cos\alpha + y\big|_{s_i} \sin\alpha) \tag{25}$$

Replacing both horizontal and vertical components of s_i , then equation (24) takes the new form:

$$\Phi_{in}(s_i) = U_\infty(z_i \cos\alpha + w_i \sin\alpha) \tag{26}$$

Where:

z_i is the horizontal distance from the point (A) to the node of the element i.

w_i is the vertical distance from the point (A) to the node of the element i.

Therefore; the dynamic condition takes the following form:

$$\phi_{s_i} = \phi_A + q_c s_i - U_\infty (z_i \cos\alpha + w_i \sin\alpha) + \Phi_{in}(A) \quad \textbf{(27)}$$

But, $\Phi_{in}(A) = 0$, then:

$$\phi_{s_i} = \phi_A + q_c s_i - U_\infty (z_i \cos\alpha + w_i \sin\alpha) \quad \textbf{(28)}$$

Also, equation (8.28) can be written as:

$$q_i = \frac{\phi_i + U_\infty (z_i \cos\alpha + w_i \sin\alpha) - \phi_A}{s_i} \quad \textbf{(29)}$$

Where [**75**]:

q_i Velocity of the element i at the cavity surface,

ϕ_i Perturbation potential of the element i at the cavity surface,

ϕ_A Perturbation potential at point A

By equation (8.29), a complete mathematical formulation for the cavitation flow problem was introduced. In the next section, a survey of previous models suggested by different researchers will be presented.

The Riabouchinsky Model

This model was proposed by Riabouchinsky [180]. The model assumed that an image plate can be placed in the flow at some point downstream from the original plate and that the flow geometry will be symmetric about the line mid-way between the plates.

The Riabouchinsky Model for Vertical Flat Plate

In this case, the domain over which Laplace's equation should be satisfied may be one of the following three cases, as shown in Fig (**8.11**).

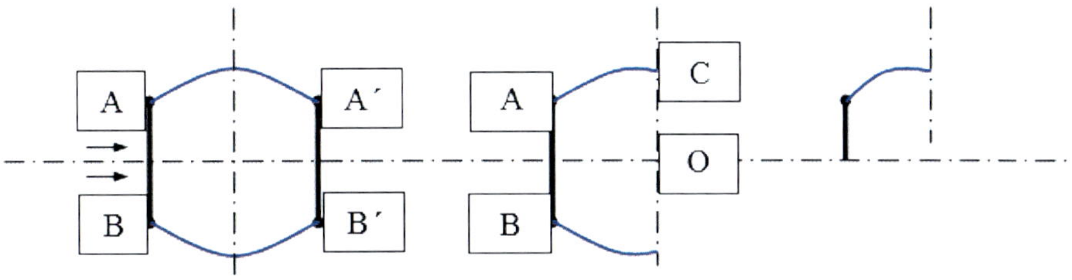

(a) Complete domain. (b) half-domain. (c) quarter-domain.

Fig. (8.11). The Riabouchinsky model for a vertical flat plate.

Without details, the governing equation with the associated boundary condition for the three cases of the domain will be as follows:

The Riabouchinsky Model for Oblique Flat Plate

In this case, a major difference will appear, that is two free surfaces encountered called upper and lower free-surfaces. Their final shape depends mainly on the angle of inclination of the flow and other parameters. The domain may be one of the two cases shown in Fig. (**8.12**). The boundary condition should take the angle of attack and inclination of the plate in the calculation and also, the iterative scheme will be modified to track the two free-surfaces simultaneously.

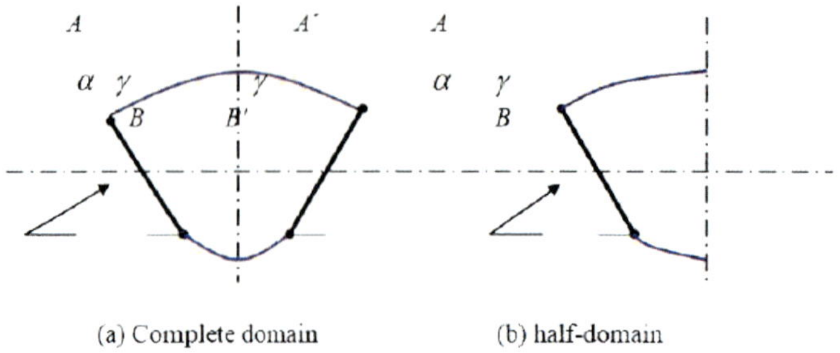

Fig. (8.12). The Riabouchinsky model for an oblique flat plate.

The Riabouchinsky Model for Wedge

In this case, the domain consists of two parts. The first one is the wedge, called wetted body and the second one is the cavity bounded by the wedge and the similar free-surface, as shown in Fig. (**8.13**).

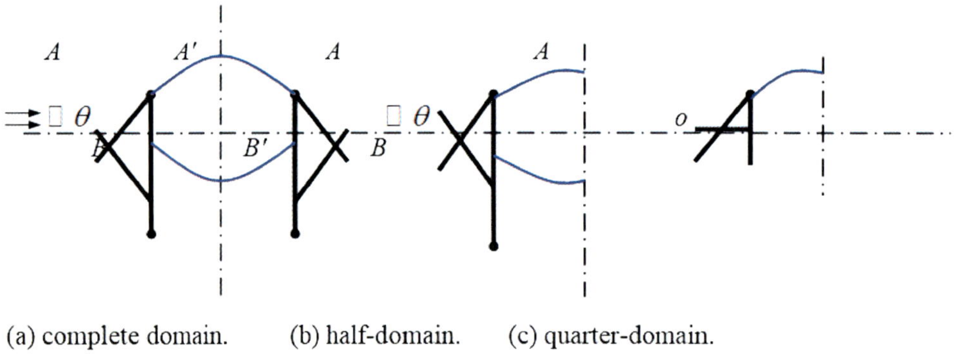

Fig. (8.13). The Riabouchinsky model for the wedge.

The governing equation and boundary conditions for the three cases of the domain will be as follows:

Complete Domain

Laplace's equation given by equation (14) was applied as a governing equation, with the kinematic boundary condition given by equation (18) was applied on the wedges and the cavity surfaces. Also, over the cavity surface, the dynamic boundary condition given by equation (29) was applied. The information in the previous case of the Riabouchinsky model for a vertical flat plate with half and quarter domain, respectively, will be taken into account if half or quarter domain of this case is used.

The Hydrofoil Cavity Model

The partially cavitating hydrofoil with chord length $c = 1$ subjected to a uniform inflow U_∞ at an angle of attack α and with ambient pressure P_∞ is considered, as shown in Fig. (**8.14**).

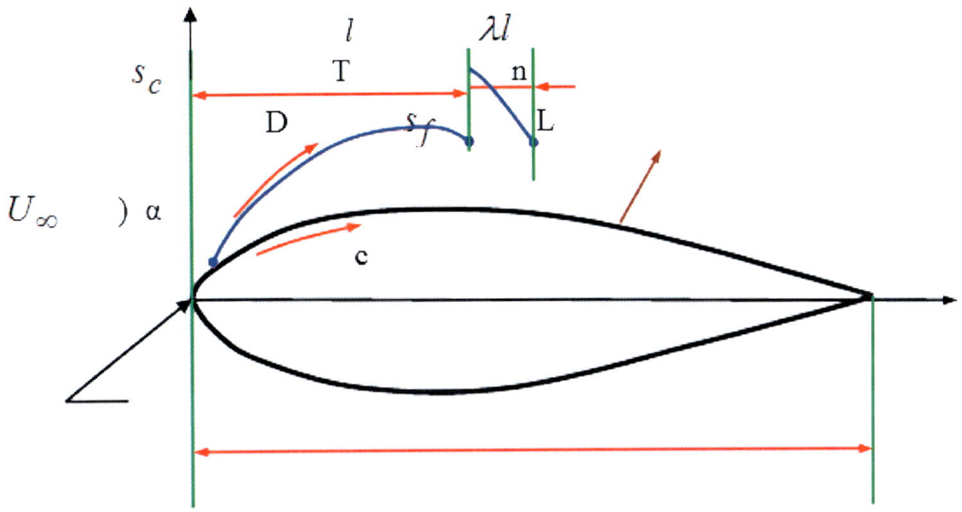

Fig. (8.14). partially cavitations on the two–dimensional hydrofoil [**53**].

The cavity detaches from the foil surface at a point D on the suction side of the foil and rejoins the foil at the point L. The point D is kept fixed, and unless otherwise mentioned, it will coincide with the foil leading edge. It is assumed that the fluid is inviscid, incompressible and that the flow is irrotational. The total velocity flow field q can be expressed in terms of either the total potential Φ, or the perturbation potential ϕ, as follows [180]:

$$q - \nabla\Phi = U_\infty + \nabla\phi \qquad \textbf{(30)}$$

The total and perturbation potentials are related as in equation (8.12), where the inflow velocity potential Φ_{in} corresponds to the uniform inflow in equation (13). The perturbation potential ϕ will satisfy Laplace's equation (8.14) in the domain outside the hydrofoil and inside the cavity. Further boundary conditions should be applied as follows:

Kinematic boundary condition

Over the wetted foil and the cavity free-surface, equation (8.18) should be applied.

Dynamic boundary condition

The pressure is required to be constant on the cavity surface and equal to P_C. Applying Bernoulli's equation given by (3) and the definition of the cavitations number (7), then the magnitude of the total velocity on the cavity surface, q_C, is found to be constant as in equation (19).

Kutta condition

This condition requires finite velocities at the foil trailing edge.

$$|\nabla\phi| < \infty \tag{31}$$

Condition at infinity

At infinity, the perturbation velocities should go to zero.

$$\nabla\phi \rightarrow 0 \tag{32}$$

Cavity termination model

A termination model must be applied at the end of the cavity. A pressure recovery termination model is employed, by which the velocity in a transition zone of length λl departs from its constant value on the cavity between D and T according to a prescribed algebraic law in the transition zone between T and L :

$$q_{tr} = U_{\infty}\sqrt{1+\sigma}\,(1-f(s_f)) \tag{33}$$

Where $f(s_f)$ is defined as follows [180]:

$$f(s_f) = \begin{cases} 0 & if \;\; s_f < s_T \\ A\left(\dfrac{s_f - s_T}{s_L - s_T}\right)^{v} & if \;\; s_T \le s_f < s_L \end{cases} \tag{34}$$

Where s_f the arc-length of the foil beneath the cavity is, measured from the cavity leading edge and $A(0 < A < 1)v(v > 0)$ is arbitrary constants. In addition to the above boundary conditions, it is assumed that the cavity height vanishes at its trailing edges, *i.e.*;

$$h(s_L) = 0 \tag{35}$$

On the wetted part of the foil that, the value of $\dfrac{\partial \phi}{\partial n}$ is known and given by the kinematic boundary condition (18), a general expression for the velocity on the cavity, including the transition zone, is given by the following algebraic law:

$$\frac{\partial \phi}{\partial s_c} + \frac{\partial \Phi_{in}}{\partial s_c} = q_c(1-f(s_f)) \tag{36}$$

Where, s_c is the arc length of the cavity, the expression for ϕ on the cavity can be found by integrating (8.36):

$$\phi(s_c) = \phi(0) - \Phi_{in}(s_c) + \Phi_{in}(0) + q_c\int_0^{s_c}(1-f(s_f))\,ds_c \tag{37}$$

Also, equation (36) can be re-written as:

$$\phi(s_C) = \phi(0) + \int_0^{s_c} (-\vec{U}_\infty \cdot \vec{S}_c + q_C (1 - f(s_f))\, ds_C \quad \textbf{(38)}$$

where, $\phi(0)$ is the potential at the leading edge of the cavity. In the numerical scheme, $\phi(0)$ is expressed in terms of the unknown potentials on the wetted part of the foil in front of the cavity. Also, the cavity shape is updated by an amount $h(s_C)$ applied normal to the cavity surface, which is determined from integrating the following ordinary differential equation:

$$q_C (1 - f(s_f)) \frac{dh}{ds} = \vec{U}_\infty \cdot \hat{n} + \frac{\partial \phi}{\partial n} \quad \textbf{(39)}$$

Then, for a given cavitation number, the thickness of the cavity , can be written as follows [**56**]:

$$\delta(s_C; \sigma) \equiv h(s_C) = \frac{1}{q_C (1 - f(s_f))} \int_0^{s_c} (\vec{U}_\infty \cdot \hat{n} + \frac{\partial \phi}{\partial n}) ds_C \quad \textbf{(40)}$$

Where, $h(s_C)$ is the cavity height, and it represents the amount by which the cavity surface is updated. The cavity closure condition at the cavity end, , will be as follows [180]:

$$\delta(l; \sigma) = 0 \quad \textbf{(41)}$$

In the next chapter, a complete boundary element formulation is developed. Then, some of the most important numerical algorithms are introduced, followed by a new algorithm developed to overcome the difficulty of estimating the potential $\phi(A)$.

BOUNDARY ELEMENT METHODS AND NUMERICAL ALGORITHMS

The basic idea to derive the integral equation is to minimize the error and deal only with the boundaries. There are two directions when deriving the integral equation, the ***direct*** and ***indirect*** approaches. In the present thesis, the first approach will be used for the derivation of the integral equation. The direct approach of formulation consists of transforming the partial differential equation describing the behavior of the unknown inside and on the boundary of the domain into an integral equation relating only boundary values and then finding out the numerical solution of the integral equation. Before deriving the boundary integral equation, it is important to make the review of some basic theorems that play an important role in the derivation of the integral equation. It is sensible to start with the ***divergence theorem*** [180] in vector calculus, which equates an integral over the volume **V** to an integral over the surface **S**.

Green's Second identity

One of the most important mathematical tools that is mostly used when deriving the boundary integral equation is Green's second identity. Consider two arbitrary functions ϕ and ϕ^* continuous at all points of the volume V and both have values that depend on the position in the volume. Consider the following product:

$$F = \phi \nabla \phi^* \tag{42}$$

Therefore, the divergence of the function F is given by:

$$\nabla \bullet F = \phi \nabla^2 \phi^* + \nabla \phi \nabla \phi^* \tag{43}$$

Applying the divergence theorem to the product $\phi \nabla \phi^*$, leads to:

$$\int_V \left(\phi \nabla^2 \phi^* + \nabla \phi \nabla \phi^* \right) dv = \int_S \left(\phi \nabla \phi^* n \right) dS \tag{44}$$

Again, apply the divergence theorem to $\phi^* \nabla\phi$, leads to:

$$\int_V \left(\phi^* \nabla^2 \phi + \nabla\phi^* \nabla\phi\right) dv = \int_S \left(\phi^* \nabla\phi n\right) dS \tag{45}$$

Subtract equation (44) from equation (45), leads to:

$$\int_V \left(\phi^* \nabla^2 \phi - \phi \nabla^2 \phi^*\right) dv = \int_S \left(\phi^* \nabla\phi - \phi \nabla\phi^*\right) n\, dS \tag{46}$$

Equation (8.46) is the starting point for the derivation of the boundary integral equation for potential flow applications. Now, it is time to start derivation of the boundary integral equation corresponding to the flow potential problems. Consider a domain Ω between the upper surface of the hydrofoil and the free-surface, see Fig. (**8.15**):

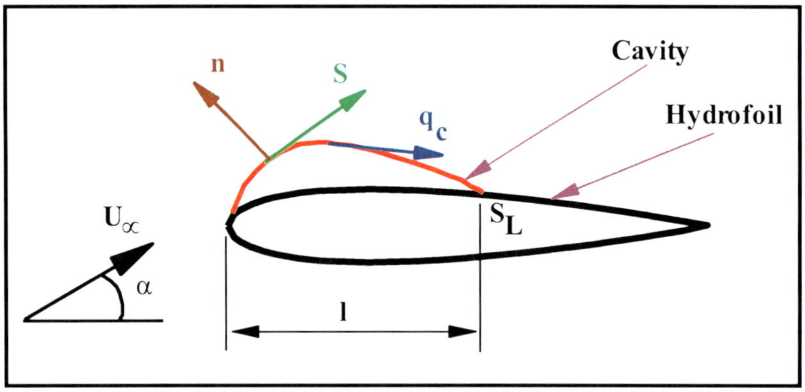

Fig. (8.15). Domain of the cavitations around a hydrofoil.

Over this domain, Laplace's equation is well defined with different associate boundary conditions [**82**]. The starting point of the formulation is Green's second identity:

$$\int_{\Omega}\left(\phi\nabla^2\phi^* - \phi^*\nabla^2\phi\right)d\Omega = \int_{\Gamma}\left(\phi\frac{\partial\phi^*}{\partial n} - \phi^*\frac{\partial\phi}{\partial n}\right)d\Gamma \tag{47}$$

In equation (47), Ω is a bounded two-dimensional region and Γ is its closed boundary curve. The functions ϕ and ϕ^* must be differentiable at least to the orders that appear in the integrals; the later function is called the fundamental solution for Laplace's equations and has two different forms according to the dimension of the problem. These forms are as follows:

$$\phi^* = -\frac{1}{2\pi}\ln r \qquad \text{For 2-D problems} \tag{48}$$

$$\phi^* = \frac{1}{4\pi r} \qquad \text{For 3-D problems} \tag{49}$$

It should be remembered that the function $\phi^*(x,y)$ is the potential at a point y called the ***field point*** due to a unit source at the point x called the ***source point***. This function and its derivative should be continuous at least up to the second order everywhere over the domain of the problem except at the source point, as shown in Fig. (**8.16**).

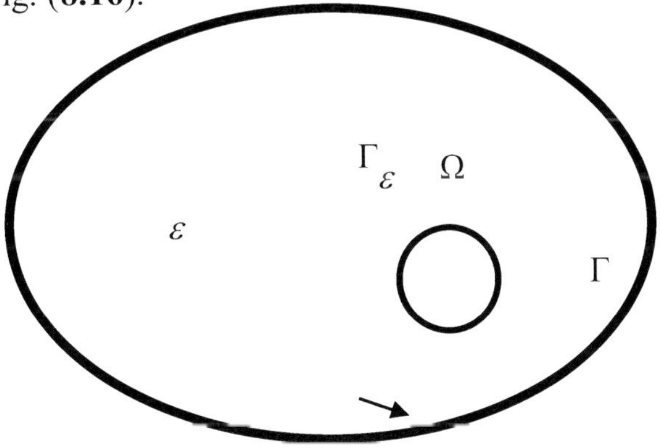

Fig. (8.16). Domain without a singular point.

Removing the singular point from the domain by removing a circle centered at the source point and having a very small radius denoted by ε, applying Green's second identity to the same domain but without the circle containing the singularity. Let the new domain is denoted by $\Omega_{new} = \Omega - \Omega_{\varepsilon}$, the application of the second Green's identity over the new domain leads to:

$$\int_{\Omega_{new}} \left(\phi\nabla^2\phi^* - \phi^*\nabla^2\phi\right) d\Omega = \int_{\Gamma} \left(\phi\frac{\partial\phi^*}{\partial n} - \phi^*\frac{\partial\phi}{\partial n}\right) d\Gamma + \int_{\Gamma_\varepsilon} \left(\phi\frac{\partial\phi^*}{\partial n} - \phi^*\frac{\partial\phi}{\partial n}\right) d\Gamma_\varepsilon \quad \textbf{(50)}$$

Also, it should be remembered that both ϕ and ϕ^* satisfies Laplace's equation over the new domain Ω_{new}. The next step of the derivation of the integral equation is to evaluate the integral $\int_{\Gamma_\varepsilon} \left(\phi\frac{\partial\phi^*}{\partial n}\right) d\Gamma_\varepsilon$, taking into consideration that the potential ϕ inside the integral is the potential at the field point y. Re-write this integral in the following form:

$$\int_{\Gamma_\varepsilon} \left(\phi(y)\frac{\partial\phi^*(x,y)}{\partial n}\right) d\Gamma_\varepsilon(y) = \int_{\Gamma_\varepsilon} \left(\left[\phi(y) - \phi(x)\right]\frac{\partial\phi^*(x,y)}{\partial n_y}\right) d\Gamma_\varepsilon(y) + \phi(x)\int_{\Gamma_\varepsilon} \left(\frac{\partial\phi^*(x,y)}{\partial n_y}\right) d\Gamma_\varepsilon(y) \quad \textbf{(51)}$$

Evaluation of the Integrals Involving Potential Derivative

The following procedure is used to evaluate the following integral:

$$I_1 = \int_{\Gamma_\varepsilon} \left(\frac{\partial\phi^*(x,y)}{\partial n_y}\right) d\Gamma_\varepsilon(y) \quad \textbf{(52)}$$

The term $\left(\frac{\partial\phi^*(x,y)}{\partial n_y}\right)$ is the normal derivative of the fundamental solution, *i.e.*;

$$\left(\frac{\partial \phi^*(x,y)}{\partial n_y}\right) = -\frac{1}{2\pi r}\frac{\partial r}{\partial n_y} \tag{53}$$

In equation (53):

r is the distance between the source and the field point.

n_y is the outward unit normal (inward the circle).

The term $\frac{\partial r}{\partial n_y}$ equals –1, therefore, equation (8.53) takes the following form:

$$\left(\frac{\partial \phi^*(x,y)}{\partial n_y}\right) = \frac{1}{2\pi r} \tag{54}$$

Because the domain under consideration is a circle, let us now transform into the polar coordinates and take the limit as the radius $\varepsilon \to 0$, and then the second integral in the right-hand side of equation (8.51) will take the following form:

$$\lim_{\varepsilon \to 0} \int_{\Gamma_\varepsilon} \left(\frac{\partial \phi^*(x,y)}{\partial n_y}\right) d\Gamma_\varepsilon(y) = \lim_{\varepsilon \to 0} \int_0^{2\pi} \frac{1}{2\pi} d\theta = 1 \tag{55}$$

The first integral in the right-hand side of equation (51) equals zero due to the same reasons that taken into consideration when evaluating equation (55), therefore;

$$\int_{\Gamma_\varepsilon} \left([\phi(y) - \phi(x)] \frac{\partial \phi^*(x,y)}{\partial n_y} \right) d\Gamma_\varepsilon(y) = 0 \tag{56}$$

Substituting from equations (55) and (56) into equation (51), leads to:

$$\lim_{\varepsilon\to 0}\int_{\Gamma_\varepsilon}\left(\phi(y)\frac{\partial\phi^*(x,y)}{\partial n}\right)d\Gamma_\varepsilon(y)=\phi(x) \tag{57}$$

Evaluation of the Integrals Involving Potential

This integral appears in the second integral on the right-hand side of equation (50) as follows:

$$I_2=\int_{\Gamma_\varepsilon}\left(\phi^*(x,y)\frac{\partial\phi(y)}{\partial n}\right)d\Gamma_\varepsilon(y) \tag{58}$$

Carry out the same procedure as done when evaluating the integral involving the potential derivative leads to the following:

$$I_{\phi^*}=\lim_{\varepsilon\to 0}\int_{\Gamma_\varepsilon}\left(\phi^*(x,y)\frac{\partial\phi(y)}{\partial n_y}\right)d\Gamma_\varepsilon(y)=-\lim_{\varepsilon\to 0}\int_0^{2\pi}\frac{1}{2\pi}\ln\varepsilon\, d\theta=0 \tag{59}$$

Back again to the first integral of the right hand side of equation (8.51), the limit of this integral over Γ_ε and due to the same procedure carried out as before is as follows:

$$\lim_{\varepsilon\to 0}\int_{\Gamma_\varepsilon}\left(\phi(y)\frac{\partial\phi^*(x,y)}{\partial n_y}-\phi^*(x,y)\frac{\partial\phi(y)}{\partial n_y}\right)d\Gamma_\varepsilon(y)=\phi(x) \tag{60}$$

Therefore, the following integral equation can be obtained:

$$\phi(x)=\int_{\Gamma}\left(\phi^*(x,y)\frac{\partial\phi(y)}{\partial n_y}-\phi(y)\frac{\partial\phi^*(x,y)}{\partial n_y}\right)d\Gamma(y) \tag{61}$$

The above procedure was the case when the source point was inside the domain of the problem. In what follows, the integral equation will be derived if the source point is moved to the boundary of the domain.

Suppose that the source point x is moved to the boundary and then take the limit when x tends to ζ on the boundary, as shown in Fig. (**8.17**).

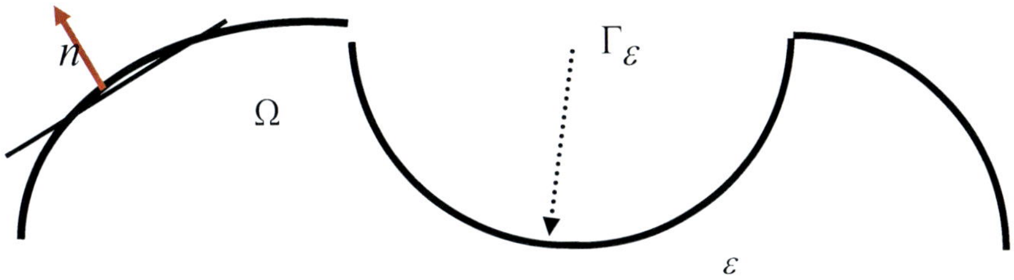

Fig. (8.17). Domain with source point on the boundary.

Again, and as done before, the point ζ should be removed by removing a semi-circle containing this point. Carry out the same procedure as done before, taking into consideration that the upper limit of integration in equation (8.55) becomes π, leads to:

$$\lim_{\varepsilon\to 0}\int_{\Gamma_\varepsilon}\left(\phi(y)\frac{\partial\phi^*(x,y)}{\partial n_y}-\phi^*(x,y)\frac{\partial\phi(y)}{\partial n_y}\right)d\Gamma_\varepsilon(y)=\frac{1}{2}\phi(\zeta) \tag{62}$$

Therefore, the following integral equation can be obtained:

$$\frac{1}{2}\phi(\zeta)=\int_{\Gamma}\left(\phi^*(\zeta,y)\frac{\partial\phi(y)}{\partial n_y}-\phi(y)\frac{\partial\phi^*(\zeta,y)}{\partial n_y}\right)d\Gamma(y) \tag{63}$$

Equation (63) is valid for every point on the smooth curve of the boundary of the domain. The general form of equation (63) can be written as follows:

$$c(\zeta)\,\phi(\zeta)=\int_{\Gamma}\phi^*(\zeta,y)\frac{\partial\phi(y)}{\partial n_y}d\Gamma(y)-\int_{\Gamma}\phi(y)\frac{\partial\phi^*(\zeta,y)}{\partial n_y}d\Gamma(y) \tag{64}$$

The terms on the right-hand side of equation (64) can take a simple form as follows:

$$q^{*} = \frac{\partial \phi^{*}(\zeta, y)}{\partial n_{y}}$$

$$q = \frac{\partial \phi(y)}{\partial n_{y}} \qquad \textbf{(65)}$$

Therefore, equation (65) can now take the following form:

$$c(\zeta)\phi(\zeta) + \int_{\Gamma} \phi(y) q^{*}(\zeta, y) d\Gamma(y) = \int_{\Gamma} \phi^{*}(\zeta, y) q(y) d\Gamma(y) \qquad \textbf{(66)}$$

The multiplier $c(\zeta)$ takes one of the following values as follows:

$c(\zeta) = 0$ If ζ outside the domain

$c(\zeta) = 1$ If ζ inside the domain

$c(\zeta) = \dfrac{\theta}{2\pi}$ On the domain **(67)**

Where θ is the angle of the boundary at the point ζ. It should be noted that at every source point, three different values are known; the multiplier $c(\zeta)$, ϕ^{*} and q^{*}. Thus, the boundary integral equation gives the potential at any source point ζ inside the domain or on the boundary as a function of the potential and normal flux density components around the boundary. The next stage shows how the integral equation may be expressed in a discrete form, leading to the boundary element method itself. Before dealing with the discretized form of the integral equation, it is important to re-write equation (66) in a discretized form as follows:

$$c(\zeta)\phi(\zeta)+\sum_{\text{element}}\int_{\Gamma}\phi(y)q^{*}(\zeta,y)d\Gamma_{\text{element}}=\sum_{\text{element}}\int_{\Gamma}\phi^{*}(\zeta,y)q(y)d\Gamma_{\text{element}} \quad (68)$$

It will be fair enough to say that no approximation has been introduced up to equation (68). Also, it is important to remember that if all integrations can be carried out analytically, an exact solution will be obtained. Unfortunately, some integrals cannot be found analytically and should be obtained numerically; this will be explained later in this chapter. Before dealing with the numerical integrations, a brief review of the discretization techniques should be introduced.

Discretization Techniques

There are three different types of discretization, the type of anyone depends on the distribution of both potential and potential derivative over the element. These types are ***constant***, ***linear*** and ***quadratic***. The accuracy, nature of the problem and the importance of the results are three parameters that control choosing the type of elements used.

Constant Element

In this case, both potential and potential derivatives are assumed to be constant along with the element, and that element has only one node denoted by i , see Fig. (**8.18**).

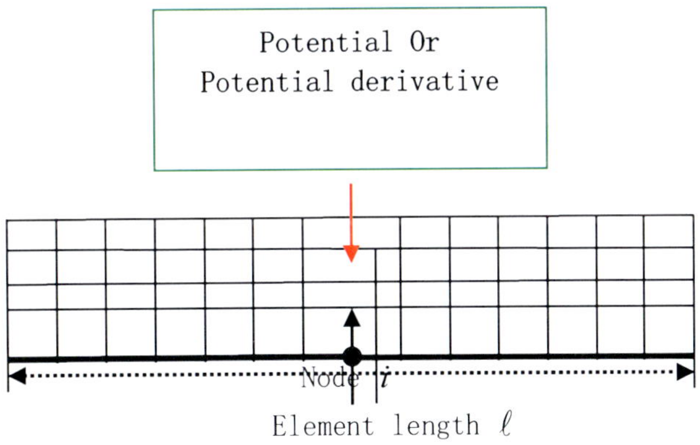

Fig. (8.18). Geometry of constant element.

Suppose that the boundary is divided into N-constant elements; therefore, the discretized form of equation (8.68) will take the following form:

$$\frac{1}{2}\phi^{i} + \sum_{j=1}^{N}\left(\int_{\Gamma} q^{*} d\Gamma_{j}\right)\phi^{j} = \sum_{j=1}^{N}\left(\int_{\Gamma} \phi^{*} d\Gamma_{j}\right) q^{j} \tag{69}$$

Where i is any boundary node, which is located in the middle of the element and j is the element number. Referring to equation (68), one can get two different integrals:

$$I_1 = \int_{\Gamma} q^{*} d\Gamma_{j} \tag{70}$$

$$I_2 = \int_{\Gamma} \phi^{*} d\Gamma_{j} \tag{71}$$

The integrals in equations (70) and (71) relate two nodes i and j where the fundamental solution is acting on the first. Carrying out these integrals on all elements of the boundary leads to the ***influence coefficients***. These coefficients can be denoted as:

$$\overline{H}^{ij} = \int_{\Gamma} q^{*} d\Gamma_{j} \tag{72}$$

$$G^{ij} = \int_{\Gamma} \phi^{*} d\Gamma_{j} \tag{73}$$

Substutiting from equations (72) and (73) into equation (69), leads to:

$$\frac{1}{2}\phi^{i} + \sum_{j=1}^{N} \overline{H}^{ij}\phi^{j} = \sum_{j=1}^{N} G^{ij} q^{j} \tag{74}$$

It should be remembered that the node at which the fundamental solution acting varies over each element on the boundary, therefore the influence coefficient given by equation (73) can take another form as follows:

$$H^{ij} = \begin{cases} \overline{H}^{ij} & i \neq j \\ \overline{H}^{ij} + \frac{1}{2} & i = j \end{cases} \quad \textbf{(75)}$$

Substuting equation (75) into equation (74), and re-arrange, leads to:

$$\sum_{j=1}^{N} H^{ij} \phi^{j} = \sum_{j=1}^{N} G^{ij} q^{j} \quad \textbf{(76)}$$

In simple matrix form, equation (76) can be re-written as:

$$H\phi = GQ \quad \textbf{(77)}$$

Referring to the influence coefficients given by equations (72) and (73), one can meet four different cases that depend on the location of the source and field points. When the source point is the field point, then a singularity will be encountered, *i.e*; H^{ii} and G^{ii}, these integrals have special treatment, while H^{ij} and G^{ij} being calculated using simple Gauss quadrature rules.

Linear Element

The first difference between the constant and linear element is the multiplier $c(\zeta)$ where in the constant element it takes 0.5 because the boundary is smooth, but now it will be used as it is in mathematics. Discretized equation (68) with N linear elements, then it takes the following form:

$$c^{i}\phi^{i} + \sum_{j=1}^{N} \int_{\Gamma} \phi q^{*} d\Gamma = \sum_{j=1}^{N} \int_{\Gamma} \phi^{*} q d\Gamma \quad \textbf{(78)}$$

In this case, it is impossible to take the potential and the potential derivative out of the two integrals $\int_{\Gamma} q^{*} d\Gamma$ and $\int_{\Gamma} \phi^{*} d\Gamma$ because over each element, these values vary linearly. Therefore, the values of ϕ and $\frac{\partial \phi}{\partial n}$ at any point over the element can be defined in terms of their nodal values and two linear interpolating functions Ψ_1 and Ψ_2 of the homogenous coordinate ζ such that:

$$\phi = [\Psi_1 \quad \Psi_2][\phi_1 \quad \phi_2]^T \tag{79}$$

$$\frac{\partial \phi}{\partial n} = [\Psi_1 \quad \Psi_2]\left[\frac{\partial \phi_1}{\partial n} \quad \frac{\partial \phi_2}{\partial n}\right]^T \tag{80}$$

In which, the functions Ψ_1 and Ψ_2 are as follows:

$$\Psi_1 = 0.5(1-\xi)$$

$$\Psi_2 = 0.5(1+\xi) \tag{81}$$

$$\int_\Gamma \left(\phi \frac{\partial \phi^*}{\partial n}\right) d\Gamma = \int_\Gamma \left([\Psi_1 \quad \Psi_2]\frac{\partial \phi^*}{\partial n}\right) d\Gamma \begin{bmatrix} \phi_1 \\ \phi_2 \end{bmatrix} = \left[h_{ij}^1 \quad h_{ij}^2\right] \begin{bmatrix} \phi_1 \\ \phi_2 \end{bmatrix} \tag{82}$$

$$\int_\Gamma \left(\phi^* \frac{\partial \phi}{\partial n}\right) d\Gamma = \int_\Gamma \left([\Psi_1 \ \Psi_2]\phi^*\right) d\Gamma \begin{bmatrix} \frac{\partial \phi_1}{\partial n} \\ \frac{\partial \phi_2}{\partial n} \end{bmatrix} = \left[g_{ij}^1 \ g_{ij}^2\right] \begin{bmatrix} \frac{\partial \phi_1}{\partial n} \\ \frac{\partial \phi_2}{\partial n} \end{bmatrix} \tag{83}$$

$$h_{ij}^1 = \int_{\Gamma_j} \left(\Psi_1 \frac{\partial \phi^*}{\partial n}\right) d\Gamma_j \qquad h_{ij}^2 = \int_{\Gamma_j} \left(\Psi_2 \frac{\partial \phi^*}{\partial n}\right) d\Gamma_j$$

$$g_{ij}^1 = \int_{\Gamma_j} \left(\Psi_1 \phi^*\right) d\Gamma_j \qquad g_{ij}^2 = \int_{\Gamma_j} \left(\Psi_2 \phi^*\right) d\Gamma_j \tag{84}$$

Therefore, the final boundary integral equation takes the following form:

$$\sum_{j=1}^{\partial\Gamma_j} H_{ij}\,\phi_j = \sum_{j=1}^{\partial\Gamma_j} G_{ij}\,\frac{\partial\phi_j}{\partial n} \tag{85}$$

Finally, as done before for a constant case, the system given by equation (85), can now be recast in a matrix form as follows:

$$H\phi = GQ \tag{86}$$

In the present and next chapters, all calculations will be performed using these two different types of elements; therefore, it is not necessarily to represent the third type Fig. (**19**).

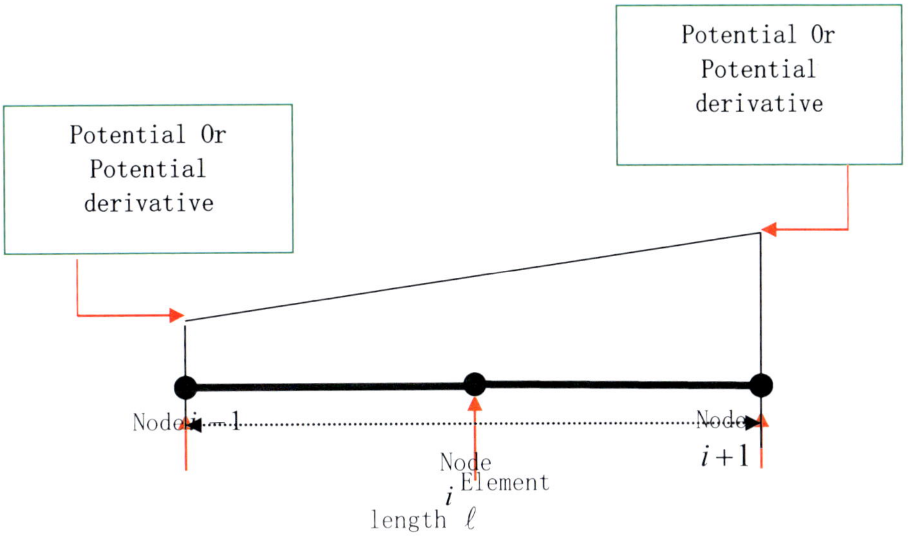

Fig. (8.19). Geometry of linear element.

After this complete formulation of the boundary element method, remaining part of the present chapter introduces different numerical algorithms for cavitation problems. These algorithms will be introduced in some detail in the next section.

NUMERICAL ALGORITHMS

Aitchison Algorithm

Aitchison [**48**] presented in his paper a numerical solution for planar and axi-symmetric Riabouchinsky cavity flow. The problem was formulated as the minimization of a function over a variable region, which was solved numerically by the variable finite elements. In the following figures, the general cavity model is followed by Riabouchinsky model in a finite domain, then finally, the region of the solution.

It is well known from Birkoff and Zarantonello [209] that for a fixed cavity length L and half-height of the plate, there is only one value of the free stream line velocity, q_c, which gives an analytical solution. From this fact, Aitchison derived the following algorithm, using the following formula for q_c, that is:

$$q_c = \sqrt{2p + q^2} \tag{87}$$

Aitchison algorithm can be summarized as follows:

(1) Solve the problem for a given value of q_c, which is known to be too low.
(2) Calculate the average pressure in each element p_j
(3) If any value of p_j is negative, increase q_c and repeat step (2), otherwise stop.

This means that the main objective of this algorithm is to find the lowest value of q_c for which the pressure is positive everywhere inside the flow.

Wrobel Algorithm

In 1992, Wrobel [180] solved cavitations behind a plate placed perpendicular to the direction of the flow in a channel of finite width. Riabouchnisky model was used for modeling the problem, see **Fig. (8.20)**.

Wrobel's algorithm can be summarized as follow

(1) Assume an initial guess for the free-surface.

(2) Use the boundary element method to find the potential derivative at each node of the free-surface.

(3) Calculate Bernoulli's constant at each node using $\beta_i = \frac{1}{2}\rho q_{c_i}^2$.

(4) Evaluate the difference between Bernoulli's constant at the tip of the plate and that value obtained in step (3).

(5) Evaluate the vertical displacement from $\Delta y_i = -\frac{\Delta\beta_i}{\rho} y_i^3$.

(6) Evaluate the following norm $Norm = \sum \frac{1}{N}\left|\frac{y_i^{k+1} - y_i^k}{y_i^k}\right|$.

(7) If the previous norm is smaller than a prescribed tolerance, stop, otherwise, guess new location using $y_i^{k+1} = y_i^k + \omega \Delta y_i^k$, and then repeat the procedure again.

Yas'ko Algorithm

In 1992, Yas'ko [180] solved plane and axi-symmetric flow and the cavity formed behind blunt-ended bodies. Yas'ko used the Riabouchinsky model in his algorithm, see **Fig. (8.20)**.

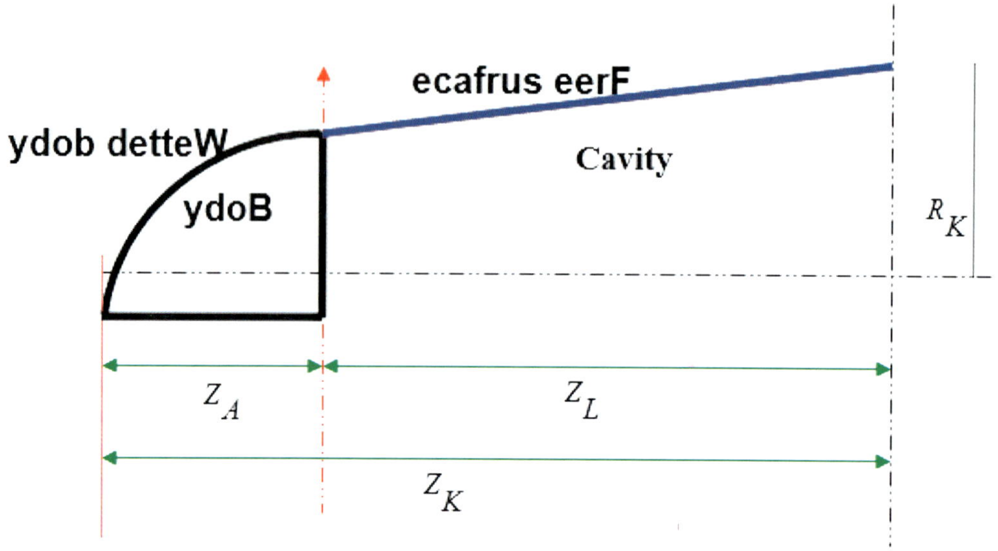

Fig. (8.20). Yas'ko problem.

Yas'ko algorithm in steps

(1) Apply the boundary integral equation at each point on the wetted body and the free-surface and get $(N+L)$ number of unknowns.

(2) Yas'ko derived an additional equation for the velocity of the free- surface.

(3) By Gauss elimination method, he solved $(N+L+1)$ equations for the unknowns. These unknowns are (L) potential on the wetted body, (N) potential derivative on the free-surface and the velocity of the free-surface.

(4) He used the kinematic boundary condition to iterate for the shape of the cavity.

(5) After iteration, he found the cavitation number as a solution parameter.

Ingber Algorithm

In this algorithm [180], flow about submerged, fully cavitating bodies at zero and non-zero angles of attack were considered. A cavity closure model was used in the algorithm and making use of the boundary element method. The domain of the problem with the closure model is shown in Fig. (**9**). Ingber assumed that he could correctly predict the point of detachment and focus on determining the correct cavity length. To do that, he proposed two iterations, the first-named as ***inner*** and the other named ***outer***.

Inner iteration

In this iteration, the point of detachment and the cavity length is given and assumed cavity shape. Then, he uses the direct boundary element method to satisfy either kinematics or dynamic condition. In steps, the inner iterations can be summarized as follows:

(1) Assume cavity length and cavity shape.

(2) Apply direct boundary element, where the kinematics boundary condition is known on both wetted body and the free-surface.

Outer iteration

In this iteration, an adjustment for the cavity length is still carried out until the dynamics condition is satisfied everywhere along the cavity surface.

Kinnas and Fine Algorithm

In 1993, Kinnas and Fine [180] suggested an algorithm for numerical non-linear analysis for the flow around two- and three-dimensional partially cavitating hydrofoil. In few steps, their algorithm can be summarized as follows:

(1) Solve the cavity problem (first iteration) with a fixed cavitation number $\sigma_O = \sigma$ to find the corresponding cavity length, ℓ_O

(2) Solve the cavity problem with fixed cavity length ℓ_o , for K-iterations and find the converged value of the cavitation number, $\sigma_o^/$.

(3) Define a new cavitations number from, $\sigma_1 = \sigma_o + \left(\sigma_o - \sigma_o^/\right)$

(4) Solve the cavity problem (first iteration) with a fixed cavitation number σ_n and find the correct cavity length ℓ_n , $n \geq 1$.

(5) Solve the cavity problem with fixed cavity length, ℓ_n for K-iterations and find the converged value of the cavitation number $\sigma_n^/, n \geq 1$.

(6) Apply Newton-Raphson iteration solution for $\sigma_n^/ - \sigma = 0$ and find a new cavitations number σ_{n+1}.

(7) Repeat steps (4-6) until $\sigma_n^/ = \sigma$ within a prescribed tolerance.

Ahmed Algorithm

Between 1994 and 1996, Ahmed *et al.* [180], solved the two-dimensional cavity flow problem, see Fig. (**8.21**), using the boundary element method. A new algorithm for that problem was derived and can be summarized in a few steps as follows:

(1) Assume an initial guess for the free-surface position.

(2) Apply the kinematics condition on both wetted and free-surface, and then use the boundary element formulation to determine the potential on both surfaces.

(3) Find the approximate velocity at each node using $q_i = \dfrac{\Phi(s_i) - \Phi(A) + x_i}{s_i}$.

(4) Evaluate the error at each point i given by the difference between q_i and $q(A)$.

(5) Verify convergence, *i.e.*, if the difference calculated in step (4) is less than a specified tolerance at all free-surface points; if this is not so, use $\Delta y_i = -y_i^2 \left(q(A) - q_i \right)$ to move the free-surface nodes and repeat steps 2-5 until convergence.

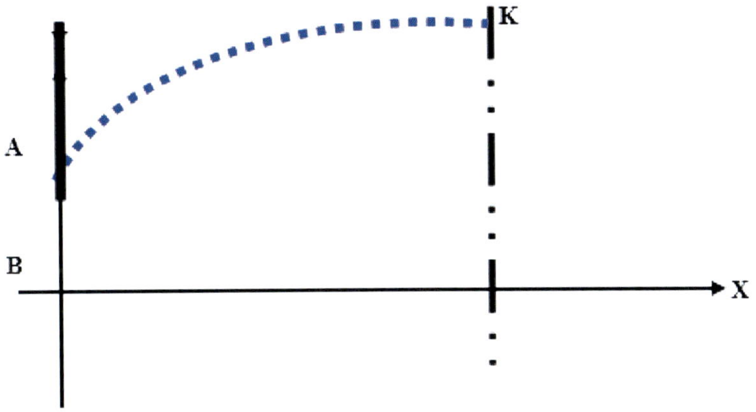

Fig. (8.21). Ahmed domain problem.

NEW (PRESENT) NUMERICAL ALGORITHM

After the previous review on the most numerical algorithms that were developed to solve cavity flow problems. ***A new numerical algorithm*** based on the ***closure model*** and uses the boundary element method will be introduced herein and listed in steps as follows [180]:

1) Solve the BE-code over domain bounded by the upper surface of the hydrofoil and the free-surface.

Over the hydrofoil surface, equation(8.55) is used as a boundary condition.

Over the free-surface, the potential in equation (75) is applied. For simplicity, $\lambda = 0$ [**56**], then, equation (75) takes the following formula:

$$\phi(s) = \phi(0) + \int_0^s \left(-\vec{U}_\infty \bullet \vec{S} + U_\infty \sqrt{1+\sigma} \right) ds \qquad \textbf{(88)}$$

Assuming that in this step the $\phi(0)$ equals zero

(1) Evaluate

$$\phi(0) = (1 - \Re) \left(\frac{\sum_{i=1}^{M} \phi(s)}{M} \right) \qquad \textbf{(89)}$$

where $\Re$ is a control parameter.

(2) Solve the BE-code once more over the same domain as in step (1) with equation (55) as a boundary condition over the hydrofoil and equation (88) using $\phi(0)$ obtained from step (2) as a boundary condition over the free-surface. This step leads to the potential derivative at each node of the free-surface $\left(\frac{\partial \phi}{\partial n} \right)_{i-node}$.

(3) Evaluate the magnitude of the total free-surface velocity.

(4) The cavity closure condition requires the thickness at the cavity trailing edge $h(s_L)$ to be zero. This is obtained by applying the following equation:

$$h(s_L) = \frac{1}{q_c} \int_0^{S_L} \left(\vec{U}_\infty \bullet \vec{n} + \left(\frac{\partial \phi}{\partial n} \right)_{i-node} \right) ds = 0 \qquad \textbf{(90)}$$

In the first iteration, the panels representing the cavity are placed on the foil surface directly under the cavity. In the following iterations, the cavity shape is updated by an amount $h(s_L)$.

Discretization Process

The present algorithm used the boundary element method with linear elements to get the required accuracy ($\delta \leq 0.0005$) that appears in Eq. (8.78) as the cavity closure condition. This accuracy is satisfied using elements that vary for different cavity lengths. However, for all cases, the linear elements that were used to discretize the domain have different lengths. Cavitation phenomenon emerges at the leading edge of the hydrofoil (at the separation point). So, the number of elements (with short length) must be increased in this area in comparison with the trailing edge area of the cavity. For example, at the cavity length l=0.6, the total number of linear elements at both the cavity surface and the hydrofoil surface equals to 9. Thus, the total number of the linear elements used within the domain is equal to 18. The length of the shortest element equals 0.0125 of chord length (C). This shortest element was placed at the leading edge. The length of the longest element equals 0.1 of chord length (C). This longest element was placed at the trailing edge of the cavity. In the next chapter, results will be presented starting with previous results obtained by other researchers but using the present algorithm to check its validity, and then new results will be listed in more detail.

RESULTS AND DISCUSSION

Cavitations problem is highly non-linear due to the inherent nonlinear conditions encountered in its mathematical formulation; therefore, analytical solutions are so difficult to obtain. Because of that, many researchers directed their efforts to numerical solutions. The aid of different mathematical models had made different numerical approaches, and some of these models had been introduced in chapter three of the present thesis. The first section of the present chapter starts by re-introducing previous results for the flow problem past vertical, oblique flat plates and over a wedge of vertex angle θ, $0 < \theta < \frac{\pi}{2}$, the purpose of re-introducing these results is to gain skills of discretization techniques, techniques of nodes numbering, and element, modification of results and any other parameters that appear when solving the problem using the boundary element methods. The second part of the present chapter introduces a numerical solution of the cavitation problem around hydrofoil using the boundary element method with a new numerical algorithm developed for this purpose. In this algorithm, the problem is solved in a completely different approach, *i.e.*, only the domain considered is bounded by the upper surface of the hydrofoil and the free- surface, therefore the difficulty in evaluating the potential at the tip of the hydrofoil becomes easy rather than before. Different cases of NACA hydrofoils are tested using the proposed technique and give good results.

RE-INTRODUCE OF SOME PREVIOUS RESULTS

Flow Over A Vertical Flat Plate

The flow over the vertical flat plate was studied before by different researchers, such as Yas'ko [180] and Ahmed *et al.*[180]. Fortunately, this problem has an analytical solution [180]. Therefore, re-introducing some previous results and the comparison with the analytical solution will give a good impression about understanding the technique of the boundary element method. The configuration of the flow problem over a vertical flat plate is shown in Fig. (**8.22**).

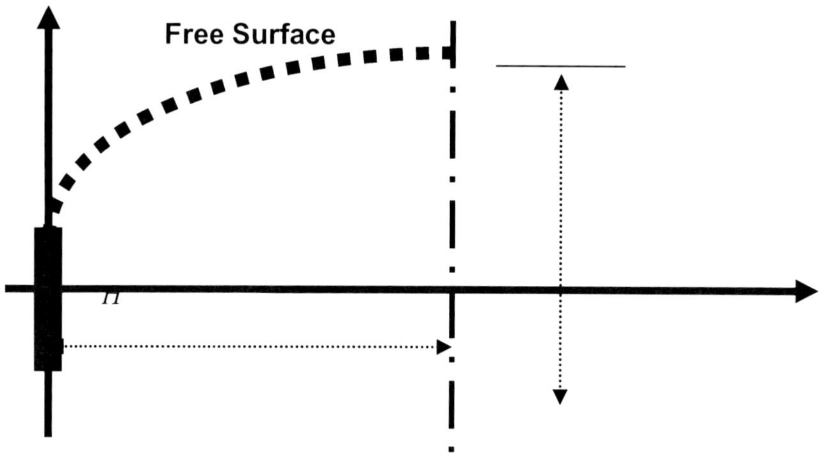

Fig. (8.22). Domain of flow problem over a vertical flat plate.

In Fig. (**8.23**):

Z Is the cavity length

H Is the vertical plate length

R Is the maximum position of the free surface.

Three different examples in [180] have been solved. The summary of these examples with previous and present calculations is listed in the table (**8.1**). The purpose of these results is to check the validity of the designed code and handle different parameters that directly affect the accuracy of the results. Another topic analyzed is the effect of the cavity length on the final shape of the free surface. Two different cases are solved for three and five cavity lengths, respectively. These results are shown in Figs. (**8.23 & 8.24**), respectively.

Table. 1. The calculations of three different examples of a vertical flat plate.

$\frac{Z}{R}$					$\frac{R}{H}$					q		
Analytical	Yas'ko	Ahmed			Analytical	Yas'ko	Ahmed			Analytical	Yas'ko	Ahmed
		Z Given	H	$\frac{Z}{R}$			R Calculated	H	$\frac{R}{H}$			
15.568	15.85	0.778	0.05	15.568	3.4847	3.491	0.17306	0.05	3.4612	1.2042	1.2042	1.2069

31.798	31.74	1.589		31.798	4.7305	4.727	0.23542		4.7084	1.1402	1.1402	1.1422
245.60	243.3	12.28		245.60	12.201	12.07	0.61293		12.258	1.0488	1.0488	1.0494

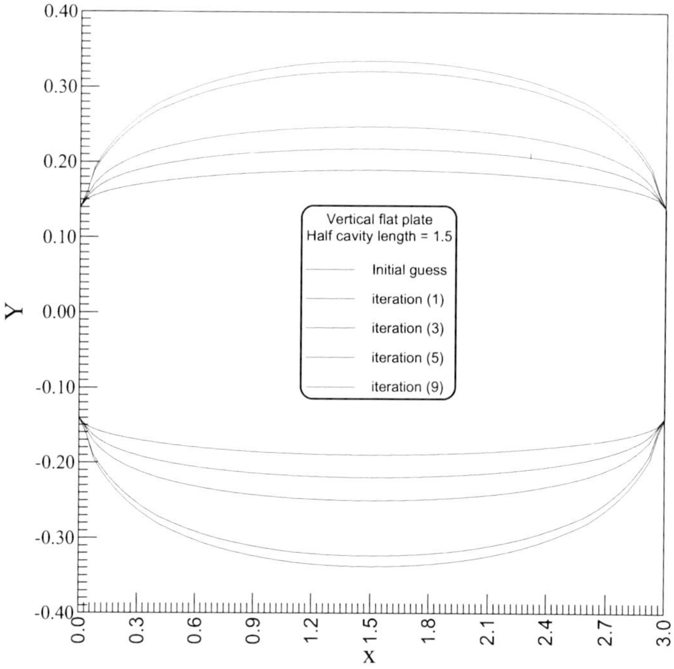

Fig. (8.23). The final free surface for three cavity length.

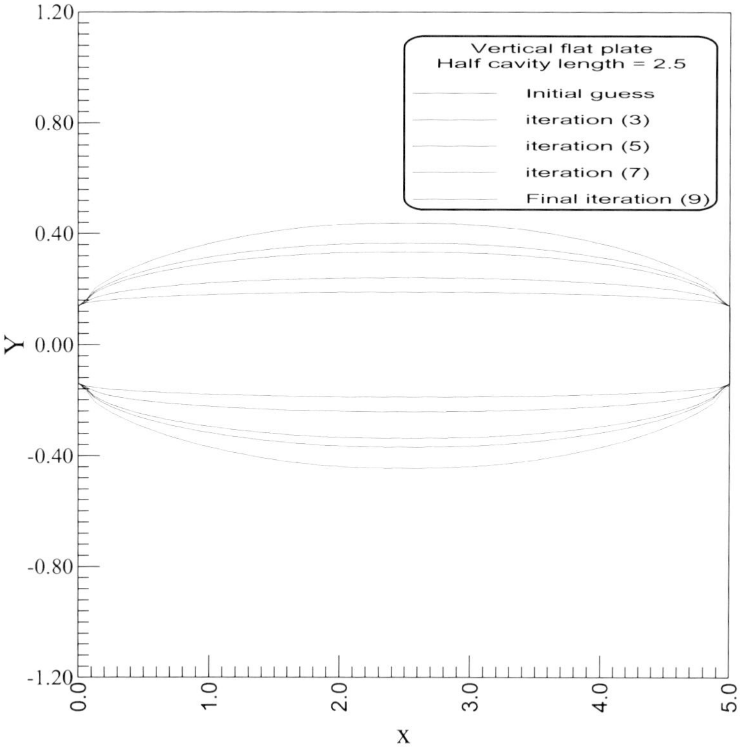

Fig. (8.24). The final free surface for five cavity lengths.

Flow over wedge

The second case that had been introduced was the flow past a wedge of a half-angle θ, $0 < \theta < \frac{\pi}{2}$, for a wide range of calculations about this case.

New results

To test the validity of the present algorithm, previous results by Kinnas *et al.* [**56**] were achieved by the present algorithm with an excellent agreement for NACA-16006. In Kinnas's case study, the cavitation number was $\sigma = 1.097$ and the cavity length was found to be $\ell = \frac{L}{c} = 0.4$. Thus, the computations were expanded to study the effect of different parameters has a direct effect on the cavity shape and length. These parameters include; hydrofoil section, angle of attack, free-stream velocity, and cavitation number. Fig. (**8.25**) shows the profiles of the different tested NACA sections (the definition of NACA sections consists of two main parts; the first part, for all NACA sections, is 00, which means that the profile is symmetric. The second part is (06, 12, 15 or 25) refers to the maximum thickness of the NACA section relative to the chord length of the hydrofoil).

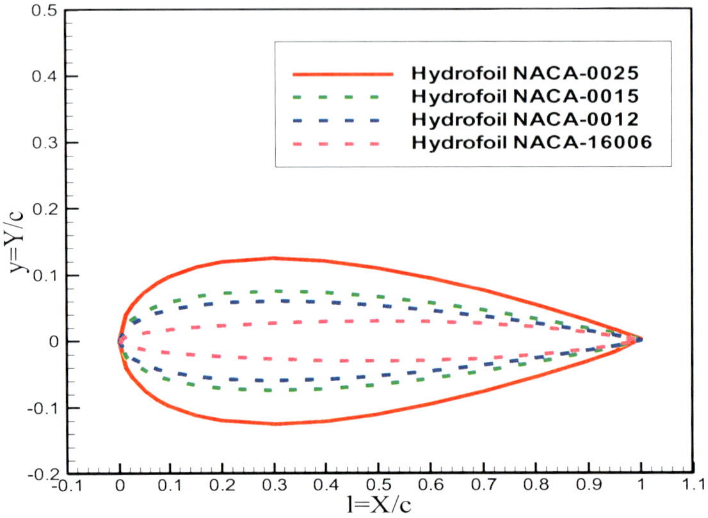

Fig. (8.25). Profiles of different tested NACA sections.

NACA-16006 Hydrofoil

Case (1)

In this case study, three parameters are prescribed and the present code is used to find the cavity length $\ell = \frac{L}{c} = 0.4$. The three parameters are; the cavitation number $\sigma = 1.097$, the inflow velocity $U_{\infty} = 1.0 m/s$ and the angle of attack $\alpha =$ 4°. It is clear that the cavity closure condition at the cavity end, δ , given in equation (8.78) is obtained at $\ell = \frac{L}{c} = 0.4$ Figs. (**26** & **27**).

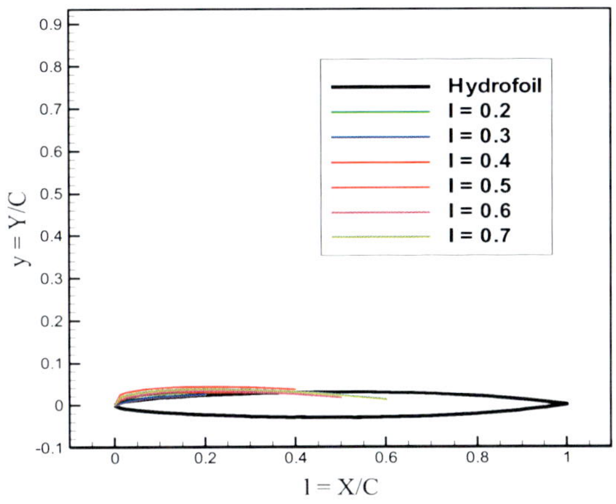

Fig. (8.26). Scaled hydrofoil section NACA-16006 at $U_{\infty} = 1.0 m/s$.

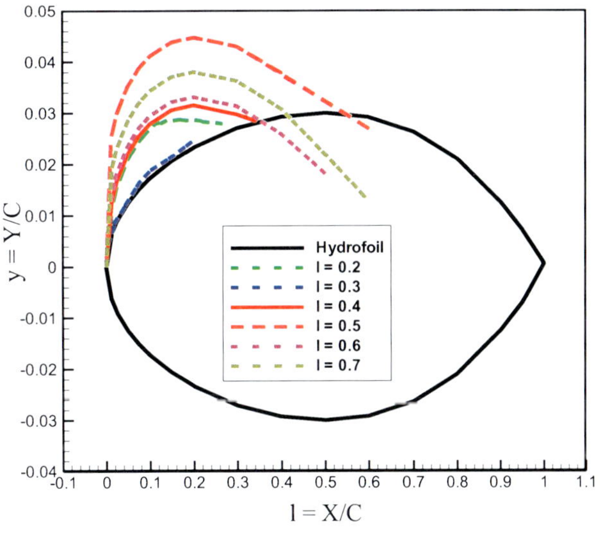

Fig. (8.27). Non-scaled hydrofoil section NACA-16006 at $U_{\infty} = 1.0 m/s$.

Case (2)

In this case, the present code is used with the same three parameters which were used in the previous case. There are two parameters that have the same value as the previous case, the cavitation number $\sigma = 1.097$ and the angle of attack $\alpha = 4^o$. However, the inflow velocity, U_∞, changed to become $U_\infty = 0.5 m/s$. The results of this case are shown in Fig. (**8.28**). It is clear that the cavity length $\ell = \frac{L}{c} = 0.425$ achieved the cavity closure condition at the cavity end δ, given in equation (8.78).

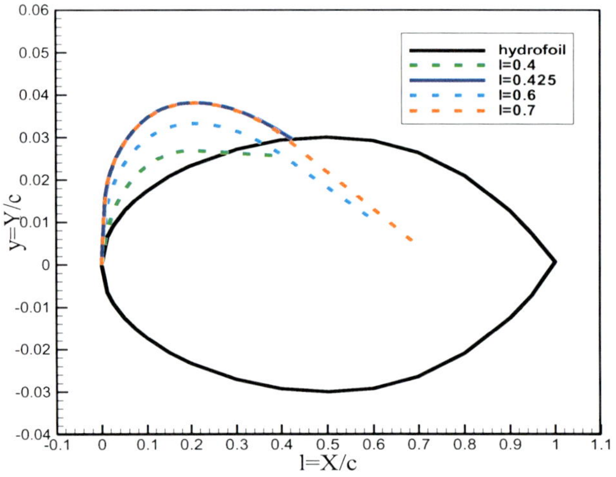

Fig. (8.28). Non-scaled hydrofoil section NACA-16006 at $U_\infty = 0.5 m/s$.

Case (3)

As in the previous two cases, the parameters with the same values are the cavitation number $\sigma = 1.097$ and the angle of attack $\alpha = 4^o$. The inflow velocity U_∞ takes a new value, $U_\infty = 2.0 m/s$. The present code finds the cavity length $\ell = \frac{L}{c} = 0.3$ that achieved the cavity closure condition at the cavity end , given in equation (78).

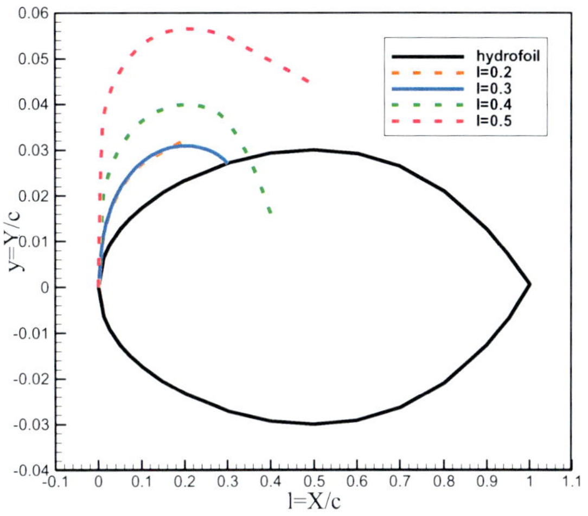

Fig. (8.29). Non-scaled hydrofoil section NACA-16006 at $U_{\infty} = 2.0 m/s$.

Case (4)

In this case study, the inflow velocity, U_{∞}, takes a new value, $U_{\infty} = 3.0 m/s$ and the other parameters still have the same values as the previous cases. In Fig. (**8.30**), all results of this case are shown, and it is clear that the cavity length, $\ell = \frac{L}{c} = 0.25$ achieved the closure condition of equation (78).

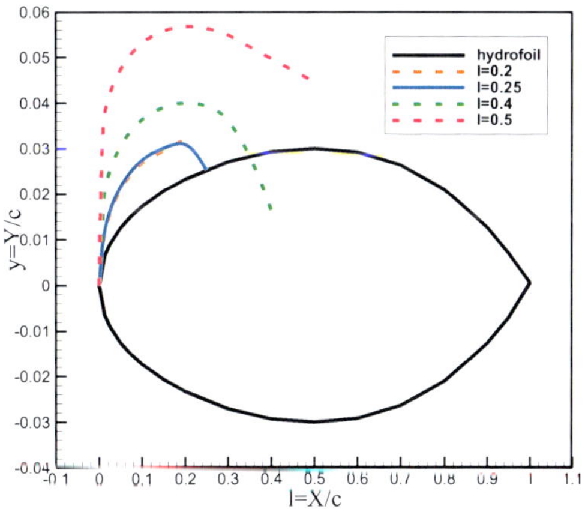

Fig. (8.30). Non-scaled hydrofoil section NACA-16006 at $U_{\infty} = 3.0 m/s$.

Case (5)

In this case, the variable parameter is the cavitation number σ and the inflow velocity U_∞ has a prescribed value. The present code is used with the three parameters, the cavitation number $\sigma = 0.852$, the angle of attack $\alpha = 4^\circ$ and the inflow velocity $U_\infty = 1.0 m/s$ to find the cavity length ℓ. The cavity length ℓ, which achieved the cavity closure condition at the cavity end, , given in equation (8.78) is obtained at $\ell = \frac{L}{c} = 0.47$. The results of this case are shown in Fig. (**8.31**).

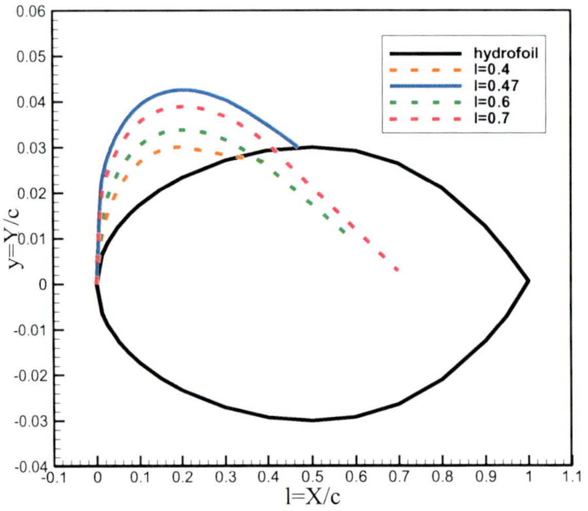

Fig. (8.31). Non-scaled hydrofoil section NACA-16006 at $\sigma = 0.852$.

Case (6)

As in the previous case, the two parameters which have the same values are the angle of attack $\alpha = 4^\circ$ and the inflow velocity $U_\infty = 1.0 m/s$. However, the cavitation number takes a new value, $\sigma = 1.26933$, the cavity length that achieved the closure condition in equation (8.78) is $\ell = \frac{L}{c} = 0.225$. The results of this case are shown in Fig. (**8.32**).

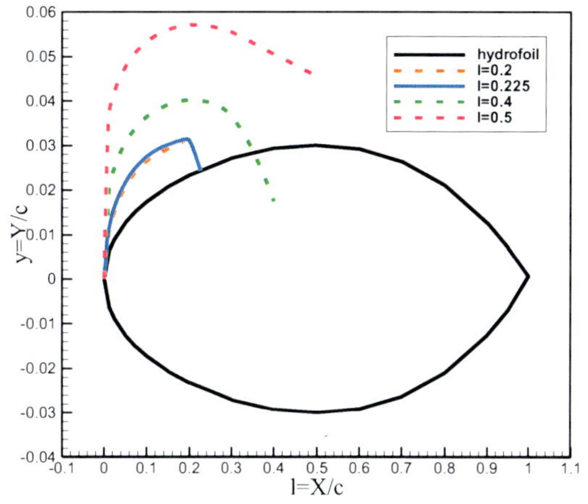

Fig. (8.32). Non-scaled hydrofoil section NACA-16006 at $\sigma = 1.26933$.

Case (7)

In the present case, the angle of attack α, is studied as a variable parameter. The other two parameters, the cavitation number $\sigma = 1.097$ and the inflow velocity $U_\infty = 1.0 m/s$ are fixed at the shown values. The present code is used with an angle of attack $\alpha = 3^o$ and gives the cavity length $\ell = \frac{L}{c} = 0.34$, which achieves the closure condition in equation (8.78). In Fig. (**8.33**) the results of this case are shown at different lengths.

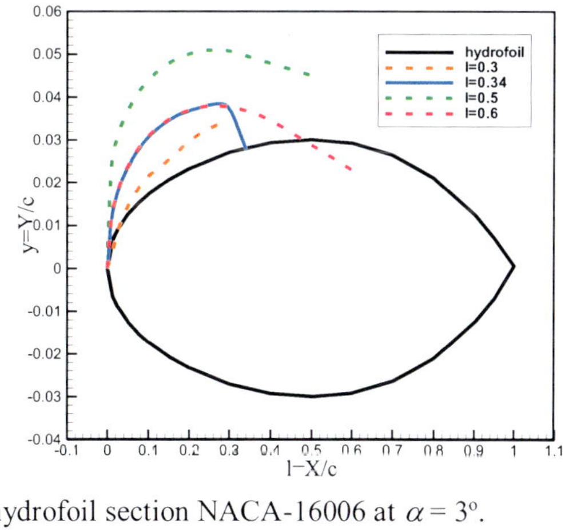

Fig. (8.33). Non-scaled hydrofoil section NACA-16006 at $\alpha = 3^o$.

Case (8)

In this case, the same two parameters, the cavitation number $\sigma = 1.097$ and the inflow velocity $U_\infty = 1.0$ are used. However, the angle of attack, α, has a new value, α = 5°. It is clear that the cavity length $\ell = \dfrac{L}{c} = 0.44$ achieved the cavity closure condition at the cavity end , given in equation (8.78). The results of this case are shown in Fig. (**8.34**).

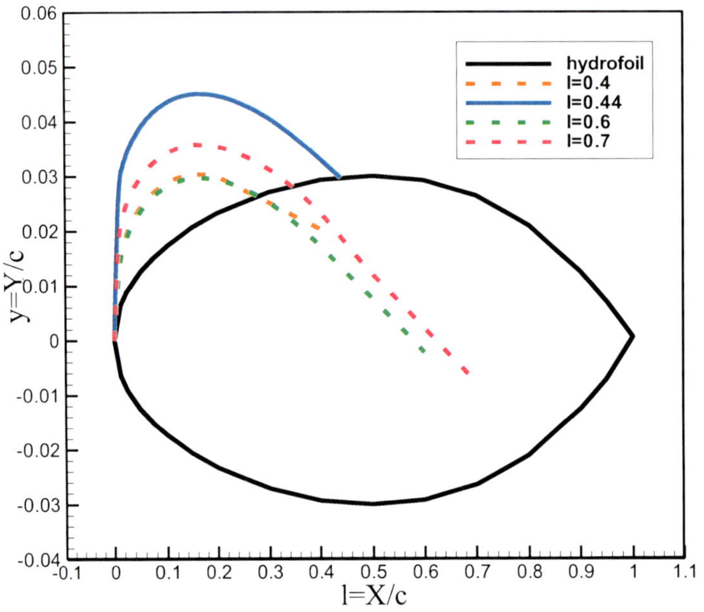

Fig. (8.34). Non-scaled hydrofoil section NACA-16006 at α= 5°.

Fig. **(8.35)** illustrates the variation of cavitation shape with free-stream velocity (U_∞). The cavity size decreases as U_∞ increases. The cavity shape is almost the same for 3 and 4 *m/s*.

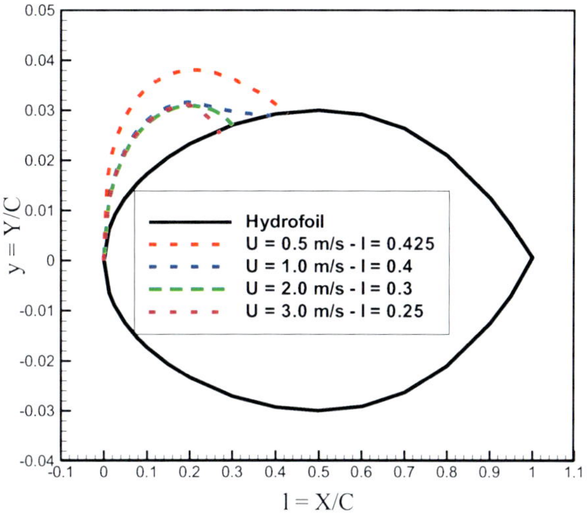

Fig. (8.35). Effect of free-stream velocity, NACA-16006.

Fig. (**8.36**) shows the effect of cavitation number on the cavity shape and length at $\alpha = 4^o$ and $U_\infty = 1.0$ *m/s*. The cavitation length, *l*, decreases as σ increases.

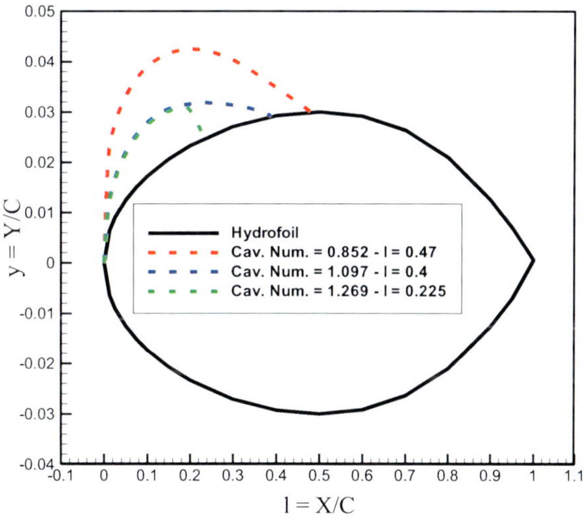

Fig. (8.36). Effect of cavitations number, NACA-16006.

The effect of the angle of attack (α) is shown in Fig. (**8.37**) for $\sigma = 1.097$ and $U_\infty = 1.0$ *m/s*. The height and length of cavitation increase with α.

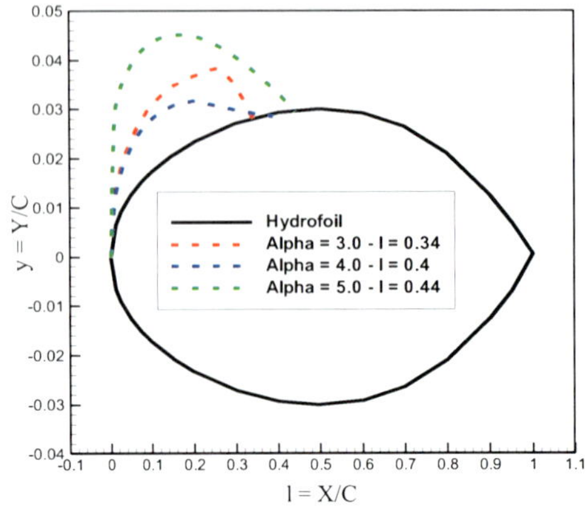

Fig. (8.37). Effect of angle of attack, NACA-16006.

From the previous results, it is shown that increase in the free stream velocity makes the cavity length decrease and the same effect occurs when the cavitation number increases. Otherwise, increase in the angle of attack makes the cavity length increase.

The summary of the previous results are listed in Table (**2**).

Table. 2. Results of NACA-16006.

Case	Free-Stream Velocity (U_∞)	Cavitation Number (σ)	Angle of Attack (α)	Cavitation Length (l)
(1)	1.0	1.097	4°	0.4
(2)	0.5	1.097	4°	0.425
(3)	2.0	1.097	4°	0.3
(4)	3.0	1.097	4°	0.25
(5)	1.0	0.852	4°	0.47
(6)	1.0	1.26933	4°	0.225

(7)	1.0	1.097	3°	0.34
(8)	1.0	1.097	5°	0.44

NACA-0012 Hydrofoil

Case (1)

In this case study, the present code is used to solve cavitation for the hydrofoil section NACA 0012 with the three different parameters which were studied with the previous hydrofoil NACA section. The new values of the three parameters are the cavitation number $\sigma = 0.9774$, the angle of attack $\alpha = 4^{\circ}$ and the inflow velocity $U_{\infty} = 0.5 m/s$. The cavity length $\ell = \dfrac{L}{c} = 0.3$ achieved the cavity closure condition at the cavity end , given in equation (8.78). Fig. (**8.38**) shows the results of this case.

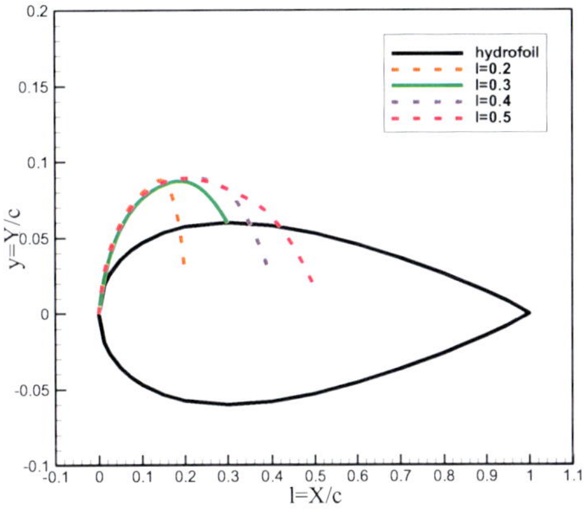

Fig. (8.38). Scaled hydrofoil section NACA-0012 at $U_{\infty} = 0.5 m/s$.

Case (2)

In this case, the present code is used with the same three parameters which were used in the previous case. There are two parameters having the same value as the previous case, the cavitation number $\sigma = 0.9774$ and the angle of attack $\alpha = 4^{o}$. However, the inflow velocity changed to become $U_{\infty} = 1.0 m/s$. The results of this case are shown in Fig. (**8.39**). It is clear that the cavity length $\ell = \frac{L}{c} = 0.3496$ achieved the cavity closure condition at the cavity end, δ , given in equation (8.78).

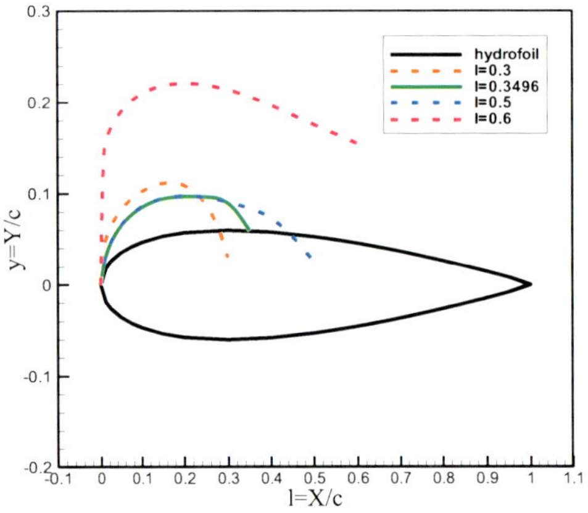

Fig. (8.39). Scaled hydrofoil section NACA-0012 at $U_{\infty} = 1.0 m/s$.

Case (3)

As the previous two cases, the parameters which have the same values are the cavitation number $\sigma = 0.9774$ and the angle of attack $\alpha = 4^{o}$. The inflow velocity takes a new value $U_{\infty} = 2.0$. The results of this case are shown in Fig. (8.41). The present code finds the cavity length $\ell = \frac{L}{c} = 0.5$ which achieved the cavity closure condition at the cavity end δ, given in equation (8.78).

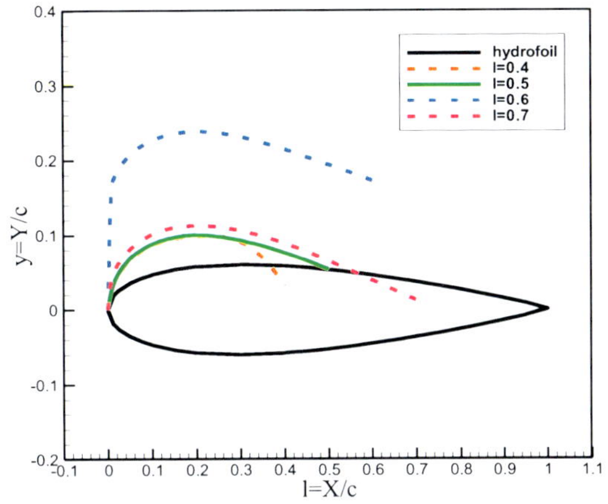

Fig. (8.40). Scaled hydrofoil section NACA-0012 at $U_{\infty} = 2.0\,m/s$.

Case (4)

In this case study, the inflow velocity takes a new value, $U_{\infty} = 4.0\,m/s$ and the other parameters still have the same values as the previous cases. In Fig. (**8.41**), all results of this case are shown and it is clear that the cavity length $\ell = \frac{L}{c} = 0.616$ achieved the same condition in equation (8.78).

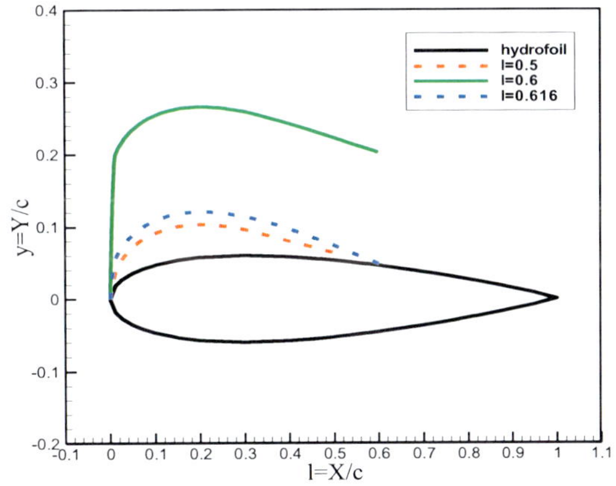

Fig. (8.41). Scaled hydrofoil section NACA-0012 at $U_{\infty} = 4.0\,m/s$.

Case (5)

In this case study, three different parameters are prescribed and the present code is used to find the cavity length $\ell = \frac{L}{c} = 0.443$. The three parameters are, the cavitation number $\sigma = 0.825$, the inflow velocity $U_{\infty} = 1.0\,m/s$ and the angle of attack $\alpha = 4^{o}$. The results of this case are shown in Fig. (**8.42**). It is clear that the cavity closure condition at the cavity end, δ , given in equation (8.78) is obtained, in all cases shown, at $\ell = \frac{L}{c} = 0.443$.

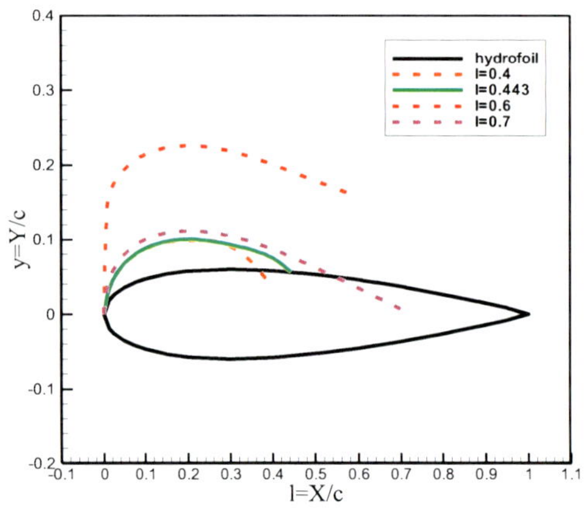

Fig. (8.42). Scaled hydrofoil section NACA-0012 at $\sigma = 0.825$.

Case (6)

As in the previous case, the two parameters which have the same values are the angle of attack $\alpha = 4^{o}$ and the inflow velocity $U_{\infty} = 1.0\,m/s$. However, the cavitation number takes a new value $\sigma = 1.1247$. The cavity length ℓ, that achieved the closure condition in equation (8.78) is $\ell = \frac{L}{c} = 0.244$. The results of this case are shown in Fig. (**8.43**).

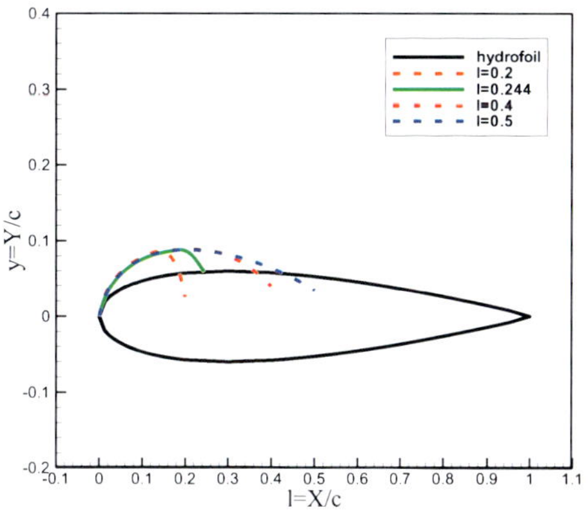

Fig. (8.43). Scaled hydrofoil section NACA-0012 at $\sigma = 1.1247$

Case (7)

In the present case, the angle of attack, α, is studied as a variable parameter. The other two parameters, the cavitation number $\sigma = 0.9774$ and the inflow velocity $U_\infty = 1.0 m/s$ are prescribed at the mentioned value. The present code is used with an angle of attack $\alpha = 3^\circ$ and gives the cavity length $\ell = \frac{L}{c} = 0.247$, which achieved the closure condition in equation (8.78). In Fig. (**8.44**), the results of this case are shown at different lengths.

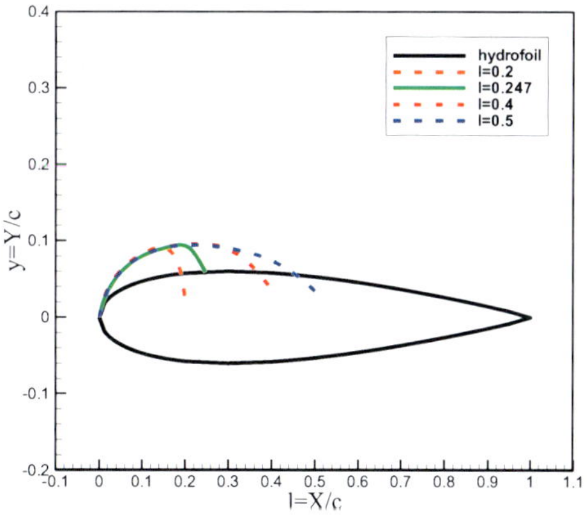

Fig. (8.44). Scaled hydrofoil section NACA-0012 at $\alpha = 3^\circ$.

Case (8)

In this case, the same two parameters, the cavitation number $\sigma = 0.9774$ and the inflow velocity $U_{\infty} = 1.0 m/s$ are used. However, the angle of attack, α, has a new value α = 5°. It is clear that the cavity length $\ell = \frac{L}{c} = 0.6$ achieved the cavity closure condition at the cavity end, , given in equation (8.78). The results of this case are shown in Fig. (**8.45**).

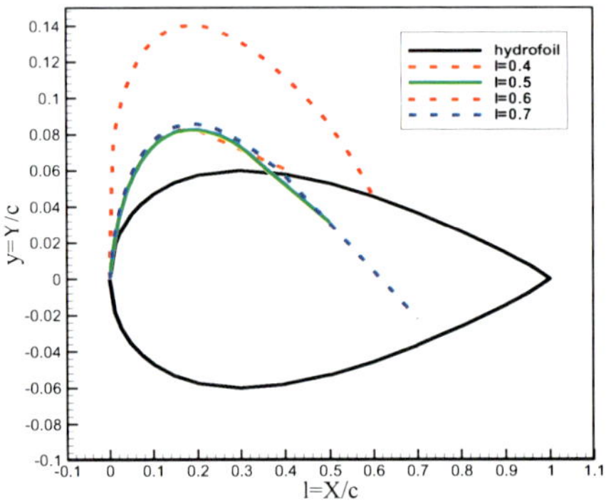

Fig. (8.45). Non-scaled hydrofoil section NACA-0012 at $\alpha = 5°$

Fig. (**8.46**) illustrates the variations in cavitation shape with free-stream velocity (U_{∞}). The cavity length increases with U_{∞} .

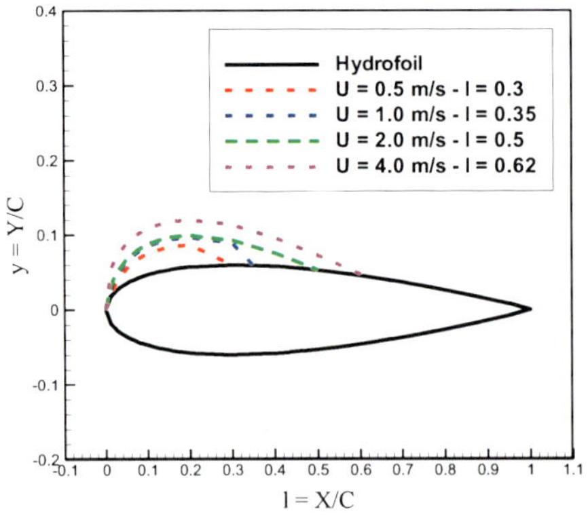

Fig. (8.46). Effect of free-stream velocity, NACA-0012.

The effect of cavitation number on the cavity shape and length at $\alpha = 4^o$ and $U_\infty = 1.0$ *m/s* is shown in Fig. (**8.47**). The cavitation length decreases as σ increases.

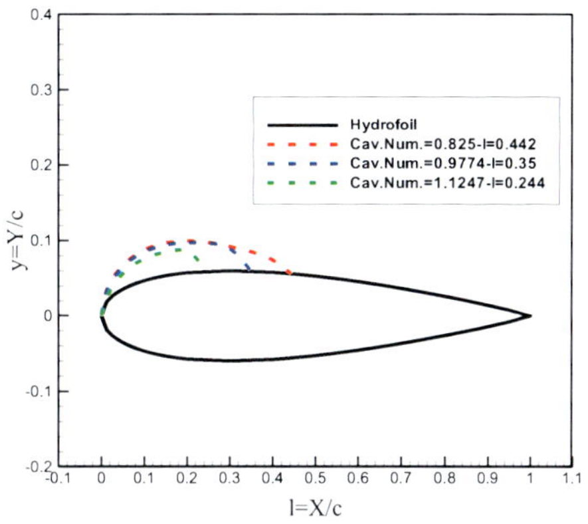

Fig. (8.47). Effect of cavitations number, NACA-0012.

Fig. (**8.48**) shows the effect of the angle of attack (α) for $\sigma = 0.977$ and $U_\infty = 1.0$ *m/s*. As expected, the height and length of cavitation increase withα.

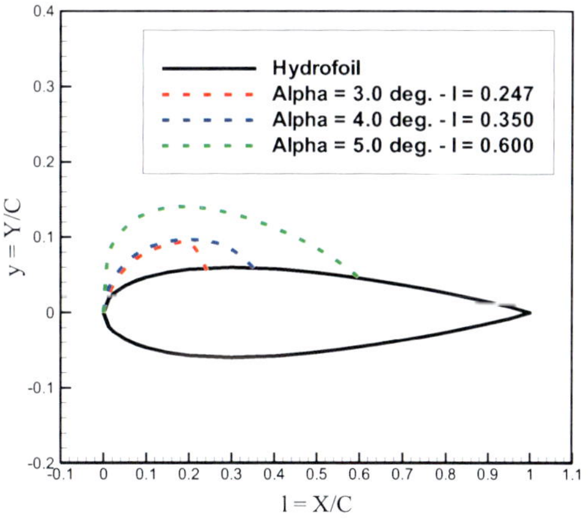

Fig. (8.48). Effect of angle of attack, NACA-0012.

From the previous results, it is shown that the free stream velocity increases the cavity length, and the same effect occurs when the angle of attack increases.

Otherwise, increase of the cavitation number makes the cavity length decrease. The summary of the previous results are listed in Table (**8.3**).

Table (8.3). Results of NACA-0012.

Case	Free-Stream	Cavitation Number (σ)	Angle of Attack (α)	Cavitation Length (l)
(1)	0.5	0.9774	4°	0.3
(2)	1.0	0.9774	4°	0.3496
(3)	2.0	0.9774	4°	0.5
(4)	4.0	0.9774	4°	0.616
(5)	1.0	0.825	4°	0.443
(6)	1.0	1.1247	4°	0.244
(7)	1.0	0.9774	3°	0.247
(8)	1.0	0.9774	5°	0.6

NACA-0015 Hydrofoil

Case (1)

In this case study, the present code is used to solve the hydrofoil section NACA 0015 with the three different parameters which were studied in the previous hydrofoil NACA section. The new values of the three parameters are the cavitation number $\sigma = 0.9774$, the angle of attack $\alpha = 5^{\circ}$ and the inflow velocity $U_{\infty} = 0.25\,m/s$. The cavity length $\ell = \frac{L}{c} = 0.251$ achieved the cavity closure condition at the cavity end δ, given in equation (8-78), Fig. (**8.49**) shows the results of this case.

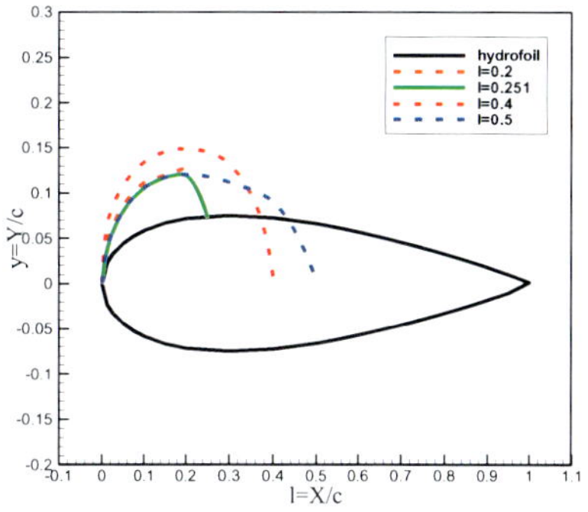

Fig. (8.49). Non-scaled hydrofoil section NACA-0015 at $U_{\infty} = 0.25 m/s$.

Case (2)

In this case, the present code is used with the same three parameters which used in the previous case. There are two parameters that have the same value as the previous case, the cavitation number $\sigma = 0.9774$ and the angle of attack $\alpha = 5^{o}$. However, the inflow velocity U_{∞}, changed to become $U_{\infty} = 0.5 m/s$. The results of this case is shown in Fig. (**8.50**). It is clear that the cavity length $\ell = \dfrac{L}{c} = 0.3$ achieved the cavity closure condition at the cavity end δ, given in equation (8-78).

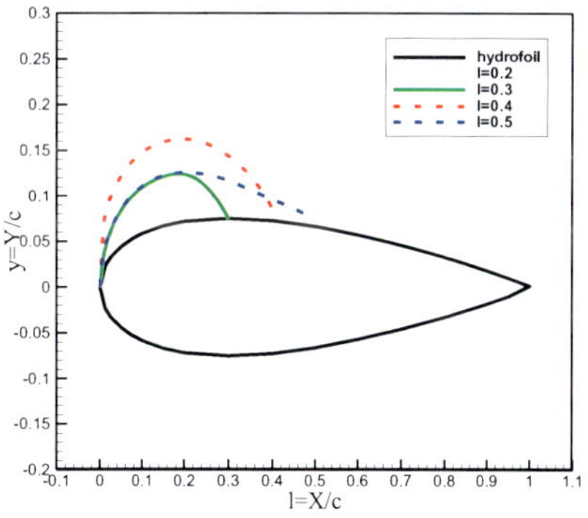

Fig. (8.50). Non-scaled hydrofoil section NACA-0015 at $U_{\infty} = 0.5 m/s$.

Case (3)

As in the previous two cases, the parameters which have the same values are the cavitation number $\sigma = 0.9774$ and the angle of attack $\alpha = 5^o$. The inflow velocity takes a new value $U_\infty = 1.0\,m/s$ the results of this case are shown in Fig. (**8.51**). The present code finds the cavity length $\ell = \frac{L}{c} = 0.4667$ which achieved the cavity closure condition at the cavity end δ, given in equation (8-78).

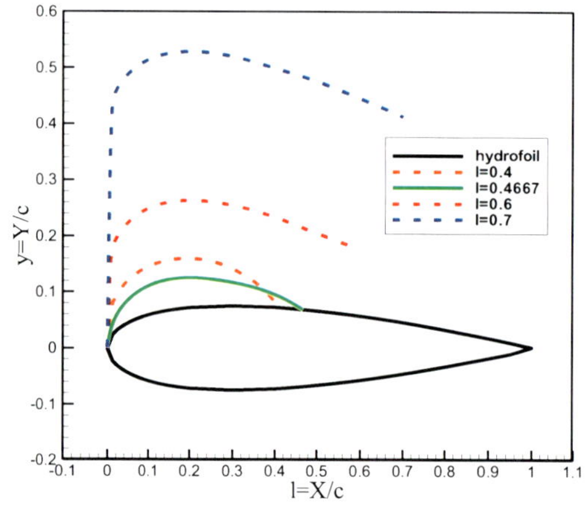

Fig. (8.51). Non-scaled hydrofoil section NACA-0015 at $U_\infty = 1.0\,m/s$.

Case (4)

In this case study, the inflow velocity takes a new value $U_\infty = 2.0\,m/s$ and the other parameters still have the same values in the previous cases. In Fig. (**8.52**), all results of this case are shown and it is clear that the cavity length, $\ell = \frac{L}{c} = 0.5$ achieved the closure condition of equation (8-78).

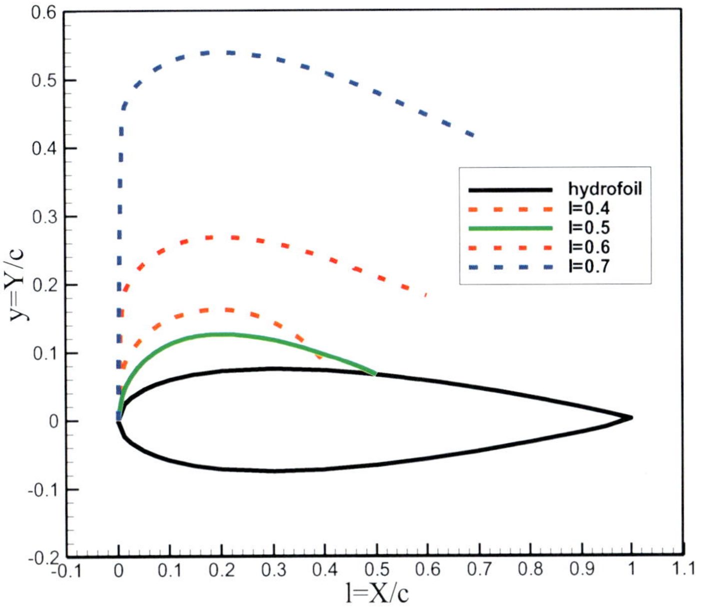

Fig. (8.52). Scaled hydrofoil section NACA-0015 at $U_{\infty} = 2.0\,m/s$.

Case (5)

In this case study, three different parameters are prescribed and the present code is used to find the cavity length, $\ell = \dfrac{L}{c} = 0.5933$, the three parameters are, the cavitation number $\sigma = 0.67585$, the inflow velocity $U_{\infty} = 1.0\,m/s$ and the angle of attack $\alpha = 4^{o}$. The results of this case are shown in Fig. (**8.53**). It is clear that the cavity closure condition at the cavity end, , given in equation (8-78) is obtained at $\ell = \dfrac{L}{c} = 0.5933$.

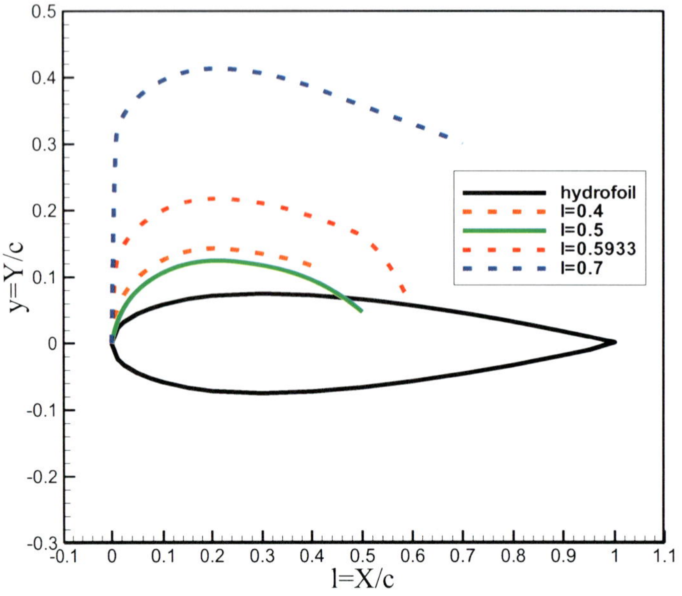

Fig. (8.53). Scaled hydrofoil section NACA-0015 at $\sigma = 0.67585$.

Case (6)

As in the previous case, the two parameters which have the same values are the angle of attack $\alpha = 4^{o}$ and the inflow velocity $U_{\infty} = 1.0\, m/s$. However, the cavitation number takes a new value $\sigma = 0.825$. The cavity length ℓ, that achieved the condition in equation (8-78) is $\ell = \dfrac{L}{c} = 0.415$. The results of this case are shown in Fig. (**8.54**).

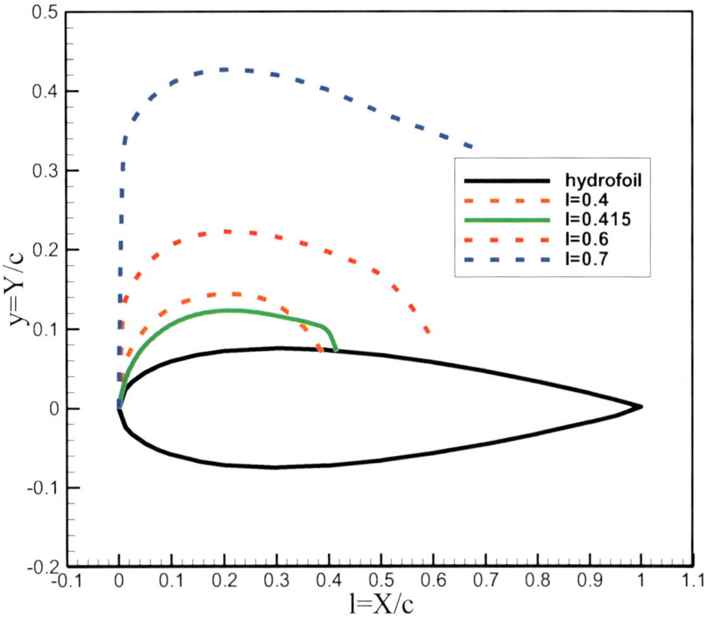

Fig. (8.54). Scaled hydrofoil section NACA-0015 at $\sigma = 0.825$.

Case (7)

In this case, the cavitation number takes a value $\sigma = 0.9774$. The two parameters, the angle of attack, α, and the inflow velocity, U_∞, still have the same values. The cavity length that achieved the condition in equation (8-78) is $\ell = \frac{L}{c} = 0.2333$. The results of this case are shown in Fig. (**8.55**).

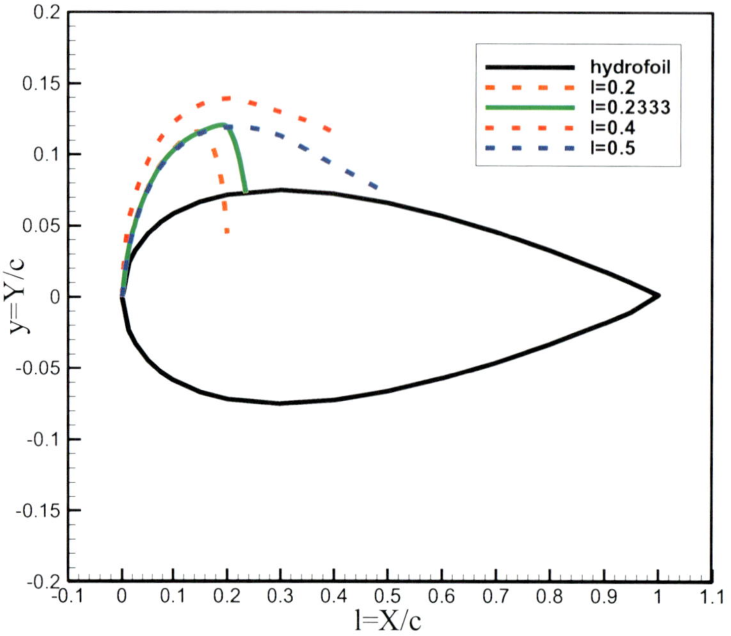

Fig. (8.55). Non-scaled hydrofoil section NACA-0015 at $\sigma = 0.9774$.

Case (8)

In the present case, the angle of attack, α is studied as a variable parameter. In this case, the other two parameters are the cavitation number $\sigma = 0.9774$ and the inflow velocity $U_{\infty} = 1.0\, m/s$. the present code is used with an angle of attack $\alpha = 6^{\circ}$ and gives the cavity length $\ell = \frac{L}{c} = 0.6$ which achieved the closure condition in equation (8-78). In Fig. (**8.56**), the results of this case are shown for different lengths.

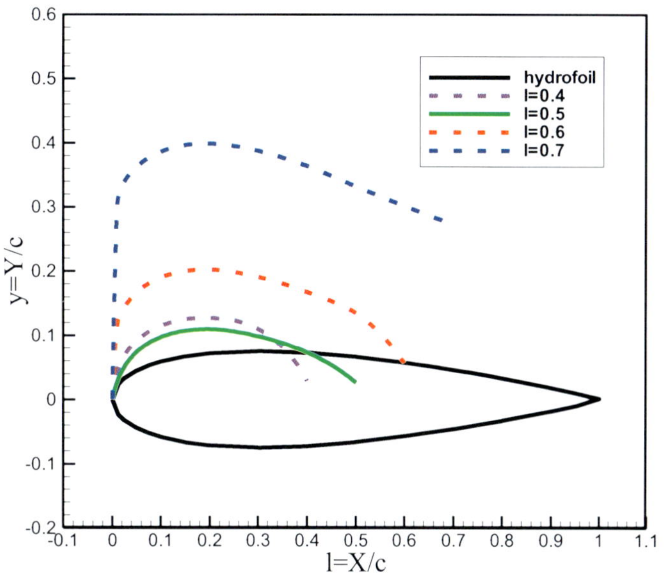

Fig. (8.56). Scaled hydrofoil section NACA-0015 at $\alpha = 6^o$.

Case (9)

In this case, the same two parameters are prescribed, the cavitation number has another value $\sigma = 0.67585$ and the inflow velocity $U_\infty = 1.0 m/s$. However, the angle of attack has a new value $\alpha = 3^o$. It is clear that the cavity length $\ell = \dfrac{L}{c} = 0.315$ achieved the cavity closure condition at the cavity end δ, given in equation (8-78). The results of this case are shown in Fig. (**8.57**).

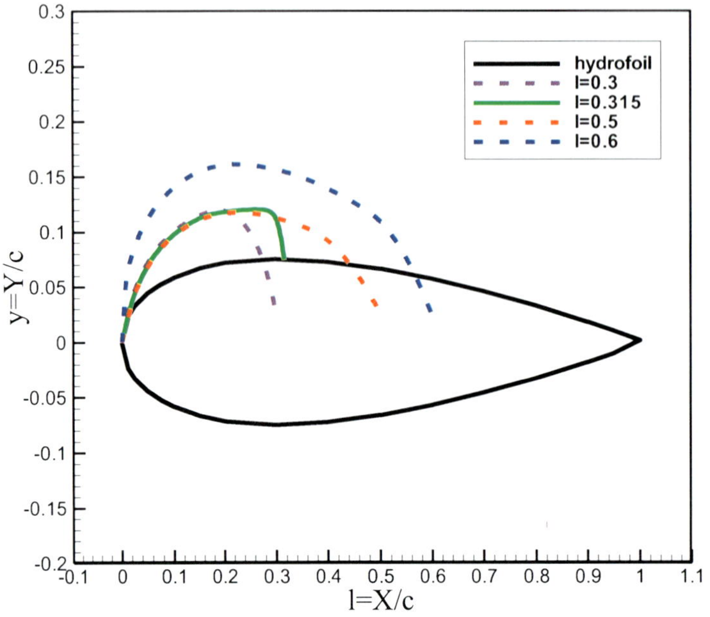

Fig. (8.57). Non-scaled hydrofoil section NACA-0015 at $\alpha = 3^o$.

Case (10)

In this case study, the previous case is applied but with a new value to the angle of attack $\alpha = 3.5^o$ and the other parameters are prescribed with the same values. The present code is used to find the cavity length ℓ. It is clear that the cavity closure condition at the cavity end δ, given in equation (8-78) is obtained at $\ell = \frac{L}{c} = 0.478$.

The results of this case are shown in Fig. (**8.58**).

Fig. (**8.59**) explains the changes of cavitation shape and length with free-stream velocity (from $U_\infty = 0.25$ to 2.0 *m/s*). The cavity height is slightly affected by the increase of U_∞. However, the cavity length increases with free-stream velocity.

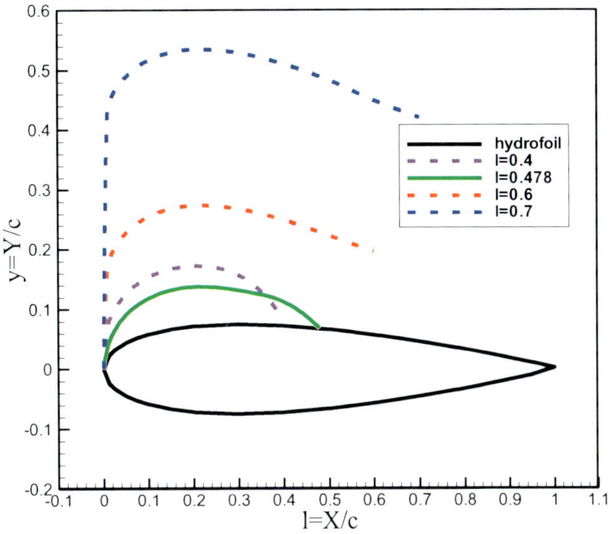

Fig. (8.58). Scaled hydrofoil section NACA-0015 at $\alpha = 3.5^{oo}$.

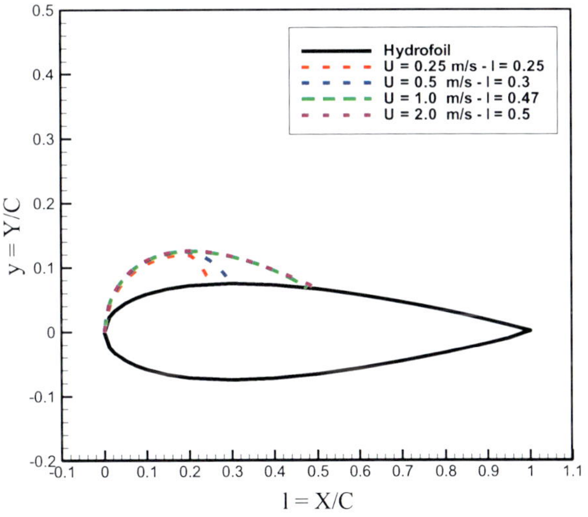

Fig. (8.59). Effect of free-stream velocity, NACA 0015.

Fig. (**8.60**) shows the effect of cavitation number on the cavity shape and length at

$\alpha = 4^o$ and $U_\infty = 1.0$ *m/s*. The cavitation height and length decrease as σ increases.

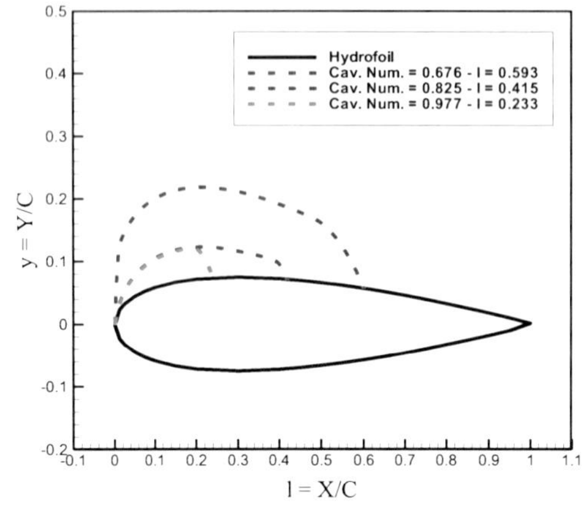

Fig. (8.60). Effect of cavitations number, NACA-0015.

Figs. (**8.61** and **8.62**) illustrate the effect of the angle of attack (α) for $\sigma = 0.67585$ and 0.9774, respectively. A range of α (from 3° to 6°) was tested for $U_\infty = 1.0$ *m/s*. Again, the height and length of cavitation increase withα.

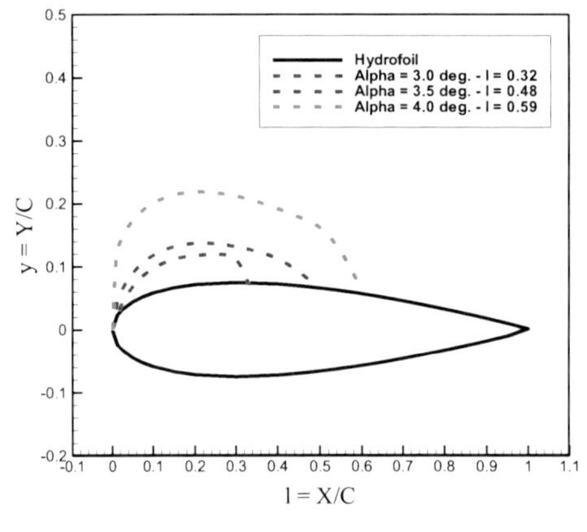

Fig. (8.61). Effect of angle of attack with cavity number = 0.67585, NACA-0015.

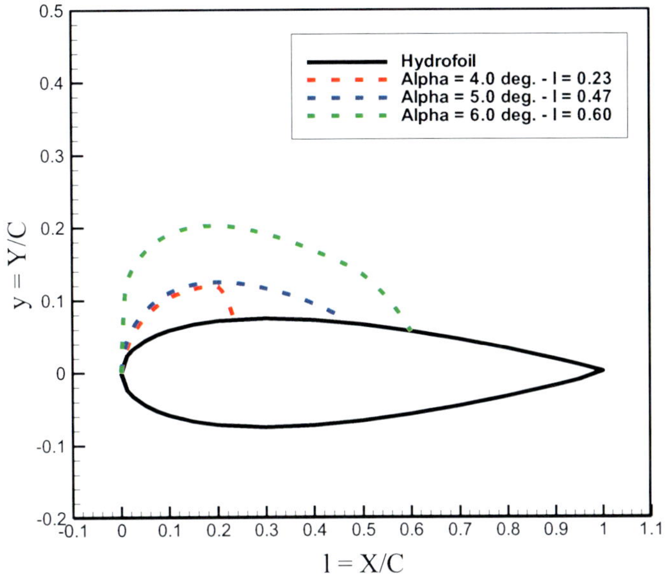

Fig. (8.62). Effect of angle of attack cavitation number = 0.9774, NACA-0015.

From the previous results, it is shown that the free stream velocity increases the cavity length, and the same effect occurs when the angle of attack increases. Otherwise, increase of the cavitation number makes the cavity length decrease. The summary of the previous results are listed in Table (**8.4**).

Table 8.4. Results of NACA-0015.

Case	**Free-Stream Velocity (U_∞)**	**Cavitation Number (σ)**	**Angle of Attack (α)**	**Cavitation Length (l)**
(1)	0.25	0.9774	5°	0.251
(2)	0.5	0.9774	5°	0.3
(3)	1.0	0.9774	5°	0.4667

(4)	2.0	0.9774	5°	0.5
(5)	1.0	0.67585	4°	0.59
(6)	1.0	0.825	4°	0.415
(7)	1.0	0.9774	4°	0.2333
(8)	1.0	0.9774	6°	0.6
(9)	1.0	0.67585	3°	0.315
(10)	1.0	0.67585	3.5°	0.478

NACA-0025 Hydrofoil

Case (1)

In this case study, the present code is used to solve the cavitation of the hydrofoil section NACA 0025 with the three different parameters studied for the previous hydrofoil NACA section. The new values of the three parameters are the cavitation number $\sigma = 0.9774$, the angle of attack $\alpha = 5^{\circ}$ and the inflow velocity $U_{\infty} = 0.5\,m/s$. The cavity length $\ell = \frac{L}{c} = 0.25$ achieved the cavity closure condition at the cavity end, δ, given in equation (8-78). Fig. (**8.63**) shows the results of this case.

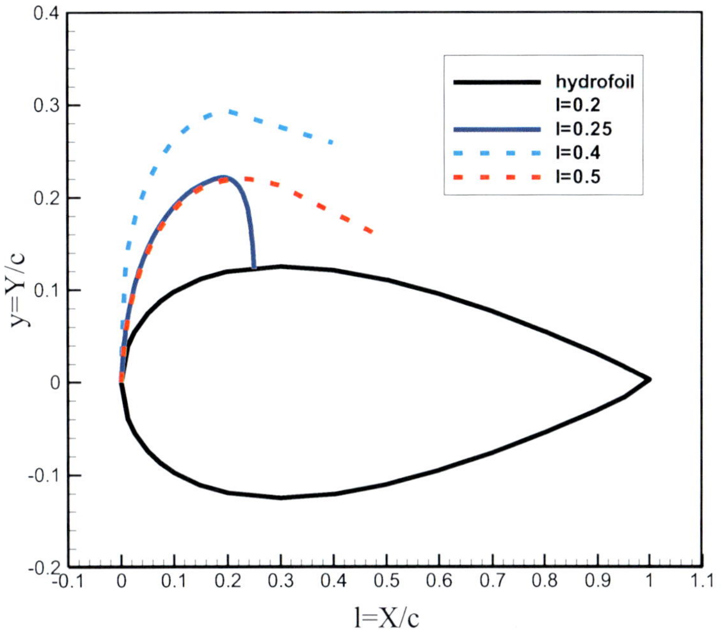

Fig. (8.63). Scaled hydrofoil section NACA-0025 at $U_{\infty} = 0.5 m/s$.

Case (2)

In this case, the present code is used with the same three parameters used in the previous case. There are two parameters that have the same value as the previous case, the cavitation number $\sigma = 0.9774$ and the angle of attack $\alpha = 5^{o}$, however, the inflow velocity changed to become $U_{\infty} = 1.0 m/s$. the results of this case are shown in Fig. (**8.64**). It is clear that the cavity length $\ell = \frac{L}{c} = 0.419$ achieved the cavity closure condition at the cavity end δ, given in equation (8-78).

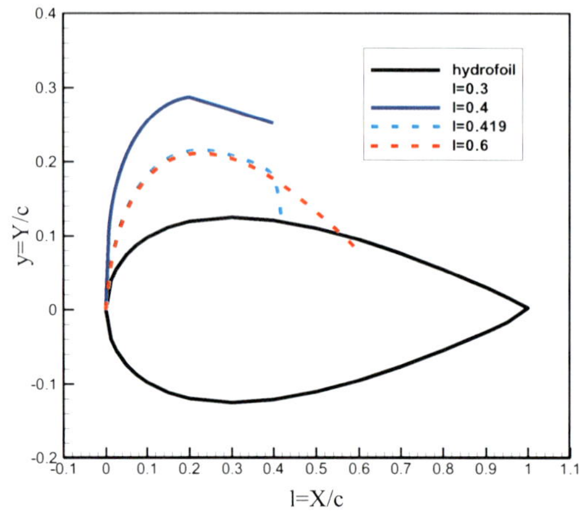

Fig. (8.64). Scaled hydrofoil section NACA-0025 at $U_\infty = 1.0 m/s$.

Case (3)

As in the previous two cases, the parameters which have the same values are the cavitation number $\sigma = 0.9774$ and the angle of attack $\alpha = 5^o$. The inflow velocity takes a new value $U_\infty = 2.0 m/s$. The results of this case are shown in Fig. (**8.65**). The present code finds the cavity length $\ell = \dfrac{L}{c} = 0.427$ which achieved the cavity closure condition at the cavity end, δ, given in equation (8-78).

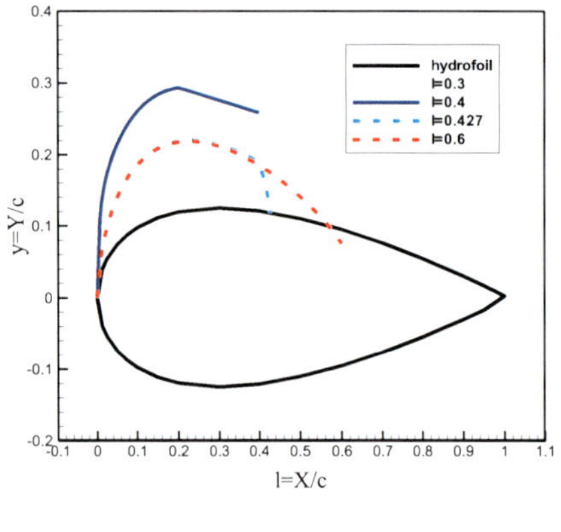

Fig. (8.65). Scaled hydrofoil section NACA-0025 at $U_\infty = 2.0 m/s$.

Case (4)

In this case study, three different parameters are prescribed and the present code is used to find the cavity length $\ell = \dfrac{L}{c} = 0.278$. The three parameters are, the cavitation number $\sigma = 0.67585$, the inflow velocity $U_\infty = 1.0\,m/s$ and the angle of attack $\alpha = 4^o$. The results of this case are shown in Fig. (**8.66**). It is clear that the cavity closure condition at the cavity end, δ, given in equation (8-78) is obtained at $\ell = \dfrac{L}{c} = 0.278$.

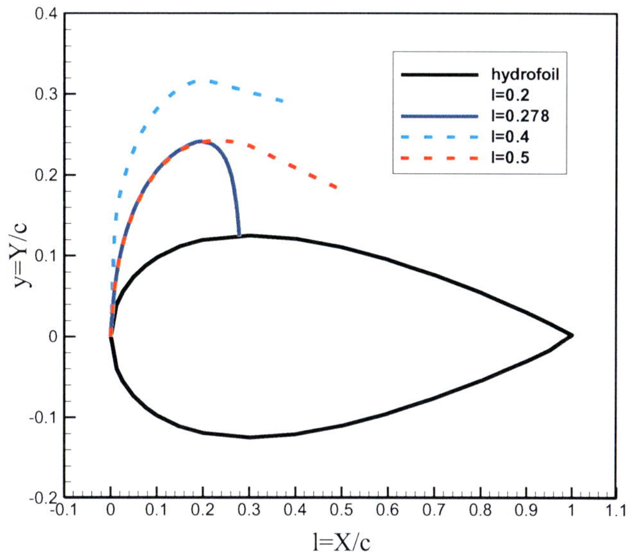

Fig. (8.66). Scaled hydrofoil section NACA-0025 at $\sigma = 0.67585$.

Case (5)

As in the previous case, the two parameters with the same values are the angle of attack $\alpha = 4^o$ and the inflow velocity $U_\infty = 1.0\,m/s$. However, the cavitation number takes a new value, $\sigma = 0.825$. The cavity length ℓ, that achieved the closure condition in equation (8-78) is $\ell = \frac{L}{c} = 0.238$. The results of this case are shown in Fig. (**8.67**).

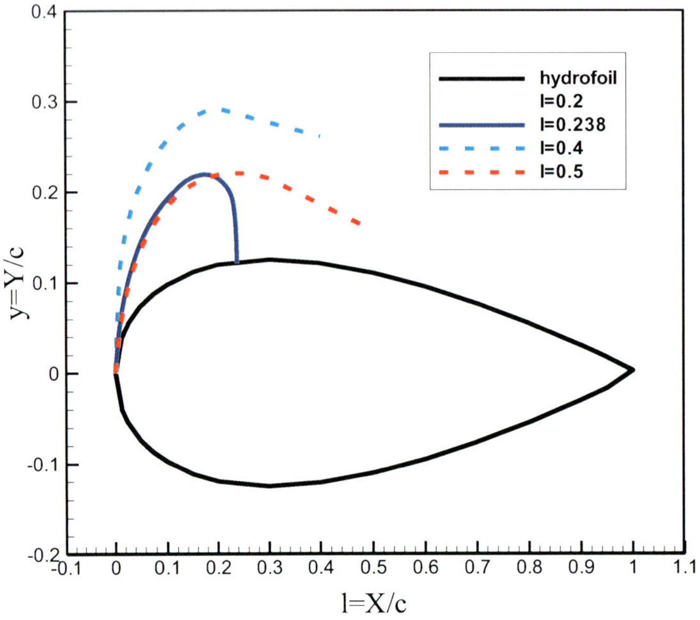

Fig. (8.67). Scaled hydrofoil section NACA-0025 at $\sigma = 0.825$.

Case (6)

In this case, the cavitation number takes a value $\sigma = 0.9774$. The two parameters, the angle of attack, α, and the inflow velocity, U_∞, still have the same values. The cavity length, ℓ, that achieved the condition in equation (8-78) is $\ell = \frac{L}{c} = 0.212$. The results of this case are shown in Fig. (**8.68**).

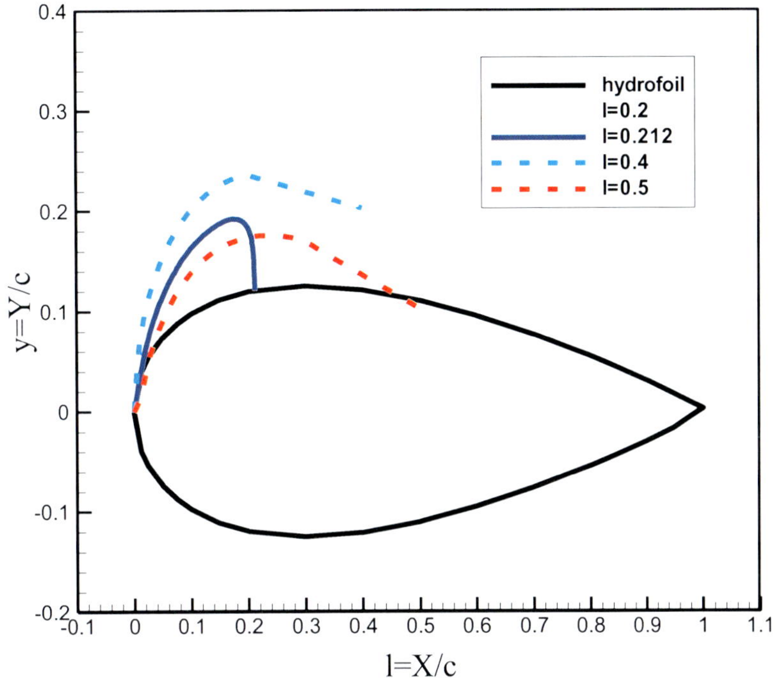

Fig. (8.68). Scaled hydrofoil section NACA-0025 at $\sigma = 0.9774$.

Case (7)

In this case, the angle of attack, α, is studied as a variable parameter. The other two parameters, the cavitation number $\sigma = 0.9774$ and the inflow velocity $U_\infty = 1.0\,m/s$ are prescribed at the shown value. The present code is used with an angle of attack $\alpha = 6^\circ$ and gives the cavity length $\ell = \dfrac{L}{c} = 0.5$, which achieves the closure condition in equation (8-78). In Fig. (**8.69**) the results of this case are shown at different lengths.

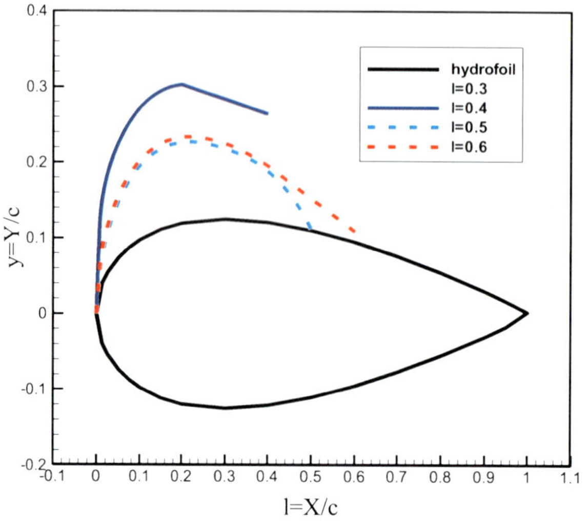

Fig. (8.69). Scaled hydrofoil section NACA-0025 at$\alpha = 6^o$.

Fig. (**8.70**) explains the changes of cavitation shape and length with free-stream velocity (from U_∞ = 0.5 to 2.0 *m/s*). The cavity height is slightly affected by the increase of U_∞. However, the cavity length increases with free-stream velocity.

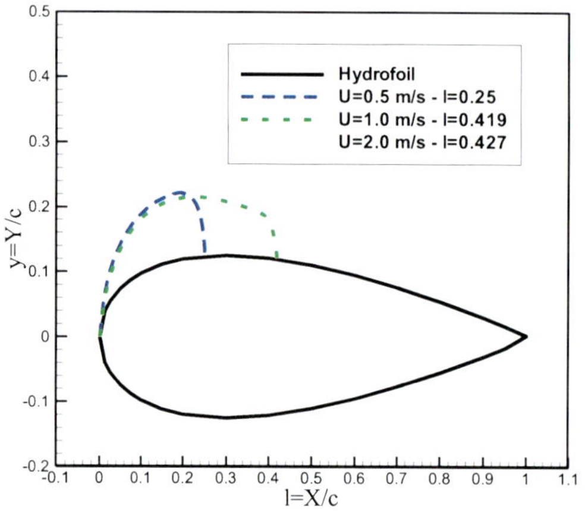

Fig. (8.70). Effect of free-stream velocity, NACA-0025.

Fig. (**8.71**) shows the effect of cavitation number on the cavity shape and length at $\alpha = 4^o$ and U_∞ = 1.0 *m/s*. The cavitation height and length decrease as σ increases.

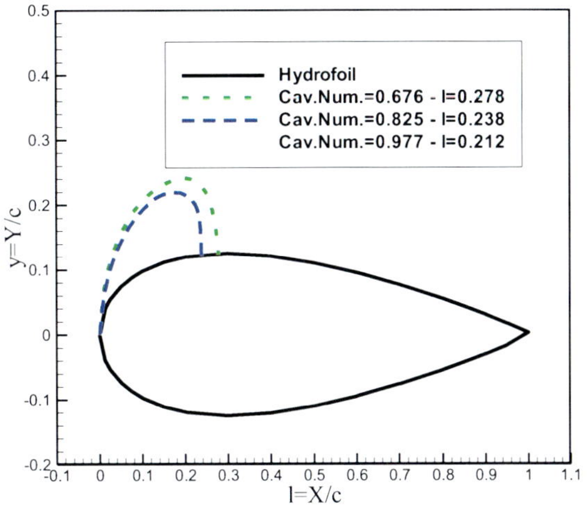

Fig. (8.71). Effect of cavitations number, NACA-0025.

Fig. (**8.72**) illustrate the effect of the angle of attack (α) for $\sigma = 0.9774$. A range of α (from 4° to 6°) was tested for U_∞ = 1.0 *m/s*. Again, the height and length of cavitation increase with α.

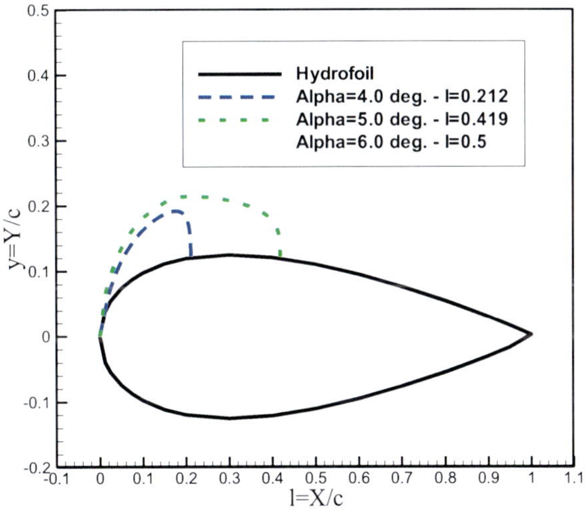

Fig. (8.72). Effect of angle of attack, NACA-0025.

From the previous results, it is shown that the free stream velocity increases the cavity length, and the same effect occurs when the angle of attack increases. Otherwise, increaseof the cavitation number makes the cavity length decrease. The summary of the previous results is listed in Table (**8.5**).

Table. 8.5. Results of NACA-0025.

Case	Free-Stream Velocity (U_∞) *m/s*	Cavitation Number (σ)	Angle of Attack (α)	Cavitation Length (l)
(1)	0.5	0.9774	5°	0.25
(2)	1.0	0.9774	5°	0.419
(3)	2.0	0.9774	5°	0.427
(4)	1.0	0.67585	4°	0.278
(5)	1.0	0.825	4°	0.238
(6)	1.0	0.9774	4°	0.212
(7)	1.0	0.9774	6°	0.5

Comparisons between Hydrofoil Sections

To identify the effect of the hydrofoil section (maximum thickness), comparisons should be made between their results at the same operating conditions. A summary of the results is shown in Tables (**8.6** – **8.8**) for the effect of cavitation number (σ), angle of attack (α) and the free-stream velocity (U_∞), respectively. It is clear that the cavitation length decreases with the hydrofoil thickness at the same value ofσ.However, it seems that the height increases with the hydrofoil thickness. For all NACA sections, it is clear that the cavitation length decreases with increasing of the cavitation number, as shown in Table (**8.6**).

Table. 8.6. Effect of σ.

NACA Section	Angle of Attack (α)	-	Cavitation Number (σ)	Cavitation Length (l)
NACA-16006	4.0°	1.0	0.852	0.47
			1.097	0.4
			1.269	0.225
NACA-0012	4.0°	1.0	0.825	0.442
			0.977	0.35
			1.125	0.244
NACA-0015	4.0°	1.0	0.676	0.593
			0.825	0.415
			0.977	0.233
NACA-0025	4.0°	1.0	0.676	0.278
			0.825	0.238
			0.977	0.212

Some figures show that similar behavior as the behavior of the cavitation number is noticed for the cavitation length. Generally, the cavitation length varies with the hydrofoil thickness at the same angle of attack (*e.g.* $\alpha = 5^o$). However, for all NACA sections, it is clear that the cavitation length increases with increasing of the angle of attack, as shown in Table (**8.7**).

Table. 8.7. Effect of α.

NACA Section	Cavitation Number (σ)	Free-stream Velocity (U_∞) *m/s*	Angle of Attack (α)	Cavitation Length (l)
NACA-16006	1.097	1.0	3.0°	0.34
			4.0°	0.4
			5.0°	0.44
NACA-0012	0.977	1.0	3.0°	0.247
			4.0°	0.35
			5.0°	0.60
NACA-0015	0.977	1.0	4.0°	0.23
			5.0°	0.47
			6.0°	0.60
	0.677	1.0	3.0°	0.32
			3.5°	0.48
			4.0°	0.59
NACA -			4.0°	0.212

0025	0.977	1.0	5.0°	0.419
			6.0°	0.5

Some other cases illustrate that the cavitation height increases with hydrofoil thickness at the same free-stream velocity ($e.g. U_\infty = 1.0\ m/s$). However, the cavitation length decreases with hydrofoil thickness. For all NACA sections, it is clear that the cavitation length increases with increasing of the free-stream velocity except the NACA 16006, as shown in Table (**8.8**).

Table. 8.8. Effect of U_∞.

NACA Section	**Angle of Attack (α)**	**Cavitation Number (σ)**	**Free-stream Velocity (U_∞) *m/s***	**Cavity Length (l)**
NACA-16006	4.0°	1.097	0.5	0.425
			1.0	0.4
			2.0	0.3
			3.0	0.25
NACA-0012	4.0°	0.977	0.5	0.3
			1.0	0.35
			2.0	0.5

			4.0	0.62
NACA-0015	5.0°	0.977	0.25	0.25
			0.5	0.3
			1.0	0.47
			2.0	0.5
NACA-0025	5.0°	0.977	0.5	0.25
			1.0	0.419
			2.0	0.427

COMPARISON WITH OTHERS

To emphasize on the validity and importance of the present boundary element algorithm, a comparison is made between the present results and those of Shimizu *et al.* [180]. They used the large eddy simulation (LES) method, which is a well-known computational fluid dynamics technique for real viscous turbulent flows, to predict cavitating flow around a two-dimensional NACA-0015 hydrofoil. They performed the computations at $Re = 1.2\times10^6$ and $\alpha = 8^o$. Their computational mesh consisted of 1.2 million finite elements. Figure (52) shows the comparison between the results of Shimizu *et al.* [180] (Fig.**51-a**) and the present results (Fig. **51-b**) for the NACA-0015 hydrofoil. The present results were carried out at a corresponding Reynolds number of 1.2×10^6 for real viscous flow. Reynolds number (Re) is calculated based on the free-stream velocity (U_∞) and chord length (c). Thus, $Re = (U_\infty c)/\nu$ and ν is the fluid kinematic viscosity. As can be seen, the present results compare well to those of [180] when concerning the dimensionless cavitation length (l). Table (**9**) shows a comparison between the present results and those of [180] for the dimensionless cavitation length (l). The maximum difference between the present results and those of [180] is 4% (This value of 4% is relative to the hydrofoil chord length, C). The main advantages of the present algorithm are its relative simplicity and its small software, hardware, and run-time requirements compared to the much more complicated real viscous flow codes.

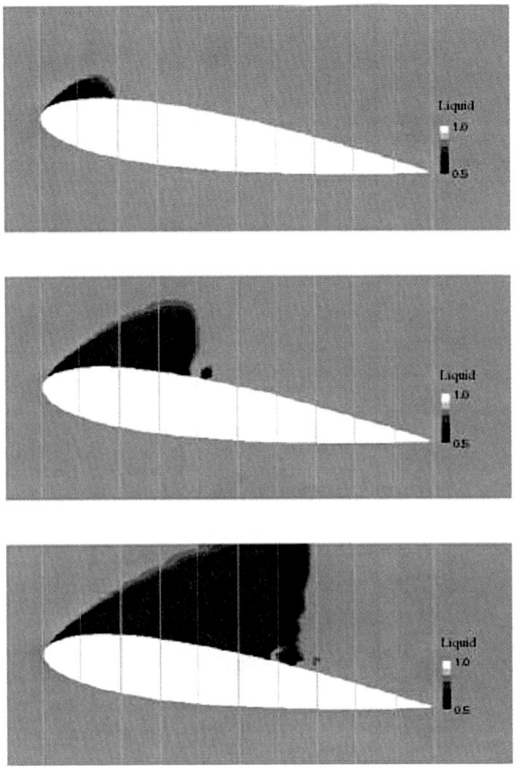

Top σ= 1.6, middle σ = 1.15, bottom σ= 0.9e

Fig. (8.72). Results of Shimizu *et al.* [180], α= 8°.

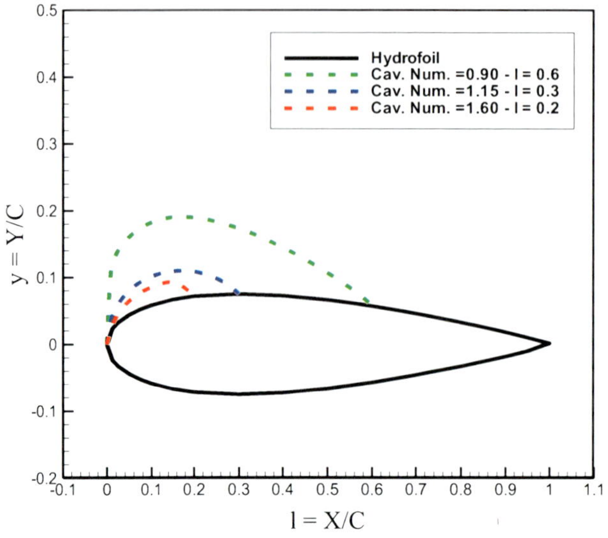

Fig. (8.73). Present results, α= 8°.

Fig. (**8.73**). Comparison between the results of Shimizu *et al.* [180] and the present results for NACA-0015 hydrofoil.

Table. 8.9. Comparison between the present results and those of [180].

Case	σ	Dimensionless cavitation length (*l*) Present model	Dimensionless cavitation length (*l*) Results of Shimizu *et al.* [85]
1	0.90	0.60	0.63
2	1.15	0.30	0.34
3	1.60	0.20	0.19

REFERENCES

[1] J. Crank, *Free and Moving Boundary Problems.* Clarendon Press: Oxford, 1984.

[2] M. Zerrokat, and C.R. Chatwin, *Computational Moving Boundary Problems.* Research Studies Press Ltd, 1994.

[3] Chang-Sup Lee Young-Dal Choi, Byoung-Kwon Ahn, *et al*, "Performance optimization of marine propellers", *Int. J. Nav. Archit,* vol. 1010; 2, no. 4, p. 211p. 216, . [http://dx.doi.org/10.2478/IJNAOE-2013-0038]

[4] L.S. Yao, and J. Prusa, "Melting and freezing", In: *Advances in Heat Transfer. 19.* Academic Press: New York, 1989. [http://dx.doi.org/10.1016/S0065-2717(08)70211-9]

[5] A. Bejan, and P.A. Tyvand, "The pressure melting of ice to an embedded cylinder", *J. Heat Transfer,* vol. 114, no. 2, pp. 532-535, 1992. [http://dx.doi.org/10.1115/1.2911311]

[6] M.K. Moallemi, B.W. Webb, and R. Viskanta, "An experimental and analytical study of close-contact melting", *J. Heat Transfer,* vol. 108, no. 4, pp. 894-899, 1986. [http://dx.doi.org/10.1115/1.3247030]

[7] G.S. Berntsen, M. Kjeldsen, and R. Muller, "Cavitations Induced Dynamics in Hydraulic Machinery", *Proc 10th International Meeting of the Work Group on The Behavior of Hydraulic Machinery under Steady Oscillatory Condition Trondheim*, Norway.

[8] S. Bal, "A Potential Based Panel Method for 2-D Hydrofoils", *Ocean Eng.,* vol. 26, no. 4, pp. 343-361, 1999. [http://dx.doi.org/10.1016/S0029-8018(97)10022-1]

[9] G. Kuiper, *Cavitations and New Blade Sections.* vol. Vol. 190. ASME FED-Cavitations and Gas-Liquid Flow in Fluid Machinery and Devices, 1994.

[10] H. Lemonnier, and A. Rowe, "Another Approach in Modelling Cavitating Flows", *J. Fluid Mech.,* vol. 195, no. 2, pp. 557-580, 1988. [http://dx.doi.org/10.1017/S0022112088002526]

[11] M. Dular, B. Bachert, B. Stoffel, and B. Sirok, "Relationship between Cavitations Structures and Cavitations Damage", *Wear,* vol. 257, no. 11, pp. 1176-1184, 2004. [http://dx.doi.org/10.1016/j.wear.2004.08.004]

[12] Z. Xie, and Y. He, "an Approach in Modeling Cavitating Flows with Gravity Effect", *Commun. Nonlinear Sci. Numer. Simul.,* vol. 2, no. 2, pp. 100-104, 1997. [http://dx.doi.org/10.1016/S1007-5704(97)90048-4]

[13] S.A. Kinnas, H. Lee, and Y.L. Young, "Boundary Element Techniques for the Prediction of Sheet and Developed Tip Vortex Cavitations", *Electron. J. Bound. Elem. BETEQ,* vol. 2001, no. 2, pp. 151-178, 2002.

[14] S.A. Kinnas, and N.E. Fine, "Fine NE. A Numerical Nonlinear Analysis of the Flow around Two- and Three-Dimensional Partially Cavitating Hydrofoils", *J. Fluid Mech.,* vol. 254, pp. 151-181, 1993. [http://dx.doi.org/10.1017/S0022112093002071]

[15] S.G. Ahmed, and L.C. Wrobel, "An Alternative BEM Approach for Solving Cavity Flow Problems", *International Conference Boundary Element Technology XI.* 1996 Hawaii, USA.

[16] M. Yasko, "Boundary element method for a hydrofoil near the free surface. J", *Eng. Anal. Bound. Elem.,* vol. 21, no. 2, pp. 191-194, 1998.

[http://dx.doi.org/10.1016/S0955-7997(98)00010-1]

[17] S. Bal, "A Potential based panel method for 2-D hydrofoils", *Ocean Eng.*, vol. 26, no. 4, pp. 343-361, 1999.
[http://dx.doi.org/10.1016/S0029-8018(97)10022-1]

[18] G. Kuiper, *Cavitation and new blade sections.* vol. Vol. 190. ASME FED-Cavitation and Gas-Liquid Flow in Fluid Machinery and Devices, 1994.

[19] H. Lemonnier, and A. Rowe, "Another approach in modeling cavitating flows", *J. Fluid Mech.*, vol. 195, no. 2, pp. 557-580, 1988.
[http://dx.doi.org/10.1017/S0022112088002526]

[20] M. Dular, B. Bachert, B. Stoffel, and B. Sirok, "Relationship between cavitation structures and cavitations damage", *Wear,* vol. 257, no. 11, pp. 1176-1184, 2004.
[http://dx.doi.org/10.1016/j.wear.2004.08.004]

[21] Z. Xie, and Y. He, "An approach in modeling cavitating flows with gravity effect", *Commun. Nonlinear Sci. Numer. Simul.*, vol. 2, no. 2, pp. 100-104, 1997.
[http://dx.doi.org/10.1016/S1007-5704(97)90048-4]

[22] S.A. Kinnas, H. Lee, and Y.L. Young, "Boundary element techniques for the prediction of sheet and developed tip vortex cavitations", *Electron. J. Bound. Elem. BETEQ,* vol. 2001, no. 2, pp. 151-178, 2002.

[23] S.A. Kinnas, and N.E. Fine, "A numerical nonlinear analysis of the flow around two- and three-dimensional partially cavitating hydrofoils", *J. Fluid Mech.,* vol. 254, pp. 151-181, 1993.
[http://dx.doi.org/10.1017/S0022112093002071]

[24] J. Anderson, *Fundamentals of Aerodynamics.* 2nd ed. McGraw–Hill: Toronto, 1991, pp. 195-208.

[25] Tsien, Hsue-shen, "Symmetrical Joukowsky airfoils in shear flow", *Quarterly of Applied Mathematics,* vol. 1, pp. 130-248, 1943.

[26] Y.L. Wu, and C. Shu, "Development of RBF-DQ method for derivative approximation and its application to simulate natural convection in concentric annuli", *Comput. Mech.,* vol. 29, no. 6, pp. 477-485, 2002.
[http://dx.doi.org/10.1007/s00466-002-0357-4]

[27] A.E. Norhan, *Mohamd, ' Boundary Elements Methods and Its Non-Linear Analysis of Hydrofoils', Supervisors: Prof. S. G. Ahmed and Prof. A. F. Abd-Elgawad, Zagazig University.* Zagazig University, 2002.

[28] TH. Yao-Tsu W, "Cavity and Wake Flows", *Annual Review of Fluid Mechanics,* 1972.annualreviews.org

[29] W. Liu, Y. Chen, S. Jun, J.S. Chen, T. Belytschko, C. Pan, R. Uras, and C. Chang, "Overview and application of the reproducing kernel particle methods", *Arch. Comput. Methods Eng.,* vol. 3, no. 1, pp. 3-80, 1996.
[http://dx.doi.org/10.1007/BF02736130]

[30] G.K. Batchelor, "Axial flow in trailing line vortices", *J. Fluid Mech.,* vol. 20, no. 4, pp. 645-658, 1964.
[http://dx.doi.org/10.1017/S0022112064001446]

[31] D. SIPP F. COPPENS & L. JACQUIN, "Theoretical and numerical analysis of wake vortices", *ESAIM: Proceedings,* vol. 7, 1999pp. 397-407 Third International Workshop on Vortex. Flows and Related Numerical Methods.

[32] C. J. Apelt, G. S. West, and A. Albin, "The effects of wake splitter plates on the flow past a circular cylinder in the range 104 <R<5×104", *Cambridge University Press,* 2006.

[33] P. Marshall, *Tulin. On the theory and modeling of real cavity flows.* Advances in Engineering Mechanics-Reflections and Outlooks, 2005, pp. 3-26.

[34] Anching Lin, *On a solution of the Lavrentiev wake model and its cascade.* Cambridge University Press, 2006.

[35] H. Lemonnier, and A. Rowe, "Another Approach in Modelling Cavitating Flows", *J. Fluid Mech.,* vol. 195, no. 2, pp. 557-580, 1988. [http://dx.doi.org/10.1017/S0022112088002526]

[36] M. Dular, B. Bachert, B. Stoffel, and B. Sirok, "Relationship between Cavitation Structures and Cavitation Damage", *Wear,* vol. 257, no. 11, pp. 1176-1184, 2004. [http://dx.doi.org/10.1016/j.wear.2004.08.004]

[37] A. D. Cox, and W. A. Clayden, *Cavitating flow about a wedge at incidence.* Cambridge University Press, 2006.

[38] W. A. Clayden, "Unpublished Ministry of Supply Report",

[39] D. Gilbarg, and D. H. Rock, "On two theories of plane potential flow with finite cavities", *U. S. Naval Ordnance Lab. Mem. No. 8718,* 1946.

[40] R.T. Knapp, *Further studies of the mechanics and damage potential of fixed type cavities,* 1956 *Proc. Symposium on Cavitations in Hydrodynamics*, London: H.M.S.O.

[41] G. Kreisel, "Cavitations with finite cavitations number", *Admiralty Research Lab. Rep. U.B.R.,* vol. C, p. 52, 1946.

[42] H. Lamb, *Hydrodynamics.* 6th ed. Cambridge University Press, 1932.

[43] L.M. Milne-Thomson, *Theoretical Hydrodynamics.* 2nd ed. Macmillan: London, 1949.

[44] L.M. Milne-Thomson, *Jacobian Elliptic Function Tables.*New York, Dover, 1950.

[45] E.H. Neville, *Jacobian Elliptic Functions.* 2nd ed. Oxford University Press, 1951.

[46] B. Perry, "Evaluation of the integrals occurring in the cavity theory of Plesset and Shaffer", *Cal. Inst. Tech. Hydrodynamics Lab. Rep. no. 21-11,* 1952.

[47] A.S. Ramsey, *A Treatise on Hydromechanics, Part II.* Bell: London, 1947.

[48] M. Shiffman, "On free boundaries of an ideal fluid, Part II", *Commun. Pure Appl. Math.,* vol. 2, no. 1, pp. 1-11, 1949. [http://dx.doi.org/10.1002/cpa.3160020101]

[49] E.T. Whittaker, and G.N. Watson, *A Course of Modern Analysis.* 4th ed. Cambridge University Press, 1940.

[50] H. Helmholtz, and H. Fripp, "On the Limits of the Optical Capacity of the Microscope", *Mon. Microsc. J,* vol. 16, no. 1, pp. 15-39, 1876. [http://dx.doi.org/10.1111/j.1365-2818.1876.tb05606.x]

[51] On the Conservation of Force, "Introduction to a Series of Lectures Delivered at Carlsruhe in the Winter of 1862–1863", *English translation,* 1895.

[52] On the Sensations of Tone as a Physiological Basis for the Theory of Music (downloadable from California Digital Library) Third Edition of English Translation, based on Fourth German Edition of 1877, By Hermann von Helmholtz, Alexander John Ellis, Published by Longmans, Green, 1895, 576 pages,

[53] On the Sensations of Tone as a Physiological Basis for the Theory of Music (downloadable from Google Books) Fourth Edition, By Hermann von Helmholtz, Alexander John Ellis, Published by Longmans, Green, 1912, 575 pages.

[54] L Rosenfeld, Biography in Dictionary of Scientific Biography (New York 1970-1990).

[55] L. Boltzmann, *Gustav Robert Kirchhoff.* Populäre Schriften: Leipzig, 1905.

[56] C. Jungnickel, and R. McCormmach, *Intellectual Mastery of Nature.*Chicago, 1986.

[57] T. Archibald, "Tension and potential from Ohm to Kirchhoff", *Centaurus,* vol. 31, no. 2, pp. 141-163, 1988.
[http://dx.doi.org/10.1111/j.1600-0498.1988.tb00684.x]

[58] E.H. Dill, "Kirchhoff's theory of rods", *Arch. Hist. Exact Sci.,* vol. 44, no. 1, pp. 1-23, 1992.
[http://dx.doi.org/10.1007/BF00379680]

[59] N.P. Khomenko, and T.M. Vyvrot, "Euler and Kirchhoff - initiators of the main directions in graph theory (Russian)", In: *On the history of the mathematical sciences 'Naukova Dumka' (Kiev, 1984),* 1984, pp. 63-69, 168.

[60] N.P. Khomenko, and T.M. Vyvrot, "Euler and Kirchhoff - initiators of the main directions in graph theory (Russian) II: Sketches on the history of mathematical physics", *Naukova Dumka (Kiev, 1985).,* vol. 183, pp. 28-34, 1985.

[61] R.C. Ackerberg, "Boundary layer separation at a free streamline. Part 1. Two-dimensional flow", *J. Fluid Mech.,* vol. 44, no. 2, pp. 211-225, 1970.
[http://dx.doi.org/10.1017/S0022112070001799]

[62] R.C. Ackerberg, "The effects of capillarity on free streamline separation", *J. Fluid Mech.,* vol. 70, no. 2, pp. 333-352, 1975.
[http://dx.doi.org/10.1017/S0022112075002066]

[63] A.J. Acosta, "A note on partial cavitations of flat plate hydrofoils", *Calif. Inst. of Tech. Hydro. Lab. Rep. E-19.9,* 1955.

[64] A.J. Acosta, "Calif. Inst. of Tech. Hydro. Lab. Rep. E-79.2", *Cavitating flow past a cascade of circular arc hydrofoils.,* 1960.

[65] A.J. Acosta, "Hydrofoils and hydrofoil craft", *Annu. Rev. Fluid Mech.,* vol. 5, no. 1, pp. 161-184, 1973.
[http://dx.doi.org/10.1146/annurev.fl.05.010173.001113]

[66] A.J. Acosta, Hollander a, "Remarks on cavitations in turbo machines", *Calif. Inst. of Tech. Eng. Div. Rep. No. 79.3,* 1959.

[67] A.J. Acosta, and R.K. DeLong, "Experimental investigation of non-steady forces on hydrofoils oscillating in heave", *Proc. IUTAM Symp. On non-steady flow of water at high speeds,* 1971pp. 95-104 Leningrad, USSR

[68] D.K. Ai, "The wall effect in cavity flows", *J. Basic Eng.,* vol. 88, no. 1, pp. 132-138, 1966.
[http://dx.doi.org/10.1115/1.3645784]

[69] V.H. Arakeri, "Viscous effects on the position of cavitations separation from smooth bodies", *J. Fluid Mech.,* vol. 68, no. 4, pp. 779-799, 1975.
[http://dx.doi.org/10.1017/S0022112075001231]

[70] A.H. Armstrong, "Abrupt and smooth separation in plane and axisymmetric flow", *Memo. Arm. Res. Est., G.B., No. 22/63,* 1953.

[71] A.H. Armstrong, and J.H. Dunham, Axisymmetric cavity flow, *Rep. Res. Est., G.B., No. 12/53.,* 1953.

[72] A.H. Armstrong, and K.G. Tadman, "Axisymmetric cavity flow about ellipsoids", Proc. Joint Admiralty-U.S. Navy Meeting on Hydro ballistics, 1954.

[73] A. Betz, and E. Petersohn, "Application of the theory of free jets", *NACA TM No. 667.*

[74] M.A. Biot, "Some simplified methods in airfoil theory", *J. Aerosp. Sci.,* vol. 9, no. 5, pp. 185-190, 1942.

[75] G. Birkhoff, and E.H. Zarantonello, *Jets, wakes, and cavities.* Academic Press, 1957.

[76] C. Brennen, "A numerical solution of axisymmetric cavity flows", *J. Fluid Mech.,* vol. 37, no. 4, pp. 671-688, 1969.

[http://dx.doi.org/10.1017/S0022112069000802]

[77] C. Brennen, "Some viscous and other real fluid effects in fully developed cavity flows", In: *Cavitations State of Knowledge.*, J.M. Robertson, G.F. Wislicenus, Eds., ASME: N.Y., 1969.

[78] C. Brennen, "Some cavitations experiments with dilute polymer solutions", *J. Fluid Mech.*, vol. 44, no. 1, pp. 51-63, 1970.
[http://dx.doi.org/10.1017/S0022112070001684]

[79] C.E. Brennen, and A.J. Acosta, "Theoretical, quasistatic analysis of cavitation compliance in turbopumps", *J. Spacecr. Rockets,* vol. 10, no. 3, pp. 175-180, 1973.
[http://dx.doi.org/10.2514/3.27748]

[80] C.E. Brennen, K. Oey, and C.D. Babcock, "On the leading edge flutter of cavitating hydrofoils", *J. Ship Res.*, vol. 24, no. 3, pp. 135-146, 1980.
[http://dx.doi.org/10.5957/jsr.1980.24.3.135]

[81] M. Brillouin, "Les surfaces de glissement de Helmholtz et la resistance des fluides", *Ann. Chim. Phys.*, vol. 23, pp. 145-230, 1911.

[82] R.V. Churchill, *Introduction to complex variables and applications.* McGraw-Hill Book Company, 1948.

[83] H. Cohen, and R. Gilbert, "Two-dimensional, steady, cavity flow about slender bodies in channels of finite width", *J. Appl. Mech.*, vol. 24, no. 2, pp. 170-176, 1957.
[http://dx.doi.org/10.1115/1.4011492]

[84] H. Cohen, C.D. Sutherland, and Y-O. Tu, "Wall effects in cavitating hydrofoil flow", *J. Ship Res.*, vol. 1, no. 3, pp. 31-39, 1957.
[http://dx.doi.org/10.5957/jsr.1957.1.4.31]

[85] E. Cumberbatch, and T.Y. Wu, "Cavity flow past a slender pointed hydrofoil", *J. Fluid Mech.*, vol. 11, no. 2, pp. 187-208, 1961.
[http://dx.doi.org/10.1017/S0022112061000469]

[86] M. Deshpande, J. Feng, and C. Merkle, "Navier-Stokes analysis of 2-D cavity flows", *ASME Cavitation and Multiphase Flow Forum FED-153,* pp. 149-155, 1993.

[87] G.A. Duller, "On the linear theory of cascades of supercavitating hydrofoils", *U.K. Nat. Eng. Lab. Rep.*, no. 218, 1966.

[88] D.A. Efros, "Hydrodynamical theory of two-dimensional flow with cavitation", *Dokl. Akad. Nauk SSSR,* vol. 51, pp. 267-270, 1946.

[89] P. Eisenberg, and H.L. Pond, "Water tunnel investigations of steady state cavities", *David Taylor Model Basin Rep. No. 668,* 1948.

[90] A.G. Fabula, "Thin airfoil theory applied to hydrofoils with a single finite cavity and arbitrary free-streamline detachment", *J. Fluid Mech.*, vol. 12, no. 2, pp. 227-240, 1962.
[http://dx.doi.org/10.1017/S0022112062000178]

[91] A.G. Fabula, "Choked flow about vented or cavitating hydrofoils", *J. Basic Eng.*, vol. 86, no. 3, pp. 561-568, 1964.
[http://dx.doi.org/10.1115/1.3653173]

[92] A. Fage, and F.C. Johansen, "On the flow of air behind an inclined flat plate of infinite span", Proc. Roy. Soc., London, Series A, 116, 170--197, 1927.

[93] J.P. Franc, and J.M. Michel, "Attached cavitation and the boundary layer: experimental investigation and numerical treatment", *J. Fluid Mech.*, vol. 154, pp. 63-90, 1985.
[http://dx.doi.org/10.1017/S0022112085001422]

[94] O. Furuya, and A.J. Acosta, "A note on the calculation of supercavitating hydrofoils with rounded noses", *J. Fluids Eng.*, vol. 95, no. 2, pp. 222-228, 1973.

[http://dx.doi.org/10.1115/1.3446994]

[95] O. Furuya, "Supercavitating linear cascades with rounded noses", *ASME J. Basic Eng., Series D,* vol. 96, pp. 35-42, 1974.

[96] O. Furuya, "Exact supercavitating cascade theory", *J. Fluids Eng.,* vol. 97, no. 4, pp. 419-429, 1975. [http://dx.doi.org/10.1115/1.3448051]

[97] O. Furuya, "Three-dimensional theory on supercavitating hydrofoils near a free surface", *J. Fluid Mech.,* vol. 71, no. 2, pp. 339-359, 1975. [http://dx.doi.org/10.1017/S0022112075002613]

[98] P.R. Garabedian, "The mathematical theory of three-dimensional cavities and jets", *Bull. Am. Math. Soc.,* vol. 62, no. 3, pp. 219-235, 1956. [http://dx.doi.org/10.1090/S0002-9904-1956-10017-7]

[99] D. Gilbarg, "A generalization of the Schwarz-Christoffel transformation", *Proc. U.S. Nat. Acad. Sci.,* vol. 35, pp. 609-612, 1949. [http://dx.doi.org/10.1073/pnas.35.10.609]

[100] D. Gilbarg, "Jets and cavities", In: *Handbuch der Physik* vol. 9. Springer-Verlag, 1960, pp. 311-445. [http://dx.doi.org/10.1007/978-3-642-45944-3_5]

[101] M.I. Gurevich, "Theory of jets in ideal fluids", In: *Academic Press, N.Y.*, 1961.

[102] N.D. Ham, "Aerodynamic loading on a two-dimensional airfoil during dynamic stall", *AIAA J.,* vol. 6, no. 10, pp. 1927-1934, 1968. [http://dx.doi.org/10.2514/3.4902]

[103] H. Helmholtz, "Uulm;ber diskontinuierliche Flüssigkeitsbewegungen", In: *Monatsber. Akad. Wiss. Berlin* vol. 23. , 1868, pp. 215-228.

[104] C.C. Hsu, "On flow past a supercavitating cascade of cambered blades", *J. Basic Eng.,* vol. 94, no. 1, pp. 163-168, 1972. [http://dx.doi.org/10.1115/1.3425356]

[105] E-Y. Hsu, and B. Perry, "Water tunnel experiments on spheres in cavity flow", *Inst. of Tech. Hydro. Lab. Rep. No. E-24.9,* 1954.

[106] V.E. Johnson, "Theoretical and experimental investigation of supercavitating hydrofoils operating near the free surface", *NASA TR R-93,* 1961.

[107] N.E. Joukowski, "I. A modification of Kirchhoff's method of determining a two-dimensional motion of a fluid given a constant velocity along an unknown streamline. II. Determination of the motion of a fluid for any condition given on a streamline", *Mat. Sbornik (Rec. Math.),* vol. 15, pp. 121-278, 1890.

[108] H.R. Kelly, "An extension of the Woods theory for unsteady cavity flows", *J. Basic Eng.,* vol. 89, no. 4, pp. 798-806, 1967. [http://dx.doi.org/10.1115/1.3609708]

[109] S.A. Kinnas, and N.E. Fine, "Non-linear analysis of the flow around partially or super-cavitating hydrofoils on a potential based panel method", *Proc. IABEM-90 Symp. Int. Assoc. for Boundary Element Methods,* pp. 289-300, 1990.

[110] G. Kirchhoff, "Zur Theorie freier Flüssigkeitsstrahlen. Z. reine Angew", *Math.,* vol. 70, pp. 289-298, 1869.

[111] G. Kreisel, "Cavitation with finite cavitation numbers", *Admiralty Res. Lab. Rep. R1/H/36,* 1946.

[112] L. Landweber, "The axially symmetric potential flow about elongated bodies of revolution", *David Taylor Model Basin Report No. 761,* 1951. [http://dx.doi.org/10.5962/bhl.title.48298]

[113] P. Leehey, "Supercavitating hydrofoil of finite span", *Proc. IUTAM Symp. on Non-steady Flow of Water at High Speeds,* 1971pp. 277-298 Leningrad.

[114] H. Lemonnier, and A. Rowe, "Another approach in modelling cavitating flows", *J. Fluid Mech.,* vol. 195, no. -1, pp. 557-580, 1988.
[http://dx.doi.org/10.1017/S0022112088002526]

[115] T. Levi-Civita, "Scie e leggi di resistenzia", *Rend. Circ. Mat. Palermo,* vol. 18, pp. 1-37, 1907.
[http://dx.doi.org/10.1007/BF03013504]

[116] M. Martin, "Unsteady lift and moment on fully cavitating hydrofoils at zero cavitation number", *J. Ship Res.,* vol. 6, no. 1, pp. 15-25, 1962.
[http://dx.doi.org/10.5957/jsr.1962.6.2.15]

[117] Y. Mimura, "The flow with wake past an oblique plate", *J. Phys. Soc. Jpn.,* vol. 13, no. 9, pp. 1048-1055, 1958.
[http://dx.doi.org/10.1143/JPSJ.13.1048]

[118] J.N. Newman, "Supercavitating flow past bodies with finite leading edge thickness", *David Taylor Model Basin Rep. No. 1081,* 1956.

[119] T. Nishiyama, "Lifting line theory of supercavitating hydrofoil of finite span", *Z. Angew. Math. Mech.,* vol. 50, no. 11, pp. 645-653, 1970.
[http://dx.doi.org/10.1002/zamm.19700501102]

[120] T. Nishiyama, and T. Ota, "Linearized potential flow models for hydrofoils in supercavitating flows", *ASME J. Basic Eng., 93, Series D, 550--564,* 1971.
[http://dx.doi.org/10.1115/1.3425309]

[121] F. Numachi, "Cavitation tests on hydrofoils designed for accelerating flow cascade: Report 1", *ASME J. Basic Eng., 83, Series D, 637--647,* 1961.

[122] F. Numachi, "Cavitation tests on hydrofoils designed for accelerating flow cascade: Report 3", *ASME J. Basic Eng., 86, Series D, 543--555,* 1964.

[123] B.R. Parkin, "Experiments on circular-arc and flat plate hydrofoils", *J. Ship Res.,* vol. 1, no. 1, pp. 34-56, 1958.
[http://dx.doi.org/10.5957/jsr.1958.2.1.34]

[124] B.R. Parkin, "Linearized theory of cavity flow in two-dimensions", *The RAND Corp. (Calif.) Rep. No. P-1745,* 1959.

[125] B.R. Parkin, "Numerical data on hydrofoil reponse to non-steady motions at zero cavitation number", *J. Ship Res.,* vol. 6, no. 1, pp. 40-42, 1962.
[http://dx.doi.org/10.5957/jsr.1962.6.4.40]

[126] W.J.M. Rankine, "On the mathematical theory of stream lines, especially those with four foci and upwards", *Phil. Trans. Roy. Soc., 267--306,* 1871.

[127] H. Reichardt, "The physical laws governing the cavitation bubbles produced behind solids of revolution in a fluid flow", *Kaiser Wilhelm Inst. Hyd. Res., Gottingen, TPA3/TIB.,* 1945.

[128] H. Reichardt, and H. Munzner, "Rotationally symmetric source-sink bodies with predominantly constant pressure distributions", *Arm. Res. Est. Trans. No. 1/50,* 1950.

[129] D. Riabouchinsky, "On steady fluid motion with a free surface", *Proc. Lond. Math. Soc.,* vol. 19, pp. 206-215, 1920.

[130] A. Roshko, "A new hodograph for free streamline theory", *NACA TN 3168.,* 1954.

[131] H. Rouse, and J.M. McNown, "Cavitation and pressure distribution: headforms at zero angles of yaw", 1948.

[132] L.I. Sedov, Plane problems in hydrodynamics and aerodynamics (in Russian). Izdat. ``Nauka", Moscow.

[133] E. Silberman, "Experimental studies of supercavitating flow about simple two-dimensional bodies in a

jet", *J. Fluid Mech.*, vol. 5, no. 3, pp. 337-354, 1959.
[http://dx.doi.org/10.1017/S0022112059000246]

[134] F. Sisto, "Linearized theory of non-stationary cascades at fully stalled or supercavitating conditions", *Z. Angew. Math. Mech.*, vol. 8, no. 8, pp. 531-542, 1967.
[http://dx.doi.org/10.1002/zamm.19670470806]

[135] R.V. Southwell, and G. Vaisey, "Fluid motions characterized by free streamlines", *Phil. Trans.*, vol. 240, pp. 117-161, 1946.

[136] R.V. Southwell, *Relaxation methods in mathematical physics.* Oxford Univ. Press, 1948.

[137] L.B. Stripling, and A.J. Acosta, "Cavitation in turbopumps---Part 1", *J. Basic Eng.*, vol. 84, no. 3, pp. 326-338, 1962.
[http://dx.doi.org/10.1115/1.3657314]

[138] H.G. Struck, "Discontinuous flows and free streamline solutions for axisymmetric bodies at zero and small angles of attack", *NASA TN D-5634,* 1970.

[139] C.D. Sutherland, and H. Cohen, "Finite cavity cascade flow", *Proc. 3rd U.S. Nat. Cong. of Appl. Math.*, pp. 837-845, 1958.

[140] R. Timman, "A general linearized theory for cavitating hydrofoils in nonsteady flow", *Proc. 2nd ONR Symp. on Naval Hydrodynamics,* 1958pp. 559-579.

[141] E. Trefftz, "Über die Kontraktion kreisföormiger Flüssigkeits-strahlen", *Z. Math. Phys.*, vol. 64, pp. 34-61, 1916.

[142] M.P. Tulin, "Steady two-dimensional cavity flows about slender bodies", *David Taylor Model Basin Rep. 834,* 1953.

[143] M.P. Tulin, "Supercavitating flow past slender delta wings", *J. Ship Res.*, vol. 3, no. 3, pp. 17-22, 1959.
[http://dx.doi.org/10.5957/jsr.1959.3.4.17]

[144] M.P. Tulin, "Supercavitating flows---small perturbation theory", *J. Ship Res.*, vol. 7, no. 3, pp. 16-37, 1964.
[http://dx.doi.org/10.5957/jsr.1964.8.1.16]

[145] J.S. Uhlman Jr, "A partially cavitated hydrofoil of finite span", *J. Fluids Eng.*, vol. 100, no. 3, pp. 353-354, 1978.
[http://dx.doi.org/10.1115/1.3448680]

[146] J.S. Uhlman, "The surface singularity method applied to partially cavitating hydrofoils", *J. Ship Res.*, vol. 31, no. 2, pp. 107-124, 1987.
[http://dx.doi.org/10.5957/jsr.1987.31.2.107]

[147] J.S. Uhlman, "The surface singularity or boundary integral method applied to supercavitating hydrofoils", *J. Ship Res.*, vol. 33, no. 1, pp. 16-20, 1989.
[http://dx.doi.org/10.5957/jsr.1989.33.1.16]

[148] H. Villat, "Sur la validite des solutions de certains problemes d'hydrodynamique", *J. Math. Pures Appl.*, vol. 10, no. 6, pp. 231-290, 1914.

[149] R.B. Wade, "Linearized theory of a partially cavitating cascade of flat plate hydrofoils", *Appl. Sci. Res.*, vol. 17, no. 3, pp. 169-188, 1967.
[http://dx.doi.org/10.1007/BF00386089]

[150] R.B. Wade, and A.J. Acosta, "Experimental observations on the flow past a planoconvex hydrofoil", *J. Basic Eng.*, vol. 88, no. 1, pp. 273-283, 1966.
[http://dx.doi.org/10.1115/1.3645828]

[151] R.B. Wade, and A.J. Acosta, "Investigation of cavitating cascades", *J. Basic Eng.*, vol. 89, no. 4, pp. 693-706, 1967.

[http://dx.doi.org/10.1115/1.3609687]

[152] D.P. Wang, and T.Y. Wu, "General formulation of a perturbation theory for unsteady cavity flows", *J. Basic Eng.*, vol. 87, no. 4, pp. 1006-1010, 1965. [http://dx.doi.org/10.1115/1.3650791]

[153] S.E. Widnall, "Unsteady loads on supercavitating hydrofoils", *J. Ship Res.*, vol. 9, no. 2, pp. 107-118, 1966. [http://dx.doi.org/10.5957/jsr.1966.10.2.107]

[154] L.C. Woods, "Aerodynamic forces on an oscillating aerofoil fitted with a spoiler", *Proc. Roy. Soc. London, Series A, 239, 328--337,* 1957.

[155] L.C. Woods, "A new relaxation treatment of flow with axial symmetry", *Q. J. Mech. Appl. Math.*, vol. 4, no. 3, pp. 358-370, 1951. [http://dx.doi.org/10.1093/qjmam/4.3.358]

[156] L.C. Woods, *The theory of subsonic plane flow.* Cambridge Univ. Press, 1961.

[157] L.C. Woods, and G.H.L. Buxton, "The theory of cascade of cavitating hydrofoils", *Q. J. Mech. Appl. Math.*, vol. 19, no. 3, pp. 387-402, 1966. [http://dx.doi.org/10.1093/qjmam/19.3.387]

[158] T.Y. Wu, "A free streamline theory for two-dimensional fully cavitated hydrofoils", *J. Math. Phys.*, vol. 35, no. 1-4, pp. 236-265, 1956. [http://dx.doi.org/10.1002/sapm1956351236]

[159] T.Y. Wu, "A linearized theory for nonsteady cavity flows", *Inst. of Tech. Eng. Div. Rep. No. 85-6,* 1957.

[160] T.Y. Wu, "A wake model for free streamline flow theory, Part 1. Fully and partially developed wake flows and cavity flows past an oblique flat plate", *J. Fluid Mech.*, vol. 13, no. 2, pp. 161-181, 1962. [http://dx.doi.org/10.1017/S0022112062000609]

[161] T.Y. Wu, "Cavity flow analysis; a review of the state of knowledge", In: *Cavitations State of Knowledge.*, J.M. Robertson, G.F. Wislicenus, Eds., ASME: N.Y., 1969.

[162] T.Y. Wu, "Cavity and wake flows", *Annu. Rev. Fluid Mech.*, vol. 4, no. 1, pp. 243-284, 1972. [http://dx.doi.org/10.1146/annurev.fl.04.010172.001331]

[163] T.Y. Wu, and D.P. Wang, "A wake model for free streamline flow theory, Part 2. Cavity flows past obstacles of arbitrary profile", *J. Fluid Mech.*, vol. 18, no. 1, pp. 65-93, 1964. [http://dx.doi.org/10.1017/S0022112064000052]

[164] T.Y. Wu, and D.P. Wang, "An approximate numerical scheme for the theory of cavity flows past obstacles of arbitrary profile", *J. Basic Eng.*, vol. 86, no. 3, pp. 556 560, 1964. [http://dx.doi.org/10.1115/1.3653172]

[165] T.Y. Wu, A.K. Whitney, and C. Brennen, "Cavity-flow wall effect and correction rules", *J. Fluid Mech.*, vol. 49, no. 2, pp. 223-256, 1971. [http://dx.doi.org/10.1017/S0022112071002039]

[166] L. Tang, J. Yin, G. Yuan, J. Du, H. Gao, X. Dong, Y. Lu, and C. Du, "General conformal transformation method based on Schwarz-Christoffel approach", *Opt. Express,* vol. 19, no. 16, pp. 15119-15126, 2011. [http://dx.doi.org/10.1364/OE.19.015119] [PMID: 21934873]

[167] C. Brennen, "A numerical solution of axisymmetric cavity flows", *J. Fluid Mech.*, vol. 37, no. 4, pp. 671-688, 1969. [http://dx.doi.org/10.1017/S0022112069000802]

[168] C. Brennen, *Cavitations and Bubble Dynamics.* Oxford University Press: New York, NY, U.S.A., 1995, pp. 238-244.

[169] J.H. Chen, "A regression relation between cavitations number and cavity length for 2-d supercavitating hydrofoils", In: *11th Conference of Naval Architecture and Marine Engineering* Tainan: Taiwan, R.O.C, 1998.

[170] J.H. Chen, and Y.C. Weng, "Analysis of flow past a two-dimensional supercavitating hydrofoil using a simple closure model", *Proc. Natl. Sci. Counc. ROC (A),* vol. 23, no. 3, pp. 411-418, 1999.

[171] T.V. Davies, "Steady two-dimensional cavity How past an aerofoil using linearized theory", *Q. J. Mech. Appl. Math.,* vol. 23, no. 1, pp. 49-76, 1970. [http://dx.doi.org/10.1093/qjmam/23.1.49]

[172] J.A. Geurst, "Linearized theory for fully cavitated hydrofoils. Int", *Shipbuilding Progress,* vol. 7, no. 65, pp. 17-27, 1960. [http://dx.doi.org/10.3233/ISP-1960-76502]

[173] Y.Z. Kehr, *On the Development and Experiment of New Section Propellers for Container Ships (I). CSBC report CSBCRD-0446, China Ship Building Corporation.* R.O.C.: Keelung, 1998.

[174] S.A. Kinnas, and N.E. Fine, "A Numerical Non-linear Analysis of the Flow around 2-D and 3-D Partially Cavitating Hydrofoils", *J. Fluid Mech.,* vol. 254, pp. 151-181, 1993. [http://dx.doi.org/10.1017/S0022112093002071]

[175] S.A. Kinnas, H. Lee, and Y.L. Young, "Boundary Element Techniques for the Prediction of Sheet and Developed Tip Vortex Cavitations", *Electron. J. Bound. Elem.,* vol. 2001, BETEQ, no. 2, pp. 151-178, 2002.

[176] G. Wright, and B. Fornberg, "scattered node compact finite difference type formulas generated from radial basis functions", *J. Comput. Phys.,* vol. 212, no. 1, pp. 99-123, 2006. [http://dx.doi.org/10.1016/j.jcp.2005.05.030]

[177] V. Gu¨lkac, "On the finite differences schemes for the numerical solution of two-dimensional moving boundary problem", *Appl. Math. Comput.,* vol. 168, no. 1, pp. 549-556, 2005. [http://dx.doi.org/10.1016/j.amc.2004.09.039]

[178] A.A. Samarskii, *Theory of Finite Difference Schemes.* Nauka: Moscow, 1983.

[179] S.G. Ahmed, "A coupled integral moment and finite difference methods for solving combined moving boundary problems", *J. Eng. Appl. Sci. (Asian Res. Publ. Netw.),* vol. 50, no. 3, pp. 433-446, 2003.

[180] N.A. Mohamed, A.F. Abdel Gawad, and S.G. Ahmed, "Computation of partial cavitation characteristcs over two dimensional symmetric hydrofoils using a newly proposed boundary element algorithm", *IeJEMTA,* vol. 1, Tiawan, 2007.http://www.iejs.org

[181] M. Jamali, *BEM Modeling of Surface Water Wave Motion with Laminar Boundary Layers*, 2005.

[182] C.A. Brebbia, and J. Dominguez, *Boundary Elements an Introductory Course.* 2nd ed. Computational Mechanics Publications: New York, Mexico, 1992.

[183] P.R. Garabedian, "The Calculation of Axially Symmetric Cavities and Jets' f ocz/", *J. AM6., v.6,* 1955.

[184] E. Amromin, "L and Ivanov, A.N. 'Axisymmetric flows round bodies in regime of the developed cavitation'", *Inf Sov. Acad. Sci. Mech. of Fluids and Gas,* no. 3, pp. 37-42, 1975.

[185] L.G. Gyzevsky, "Numerical analysis of cavitational flows", *Sub. Div of Sov. Acad Sci. Int Thermophysics Preprint No 40, 1979 (into Russian),* 1979.

[186] L.A. Kojouro, "Calculation of Axisymmetric Stream Flows Round of the Bodies by Ryabouchinsky scheme", *Sci. Note of CAHI,* vol. 11, no. 5, pp. 109-115, 1980.

[187] D. Riabouchinsky, "On Steady Fluid Motion with Free Surface", In: *Proc. lomd. AM/t. ,%c.* vol. 19. , 1920, pp. 202-212.

[188] M. Abramovitz, I.A. Stegun, Ed., *Handbook of Mathematical Functions New York.*Dover, 1974.

[189] C.A. Brebbia, J.C.F. Telles, and L.C. Wrobel, *Boundary Element Techniques Theory and Applications in Engineering Springer-Verlag.* Berlin, New York, 1984.

[190] H. Lemonnier, and A. Rowe, "Another approach in modeling cavitating flows", *J. Fluid Mech,* vol. 195, no. -1, pp. 557-580, 1988.

[191] C.A. Brebbia, J.C.F. Telles, and L.C. Wrobel, *' Boundary Element Techniques.* Springer-Verlag: Berlin, Heidelberg, New York, Tokyo, 1984. [http://dx.doi.org/10.1007/978-3-642-48860-3]

[192] M. Joyce, "Aitchison, Andreas Karageorghis. Numerical solution of a free surface problem by a boundary element method", *Int. J. Numer. Methods Fluids,* vol. 8, no. 1, pp. 91-96, 1988. [http://dx.doi.org/10.1002/fld.1650080108]

[193] M.S. Ingber, and C.E. Hailey, "A Numerical Approach for Modelling Cavitating Flows", *De Gruyter,* 1991. [http://dx.doi.org/10.1515/9783110871906-013]

[194] S.G. Ahmed, and L.C. Wrobel, "Numerical solution of solidification problems using integral equations", *Third Int. Conference of free and moving boundary problems,* 1995 Bled, Slovenia.

[195] N.N. Yas'ko, "Calculation of the plane and axisymmetric Riabouchinsky cavitational flows by the direct boundary element method, Mechanical and Mathematical Faculty, Dniepropetrovsk University, 320625 Dnepropetrovsk-10, Ukraine", *Transactions on Modelling and Simulation,* vol. 6, 1993. www.witpress.com

[196] A. Spyros, "Kinnas, Ching-Yeh Hsin. Boundary element method for the analysis of the unsteady flow around extreme propeller geometries", *AIAA J.,* 2012. [http://dx.doi.org/10.2514/3.10973]

[197] G. Vaz, and W. Marin, "Modelling three dimensional sheet cavitation on marine propellers using a boundary element method", In: *Symposium on Cavitation CAV2006.* Wageningen: The Netherlands, 2006.

[198] D.E. Edmunds, and W.D. Evans, *Spectral theory and differential operators.* Oxford University Press, 1987.

[199] J. Berthier, *Micro-Drops and Digital Microfluidics (Second Edition)*, 2013.

[200] M.A. Jaswon, and G.T. Symm, "Integral equation methods in potential theory and elastostatics", In: *Lynge & Søn ILAB-ABF (Copenhagen, Denmark)* Academic Press: London, 1977.

[201] Graeme Fairweather, Andreas Karageorghis, "The method of fundamental solutions for elliptic boundary value problems", *Advances in Computational Mathematics 9,* pp. 69-95 69, 1998.

[202] S. Bal, "A Potential Based Panel Method for 2-D Hydrofoils", *J. Ocean Eng.,* vol. 26, no. 4, pp. 343-361, 2010.

[203] H. Yamaguchi, H. Kato, M. Tanimura, and M. Toyoda, "A Study on Blade Sections for a Trans-Cavitating Propeller", *J. Soc. Nav. Archi. Jpn.,* vol. 182, no. 182, pp. 49-57, 1997. [http://dx.doi.org/10.2534/jjasnaoe1968.1997.182_49]

[204] S.A. Kinnas, and N.E. Fine, "A Numerical Non-linear Analysis of the Flow around 2-D and 3-D Partially Cavitating Hydrofoils", *J. Fluid Mech.,* vol. 254, pp. 151-181, 1993. [http://dx.doi.org/10.1017/S0022112093002071]

[205] Z. Xie, and Y. He, "An Approach in Modeling Cavitating Flows with Gravity Effect", *Commun. Nonlinear Sci. Numer. Simul.,* vol. 2, no. 2, pp. 100-104, 1997. [http://dx.doi.org/10.1016/S1007-5704(97)90048-4]

[206] G.S. Berntsen, M. Kjeldsen, and R. Muller, "Cavitation Induced Dynamics in Hydraulic Machinery", *Proc. 10th International Meeting of the Work Group on "The Behaviour of Hydraulic Machinery under Steady Oscillatory Condition.* June 26-28 2001 Trondheim, Norway.

[207] M. Dular, B. Bachert, B. Stoffel, and B. Sirok, "Relationship between Cavitation Structures and Cavitation Damage", *Wear,* vol. 257, no. 11, pp. 1176-1184, 2004. [http://dx.doi.org/10.1016/j.wear.2004.08.004]

[208] Y.A. Çengel, *Heat transfer: a practical approach.* 2nd ed. McGraw-Hill: Boston, 2003.

[209] E. Yao, H. Wang, L. Pan, X. Wang, and R. Woding, "Vertical Water-Entry of Bullet-Shaped Projectiles", *Z. Angew. Math. Phys.,* vol. 2, no. 6, 2014.

APPENDICES

APPENDIX –A

The Divergence Theorem

Define a vector quantity $\vec{F}$ in its components in the x , y , and z directions as follows:

$$\vec{F} = F_x\, i + F_y\, j + F_z\, k \qquad \textbf{(A-1)}$$

In which, i , j , and k are the unit vectors in the directions x , y , and z , respectively. The divergence theorem states that:

$$\int_V \left(\frac{\partial F_x}{\partial x} + \frac{\partial F_y}{\partial y} + \frac{\partial F_z}{\partial z} \right) dv = \int_S \left(\vec{F} \bullet \vec{n} \right) dS \qquad \textbf{(A-2)}$$

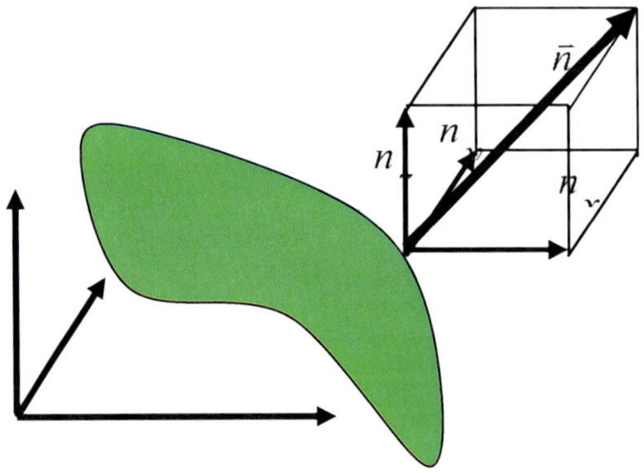

Fig. (1). The unit normal vector.

The vector n can be written as:

$$\vec{n} = n_x\, i + n_y\, j + n_z\, k \qquad \textbf{(A-3)}$$

S. G. Ahmed

The left hand side of equation (2) is called the ***divergence of F*** and written as $\nabla \bullet F$, therefore equation (2) can be re-written as follows:

$$\int_V (\nabla \bullet F) dv = \int_S (\vec{F} \bullet \vec{n}) dS \qquad \textbf{(A-4)}$$

Green's Second Identity

One of the most important mathematical tools used when deriving the boundary integral equation is Green's second identity. Consider two arbitrary functions ϕ and ϕ^* continuous at all points of the volume V and both have values that depend on the position in the volume. Consider the following product **[49]**:

$$F = \phi(\nabla \bullet \phi^*) \qquad \textbf{(A-5)}$$

Therefore, the divergence of the function F is given by:

$$\nabla \bullet F = \phi(\nabla^2 \bullet \phi^*) + (\nabla \bullet \phi)(\nabla \bullet \phi^*) \qquad \textbf{(A-6)}$$

Applying the divergence theorem to the product $\phi(\nabla \bullet \phi^*)$ leads to:

$$\int_V (\phi(\nabla^2 \bullet \phi^*) + (\nabla \bullet \phi)(\nabla \bullet \phi^*)) dV = \int_S ((\phi(\nabla \bullet \phi^*)) \bullet \bar{n}) ds \qquad \textbf{(A-7)}$$

Again, applying the divergence theorem to $\phi^* (\nabla \bullet \phi)$ leads to:

$$\int_V (\phi^* (\nabla^2 \bullet \phi) + (\nabla \bullet \phi^*)(\nabla \bullet \phi)) dV = \int_S ((\phi^* (\nabla \bullet \phi)) \bullet \bar{n}) ds \qquad \textbf{(A-a8)}$$

Subtracting equation (7) from equation (8) leads to:

$$\int_V (\phi^* (\nabla^2 \bullet \phi) - \phi (\nabla^2 \bullet \phi^*)) dV = \int_S (\phi^* (\nabla \bullet \phi) - \phi (\nabla \bullet \phi^*)) \bullet \bar{n}) ds \quad \textbf{(A-9)}$$

APPENDIX –B

Dilatation

If the coordinate system is changed from r_0 to coordinate r, then the element of volume changes by the formula:

$$d\Omega = \frac{\partial(x, y, z)}{\partial(x_0, y_0, z_0)} = dx_0 dy_0 dz_0 = J d\Omega_0 \quad \textbf{(B-1)}$$

If we consider r_0 as one of the material coordinates, they are Cartesian coordinates at time equals to zero, considering $dx_0 dy_0 dz_0$ as the volume Ω_0 of an elementary parallelepiped. Consider this elementary parallelepiped at a given point r_0 initially, as shown in Fig. (**1**). By the motion, this parallelepiped is moved and distorted. Since the motion is continuous, therefore, it can not break up, and at times, t is considered closer to the point $r = r(r_0, t)$. By applying equation (B-1), its volume is determined as $d\Omega = J d\Omega_0$ and hence

$$J = \frac{d\Omega}{d\Omega_0} \quad \textbf{(B-2)}$$

is defined as the ratio of an elementary material volume to its initial volume, and also called ***dilatation*** or ***expansion***.

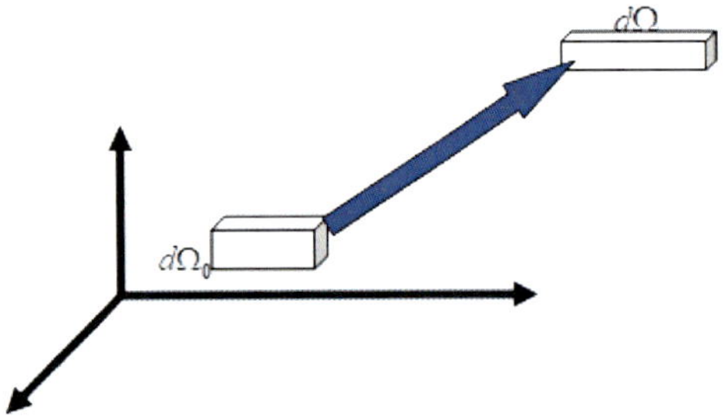

Fig. (2). Elementary paraellelepiped as it moves and distorts during the time interval from zero to any time t.

APPENDIX – C

Reynolds Transport Theorem

Let $f(\mathrm{r},t)$ be any function and $\Omega(\mathrm{t})$ be a closed volume moving with fluid that comprising of the same fluid particles, then:

$$F(t) = \int_{\Omega(t)} f(\mathrm{r},t)d\Omega \tag{C-1}$$

Equation (C-1) is a function of time that can be calculated. In order to find its material derivative dF/dt, the integral over a varying volume $\Omega(\mathrm{t})$ can not be found bydifferentiation through the integral sign. For integration with respect to r_0 -space (Fig. **1** in appendix B), it would be possible to interchange differentiation and integration since d/dt is differentiable with respect to time, keeping r_0 constant. As the transformation $r = r(r_0, t)$ and $d\Omega = Jd\Omega_0$ proceeds, $\Omega(t)$ has been defined as a moving material volume and so has derivedfrom some fixed volume Ω_0 at the time $t = \mathbf{0}$, now it is possible to write:

$$\begin{aligned}
\frac{d}{dt}\int_{\Omega(t)} f(\mathrm{r},t)d\Omega &= \frac{d}{dt}\int_{\Omega_0} f(\mathrm{r}(\mathrm{r}_0,t),t)Jd\Omega_0 \\
&= \int_{\Omega_0} \frac{df}{dt}J + f\frac{dJ}{dt}d\Omega_0 \\
&= \int_{\Omega_0} \left\{\frac{df}{dt} + f(\nabla \bullet \mathrm{v})\right\}Jd\Omega_0 \\
&= \int_{\Omega(t)} \left\{\frac{df}{dt} + f(\nabla \bullet \mathrm{v})\right\}d\Omega
\end{aligned} \tag{C-2}$$

Since $d/dt = (\partial/\partial t) + v \bullet \nabla$ it is possible to put this formula into a number of different shapes. By substituting the material derivative and collecting the gradient terms, one can get the following:

$$\frac{d}{dt}\int_{\Omega(t)} f(\mathrm{r},t)d\Omega = \int_{\Omega(t)} \left\{\frac{df}{dt} + \nabla \bullet (fv)\right\} d\Omega \qquad \textbf{(C-3)}$$

Now applying Green's theorem to the second integrand of equation (C-3), one finally arrives at:

$$\frac{d}{dt}\int_{\Omega(t)} f(\mathrm{r},t)d\Omega = \int_{\Omega(t)} \left\{\frac{df}{dt}\right\} d\Omega + \int_{\Gamma(t)} \{fv \bullet n\} d\Gamma \qquad \textbf{(C-4)}$$

In equation (C-4), $\Gamma(t)$ is the surface of $\Omega(t)$ and the factor $v \bullet n$ is called the ***normal component of the velocity of the boundary***. In phase change problems, the factor $v \bullet n$ represents ***the rate of progress of the phase front***.

APPENDIX –D

Computer Codes

```
C  THIS PROGRAM IS USED FOR SOLVING THE DIFFUSION
C  PROBLEM WITH FIXIED BOUNDARIES USING B.E.M.
C-----------------------------------------------------------------
      CHARACTER*10 FILEIN,FILEOUT
      IMPLICIT DOUBLE PRECISION (A-H,O-Z)
      DIMENSION T(0:200),A(0:100),X(100),XJ(100),X1(100)
      DIMENSION U(200,200),G1(100),G2(100)
      DIMENSION B1(100),B2(100),B11(100),B22(100)
      DIMENSION Q1(100),Q2(100)
      DIMENSION BB(100,100),R2(100)
      DIMENSION A2(0:100,100),WW(100,100),ZB(100)
```

```
      DIMENSION W1(100,100),R11(100),A3(0:100)
      DIMENSION W2(100),B10(100),W3(100,100)
      DIMENSION U1(100,100),F(100,100)
      DIMENSION R23(100),ZBB(100),R15(100)
      DIMENSION R16(100),F1(100,100),F2(100,100)
      DIMENSION RO(100),A5(0:100,100),W4(100,100)
      DIMENSION POT1(100),POT2(100)
      DIMENSION TM1(2000),POT1A(100),SUM1(2000)
C---------------------------------------------------------------
 973  FORMAT(2X,2(2X,E12.5))
 4341 FORMAT(2X,'TEMP. AT INTERNAL POINTS')
 297  FORMAT(2X,'J',I3,2X,'K3',I3,2X,'BB(J,K2)',E12.5,'WW',E12.5)
 9700 FORMAT(2X,'I',I3,3X,'ZB(I)',E12.5)
 9701 FORMAT(2X,'I11',I3,3X,'ZBB(I11)',E12.5)
 9702 FORMAT(2X,'J',I3,3X,'I5',I3,3X,'F1(J,I5)',E12.5)
 9703 FORMAT(5X,'--------------------------------------------')
 9704 FORMAT(2X,'J',I3,3X,'I6',I3,3X,'F2(J,I6)',E12.5)
 9710 FORMAT(2X,'W2(KaK3)',E12.5)
 9711 FORMAT(2X,'B10(K3)',E12.5)
 9712 FORMAT(2X,'W4(J,KK3)',E12.5)
```

```
9713 FORMAT(2X,'J',I3,2X,'K1',I3,2X,'BB(J,K1)',E12.5)

2976 FORMAT(2X,'J',I3,2X,'K3',I3,2X,'U1(J,K3)',E12.5)

9739 FORMAT(2X,'POT1',E12.5,2X,'POT2',E12.5)

8700 FORMAT(2X,5(3X,E12.5))

8904 FORMAT(2X,'K10',I3,5X,'ANAL.SOL',E12.5,5X,'B.E.SOL',E12.5)

C-----------------------------------------------------------------

   COMMON INP,IPR

   INP=20

   IPR=30

   WRITE(*, ' (A) ') 'NAME OF INPUT FILE (MAX. 10 CHART.)'

   READ(*,' (A) ')FILEIN

   OPEN(INP,FILE=FILEIN,STATUS='OLD')

   WRITE(*,' (A) ') 'NAME OF OUTPUT FILE (MAX. 10 CHART.)'

   READ(*,' (A) ')FILEOUT

   OPEN(IPR,FILE=FILEOUT,STATUS='NEW')

C-----------------------------------------------------------------

CM   :  NUMBER OF EXTREME POINTS

CM2  :  NUMBER OF INTERNAL POINTS

CN   :  NUMBER OF TIME INTERVALS

CDT  :  TIME STEP
```

```
CUO  :  INITIAL TEMPERATURE (TILL NOW UO IS CONSTANT)
CU11 :  BOUNDARY CONDITION AT X=0.0
CU22 :  BOUNDARY CONDITION AT X=L
CX( ) :  EXTREME POINTS
CX1( ):  INTERNAL POINTS
CXJ( ):  SOURCE POINTS
C    Q11 :  BOUNDARY CONDITION AT X=0.0
C    Q22 :  BOUNDARY CONDITION AT X=L
C    L  :  NUMBER SPECIFIES THE TYPE OF B.C.
CL=1 :  MEANS B.C. ARE POTENTIAL AT EXT.POINTS
CL=2 :  MEANS B.C. ARE FLUX AT EXT.POINTS
CFO  :  FLUX APPLIED AT X=L AS B.C.
C
```

```
    C1=0.278393
    C2=0.230389
    C3=0.000972
    C4=0.078108
    READ(INP,*) FO
    READ(INP,*)L
    IF(L.EQ.1) GO TO 1
```

```
      GO TO 2
CREAD THE INPUT DATA
   1  READ(INP,*)M,N,M2,DT,UO,U11,U22
C--------------------------------------------------------------
CREAD THE EXTREME POINTS
      DO 10 I=1,M
      READ(INP,*) X(I)
  10  CONTINUE
C--------------------------------------------------------------
CREAD THE SOURCE POINTS
      DO 20 I=1,M
      READ(INP,*)XJ(I)
  20  CONTINUE
C--------------------------------------------------------------
CREAD THE INTERNAL POINTS
      IF(M2.EQ.0) GO TO 103
      DO 21 I=1,M2
      READ(INP,*) X1(I)
  21  CONTINUE
C--------------------------------------------------------------
```

```
 103  DO 25 K=1,1
      T(K-1)=0.0
  25  CONTINUE
C-----------------------------------------------------------------
CCALCULATION THE TIME STEPS.
      DO 100 K1=1,N
      T(K1)=T(K1-1)+DT
  100 CONTINUE
C-----------------------------------------------------------------
CCALCULATION THE POTENTIAL & FLUXES MATRICES
CAT THE FIRST TIME STEP.
      K=1
      A(K-1)=1./(4.*(T(K)-T(K-1)))
      DO 40 J=1,M
      DO 50 I=1,M
      IF(X(I).EQ.XJ(J)) GO TO 808
      D=DABS(XJ(J)-X(I))
      Y=DSQRT(A(K-1))
      IF(Y.GT.2.0) GO TO 17
      Z=(1.+C1*Y+C2*Y**2+C3*Y**3+C4*Y**4)**4
```

```
      ERFY=(1.-(1./Z))
      GO TO 15
   17 ERFY=1.0
   15 U(J,I)=2.*((-1)**I)*((.5*0.5641895835*DEXP(-A(K-1))/Y)
     1+(.5*(ERFY-1.0)))
      GO TO 50
  808 U(J,I)=2.*((-1)**I)*(0.5641895835)*(DSQRT(T(K)-T(K-1)))
   50 CONTINUE
   40 CONTINUE
      DO 41 J=1,M
      DO 51 I=1,M
      IF(XJ(J).EQ.X(I)) GO TO 11
      DR=DABS(XJ(J)-X(I))
      DY=DSQRT(A(K-1))
      IF(DY.GT.2.0) GO TO 12
      DY1=(1.+C1*DY+C2*DY**2+C3*DY**3+C4*DY**4)**4
      ERFDY=1.-(1./DY1)
      GO TO 13
   12 ERFDY=1.0
   13 IF(J.LT.I) GO TO 14
```

```
      SGN=-1.0
      GO TO 16
   14 SGN=1.0
   16 F(J,I)=SGN*(1.0-ERFDY)
      GO TO 51
   11 F(J,I)=0.0
   51 CONTINUE
   41 CONTINUE
C
      B11(K)=U11-UO-(U11*F(1,1))+(U22*F(1,2))
      B22(K)=U22-UO-(U11*F(2,1))+(U22*F(2,2))
CCALCULATION THE FLUX AT THE
CFIRST TIME STEP.
C______________________________
      XX=(U(1,1)*U(2,2)-U(2,1)*U(1,2))
      Q1(K)=(B11(K)*U(2,2)-B22(K)*U(1,2))/XX
      Q2(K)=(B11(K)*U(1,1)-B22(K)*U(2,1))/XX
      WRITE(IPR,8700) B11(K),B22(K),XX,Q1(K),Q2(K)
C---------------------------------------------------------------
CCALCULATION THE POTENTIAL & FLUXES MATRICES
```

```
CAT THE INTERMEDIATE TIME STEPS.
      DO 200 K=2,N
      DO 201 J=1,M
      DO 202 I=1,M
      U(J,I)=0.0
 202  CONTINUE
 201  CONTINUE
C
C
      DO 2019 J=1,M
      DO 2029 I=1,M
      F(J,I)=0.0
 2029 CONTINUE
 2019 CONTINUE
      G1(1)=U11-UO
      G2(1)=U22-UO
      DO 300 KK=1,K-1
      A(KK-1)=1./(4.*(T(K)-T(KK-1)))
      A(KK)=1./(4.*(T(K)-T(KK)))
      DO 600 J=1,M
```

```
   DO 700 I=1,M
   IF (X(I).EQ.XJ(J)) GO TO 909
   D=DABS(XJ(J)-X(I))
   Y1=DSQRT(A(KK-1))
   Y2=DSQRT(A(KK))
   IF(Y1.GT.2.0) GO TO 37
   Y11=(1.+C1*Y1+C2*Y1**2+C3*Y1**3+C4*Y1**4)**4
   ERFY1=(1.-(1./Y11))
   GO TO 39
37 ERFY1=1.0
39 IF(Y2.GT.2.0) GO TO 377
   Y22=(1.+C1*Y2+C2*Y2**2+C3*Y2**3+C4*Y2**4)**4
   ERFY2=(1.-(1./Y22))
   GO TO 224
377 ERFY2=1.0
224 U(J,I)=2.*(.5*.5641895835*((DEXP(-A(KK-1))/Y1)-(DEXP
   4(-A(KK))/Y2))+(.5*(ERFY1-ERFY2)))
   GO TO 700
909 U(J,I)=2.*0.5641895835*(DSQRT(T(K)-T(KK-1))-DSQRT(
   5T(K)-T(KK)))
```

```
700 CONTINUE
600 CONTINUE
    DO 601 J=1,M
    DO 701 I=1,M
    IF(XJ(J).EQ.X(I)) GO TO 31
    DF=DABS(XJ(J)-X(I))
    DY10=DSQRT(A(KK-1))
    IF(DY10.GT.2.0) GO TO 32
    DY1F=(1.+C1*DY10+C2*DY10**2+C3*DY10**3+C4*DY10**4)**4
    ERFDY1=1.-(1./DY1F)
    GO TO 33
 32 ERFDY1=1.0
 33 DY20=DSQRT(A(KK))
    IF(DY20.GT.2.0) GO TO 34
    DY2F=(1.+C1*DY20+C2*DY20**2+C3*DY20**3+C4*DY20**4)**4
    ERFDY2=1.-(1./DY2F)
    GO TO 36
 34 ERFDY2=1.0
 36 IF(J.LT.I) GO TO 38
    SGN=-1.0
```

```
    GO TO 418
 38  SGN=1.0
 418  F(J,I)=(SGN)*(ERFDY2-ERFDY1)
    GO TO 701
 31  F(J,I)=0.0
 701 CONTINUE
 601 CONTINUE
C---------------------------------------------------------------
CFORMULATION THE RIGHT HAND VECTOR
CFOR THE SYSTEM OF EQUATIONS.
C___________________________________
     G1(KK)=G1(KK)+(Q1(KK)*U(1,1)-Q2(KK)*U(1,2))
    1-(U11*F(1,1))+(U22*F(1,2))
     G2(KK)=G2(KK)+(Q1(KK)*U(2,1)-Q2(KK)*U(2,2))
    2-(U11*F(2,1))+(U22*F(2,2))
C
     DO 45 J=1,M
     DO 47 I=1,M
     U(J,I)=0.0
  47  CONTINUE
```

```
 45  CONTINUE
     DO 459 J=1,M
     DO 479 I=1,M
     F(J,I)=0.0
 479  CONTINUE
 459  CONTINUE
     G1(KK+1)=G1(KK)
     G2(KK+1)=G2(KK)
 300  CONTINUE
C-------------------------------------------------------------
CCALCULATION THE POTENTIAL MATRIX AT
CTHE LAST TIME STEP.
     KK=K
     DO 480 J3=1,M
     DO 490 I3=1,M
     U(J3,I3)=0.0
 490  CONTINUE
 480  CONTINUE
     DO 4809 J=1,M
     DO 4909 I=1,M
```

```
      F(J,I)=0.0
 4909 CONTINUE
 4809 CONTINUE
      A(KK-1)=1./(4.*(T(K)-T(KK-1)))
      DO 1000 J=1,M
      DO 2000 I=1,M
      IF(X(I).EQ.XJ(J)) GO TO 1001
      D=DABS(X(I)-XJ(J))
      V1=DSQRT(A(KK-1))
      IF(V1.GT.2.0) GO TO 858
      V11=(1.+C1*V1+C2*V1**2+C3*V1**3+C4*V1**4)**4
      ERFV1=(1.-(1./V11))
      GO TO 9061
 858  ERFV1=1.0
 9061 U(J,I)=2.*((-1)**I)*((.5*.5641895835*(DEXP(-A(KK-1))/V1))
     8+(.5*(ERFV1-1.)))
      GO TO 2000
 1001 U(J,I)=(2.*((-1)**I)*(.5641895835))*(DSQRT(T(K)-T(KK-1)))
 2000 CONTINUE
 1000 CONTINUE
```

```
C----------------------------------------------------------------
      DO 971 J=1,M
      DO 972 I=1,M
      IF(XJ(J).EQ.X(I)) GO TO 604
      SZ=DABS(XJ(J)-X(I))
      SZ1=DSQRT(A(KK-1))
      IF(SZ1.GT.2.0) GO TO 704
      SZ11=(1.+C1*SZ1+C2*SZ1**2+C3*SZ1**3+C4*SZ1**4)**4
      ERFSZ1=1.-(1./SZ11)
      GO TO 7041
 704  ERFSZ1=1.0
 7041 IF(J.LT.I) GO TO 7042
      SGN=-1.0
      GO TO 7043
 7042 SGN=1.0
 7043 F(J,I)=(SGN)*(1.0-ERFSZ1)
      GO TO 972
 604  F(J,I)=0.0
 972  CONTINUE
 971  CONTINUE
```

```
      B1(K)=G1(K)-(U11*F(1,1))+(U22*F(1,2))
      B2(K)=G2(K)-(U11*F(2,1))+(U22*F(2,2))
      XX1=(U(1,1)*U(2,2)-U(2,1)*U(1,2))
      Q1(KK)=(B1(K)*U(2,2)-B2(K)*U(1,2))/XX1
      Q2(KK)=(B2(K)*U(1,1)-B1(K)*U(2,1))/XX1
  200 CONTINUE
C---------------------------------------------------------------
C
      DO 9002 K1=1,N
      WRITE(IPR,973) Q1(K1),Q2(K1)
 9002 CONTINUE
      WRITE(IPR,4341)
C---------------------------------------------------------------
      IF(M2.EQ.0) GO TO 9339
C---------------------------------------------------------------
CCALCULATION THE FREE TERM AT EACH TIME STEP.
C______________________________________________
      DO 260 J=1,M2
      DO 226 K1=1,N
      DO 141 I2=1,1
```

```
      R1=DABS(X1(J)-X(I2))
      RR=R1/(2.*DSQRT(T(K1)))
      RY=(1.+C1*RR+C2*RR**2+C3*RR**3+C4*RR**4)**4
      IF(RR.GT.2.0) GO TO 91
      ERFRY=(1.-(1./RY))
      GO TO 81
91    ERFRY=1.0
81    D6=X1(J)-X(I2)
      IF(D6.LT.0.) GO TO 905
      GO TO 1091
905   ERFRY=-ERFRY
1091  R21=DABS(X1(J)-X(I2+1))
      RV=R21/(2.*DSQRT(T(K1)))
      RVY=(1.+C1*RV+C2*RV**2+C3*RV**3+C4*RV**4)**4
      IF(RV.GT.2.0) GO TO 9777
      ERFRVY=(1.-(1./RVY))
      GO TO 811
9777  ERFRVY=1.0
811   D7=X1(J)-X(I2+1)
      IF(D7.LT.0.) GO TO 9005
```

```
      GO TO 1090
 9005 ERFRVY=-ERFRVY
 1090 BB(J,K1)=(UO/2.)*(ERFRY-ERFRVY)
      WRITE(IPR,9713) J,K1,BB(J,K1)
 141  CONTINUE
 226  CONTINUE
 260  CONTINUE
C--------------------------------------------------------------
CCALCULATION THE TEMP. AT INTERNAL POINTS.
      DO 26 J=1,M2
      K2=1
      DO 142 I=1,M
      R2(I)=DABS(X1(J)-X(I))
      A2(K2-1,I)=(R2(I)**2)/(4.*(T(K2)-T(K2-1)))
      D8=DSQRT(A2(K2-1,I))
      IF(D8.GT.2.0) GO TO 6160
      D88=(1.+C1*D8+C2*D8**2+C3*D8**3+C4*D8**4)**4
      ERFD8=1.-(1./D88)
      GO TO 6510
 6160 ERFD8–1.0
```

```
6510 ZB(I)=(R2(I)/3.5449077)*(DEXP(-A2(K2-1,I))/D8)
  2+(R2(I)/2.0)*(ERFD8-1.0)
  WRITE(IPR,9700) I,ZB(I)
142  CONTINUE
  DO 1420 I11=1,M
  RO(I11)=DABS(X1(J)-X(I11))
  A5(K2-1,I11)=(RO(I11)**2)/(4.*(T(K2)-T(K2-1)))
  D15=DSQRT(A5(K2-1,I11))
  IF(D15.GT.2.0) GO TO 301
  D155=(1.+C1*D15+C2*D15**2+C3*D15**3+C4*D15**4)**4
  ERFD15=1.-(1./D155)
  GO TO 302
301  ERFD15=1.0
302  R23(I11)=X1(J)-X(I11)
  IF(R23(I11).GT.0.0) GO TO 651
  SGN=-1.0
  GO TO 616
651  SGN=1.0
616  ZBB(I11)=(.5*SGN)*(1.0-ERFD15)
  WRITE(IPR,9701) I11,ZBB(I11)
1420 CONTINUE
```

```
      WW(J,K2)=BB(J,K2)+(Q2(K2)*ZB(1))-(Q1(K2)*ZB(2))
     1+(U11*ZBB(1))-(U22*ZBB(2))
      WRITE(IPR,297) J,K2,BB(J,K2),WW(J,K2)
      DO 143 K3=2,N
      DO 1431 J9=1,M2
      DO 1432 I9=1,M
      W1(J9,I9)=0.0
 1432 CONTINUE
 1431 CONTINUE
      W2(1)=0.0
      DO 1461 I=1,M
      R11(I)=DABS(X1(J)-X(I))
 1461 CONTINUE
      DO 146 KK3=1,K3-1
      DO 1462 I5=1,M
      A3(KK3-1)=(R11(I5)**2)/(4.*(T(K3)-T(KK3-1)))
      A3(KK3)=(R11(I5)**2)/(4.*(T(K3)-T(KK3)))
      Y31=DSQRT(A3(KK3-1))
      Y32=DSQRT(A3(KK3))
      IF(Y31.GT.2.0) GO TO 3333
```

```
    Y311=(1.+C1*Y31+C2*Y31**2+C3*Y31**3+C4*Y31**4)**4
    ERFY31=1.-(1./Y311)
    GO TO 4444
3333 ERFY31=1.0
4444 IF(Y32.GT.2.0) GO TO 5555
    Y322=(1.+C1*Y32+C2*Y32**2+C3*Y32**3+C4*Y32**4)**4
    ERFY32=1.-(1./Y322)
    GO TO 6666
5555 ERFY32=1.0
6666 W1(J,I5)=(R11(I5)/3.5449077)*((DEXP(-A3(KK3-1)))/Y31
   3-(DEXP(-A3(KK3)))/Y32)+((R11(I5)/2.)*(ERFY31-ERFY32))
    R15(I5)=X1(J)-X(I5)
    IF(R15(I5).GT.0.) GO TO 652
    SGN=-1.0
    GO TO 4433
652  SGN=1.0
4433 F1(J,I5)=(.5*SGN)*(ERFY32-ERFY31)
    WRITE(IPR,9702) J,I5,F1(J,I5)
1462 CONTINUE
    DO 1463 I6=1,1
```

```
      W2(KK3)=W2(KK3)-(Q1(KK3)*W1(J,I6))+(Q2(KK3)*W1(J,I6+1))
     3+(U11*F1(J,I6))-(U22*F1(J,I6+1))
      W2(KK3+1)=W2(KK3)
      WRITE(IPR,9710) W2(KK3)
 1463 CONTINUE
      WRITE(IPR,9703)
 146  CONTINUE
      KK3=K3
      DO 1464 J8=1,M2
      DO 1465 I8=1,M
      W1(J8,I8)=0.0
 1465 CONTINUE
 1464 CONTINUE
      B10(K3)=W2(K3)
      WRITE(IPR,9711) B10(K3)
      DO 1466 I6=1,M
      A3(KK3)=(R11(I6)**2)/(4.*(T(K3)-T(KK3-1)))
      Y41=DSQRT(A3(KK3))
      IF(Y41.GT.2.0) GO TO 8888
      Y411=(1.+C1*Y41**2+C2*Y41**2+C3*Y41**3+C4*Y41**4)**4
```

```
      ERFY41=1.-(1./Y411)
      GO TO 9999
8888 ERFY41=1.0
9999 W3(J,I6)=(R11(I6)/3.5449077)*((DEXP(-A3(KK3)))/Y41)
     3+((R11(I6)/2.)*(ERFY41-1.0))
      R16(I6)=X1(J)-X(I)
      IF(R16(I6).GT.0.) GO TO 657
      SGN=-1.0
      GO TO 5540
657  SGN=1.0
5540 F2(J,I6)=(.5*SGN)*(1.0-ERFY41)
      WRITE(IPR,9704) J,I6,F2(J,I6)
1466 CONTINUE
      DO 1467 I7=1,1
      W4(J,KK3)=B10(K3)-(Q1(KK3)*W3(J,I7))+(Q2(KK3)
     6*W3(J,I7+1))+(U11*F2(J,I7))-(U22*F2(J,I7+1))
      WRITE(IPR,9712) W4(J,KK3)
1467 CONTINUE
      U1(J,K3)=BB(J,K3)+W4(J,KK3)
      WRITE(IPR,2976) J,K3,U1(J,K3)
```

```
 143  CONTINUE
 26   CONTINUE
      GO TO 9339
C ************************************************************
C
  2   READ(INP,*)M,N,M2,DT,UO,Q11,Q22
CREAD THE EXTREME POINTS
      DO 101 I=1,M
      READ(INP,*) X(I)
 101 CONTINUE
C------------------------------------
CREAD THE SOURCE POINTS
      DO 102 I=1,M
      READ(INP,*)XJ(I)
 102  CONTINUE
C------------------------------------
CREAD THE INTERNAL POINTS
      IF(M2.EQ.0) GO TO 1031
      DO 210 I=1,M2
      READ(INP,*) X1(I)
```

```
 210 CONTINUE
C-----------------------------------
 1031 DO 251 K=1,1
     T(K-1)=0.0
 251 CONTINUE
CCALCULATION THE TIME STEPS.
C      ________________________________
     DO 199 K1=1,N
     T(K1)=T(K1-1)+DT
 199 CONTINUE
C -----------------------------------------------------------------------------
CCALCULATION THE POTENTIAL & FLUXES MATRICES
CAT THE FIRST TIME STEP.
     DO 6900 J=1,M
     DO 6901 I=1,M
     U(J,I)=0.0
     F(J,I)=0.0
 6901 CONTINUE
 6900 CONTINUE
     K=1
```

```
   A(K-1)=1./(4.*(T(K)-T(K-1)))
   DO 401 J=1,M
   DO 501 I=1,M
   IF(X(I).EQ.XJ(J)) GO TO 8080
   D=DABS(XJ(J)-X(I))
   Y=DSQRT(A(K-1))
   IF(Y.GT.2.0) GO TO 171
   Z=(1.+C1*Y+C2*Y**2+C3*Y**3+C4*Y**4)**4
   ERFY=(1.-(1./Z))
   GO TO 151
 171 ERFY=1.0
 151 U(J,I)=2.*((.5*0.5641895835*DEXP(-A(K-1))/Y)
  1+(.5*(ERFY-1.0)))
   GO TO 501
8080 U(J,I)=2.*.5641895835*(DSQRT(T(K)-T(K-1)))
 501 CONTINUE
 401 CONTINUE
   DO 410 J=1,M
   DO 510 I=1,M
   IF(XJ(J).EQ.X(I)) GO TO 110
```

```
      DR=DABS(XJ(J)-X(I))
      DY=DSQRT(A(K-1))
      IF(DY.GT.2.0) GO TO 128
      DY1=(1.+C1*DY+C2*DY**2+C3*DY**3+C4*DY**4)**4
      ERFDY=1.-(1./DY1)
      GO TO 138
  128 ERFDY=1.0
  138 IF(J.LT.I) GO TO 148
      SGN=1.0
      GO TO 168
  148 SGN=-1.0
  168 F(J,I)=SGN*(1.0-ERFDY)
      GO TO 510
  110 F(J,I)=0.0
  510 CONTINUE
  410 CONTINUE
C
      B11(K)=-UO+(Q11*U(1,1))-(Q22*U(1,2))
      B22(K)=-UO+(Q11*U(2,1))-(Q22*U(2,2))
C
```

```
CCALCULATION THE POTENTIAL AT THE
CFIRST TIME STEP.
      XX=((1.+F(2,2))*(1.-F(1,1))+(F(2,1)*F(1,2)))
      POT1(K)=((-B11(K)*(1.+F(2,2)))+(B22(K)*F(1,2)))/XX
      POT2(K)=((B22(K)*(F(1,1)-1.))-(B11(K)*F(2,1)))/XX
C
C
CCALCULATION THE POTENTIAL & FLUXES MATRICES
CAT THE INTERMEDIATE TIME STEPS.
      DO 203 K=2,N
      DO 2015 J33=1,M
      DO 2025 I33=1,M
      U(J33,I33)=0.0
 2025 CONTINUE
 2015 CONTINUE
      DO 2010 J44=1,M
      DO 2020 I44=1,M
      F(J44,I44)=0.0
 2020 CONTINUE
 2010 CONTINUE
```

```
      G1(1)=-UO
      G2(1)=-UO
      DO 3001 KK=1,K-1
      A(KK-1)=1./(4.*(T(K)-T(KK-1)))
      A(KK)=1./(4.*(T(K)-T(KK)))
      DO 6001 J12=1,M
      DO 7001 I=1,M
      IF (X(I).EQ.XJ(J12)) GO TO 9091
      D=DABS(XJ(J12)-X(I))
      Y1=DSQRT(A(KK-1))
      Y2=DSQRT(A(KK))
      IF(Y1.GT.2.0) GO TO 307
      Y11=(1.+C1*Y1+C2*Y1**2+C3*Y1**3+C4*Y1**4)**4
      ERFY1=(1.-(1./Y11))
      GO TO 309
307   ERFY1=1.0
309   IF(Y2.GT.2.0) GO TO 3771
      Y22=(1.+C1*Y2+C2*Y2**2+C3*Y2**3+C4*Y2**4)**4
      ERFY2=(1.-(1./Y22))
      GO TO 2240
```

```
 3771 ERFY2=1.0
 2240 U(J12,I)=2.*(.5*.5641895835*((DEXP(-A(KK-1))/Y1)-(DEXP
     4(-A(KK))/Y2))+(.5*(ERFY1-ERFY2)))
      GO TO 7001
 9091 U(J12,I)=2.*.5641895835*(DSQRT(T(K)-T(KK-1))-DSQRT(
     5T(K)-T(KK)))
 7001 CONTINUE
 6001 CONTINUE
C
      DO 6010 J13=1,M
      DO 7010 I=1,M
      IF(XJ(J13).EQ.X(I)) GO TO 3101
      DF=DABS(XJ(J13)-X(I))
      DY10=DSQRT(A(KK-1))
      IF(DY10.GT.2.0) GO TO 3201
      DY1F=(1.+C1*DY10+C2*DY10**2+C3*DY10**3+C4*DY10**4)**4
      ERFDY1=1.-(1./DY1F)
      GO TO 3301
 3201 ERFDY1=1.0
 3301 DY20=DSQRT(A(KK))
```

```
      IF(DY20.GT.2.0) GO TO 3401
      DY2F=(1.+C1*DY20+C2*DY20**2+C3*DY20**3+C4*DY20**4)**4
      ERFDY2=1.-(1./DY2F)
      GO TO 3601
 3401 ERFDY2=1.0
 3601 IF(J13.LT.I) GO TO 3801
      SGN=1.0
      GO TO 4180
 3801 SGN=-1.0
 4180 F(J13,I)=(SGN)*(ERFDY2-ERFDY1)
      GO TO 7010
 3101 F(J13,I)=0.0
 7010 CONTINUE
 6010 CONTINUE
C
CFORMULATION THE RIGHT HAND VECTOR
CFOR THE SYSTEM OF EQUATIONS.
      G1(KK)=G1(KK)+(Q11*U(1,1)-Q22*U(1,2))
     1-(POT1(KK)*F(1,1))+(POT2(KK)*F(1,2))
      G2(KK)=G2(KK)+(Q11*U(2,1)-Q22*U(2,2))
```

```
     2-(POT1(KK)*F(2,1))+(POT2(KK)*F(2,2))
C
      DO 4501 J66=1,M
      DO 4701 I66=1,M
      U(J66,I66)=0.0
 4701 CONTINUE
 4501 CONTINUE
      DO 4590 J77=1,M
      DO 4799 I77=1,M
      F(J77,I77)=0.0
 4799 CONTINUE
 4590 CONTINUE
      G1(KK+1)=G1(KK)
      G2(KK+1)=G2(KK)
 3001 CONTINUE
C
CCALCULATION THE POTENTIAL MATRIX AT
CTHE LAST TIME STEP.
      KK=K
      DO 4800 J3=1,M
```

```
      DO 4900 I3=1,M
      U(J3,I3)=0.0
4900 CONTINUE
4800 CONTINUE
      DO 4819 J15=1,M
      DO 4919 I15=1,M
      F(J15,I15)=0.0
4919 CONTINUE
4819 CONTINUE
      A(KK-1)=1./(4.*(T(K)-T(KK-1)))
      DO 1999 J20=1,M
      DO 2999 I20=1,M
      IF(X(I20).EQ.XJ(J20)) GO TO 1991
      D=DABS(X(I20)-XJ(J20))
      V1=DSQRT(A(KK-1))
      IF(V1.GT.2.0) GO TO 1858
      V11=(1.+C1*V1+C2*V1**2+C3*V1**3+C4*V1**4)**4
      ERFV1=(1.-(1./V11))
      GO TO 9961
1858 ERFV1=1.0
```

```
 9961 U(J20,I20)=2.*((.5*.5641895835*(DEXP(-A(KK-1))/V1))
    8+(.5*(ERFV1-1.)))
    GO TO 2999
 1991 U(J20,I20)=(2.*0.5641895835)*(DSQRT(T(K)-T(KK-1)))
 2999 CONTINUE
 1999 CONTINUE
C--------------------------------------------------------------
    DO 9719 J16=1,M
    DO 9729 I16=1,M
    IF(XJ(J16).EQ.X(I16)) GO TO 6049
    SZ=DABS(XJ(J16)-X(I16))
    SZ1=DSQRT(A(KK-1))
    IF(SZ1.GT.2.0) GO TO 7049
    SZ11=(1.+C1*SZ1+C2*SZ1**2+C3*SZ1**3+C4*SZ1**4)**4
    ERFSZ1=1.-(1./SZ11)
    GO TO 7941
 7049 ERFSZ1=1.0
 7941 IF(J16.LT.I16) GO TO 7942
    SGN=1.0
    GO TO 7943
```

```
 7942 SGN=-1.0
 7943 F(J16,I16)=(SGN)*(1.0-ERFSZ1)
      GO TO 9729
 6049 F(J16,I16)=0.0
 9729 CONTINUE
 9719 CONTINUE
C
      B1(K)=G1(K)+(Q11*U(1,1))-(Q22*U(1,2))
      B2(K)=G2(K)+(Q11*U(2,1))-(Q22*U(2,2))
C
C
      XX1=((1.-F(1,1))*(1.+F(2,2)))+(F(1,2)*F(2,1))
      POT1(KK)=((-B1(K)*(1.+F(2,2)))+(B2(K)*F(1,2)))/XX1
      POT2(KK)=((B2(K)*(F(1,1)-1.))-(B1(K)*F(2,1)))/XX1
  203 CONTINUE
      DO 9222 K1=1,N
      WRITE(IPR,9739) POT1(K1),POT2(K1)
 9222 CONTINUE
C--------------------------------------------------------------
CTHIS PART COMPUTE THE ANALYTICAL SOL. THIS ANALYTICAL
```

```
CSOLUTION TAKEN FROM "HEAT CONDUCTION IN SOLIDS"
CBY : CARSLAW & JAEGER ---- PAGE 112
C____________________________________________________________
      PI=3.14149265
      PI2=PI*PI
      DO 8900 K9=1,N
      SUM1(1)=0.0
      DO 8901 I9=1,1000
      T9=FLOAT(I9)
      TM1(I9)=(((-1)**(I9))/(T9**2))*DEXP(-T(K9)*(T9**2)*PI2)
      SUM1(K9)=SUM1(K9)+TM1(I9)
 8901 CONTINUE
      POT1A(K9)=(FO*T(K9))-FO*((1./6.)+(.2026423673*SUM1(K9)))
 8900 CONTINUE
      DO 8903 K10=1,N
      WRITE(IPR,8904) K10,POT1A(K10),POT1(K10)
 8903 CONTINUE
 9339 CLOSE (INP)
      CLOSE (IPR)
      STOP
```

```
    END
```

Program -2

```
C--------------------------------------------------------------
C     PROGRAM POCONBE
C
C  PROGRAM 1
C ---------------------------------------------------------------
c     IMPORTANT NOTE : IF YOU NUMBERED THE POINTS COUNTERCLOCK
C              WISE DIRECTION YOU SHOULD TAKE THE FOLLOWING
C              EXPRESSION FOR H IN SUBROUTINE (4) AS:
C              H=H+(---
C              BUT IF YOU NUMBERED CLOCK WISE DIRECTION
C              TAKE H AS:
C              H=H-(----
C ---------------------------------------------------------------
C     THIS PROGRAM SOLVES TWO DIMENSIONAL (PO)TENTIAL PROBLEMS
C  USING (CON)STANT (B)OUNDARY (E)LEMENTS
      CHARACTER*10 FILEIN,FILEOUT
      DIMENSION X(101),Y(101),XM(100),YM(100),FI(100),DFI(100)
```

```
    DIMENSION KODE(100),CX(20),CY(20),SOL(20)
    COMMON/MATG/ G(100,100)
    COMMON/MATH/ H(100,100)
    COMMON    N,L,INP,IPR
    INP=20
    IPR=30
    WRITE(*,' (A) ') 'NAME OF INPUT FILE (MAX. 10 CHART.)'
    READ(*,' (A) ')FILEIN
    OPEN(INP,FILE=FILEIN,STATUS='OLD')
    WRITE(*,' (A) ') 'NAME OUTPUT FILE (MAX. 10 CHART)'
    READ(*,' (A) ')FILEOUT
    OPEN(IPR,FILE=FILEOUT,STATUS='NEW')
C   INITIALIZATION THE PARAMETERS OF THE PROGRAME
C
C   SET MAXIMUN DIMENSION OF THE SYSTEM OF EQUATIONS (NX)
C       (THIS NUMBER MUST BE EQUAL OR SMALLER THAN THE
DIMENSION OF
C
    NX=100
C ASSIGN NUMBERS FOR INPUT AND OUTPUT FILE
C
```

```
      CALL INPUT(CX,CY,X,Y,KODE,FI)
C
C  COMPUTE H AND G MATRICES AND FORM SYSTEM (A X = F)
C
      CALL GHMAT(X,Y,XM,YM,G,H,FI,DFI,KODE,NX)
C
C  SOLVE SYSTEM OF EQUATIONS
C
      CALL SLNPD(G,DFI,D,N,NX)
C
C  COMPUTE  POTENTIAL VALUES AT INTERNAL POINTS
C
      CALL INTER(FI,DFI,KODE,CX,CY,X,Y,SOL)
C
C  PRINT RESULTS AT BOUNDARY NODES AND INTERNAL POINTS
C
      CALL OUTPT(XM,YM,FI,DFI,CX,CY,SOL)
C
      CLOSE(INP)
      CLOSE(IPR)
```

```
      STOP
      END
C---------------------------------------------------------------
      SUBROUTINE INPUT(CX,CY,X,Y,KODE,FI)
C
C  PROGRAM 2
C
      COMMON N,L,INP,IPR
      DIMENSION CX(1),CY(1),X(1),Y(1),KODE(1),FI(1)
C
C  N= NUMBER OF BOUNDARY NODES (=NUMBER OF ELEMENTS)
C   L= NUMBER OF INTERNAL POINTS WHERE THE FUNCTION IS
CALCULATED
C
      WRITE(IPR,100)
  100 FORMAT(' ',79('*'))
      READ(INP,*)ID
      WRITE(IPR,434)ID
 434  FORMAT(2X,'NO. OF GAUSSIAN POINTS=',I2)
C
C  READ NUMBER OF BOUNDARY ELEMENTS AND INTERNAL POINTS
```

```
C

      READ(INP,*)N,L

      WRITE(IPR,300)N,L

  300 FORMAT(//' DATA'//2X,'NUMBER OF BOUNDARY ELEMENTS =',I3/2X,'NUMBER

     1 OF INTERNAL POINTS WHERE THE FUNCTION IS CALCULATED =',I3)

C

C

C  READ COORDINATES OF EXTREME POINTS OF THE BOUNDARY ELEMENTS

C  IN ARRAYS X AND Y

C

      WRITE(IPR,500)

  500 FORMAT(//2X,'COORDINATES OF THE EXTREME POINTS OF THE BOUNDARY ELE

     1MENTS',//1X,'POINT',7X,'X',15X,'Y')

      DO 10 I=1,N

      READ(INP,*) X(I),Y(I)

      WRITE(IPR,700)I,X(I),Y(I)

   10 CONTINUE

  700 FORMAT(2X,I3,2(2X,E14.5))
```

```
C
C  READ BOUNDARY CONDITIONS IN FI(I) VECTOR, IF KODE(I)=0 THE FI(I)
C  VALUE IS A KNOWN POTENTIAL;IF KODE(I)=1 THE FI(I) VALUE IS A
C  KNOWN POTENTIAL DERIVATIVE (FLUX).
C
      WRITE(IPR,800)
  800 FORMAT(//2X,'BOUNDARY CONDITIONS'//2X,'NODE',6X,'CODE',7X,'PRESCRI
     1BED VALUE')
      DO 20 I=1,N
      READ(INP,*) KODE(I),FI(I)
   20 WRITE(IPR,950)I,KODE(I),FI(I)
  950 FORMAT(2X,I3,9X,I1,8X,E14.5)
C
C  READ COORDINATES OF THE INTERNAL POINTS
C
      IF(L.EQ.0) GO TO 30
      DO 65 I=1,L
      READ(INP,*) CX(I),CY(I)
   65 CONTINUE
```

```
  30 RETURN

     END

C-------------------------------------------------------------

     SUBROUTINE GHMAT(X,Y,XM,YM,G,H,FI,DFI,KODE,NX)

C

C  PROGRAM 3

C

C  THIS SUBROUTINE COMPUTES THE G AND H MATRICES

C  AND FORMS THE SYSTEM OF EQUATIONS A X = F

C
     COMMON N,L,INP,IPR

     DIMENSION X(1),Y(1),XM(1),YM(1),FI(1),KODE(1)

     DIMENSION DFI(1),G(NX,NX),H(NX,NX)

C

C  COMPUTE THE NODAL COORDINATES AND STORE IN ARRAYS XM
AND YM

C

     X(N+1)=X(1)

     Y(N+1)=Y(1)

     DO 10 I=1,N

     XM(I)=(X(I)+X(I+1))/2.
```

```
 10 YM(I)=(Y(I)+Y(I+1))/2.
C
C  COMPUTE THE COEFFICIENTS OF G AND H MATRICES
C
      DO 30 I=1,N
      DO 30 J=1,N
      KK=J+1
      IF(I-J)20,25,20
 20 CALL EXTIN(XM(I),YM(I),X(J),Y(J),X(KK),Y(KK),H(I,J),G(I,J))
      GO TO 30
 25 CALL LOCIN(X(J),Y(J),X(KK),Y(KK),G(I,J))
      H(I,J)=3.1415926
 30 CONTINUE
C
C   REORDER THE COLUMNS OF THE SYSTEM OF EQUATIONS IN
ACCORDANCE
C  WITH THE BOUNDARY CONDITIONS AND FORM SYSTEM MATRIX A
WHICH
C  IS STORED IN G
C
      DO 55 J=1,N
```

```
      IF(KODE(J))55,55,40
   40 DO 50 I=1,N
      CH=G(I,J)
      G(I,J)=-H(I,J)
      H(I,J)=-CH
   50 CONTINUE
   55 CONTINUE
C
C  FORM THE RIGHT HAND SIDE VECTOR F WHICH IS STORED IN DFI
C
      DO 60 I=1,N
      DFI(I)=0.
      DO 60 J=1,N
      DFI(I)=DFI(I)+H(I,J)*FI(J)
   60 CONTINUE
      RETURN
      END
C--------------------------------------------------------------
      SUBROUTINE EXTIN(XP,YP,X1,Y1,X2,Y2,H,G)
C  PROGRAM 4
```

```
C  THIS SUBROUTINE COMPUTES THE OFF DIAGONAL COEFFICIENTS
OF

C  H AND G MATRICES USING GAUSS QUADRATURE

C

C  RA=DISTANCE FROM THE COLLOCATION POINT TO THE

C  GAUSS INTEGRATION POINTS ON THE BOUNDARY ELEMENT

C  DIST=DISTANCE FROM THE COLLOCATION POINT TO THE LINE

C  TANGENT TO THE ELEMENT.

      DIMENSION XCO(6),YCO(6),GI(6),OME(6)

      GI(1)=.93246951

      GI(2)=-GI(1)

      GI(3)=.66120938

      GI(4)=-GI(3)

      GI(5)=.23861918

      GI(6)=-GI(5)

      OME(1)=.17132449

      OME(2)=OME(1)

      OME(3)=.36076157

      OME(4)=OME(3)

      OME(5)=.46791393

      OME(6)=OME(5)
```

```
C
      AX=(X2-X1)/2.
      BX=(X2+X1)/2.
      AY=(Y2-Y1)/2.
      BY=(Y2+Y1)/2.
      XJ=SQRT(AX**2+AY**2)
      ENX=-AY/XJ
      ENY=AX/XJ
C  COMPUTE THE DISTANCE FROM THE POINT TO THE LINE OF THE ELEMENT
C COMPUTE G AND H COEFFITIENTS
      H=0.0
      G=0.0
      DO 40 I=1,6
      XCO(I)=AX*GI(I)+BX
      YCO(I)=AY*GI(I)+BY
      RA=SQRT((XP-XCO(I))**2+(YP-YCO(I))**2)
      G=G+ALOG(1/RA)*OME(I)*XJ
      H=H+((XCO(I)-XP)*ENX+(YCO(I)-YP)*ENY)*OME(I)*XJ/RA**2
   40 CONTINUE
      RETURN
```

```
      END
C---------------------------------------------------------------
      SUBROUTINE LOCIN(X1,Y1,X2,Y2,G)
C
C  PROGRAM 5
C
C  THIS SUBROUTINE COMPUTES THE VALUES OF THE DIAGONAL
C  COEFFITIENTS OF THE G MATRIX
      AX=(X2-X1)/2.
      AY=(Y2-Y1)/2.
      SR=SQRT(AX**2+AY**2)
      G=2.*SR*(1.-ALOG(SR))
      RETURN
      END
C---------------------------------------------------------------
      SUBROUTINE SLNPD(A,B,D,N,NX)
C
C  PROGRAM 6
C
C SOLUTION OF LINEAR SYSTEMS OF EQUATIONS
```

```
C BY THE GAUSS ELIMINATION METHOD PROVIDING
C FOR INTERCHANGING ROWS WHEN ENCOUNTERING A
C ZERO DIAGONAL COEFICIENT
C
C A : SYSTEM MATRIX
C B : ORIGINALLY IT CONTAINS THE INDEPENDENT
C    COEFFICIENTS. AFTER SOLUTION IT CONTAINS
C    THE VALUES OF THE SYSTEM UNKNOWNS.
C
C N : ACTUAL NUMBER OF UNKNOWNS
C NX: ROW AND COLUMN DIMENSION OF A
C
      DIMENSION B(NX),A(NX,NX)
C
      TOL=1.E-6
C
      N1=N-1
      DO 100 K=1,N1
      K1=K+1
      C=A(K,K)
```

```
      IF(ABS(C)-TOL)1,1,3
    1 DO 7 J=K1,N
C
C TRY TO INTERCHANGE ROWS TO GET NON ZERO DIAGONAL COEFFICIENT
C
      IF(ABS((A(J,K)))-TOL)7,7,5
    5 DO 6 L=K,N
      C=A(K,L)
      A(K,L)=A(J,L)
    6 A(J,L)=C
      C=B(K)
      B(K)=B(J)
      B(J)=C
      C=A(K,K)
      GO TO 3
    7 CONTINUE
      GO TO 8
C
C DIVIDE ROW BY DIAGONAL COEFFICIENT
C
```

```
   3 C=A(K,K)
     DO 4 J=K1,N
   4 A(K,J)=A(K,J)/C
     B(K)=B(K)/C
C
C ELIMINATE UNKNOWN X(K) FROM ROW I
C
     DO 10 I=K1,N
     C=A(I,K)
     DO 9 J=K1,N
   9 A(I,J)=A(I,J)-C*A(K,J)
  10 B(I)=B(I)-C*B(K)
 100 CONTINUE
C
C COMPUTE LAST UNKNOWN
C
     IF(ABS((A(N,N)))-TOL)8,8,101
 101 B(N)=B(N)/A(N,N)
C
C APPLY BACKSUBSTITUTION PROCESS TO COMPUTE REMAINING
UNKNOWNS
```

```
C
      DO 200 L=1,N1
      K=N-L
      K1=K+1
      DO 200 J=K1,N
  200 B(K)=B(K)-A(K,J)*B(J)
C
C COMPUTE VALUE OF DETERMINANT
C
      D=1.
      DO 250 I=1,N
  250 D=D*A(I,I)
      GO TO 300
    8 WRITE(IPR,2) K
    2 FORMAT(' **** SINGULARITY IN ROW',I5)
      D=0.0
  300 RETURN
      END
C-----------------------------------------------------------------
      SUBROUTINE INTER(FI,DFI,KODE,CX,CY,X,Y,SOL)
```

```
C

C  PROGRAM 7

C

C  THIS SUBROUTINE COMPUTES THE VALUES OF POTENTIAL AT
INTERNAL POINTS

C

      COMMON N,L,INP,IPR

      DIMENSION FI(1),DFI(1),KODE(1),CX(1),CY(1),X(1),Y(1),SOL(1)

C

C  REARRANGE THE FI AND DFI ARRAYS TO STORE ALL THE VALUES
OF THE

C  POTENTIAL IN FI AND ALL THE VALUES OF THE DERIVATIVE IN DFI

C

      DO 20 I=1,N

      IF(KODE(I)) 20,20,10

   10 CH=FI(I)

      FI(I)=DFI(I)

      DFI(I)=CH

   20 CONTINUE

C

C  COMPUTE THE VALUES OF POTENTIAL AT INTERNAL POINTS
```

```
C
      IF(L.EQ.0) GO TO 50
      DO 40 K=1,L
      SOL(K)=0.
      DO 30 J=1,N
      KK=J+1
      CALL EXTIN(CX(K),CY(K),X(J),Y(J),X(KK),Y(KK),A,B)
   30 SOL(K)=SOL(K)+DFI(J)*B-FI(J)*A
   40 SOL(K)=SOL(K)/(2*3.1415926)
   50 RETURN
      END
C----------------------------------------------------------------
      SUBROUTINE OUTPT(XM,YM,FI,DFI,CX,CY,SOL)
C
C  PROGRAM 8
C
C  THIS SUBROUTINE PRINTS THE VALUES OF THE POTENTIAL AND ITS NORMAL
C  DERIVATIVE AT BOUNDARY NODES. IT ALSO PRINTS THE VALUES OF THE
C  POTENTIAL AT INTERNAL POINTS
```

```
      COMMON N,L,INP,IPR
      DIMENSION XM(1),YM(1),FI(1),DFI(1),CX(1),CY(1),SOL(1)
      WRITE(IPR,100)
  100 FORMAT(' ',79('*')//1X,'RESULTS'//2X,'BOUNDARY NODES'//8X,'X',15
     1X,'Y',13X,'POTENTIAL',3X,'POTENTIAL DERIVATIVE'/)
      DO 10 I=1,N
   10 WRITE(IPR,200) XM(I),YM(I),FI(I),DFI(I)
  200 FORMAT(4(2X,E14.5))
C
      IF(L.EQ.0) GO TO 30
      WRITE(IPR,300)
  300 FORMAT(//,2X,'INTERNAL POINTS',//8X,'X',15X,'Y',13X,'POTENTIAL',/
     1)
      DO 20 K=1,L
   20 WRITE(IPR,400)CX(K),CY(K),SOL(K)
  400 FORMAT(3(2X,E14.5))
   30 WRITE(IPR,500)
  500 FORMAT(' ',79('*'))
      RETURN
      END
```

Program -3-

```
C     ********************************************************
C
C     PROGRAM POLINBE    ****
C
C  PROGRAM 9
C------------------------------------------------------------
CTHIS PROGRAM IS TO MODIFIED TO CALCULATE FLUX AT INTERNAL
CPOINTS.
C------------------------------------------------------------
C
C    THIS PROGRAM SOLVES TWO DIMENSIONAL (PO)TENTIAL PROBLEMS
C  USING (LIN)EAR (B)OUNDARY (E)LEMENTS
C
      CHARACTER*10 FILEIN,FILEOUT
      COMMON/MATG/ G(80,160)
      COMMON/MATH/ H(80,80)
      COMMON N,L,INP,IPR
      DIMENSION X(81),Y(81),FI(80),DFI(160)
```

```
      DIMENSION KODE(160),CX(20),CY(20),POT(20),FLUX1(20),FLUX2(20)
C
C  SET MAXIMUN DIMENSION OF THE SYSTEM OF EQUATIONS (NX)
C  NX   = MAXIMUN NUMBER OF NODES = MAXIMUN NUMBER OF
ELEMENTS
C
      NX=80
      NX1=2*NX
C
C  ASSIGN NUMBERS FOR INPUT AND OUTPUT FILES
C
      INP=20
      IPR=30
C
C  READ NAMES AND OPEN FILES FOR INPUT AND OUTPUT
C
      WRITE(*,' (A) ') ' NAME OF THE INPUT FILE (MAX. 10 CHART.)'
      READ(*,' (A) ') FILEIN
      OPEN(INP,FILE=FILEIN,STATUS='OLD')
      WRITE(*,' (A) ') ' NAME OF THE OUTPUT FILE (MAX.10 CHART.)'
      READ(*,' (A) ') FILEOUT
```

```
      OPEN(IPR,FILE=FILEOUT,STATUS='NEW')
C
C  READ DATA
C
      CALL INPUTPL(CX,CY,X,Y,KODE,DFI)
C
C  COMPUTE G AND H MATRICES AND FORM SYSTEM (A X = F)
C
      CALL GHMATPL(X,Y,G,H,FI,DFI,KODE,NX,NX1)
C
C  SOLVE SYSTEM OF EQUATIONS
C
      CALL SLNPD(H,FI,D,N,NX)
C
C  COMPUTE POTENTIAL VALUES AT INTERNAL POINTS
C
      CALL INTERPL(FI,DFI,KODE,CX,CY,X,Y,POT,FLUX1,FLUX2)
C
C  PRINT RESULTS AT BOUNDARY NODES AND INTERNAL POINTS
C
```

```
      CALL OUTPTPL(X,Y,FI,DFI,CX,CY,POT,FLUX1,FLUX2)
C
      CLOSE (INP)
      CLOSE (IPR)
      STOP
      END
C-------------------------------------------------------------------
      SUBROUTINE INPUTPL(CX,CY,X,Y,KODE,DFI)
C
C  PROGRAM 10
C
C  N= NUMBER OF BOUNDARY ELEMENTS
C  L= NUMBER OF INTERNAL POINTS
C
      COMMON N,L,INP,IPR
      DIMENSION CX(1),CY(1),X(1),Y(1),KODE(1),DFI(1)
      WRITE(IPR,100)
  100 FORMAT(' ',79('*'))
C
C
```

```
C  READ NUMBER OF ELEMENTS AND INTERNAL POINTS

    READ(INP,*)IH

    WRITE(IPR,888) IH

 888  FORMAT(2X,'NO. OF GAUSSIAN POINTS',I2)

C

    READ(INP,*)N,L

    WRITE(IPR,300)N,L

 300 FORMAT(//' DATA'//2X,'NUMBER OF BOUNDARY ELEMENTS
=',I3/2X,'NUMBER

   1 OF INTERNAL POINTS WHERE THE FUNCTION IS CALCULATED
=',I3)

C

C  READ BOUNDARY NODES COORDINATES IN ARRAYS X AND Y

C

    WRITE(IPR,500)

 500 FORMAT(//2X,'COORDINATES OF THE EXTREME POINTS OF THE
BOUNDARY ELE

   1MENTS',//2X,'POINT',10X,'X',18X,'Y')

    DO 444 I=1,N

 444 READ(INP,*) X(I),Y(I)

    DO 10 I=1,N

  10 WRITE(IPR,700)I,X(I),Y(I)
```

```
 700 FORMAT(3X,I3,2(5X,E14.5))
C
C  READ BOUNDARY CONDITIONS IN DFI(I) VECTOR.IF KODE(I)=0
C  THE DFI(I) VALUE IS A KNOWN POTENTIAL; IF KODE(I)=1 THE
C  DFI(I) VALUE IS A KNOWN POTENTIAL DERIVATIVE (FLUX).
C  TWO BOUNDARY CONDITIONS ARE READ PER ELEMENT.
C  ONE NODE MAY HAVE TWO DIFFERENT VALUES OF THE
C  POTENTIAL DERIVATIVE BUT ONLY ONE VALUE OF THE
POTENTIAL
C
     WRITE(IPR,800)
 800 FORMAT(//2X,'BOUNDARY CONDITIONS'//15X,'------FIRST NODE----
-',
    19X,'-----SECOND NODE-----'/17X,'PRESCRIBED',20X,'PRESCRIBED'/
    2,1X,'ELEMENT',12X,'VALUE',7X,'CODE',14X,'VALUE',7X,'CODE')
     DO 20 I=1,N
     READ(INP,*) KODE(2*I-1),DFI(2*I-1),KODE(2*I),DFI(2*I)
  20 WRITE(IPR,950)I,DFI(2*I-1),KODE(2*I-1),DFI(2*I),KODE(2*I)
 950 FORMAT(2X,I3,2(10X,E14.7,5X,I1))
C
C  READ COORDINATES OF THE INTERNAL POINTS
```

```
C
      IF(L.EQ.0) GO TO 30
      DO 777 I=1,L
  777 READ(INP,*) CX(I),CY(I)
   30 RETURN
      END
C-----------------------------------------------------------------------
      SUBROUTINE GHMATPL(X,Y,G,H,FI,DFI,KODE,NX,NX1)
C
C  PROGRAM 11
C
C  THIS SUBROUTINE COMPUTES THE G AND H MATRICES
C  AND FORMS THE SYSTEM OF EQUATIONS A X = F
C  H IS A SQUARE MATRIX (N,N); G IS RECTANGULAR (N,2*N)
C
      COMMON N,L,INP,IPR
      DIMENSION X(1),Y(1),G(NX,NX1),H(NX,NX),FI(1),KODE(1),DFI(1)
      NN=2*N
      DO 10 I=1,N
      DO 6 J=1,N
```

```
 6 H(I,J)=0.
   DO 10 J=1,NN
10 G(I,J)=0.
C
C  COMPUTE THE COEFFICIENTS OF G AND H MATRICES
C
   X(N+1)=X(1)
   Y(N+1)=Y(1)
   DO 100 I=1,N
   NF=I+1
   NS=I+N-2
   DO 50 JJ=NF,NS
   IF(JJ-N)30,30,20
20 J=JJ-N
   GO TO 40
30 J=JJ
40 CALL EXTINPL(X(I),Y(I),X(J),Y(J),X(J+1),Y(J+1),A1,A2,B1,B2)
   IF(J-N)42,43,43
42 H(I,J+1)=H(I,J+1)+A2
   GO TO 44
```

```
 43 H(I,1)=H(I,1)+A2
 44 H(I,J)=H(I,J)+A1
    G(I,2*J-1)=B1
    G(I,2*J)=B2
 50 H(I,I)=H(I,I)-A1-A2
    NF=I+N-1
    NS=I+N
    DO 95 JJ=NF,NS
    IF(JJ-N)70,70,60
 60 J=JJ-N
    GO TO 80
 70 J=JJ
 80 CALL LOCINPL(X(J),Y(J),X(J+1),Y(J+1),B1,B2)
    IF(JJ-NF)82,82,83
 82 CH=B1
    B1=B2
    B2=CH
 83 G(I,2*J-1)=B1
 95 G(I,2*J)=B2
C
```

```
C  ADD ONE TO THE DIAGONAL COEFFICIENTS
C  FOR EXTERNAL PROBLEMS.
C
      IF(H(I,I)) 98,100,100
   98 H(I,I)=6.2831852+H(I,I)
  100 CONTINUE
C
C  REORDER THE COLUMNS OF THE SYSTEM OF EQUATIONS IN ACCORDANCE
C  WITH THE BOUNDARY CONDITIONS AND FORM THE SYSTEM MATRIX A
C  WHICH IS STORED IN H
C
      DO 155 I=1,N
      DO 150 J=1,2
      IF(KODE(2*I-2+J))110,110,150
  110 IF(I.NE.N .OR. J.NE.2) GO TO 125
      IF(KODE(1)) 115,115,113
  113 DO 114 K=1,N
      CH=H(K,1)
      H(K,1)=-G(K,2*N)
```

```
 114 G(K,2*N)=-CH
     GO TO 150
 115 DO 116 K=1,N
     H(K,1)=H(K,1)-G(K,2*N)
 116 G(K,2*N)=0.
     GO TO 150
 125 IF(I.EQ.1 .OR. J.GT.1 .OR. KODE(2*I-2).EQ.1) GO TO 130
     DO 129 K=1,N
     H(K,I)=H(K,I)-G(K,2*I-1)
 129 G(K,2*I-1)=0.
     GO TO 150
 130 DO 132 K=1,N
     CH=H(K,I-1+J)
     H(K,I-1+J)=-G(K,2*I-2+J)
 132 G(K,2*I-2+J)=-CH
 150 CONTINUE
 155 CONTINUE
C
C  FORM THE RIGHT HAND SIDE VECTOR F WHICH IS STORED IN FI
C
```

```
      DO 160 I=1,N
      FI(I)=0.
      DO 160 J=1,NN
      FI(I)=FI(I)+G(I,J)*DFI(J)
  160 CONTINUE
      RETURN
      END
C-----------------------------------------------------------------------
      SUBROUTINE EXTINPL(XP,YP,X1,Y1,X2,Y2,A1,A2,B1,B2
     1,DQ1F1,DQ2F1,DQ1F2,DQ2F2,DU1F1,DU2F1,DU1F2,DU2F2,K)
C
C  PROGRAM 12
CTHIS SUBROUTINE COMPUTES THE INTEGRAL OF SEVERAL NON-SINGULAR
CFUNCTIONS ALONG THE BOUNDARY ELEMENTS USING A FOUR POINTS
CGAUSSIAN POINTS.
CWHEN K=0 THE OFF DIAGIONAL COEFFICIENTS OF H&G ARE COMPUTED
CWHEN K=1 ALL COFFICIENTS ARE NEEDED FOR CAL. OF POT. AND FLUX
CAT INTERNAL POINTS ARE COMPUTED
```

```
C(A1,A2,B1,B2,F11,F21,F12,F22,E11,E21,E12,E22)
CRA : DIST FROM THE COLLOCATION POINT TO GAUSSIAN POINTS
CETA1 , ETA2 :COMPONENTS OF THE UNIT NORMAL TO THE ELEMENT
CRD1,RD2,RDN : RADIUS DERIVATIVES
C-----------------------------------------------------------------------
      DIMENSION XCO(4),YCO(4),GI(4),OME(4)
      GI(1)=.86113631159
      GI(2)=-GI(1)
      GI(3)=.33998104358
      GI(4)=-GI(3)
      OME(1)=.3478548451
      OME(2)=OME(1)
      OME(3)=.6521451548
      OME(4)=OME(3)
C
      AX=(X2-X1)/2.
      BX=(X2+X1)/2.
      AY=(Y2-Y1)/2.
      BY=(Y2+Y1)/2.
      SL=SQRT(AX**2+AY**2)
```

```
      ETA1=AY/SL
      ETA2=-AX/SL
      A1=0.
      A2=0.
      B1=0.
      B2=0.
      DQ1F1=0.
      DQ2F1=0.
      DQ1F2=0.
      DQ2F2=0.
      DU1F1=0.
      DU2F1=0.
      DU1F2=0.
      DU2F2=0.
CCOMPUTE THE TERMS TO BE INCLINED IN THE G & H MATRICES OR
THE
CTERMS NEEDS FOR COMPUTATION OF THE POTENTIAL AND FLUXES
AT
CINTERNAL POINTS.
      DO 40 I=1,4
      XCO(I)=AX*GI(I)+BX
```

```
      YCO(I)=AY*GI(I)+BY
      RA=SQRT((XP-XCO(I))**2+(YP-YCO(I))**2)
      RD1=(XCO(I)-XP)/RA
      RD2=(YCO(I)-YP)/RA
      RDN=RD1*ETA1+RD2*ETA2
      F1=(1.-GI(I))/2.
      F2=(1.+GI(I))/2.
      IF(K) 30,30,10
   10 DU1=RD1*OME(I)*SL/RA
      DU2=RD2*OME(I)*SL/RA
      DQ1=-((2.*RD1**2-1.)*ETA1+2.*RD1*RD2*ETA2)*OME(I)*SL/RA**2
      DQ2=-((2.*RD2**2-1.)*ETA2+2.*RD1*RD2*ETA1)*OME(I)*SL/RA**2
      DQ1F1=DQ1F1+DQ1*F1
      DQ1F2=DQ1F2+DQ1*F2
      DQ2F1=DQ2F1+DQ2*F1
      DQ2F2=DQ2F2+DQ2*F2
      DU1F1=DU1F1+DU1*F1
      DU1F2=DU1F2+DU1*F2
      DU2F1=DU2F1+DU2*F1
      DU2F2=DU2F2+DU2*F2
```

```
 30  H=RDN*OME(I)*SL/RA

     G=ALOG(1./RA)*OME(I)*SL

     A1=A1-F1*H

     A2=A2-F2*H

     B1=B1+F1*G

 40  B2=B2+F2*G

     RETURN

     END

C----------------------------------------------------------------------

     SUBROUTINE LOCINPL(X1,Y1,X2,Y2,B1,B2)

C

C  PROGRAM 13

C

C  THIS SUBROUTINE COMPUTES THE PARTS OF THE G MATRIX

C  COEFFICIENTS CORRESPONDING TO INTEGRALS ALONG AN
ELEMENT

C  THAT INCLUDES THE COLLOCATION POINT.

C

     SEP=SQRT((X2-X1)**2+(Y2-Y1)**2)

     B1=SEP*(1.5-ALOG(SEP))/2.

     B2=SEP*(0.5-ALOG(SEP))/2.
```

```
      RETURN
      END
C----------------------------------------------------------------------
      SUBROUTINE SLNPD(A,B,D,N,NX)
C
C PROGRAM 6
C
C SOLUTION OF LINEAR SYSTEMS OF EQUATIONS
C BY THE GAUSS ELIMINATION METHOD PROVIDING
C FOR INTERCHANGING ROWS WHEN ENCOUNTERING A
C ZERO DIAGONAL COEFICIENT
C
C A : SYSTEM MATRIX
C B : ORIGINALLY IT CONTAINS THE INDEPENDENT
C    COEFFICIENTS. AFTER SOLUTION IT CONTAINS
C    THE VALUES OF THE SYSTEM UNKNOWNS.
C
C N : ACTUAL NUMBER OF UNKNOWNS
C NX: ROW AND COLUMN DIMENSION OF A
C
```

```
      COMMON NMUDO,LMUDO,INP,IPR
      DIMENSION A(NX,NX),B(NX)
C
      TOL=1.E-6
C
      N1=N-1
      DO 100 K=1,N1
      K1=K+1
      C=A(K,K)
      IF(ABS(C)-TOL)1,1,3
    1 DO 7 J=K1,N
C
C TRY TO INTERCHANGE ROWS TO GET NON ZERO DIAGONAL
COEFFICIENT
C
      IF(ABS(A(J,K)-TOL))7,7,5
    5 DO 6 L=K,N
      C=A(K,L)
      A(K,L)=A(J,L)
    6 A(J,L)=C
      C=B(K)
```

```
      B(K)=B(J)
      B(J)=C
      C=A(K,K)
      GO TO 3
    7 CONTINUE
C
C DIVIDE ROW BY DIAGONAL COEFFICIENT
C
    3 C=A(K,K)
      DO 4 J=K1,N
    4 A(K,J)=A(K,J)/C
      B(K)=B(K)/C
C
C ELIMINATE UNKNOWN X(K) FROM ROW I
C
      DO 10 I=K1,N
      C=A(I,K)
      DO 9 J=K1,N
    9 A(I,J)=A(I,J)-C*A(K,J)
   10 B(I)=B(I)-C*B(K)
```

```
 100 CONTINUE
C
C COMPUTE LAST UNKNOWN
C
    IF(ABS(A(N,N))-TOL)8,8,101
 101 B(N)=B(N)/A(N,N)
C
C APPLY BACKSUBSTITUTION PROCESS TO COMPUTE REMAINING
UNKNOWNS
C
    DO 200 L=1,N1
    K=N-L
    K1=K+1
    DO 200 J=K1,N
 200 B(K)=B(K)-A(K,J)*B(J)
C
C COMPUTE VALUE OF DETERMINANT
C
    D=1.
    DO 250 I=1,N
 250 D=D*A(I,I)
```

```
      GO TO 300
    8 WRITE(IPR,2) K
    2 FORMAT(' **** SINGULARITY IN ROW',I5)
      D=0.
  300 RETURN
      END
C---------------------------------------------------------------------
      SUBROUTINE INTERPL(FI,DFI,KODE,CX,CY,X,Y,POT,FLUX1,FLUX2)
C
C  PROGRAM 14
C
C  THIS SUBROUTINE COMPUTES THE VALUES OF POTENTIAL AND
FLUXES AT
C  INTERNAL POINTS
C
      COMMON N,L,INP,IPR
      DIMENSION FI(1),DFI(1),KODE(1),CX(1),CY(1),X(1),Y(1)
      DIMENSION POT(1),FLUX1(1),FLUX2(11)
C
C  REARRANGE THE FI AND DFI ARRAYS TO STORE ALL THE VALUES
OF THE
```

```
C  POTENTIAL IN FI AND ALL THE VALUES OF THE DERIVATIVE IN DFI
C
      DO 155 I=1,N
      DO 150 J=1,2
      IF(KODE(2*I-2+J))110,110,150
  110 IF(I.NE.N .OR. J.NE.2) GO TO 125
      IF(KODE(1)) 114,114,113
  113 CH=FI(1)
      FI(1)=DFI(2*N)
      DFI(2*N)=CH
      GO TO 150
  114 DFI(2*N)=DFI(1)
      GO TO 150
  125 IF(I.EQ.1 .OR. J.EQ.2 .OR. KODE(2*I-2).EQ.1) GO TO 130
      DFI(2*I-1)=DFI(2*I-2)
      GO TO 150
  130 CH=FI(I-1+J)
      FI(I-1+J)=DFI(2*I-2+J)
      DFI(2*I-2+J)=CH
  150 CONTINUE
```

```
 155 CONTINUE
C
C  COMPUTE THE VALUES OF POTENTIAL AND FLUXES AT INTERNAL
POINTS
C
      IF(L.EQ.0) GO TO 50
      DO 40 K=1,L
      POT(K)=0.
      FLUX1(K)=0.
      FLUX2(K)=0.
      DO 30 J=1,N
      CALL EXTINPL(CX(K),CY(K),X(J),Y(J),X(J+1),Y(J+1),A1,A2,B1,B2
     1,DQ1F1,DQ2F1,DQ1F2,DQ2F2,DU1F1,DU2F1,DU1F2,DU2F2,1)
      IF(J-N)32,33,33
   32 POT(K)=POT(K)+DFI(2*J-1)*B1+DFI(2*J)*B2-FI(J)*A1-FI(J+1)*A2
      FLUX1(K)=FLUX1(K)+DFI(2*J-1)*DU1F1+DFI(2*J)*DU1F2
     1-FI(J)*DQ1F1-FI(J+1)*DQ1F2
      FLUX2(K)=FLUX2(K)+DFI(2*J-1)*DU2F1+DFI(2*J)*DU2F2
     1-FI(J)*DQ2F1-FI(J+1)*DQ2F2
      GO TO 30
   33 POT(K)=POT(K)+DFI(2*J-1)*B1+DFI(2*J)*B2-FI(J)*A1-FI(1)*A2
```

```
      FLUX1(K)=FLUX1(K)+DFI(2*J-1)*DU1F1+DFI(2*J)*DU1F2
     1-FI(J)*DQ1F1-FI(1)*DQ1F2
      FLUX2(K)=FLUX2(K)+DFI(2*J-1)*DU2F1+DFI(2*J)*DU2F2
     1-FI(J)*DQ2F1-FI(1)*DQ2F2
   30 CONTINUE
      POT(K)=POT(K)/(2*3.1415926)
      FLUX1(K)=FLUX1(K)/(2.*3.1415926)
   40 FLUX2(K)=FLUX2(K)/(2.*3.1415926)
   50 RETURN
      END
C----------------------------------------------------------------------
      SUBROUTINE OUTPTPL(X,Y,FI,DFI,CX,CY,POT,FLUX1,FLUX2)
C
C  PROGRAM 15
C
C  THIS SUROUTINE PRINTS THE VALUES OF THE POTENTIAL AND ITS
C  NORMAL DERIVATIVE AT BOUNDARY NODES. IT ALSO PRINTS THE
C  VALUES OF THE POTENTIAL AT INTERNAL POINTS.
C
      COMMON N,L,INP,IPR
```

```
      DIMENSION X(1),Y(1),FI(1),DFI(1),CX(1),CY(1)

      DIMENSION POT(1),FLUX1(1),FLUX2(1)

C

      WRITE(IPR,100)

  100 FORMAT(' ',79('*')//2X,'RESULTS'//2X,'BOUNDARY NODES'//

     156X,'POTENTIAL DERIVATIVE'/

     29X,'X',15X,'Y',12X,'POTENTIAL',6X,'BEFORE NODE',6X,'AFTER
NODE'/)

      WRITE(IPR,200) X(1),Y(1),FI(1),DFI(2*N),DFI(1)

      DO 10 I=2,N

   10 WRITE(IPR,200) X(I),Y(I),FI(I),DFI(2*I-2),DFI(2*I-1)

  200 FORMAT(5(2X,E14.5))

      IF(L.EQ.0) GO TO 30

      WRITE(IPR,300)

  300 FORMAT(//,2X,'INTERNAL
POINTS',//9X,'X',15X,'Y',12X,'POTENTIAL',

     19X,'FLUX X',10X,'FLUX Y'/)

      DO 20 K=1,L

   20 WRITE(IPR,400)CX(K),CY(K),POT(K),FLUX1(K),FLUX2(K)

  400 FORMAT(5(2X,E14.5))

   30 WRITE(IPR,500)
```

```
 500 FORMAT(' ',79('*'))
     RETURN
     END
C-------------------------------------------------------------------
C***************
C        CLASSIC PROBLEM
C***************
CSOLVED BY (1) CITRON (1960(
C (2)  WALYDER(1994(
C*****************************************
C*****************************************
C*****************************************
Cheat diffusion moving boundary problem.
CThe code is denoted by : DFCLASSBE.F:--------
C*****************************************
C*****************************************
CDF   : means DIFFUSION PROBLEB
CCLASS: means CLASSIC PROBLEM SOLVED BEFORE.
CBE   : means BOUNDARY ELEMENT.
C*****************************************
```

```
C*****************************************
CTHE BOUNDARY CONDITIONS AT THE EXTREME BOUNDRIES ARE:
CQ11=FLUX AT X=0 & Q22=FLUX AT X=L .
C-------------------------------------------------------------
      IMPLICIT DOUBLE PRECISION (A-H,O-Z(
      CHARACTER*10 FILEIN,FILEOUT
C-------------------------------------------------------------
      DIMENSION X(1000),XJ(1000),T(0:1000),U(1000,1000)
      DIMENSION F(1000,1000),A(0:1000)
      DIMENSION POT1(1000),POT2(1000),G1(1000),G2(1000)
      DIMENSION B1(1000),B2(1000)
C-------------------------------------------------------------
      DIMENSION A02(100,0:1000),A01(100,0:1000(
      DIMENSION RI1(1000),RI2(1000(
      DIMENSION SOLX1(100,0:1000),B11(1000),B22(1000)
C-------------------------------------------------------------
      DIMENSION COLYY1(100,1000),COLYY2(100,1000(
      DIMENSION COLVV1(100,1000),COLVV2(100,1000(
      DIMENSION UMOV(1000),COLYY3(100,1000(
      DIMENSION XI(1000),DINT(1000)
```

```
    DIMENSION AW1(100,1000),GW1(100,1000(
    DIMENSION QW(100,1000),SI(1000),DFI(1000)
    DIMENSION DW1(100,1000),FW1(100,1000(
C--------------------------------------------------------------
    DIMENSION XCOR(1000),VS10(1000),VS20(1000(
    DIMENSION AA0(100,0:1000(
    DIMENSION GG1(1000),GG2(1000)
    DIMENSION GG3(1000),GG4(1000)
    DIMENSION GG5(1000),GG6(1000)
    DIMENSION GG8(1000(
    DIMENSION GG9(1000),GG7(100,1000)
C--------------------------------------------------------------
 4466 FORMAT(2X,'POT1(K)',3X,E16.7,3X,'POT2(K)',3x,E16.7(
 2247 FORMAT(2X,2(3X,E16.7((
 2246 format(2x,E16.7(
 3311 FORMAT(2X,I3,2X,3(3X,E16.7((
 8050 FORMAT(2X,'N1',I3,3X,'T(N1)',2X,E16.7(
 5028 FORMAT(2X,5(4X,E16.7((
 4420 FORMAT(2X,2(4X,E16.7((
 8811 FORMAT(2X,4(3X,E16.7((
```

```
1717 FORMAT(2X('****************************************',
4421 FORMAT(2X,3(3X,E16.7((
4432 FORMAT(2X,I3,4X,E16.7(
8812 FORMAT(2X,3(4X,E16.7((
1780 format(2x,'iteration number('
1790 format(2x('----------------------------------------',
3155 FORMAT(5X,E16.7(
8813 FORMAT(5X,'II',I3,5X,'XCOR(I)',E16.7(
8814 format(3x,'I am here to ckeck('
8815 format(3x,'end of the ckecked part('
1244 FORMAT(5X,'FINIAL RESULTS ARE('-:
1243 format(3x,'ask me to write these coeff('.
4490 format(3x,i3,4(3x,E16.7((
C---------------------------------------------------------------
CSOME NOTATIONS USED IN THE CODE:
Ce :DENSITY ....................=RO
CL :LATENT HEAT OF DIFFUSION....=HLAT
CUa:AMBIENT TEMPERATURE.........=HAMB
CUm:MELTING TEMPERATURE.........=HMLT
Ch :COEF. OF HEAT TRANSFER......=HCOF
```

```
Cq :FLUX........................=DFI(...)

Ck :CONDUCTIVITY................=COND

CSi:RATE OF CHANGE OF M.B.......=SI(...)

CUo:INITIAL TEMPERATURE.........=UO

CC :CAPICITY....................=C

Cqo:FLUX AT X=0.................=Q11

Cql:FLUX AT X=L.................=Q22

CM :NUMBER OF EXTREME POINTS....=M

CN :NUMBER OF TIME INTERVALS....=N

CDT:TIME STEP...................=DT

CX :EXTREME POINTS.............=X(...)

CXJ:SOURCE POINTS..............=XJ(...)

CXI:MOVING BOUNDARY LOCATION....=XI(...)

CNN1: NUMBER OF TIME INTERVALS AFTER THE MELTING TIME.

CSSQ: FLUX INTERED TO GUESS THE LOCATION OF THE M.B.

C-----------------------------------------------------------------

      COMMON INP,IPR

      INP=20

      IPR=30

C -----------------------------------------------------------------
```

```
    WRITE(*, ' (A) ') 'NAME OF INPUT FILE (MAX. 10 CHART'(.
    READ(*,' (A) ')FILEIN
    OPEN(INP,FILE=FILEIN,STATUS='OLD('
    WRITE(*,' (A) ') 'NAME OF OUTPUT FILE (MAX. 10 CHART'(.
    READ(*,' (A) ')FILEOUT
    OPEN(IPR,FILE=FILEOUT,STATUS='UNKNOWN('
Cthese cofficients are used for evaluating any error function
Cwhere the error function defined by-:
Cerf(x)=1-(1/(1+c1*x+c2*x**2+c3*x**3+c4*x**4)**4)
C*********************************************************
    C1=0.278393
    C2=0.230389
    C3=0.000972
    C4=0.078108
C-----------------------------------------------------------------
    Z1=0.0705230784
    Z2=0.0422820123
    Z3=0.0092705272
    Z4=0.0001520143
    Z5=0.0002765672
```

```
      Z6=0.0000430638
C
```

```
CREAD THE INPUT DATA
      READ(INP,*)M,N,DT,UO,Q11,Q22,HMLT,HAMB,SSQ,NN1,exx
C----------------------------------------
CREAD THE EXTREME POINTS
C----------------------
      DO 101 I=1,M
      READ(INP,*) X(I)
 101  CONTINUE
C-----------------------------------
CREAD THE SOURCE POINTS
C-----------------------
      DO 102 I=1,M
      READ(INP,*)XJ(I(
 102 CONTINUE
C   ----------------------------------
      DO 251 K=1,1
      T(K-1)=0.0
 251  CONTINUE
```

```
C---------------------------------------------------------------
CCALCULATION THE TIME STEPS.
C______________________________
    DO 199 K=1,N
    T(K)=T(K-1)+DT
 199  CONTINUE
C---------------------------------------------------------------
CCALCULATION THE POTENTIAL & FLUXES MATRICES
CAT THE FIRST TIME STEP.
C_______________________________________________
    DO 6900 J=1,M
    DO 6901 I=1,M
    U(J,I)=0.0
    F(J,I)=0.0
 6901 CONTINUE
 6900 CONTINUE
    K=1
    A(K-1)=1./(4.*(T(K)-T(K-1(((
    DO 401 J=1,M
    DO 501 I=1,M
```

```
      IF(X(I).EQ.XJ(J)) GO TO 8080
      D=DABS(XJ(J)-X(I((
      Y=DSQRT(A(K-1((
C
      Z=(1.+Z1*Y+Z2*Y**2+Z3*Y**3+Z4*Y**4
+*      Z5*Y**5+Z6*Y**6)**16
      ERFY=(1.-(1./Z((
C
      U(J,I)=2.*((.5*0.5641895835*DEXP(-A(K-1))/Y(
)*5.)+1     ERFY-1.0(((
      GO TO 501
 8080 U(J,I)=2.*.5641895835*(DSQRT(T(K)-T(K-1(((
 501  CONTINUE
 401  CONTINUE
C     DO 410 J=1,M
      DO 510 I=1,M
      IF(XJ(J).EQ.X(I)) GO TO 110
      DR=DABS(XJ(J)-X(I((
      DY=DSQRT(A(K-1((
      DY1=(1.+Z1*DY+Z2*DY**2+Z3*DY**3+Z4*DY**4
```

```
+*    Z5*DY**5+Z6*DY**6)**16
      ERFDY=1.-(1./DY1(
      IF(J.LT.I) GO TO 148
      SGN=1.0
      GO TO 168
  148 SGN=-1.0
  168 F(J,I)=SGN*(1.0-ERFDY(
      GO TO 510
  110 F(J,I)=0.0
  510 CONTINUE
  410 CONTINUE
C
      B11(K)=-UO+(Q11*U(1,1))-(Q22*U(1,2((
      B22(K)=-UO+(Q11*U(2,1))-(Q22*U(2,2((
C
CCALCULATION THE POTENTIAL AT THE
CFIRST TIME STEP.
C______________________________
C
C
```

```
      XX=((1.+F(2,2))*(1.-F(1,1))+(F(2,1)*F(1,2(((
      POT1(K)=((-B11(K)*(1.+F(2,2)))+(B22(K)*F(1,2)))/XX
      POT2(K)=((B22(K)*(F(1,1)-1.))-(B11(K)*F(2,1)))/XX
      IF(POT2(K).GT.HMLT) GO TO 1111
C
C
CCALCULATION THE POTENTIAL & FLUXES MATRICES
CAT THE INTERMEDIATE TIME STEPS.
C_______________________________________________
      DO 203 K=2,N
      DO 2015 J33=1,M
      DO 2025 I33=1,M
      U(J33,I33)=0.0
 2025 CONTINUE
 2015 CONTINUE
C
C
      DO 2010 J44=1,M
      DO 2020 I44=1,M
      F(J44,I44)=0.0
```

```
 2020 CONTINUE

 2010 CONTINUE

      G1(1)=-UO

      G2(1)=-UO

      DO 3001 KK=1,K-1

      A(KK-1)=1./(4.*(T(K)-T(KK-1(((

      A(KK)=1./(4.*(T(K)-T(KK(((

      DO 6001 J12=1,M

      DO 7001 I=1,M

      IF (X(I).EQ.XJ(J12)) GO TO 9091

      D=DABS(XJ(J12)-X(I((

      Y1=DSQRT(A(KK-1((

      Y2=DSQRT(A(KK((

C

      Y11=(1.+Z1*Y1+Z2*Y1**2+Z3*Y1**3+Z4*Y1**4

+*     Z5*Y1**5+Z6*Y1**6)**16

      ERFY1=(1.-(1./Y11((

C

      Y22=(1.+Z1*Y2+Z2*Y2**2+Z3*Y2**3+Z4*Y2**4

+*     Z5*Y2**5+Z6*Y2**6)**16
```

```
      ERFY2=(1.-(1./Y22((
C
      U(J12,I)=2.*(.5*.5641895835*((DEXP(-A(KK-1))/Y1)-(DEXP
-)4     A(KK))/Y2))+(.5*(ERFY1-ERFY2(((
      GO TO 7001
 9091 U(J12,I)=2.*.5641895835*(DSQRT(T(K)-T(KK-1))-DSQRT)
5     T(K)-T(KK(((
 7001 CONTINUE
 6001 CONTINUE
C
C
      DO 6010 J13=1,M
      DO 7010 I=1,M
      IF(XJ(J13).EQ.X(I)) GO TO 3101
      DF=DABS(XJ(J13)-X(I((
      DY10=DSQRT(A(KK-1((
C
      DY1F=(1.+Z1*DY10+Z2*DY10**2+Z3*DY10**3+Z4*DY10**4
+*      Z5*DY10**5+Z6*DY10**6)**16
      ERFDY1=1.-(1./DY1F(
```

```
C
     DY20=DSQRT(A(KK((
C
     DY2F=(1.+Z1*DY20+Z2*DY20**2+Z3*DY20**3+Z4*DY20**4
+*     Z5*DY20**5+Z6*DY20**6)**16
     ERFDY2=1.-(1./DY2F(
C
     IF(J13.LT.I) GO TO 3801
     SGN=1.0
     GO TO 4180
 3801 SGN=-1.0
 4180 F(J13,I)=(SGN)*(ERFDY2-ERFDY1(
     GO TO 7010
 3101 F(J13,I)=0.0
 7010 CONTINUE
 6010 CONTINUE
C
CFORMULATION THE RIGHT HAND VECTOR
CFOR THE SYSTEM OF EQUATIONS.
C__________________________      ____________
```

```
      G1(KK)=G1(KK)+(Q11*U(1,1)-Q22*U(1,2((
)-1    POT1(KK)*F(1,1))+(POT2(KK)*F(1,2((
      G2(KK)=G2(KK)+(Q11*U(2,1)-Q22*U(2,2((
)-2    POT1(KK)*F(2,1))+(POT2(KK)*F(2,2((
C
C
      DO 4501 J66=1,M
      DO 4701 I66=1,M
      U(J66,I66)=0.0
 4701 CONTINUE
 4501 CONTINUE
      DO 4590 J77=1,M
      DO 4799 I77=1,M
      F(J77,I77)=0.0
 4799 CONTINUE
 4590 CONTINUE
      G1(KK+1)=G1(KK(
      G2(KK+1)=G2(KK(
 3001 CONTINUE
C
```

```
C
CCALCULATION THE POTENTIAL MATRIX AT
CTHE LAST TIME STEP.
C____________________________________
     KK=K
     DO 4800 J3=1,M
     DO 4900 I3=1,M
     U(J3,I3)=0.0
 4900 CONTINUE
 4800 CONTINUE
     DO 4819 J15=1,M
     DO 4919 I15=1,M
     F(J15,I15)=0.0
 4919 CONTINUE
 4819 CONTINUE
     A(KK-1)=1./(4.*(T(K)-T(KK-1(((
     DO 1999 J20=1,M
     DO 2999 I20=1,M
     IF(X(I20).EQ.XJ(J20)) GO TO 1991
     D=DABS(X(I20)-XJ(J20((
```

```
    V1=DSQRT(A(KK-1((
C
    V11=(1.+Z1*V1+Z2*V1**2+Z3*V1**3+Z4*V1**4
+*    Z5*V1**5+Z6*V1**6)**16
C
    ERFV1=(1.-(1./V11((
    U(J20,I20)=2.*((.5*.5641895835*(DEXP(-A(KK-1))/V1((
)*5.)+8    ERFV1-1(((.
    GO TO 2999
 1991 U(J20,I20)=(2.*0.5641895835)*(DSQRT(T(K)-T(KK-1(((
 2999 CONTINUE
 1999 CONTINUE
C---------------------------------------------------------------
    DO 9719 J16=1,M
    DO 9729 I16=1,M
    IF(XJ(J16).EQ.X(I16)) GO TO 6049
    SZ=DABS(XJ(J16)-X(I16((
    SZ1=DSQRT(A(KK-1((
C
    SZ11=(1.+Z1*SZ1+Z2*SZ1**2+Z3*SZ1**3+Z4*SZ1**4
```

```
+*    Z5*SZ1**5+Z6*SZ1**6)**16
      ERFSZ1=1.-(1./SZ11(
C
      IF(J16.LT.I16) GO TO 7942
      SGN=1.0
      GO TO 7943
 7942 SGN=-1.0
 7943 F(J16,I16)=(SGN)*(1.0-ERFSZ1(
      GO TO 9729
 6049 F(J16,I16)=0.0
 9729 CONTINUE
 9719 CONTINUE
C
C
      B1(K)=G1(K)+(Q11*U(1,1))-(Q22*U(1,2((
      B2(K)=G2(K)+(Q11*U(2,1))-(Q22*U(2,2((
C
      XX1=((1.-F(1,1))*(1.+F(2,2)))+(F(1,2)*F(2,1((
      POT1(KK)=((-B1(K)*(1.+F(2,2)))+(B2(K)*F(1,2)))/XX1
      POT2(KK)=((B2(K)*(F(1,1)-1.))-(B1(K)*F(2,1)))/XX1
```

```
      IF(POT2(KK).GT.HMLT) GO TO 1111
 203  CONTINUE
C***********************************************************
 1111 N1=K-1
C------------------------------------------------------------
      N2=N1+NN1
      DO 5999 KI2=N1+1,N2
      T(KI2)=T(KI2-1)+DT
 5999 CONTINUE
      II=N1+1
      XI(II-1)=1.0
      SI(II)=SSQ
      DFI(II)=Q22-SI(II(
      DO 5988 KI2=N1+1,N2
      COND=1.0
C------------------------------------------------------------
      JJ=1
      XI(KI2)=XI(KI2-1)-(SI(II)*(T(KI2)-T(KI2-1(((
 2037 xi(ii)=xi(ki2 (
      DO 9019 I=1,1
```

```
      XCOR(I)=0.0
      XCOR(I+1)=XI(II(
 9019 CONTINUE
C
      DO 8019 I=1,2
      VS10(I)=(XCOR(I))/(2.*DSQRT(T(KI2(((
      VS20(I)=(XCOR(I)-1.)/(2.*DSQRT(T(KI2(((
      VS1=DABS(VS10(I((
      VS2=DABS(VS20(I((
      VS11=(1.+C1*VS1+C2*VS1**2+C3*VS1**3+C4*VS1**4)**4
      VS22=(1.+C1*VS2+C2*VS2**2+C3*VS2**3+C4*VS2**4)**4
      ERFVS1=1.-(1./VS11(
      ERFVS2=1.-(1./VS22(
      IF(VS10(i).LT.0.0) GO TO 4001
      GO TO 4002
 4001 ERFVS1=-ERFVS1
 4002 IF(VS20(i).LT.0.0) GO TO 4003
      GO TO 4004
 4003 ERFVS2=-ERFVS2
 4004 DINT(I)=(UO)*(ERFVS1-ERFVS2(
```

```
 8019 CONTINUE
C*************************************************************
Ccalculation the domain integral from 0 to x(t(
C*********************************************
Cthe statement 6119 for the case singularity of
Ctime appears
C*********************************************
      COLVV2(1,N1)=0.0
      COLVV2(2,N1)=0.0
C
      DO 1215 IM=N1+1,KI2
      IF(IM.EQ.ki2) GO TO 6119
      GO TO 6229
 6119 DO 7019 I=1,2
      AW1(I,IM)=(XCOR(I)-XI(II((
      DW1(I,IM)=0.0
      FW1(I,IM)=0.0
      GW1(I,IM)=(T(KI2)-T(IM-1))*SI(II(
      QW(I,IM)=(4.*COND/SI(II((
      TAG1=DABS(GW1(I,IM)-AW1(I,IM))/(DSQRT(DABS(QW(I,IM(
```

```
**    GW1(I,IM((((
    TAG2=DABS(GW1(I,IM)+AW1(I,IM))/(DSQRT(DABS(QW(I,IM(
**    GW1(I,IM((((
    TAG10=(1.+Z1*TAG1+Z2*TAG1**2+Z3*TAG1**3+Z4*TAG1**4
+*    Z5*TAG1**5+Z6*TAG1**6)**16
    TAG20=(1.+Z1*TAG2+Z2*TAG2**2+Z3*TAG2**3+Z4*TAG2**4
+*    Z5*TAG2**5+Z6*TAG2**6)**16
    EFTAG1=1.-(1./TAG10(
    EFTAG2=1.-(1./TAG20(
    TAG101=(GW1(I,IM)-AW1(I,IM((
    IF(TAG101.LT.0.0) GO TO 8152
    GO TO 8153
 8152 EFTAG1=-EFTAG1
C
 8153 if(aw1(i,im).eq.0.0) go to 6412
    IF(AW1(I,IM).LT.0.0) GO TO 9152
    GO TO 9153
 9152 EFTAG2=-EFTAG2
    COLVV2(I,IM)=.5*2.*((1.-EFTAG1)+(DEXP(4.*AW1
)*    I,IM)/QW(I,IM)))*( 1.0 EFTAG2((
```

```
    go to 7019
 9153 COLVV2(I,IM)=.5*2.*((-1.-EFTAG1)+(DEXP(4.*AW1
)*    I,IM)/QW(I,IM)))*(1.0-EFTAG2((
    go to 7019
c
 6412 COLVV2(I,IM)=.5*2.*((-EFTAG1)+(DEXP(4.*AW1
)*    I,IM)/QW(I,IM)))*(-EFTAG2((
 7019 CONTINUE
    GO TO 1215
C*********************************************
 6229 GG7(1,N1)=0.0
    GG7(2,N1)=0.0
    DO 4019 I=1,2
    AW1(I,IM)=((T(ki2)-T(IM-1))*SI(IM))+(XCOR(I)-XI(IM-1((
    GW1(I,IM)=((T(ki2)-T(IM-1))*SI(IM((
    DW1(I,IM)=((T(ki2)-T(IM-1))*SI(IM))+(XI(IM)-XI(IM-1((
    FW1(I,IM)=(T(ki2)-T(IM))*SI(IM(
    QW(I,IM)=(4.*COND/SI(IM((
    FW10=(FW1(I,IM)+AW1(I,IM))/(DSQRT(QW(I,IM*(
*    FW1(I,IM(((
```

```
     FW20=(FW1(I,IM)-AW1(I,IM))/(DSQRT(QW(I,IM*(
*     FW1(I,IM(((
     FW11=DABS(FW10(
     FW22=DABS(FW20(
     FW111=(1.+Z1*FW11+Z2*FW11**2+Z3*FW11**3+Z4*FW11**4
+*    Z5*FW11**5+Z6*FW11**6)**16
     FW222=(1.+Z1*FW22+Z2*FW22**2+Z3*FW22**3+Z4*FW22**4
+*    Z5*FW22**5+Z6*FW22**6)**16
     ERFFW1=1.-(1./FW111(
     ERFFW2=1.-(1./FW222(
     IF(FW20.LT.0.0)GO TO 2398
     GO TO 6872
 2398 ERFFW2=-ERFFW2
C--------------------------------------------------------------
 6872 if(fw10.lt.0.0) go to 5872
     go to 5873
 5872 erffw1=-erffw1
 5873 TX30=(GW1(I,IM)+AW1(I,IM))/(DSQRT(QW(I,IM*(
*     GW1(I,IM(((
     WL10=(DW1(I,IM)+AW1(I,IM))/(DSQRT(QW(I,IM*(
```

```
*     DW1(I,IM(((
      WL30=(DW1(I,IM)-AW1(I,IM))/(DSQRT(QW(I,IM*(
*     DW1(I,IM(((
      WL40=(GW1(I,IM)-AW1(I,IM))/(DSQRT(QW(I,IM*(
*     GW1(I,IM(((
C*************************************************************
*
      TX3=DABS(TX30(
      WL1=DABS(WL10(
      WL3=DABS(WL30(
      WL4=DABS(WL40(
      TX33=(1.+Z1*TX3+Z2*TX3**2+Z3*TX3**3+Z4*TX3**4
+*     Z5*TX3**5+Z6*TX3**6)**16
      WL11=(1.+Z1*WL1+Z2*WL1**2+Z3*WL1**3+Z4*WL1**4
+*     Z5*WL1**5+Z6*WL1**6)**16
      WL33=(1.+Z1*WL3+Z2*WL3**2+Z3*WL3**3+Z4*WL3**4
+*     Z5*WL3**5+Z6*WL3**6)**16
      WL44=(1.+Z1*WL4+Z2*WL4**2+Z3*WL4**3+Z4*WL4**4
+*     Z5*WL4**5+Z6*WL4**6)**16
C---------------------------------------------------------------
      TXX1=DEXP(4.*AW1(I,IM)/QW(I,IM((
```

```
C-------------------------------------------------------------
     ERFTX3=1.-(1./TX33(
     ERFWL1=1.-(1./WL11(
     ERFWL3=1.-(1./WL33(
     ERFWL4=1.-(1./WL44(
C-------------------------------------------------------------
     IF(WL30.LT.0.0) GO TO 4359
     GO TO 4906
 4359 ERFWL3=-ERFWL3
 4906 IF(WL40.LT.0.0) GO TO 4801
     GO TO 416
 4801 ERFWL4=-ERFWL4
C
 416  if(tx30.lt.0.0) go to 6511
     go to 6512
 6511 erftx3=-erftx3
 6512 if(wl10.lt.0.0) go to 6513
     go to 6514
 6513 erfwl1=-erfwl1
 6514 GG7(I,JM)=GG7(I,JM-1)+HMLT*((TXX1*(ERFWL1
```

```
-*     ERFTX3))+(ERFWL3-ERFWL4((
C
 4019 CONTINUE
 1215 CONTINUE
C
C-------------------------------------------------------------
CCAL.THE TIME INTEGRALS IN THE INTERVAL 0 TO Tm
C*********************************************
 1234 SOLX1(1,0)=0.0
     SOLX1(2,0)=0.0
     DO 3903 KI=1,N1
     SOLX1(1,KI)=0.0
     SOLX1(2,KI)=0.0
 3903 CONTINUE
     DO 8003 KI=1,N1
     DO 8004 I=1,2
C************************************************
Cthe statement 8993 means that the source point at
CX=0.0 , and the field point at X=0.0 & X=L.
C
```

```
CNOTE : when source at X=0.0 and field point at x=0
C  ----we have two integrals & and also for source
C      point at X=X(t(
C***********************************************
     IF(I.EQ.1) GO TO 8993
     GO TO 8994
 8993 SMR1=0.56418958*(DSQRT(T(KI2)-T(KI-1))-DSQRT(T(KI2(
-*    T(KI(((
     SMR3=0.0
C
     RI1(I)=DABS(XCOR(I)-1.0(
     A01(I,KI-1)=RI1(I)**2/(4.*(T(KI2)-T(KI-1(((
     A01(I,KI)=RI1(I)**2/(4.*(T(KI2)-T(KI(((
     QA1=DEXP(-A01(I,KI-1((
     QA2=DEXP(-A01(I,KI((
     QA3=DSQRT(A01(I,KI-1((
     QA4=DSQRT(A01(I,KI((
C*************************************
C*************************************
     TECT1=QA1/QA3
```

```
      TECT2=QA2/QA4
      TECT3=TECT1-TECT2
      TECT4=(.28209479*RI1(I))*TECT3
      QA5=(1.+Z1*QA3+Z2*QA3**2+Z3*QA3**3+Z4*QA3**4
+*      Z5*QA3**5+Z6*QA3**6)**16
      QA6=(1.+Z1*QA4+Z2*QA4**2+Z3*QA4**3+Z4*QA4**4
+*      Z5*QA4**5+Z6*QA4**6)**16
      ERFQA5=1.-(1./QA5(
      ERFQA6=1.-(1./QA6(
      TECT5=(0.5*RI1(I))*(ERFQA5-ERFQA6(
      SMR2=TECT4+TECT5
      RII=(XCOR(I)-1.0(
      IF(RII.LT.0.0) GO TO 8221
      SGN=1.0
      GO TO 8222
 8221 SGN=-1.0
 8222 SMR4=(0.5*SGN)*(ERFQA6-ERFQA5(
C
C
      SOLX1(I,KI)=SOLX1(I,KI-1)+(2.*(Q22*SMR2-Q11*SMR1
```

```
-*    POT2(KI)*SMR4+POT1(KI)*SMR3))
      GO TO 8004
C
C**************************************************
Cthe statement 8994 means the source point at x=x(t(
Cand the field point at X=0.0 & X=L.
C
CNOTE : when source at  x(t)  and the field  point  at
C   ----x=0 or x=l ,in each case we have two integrals
C**************************************************
C
 8994 RI1(I)=DABS(XCOR(I)-1.0(
      RI2(I)=DABS(XCOR(I((
      A01(I,KI-1)=RI1(I)**2/(4.*(T(KI2)-T(KI-1(((
      A01(I,KI)=RI1(I)**2/(4.*(T(KI2)-T(KI(((
      A02(I,KI-1)=RI2(I)**2/(4.*(T(KI2)-T(KI-1(((
      A02(I,KI)=RI2(I)**2/(4.*(T(KI2)-T(KI(((
      DS1=DEXP(-A01(I,KI-1((
      DS2=DEXP(-A01(I,KI((
      DS3=DEXP(-A02(I,KI-1((
```

```
      DS4=DEXP(-A02(I,KI((
      FS1=DSQRT(A01(I,KI-1((
      FS2=DSQRT(A01(I,KI((
      FS3=DSQRT(A02(I,KI-1((
      FS4=DSQRT(A02(I,KI((
C*************************
C*************************
      GS1=DS1/FS1
      GS2=DS2/FS2
      GS3=DS3/FS3
      GS4=DS4/FS4
      GEST1=GS3-GS4
      GEST2=GS1-GS2
      TOF1=(1.+Z1*FS1+Z2*FS1**2+Z3*FS1**3+Z4*FS1**4
+*    Z5*FS1**5+Z6*FS1**6)**16
      TOF2=(1.+Z1*FS2+Z2*FS2**2+Z3*FS2**3+Z4*FS2**4
+*    Z5*FS2**5+Z6*FS2**6)**16
      TOF3=(1.+Z1*FS3+Z2*FS3**2+Z3*FS3**3+Z4*FS3**4
+*    Z5*FS3**5+Z6*FS3**6)**16
      TOF4=(1.+Z1*FS4+Z2*FS4**2+Z3*FS4**3+Z4*FS4**4
```

```
+*     Z5*FS4**5+Z6*FS4**6)**16
       EFTOF1=1.-(1./TOF1   (
       EFTOF2=1.-(1./TOF2(
       EFTOF3=1.-(1./TOF3(
       EFTOF4=1.-(1./TOF4(
       SEX1=EFTOF3-EFTOF4
       SEX2=EFTOF1-EFTOF2
       SEX3=EFTOF4-EFTOF3
       SEX4=EFTOF2-EFTOF1
       EXM1=XCOR(I(
       IF(EXM1.LT.0.0) GO TO 9181
       SGNR1=1.0
       GO TO 9182
 9181 SGNR1=-1.0
 9182 SMR11=(RI2(I)*.28209479*GEST1)+(.5*RI2(I)*SEX1(
       SMR33=.5*SGNR1*SEX3
       SMR22=(RI1(I)*.28209479*GEST2)+(.5*RI1(I)*SEX2)
       EXM2=XCOR(I)-1.0
       IF(EXM2.LT.0.0) GO TO 8223
       SGNN=1.0
```

```
     GO TO 8224
 8223 SGNN=-1.0
 8224 SMR44=.5*SGNN*SEX4
C^^^^^^^^^^^^^^^^^^^^^^^^^^^^^^^^^^^^^^^^
C^^^^^^^^^^^^^^^^^^^^^^^^^^^^^^^^^^^^^^^^
C
     SOLX1(I,KI)=SOLX1(I,KI-1)+(2.*(Q22*SMR22-Q11*SMR11-
*    POT2(KI)*SMR44+POT1(KI)*SMR33))
C
 8004 CONTINUE
 8003 CONTINUE
C********************************************************
CCAL. THE INTEGRALS FROM Tm to Tn
C********
C      1010 : SOURCE AT X=0.0 & FIELD AT X=0.0
C : 6300SOURCE AT X=X(t)& FIELD AT X=0.0
C********************************************************
     DO 4 I=1,2
     COLYY1(I,N1)=0.0
     COLYY2(I,N1)=0.0
```

```
      COLVV1(I,N1)=0.0
      COLYY3(I,N1)=0.0
    4 CONTINUE
C*****************
C*****************
      DO 1815 I=N1+1,KI2
      IF(I.EQ.ki2) GO TO 8062
      GO TO 8072
 8062 DO 8005 IN=1,2
      IF(IN.EQ.1) GO TO 1010
      GO TO 6300
 1010 COLYY1(IN,I)=-2.*Q11*(.5641896*(DSQRT(T(KI2)-T(I-1((((
C
      COLYY3(IN,I)=0.0
C
      GO TO 8005
C
 6300 RI1(IN)=DABS(XCOR(IN((
      AA0(IN,I-1)=(RI1(IN)**2)/(4.*(T(KI2)-T(I-1(((
C
```

```
      XBX1=DSQRT(AA0(IN,I-1((
C******************************
C******************************
C
      XBX3=DEXP(-AA0(IN,I-1((
C
      XBX5=XBX3/XBX1
C
      XBX7=(1.+Z1*XBX1+Z2*XBX1**2+Z3*XBX1**3+Z4*XBX1**4
+*     Z5*XBX1**5+Z6*XBX1**6)**16
C
      EFXBX1=1.-(1./XBX7(
C
      COLYY1(IN,I)=-2.*Q11*(((.28209479*XBX5
*1     RI1(IN))+(.5*RI1(IN)*(EFXBX1-1.0((((
C
      EXAMA=XCOR(IN(
      IF(EXAMA.LT.0.0) GO TO 6380
      SGNN=1.0
      GO TO 6338
```

```
 6380 SGNN=-1.0

 6338 COLYY3(IN,I)=2.*(.5*SGNN*(1.0-EFXBX1((


C*****************************************************

C*****************************************************

 8005 CONTINUE

    GO TO 1815

C*******************************************

C : 1212SOURCE AT X=0.0

C : 1313SOURCE AT X=X(t(

C*******************************************

CGG1 :sum of pot. int. with source  at  x=x(t (

C     and field x=0.0 for times t(m) to t(n-1(

CGG3 :sum of pot. int. with source at  x=0.0&

C     field at x=0.0 for times t(m) to  t(n-1(

CGG2 :sum of flux int. with source at x=x(t& (

C     field at x=0.0 for times t(m) to  t(n-1(

CGG4 :sum of flux int. with source at x=0.0&

C     field at x=0.0 for times t(m) to  t(n-1(

C  *******************************************
```

```
C
 8072 GG1(N1)=0.0
      GG2(N1)=0.0
      GG3(N1)=0.0
      GG4(N1)=0.0
C
      DO 8115 IN=1,2
      IF(IN.EQ.1) GO TO 1212
      GO TO 1313
 1212 GG3(I)=GG3(I-1)-2.*(.5641896*Q11*(DSQRT(T(KI2-(
*     T(I-1))-DSQRT(T(KI2)-T(I((((
      GG4(I)=GG4(I-1)+0.0
      GO TO 8115
C
 1313 RI1(IN)=DABS(XCOR(IN((
      AA0(IN,I-1)=RI1(IN)**2/(4.*(T(KI2)-T(I-1(((
      AA0(IN,I)=RI1(IN)**2/(4.*(T(KI2)-T(I(((
C
      XBX1=DSQRT(AA0(IN,I-1((
      XBX2=DSQRT(AA0(IN,I((
```

```
C*****************************
C*****************************
C
     XBX3=DEXP(-AA0(IN,I-1((
     XBX4=DEXP(-AA0(IN,I((
C
     XBX5=XBX3/XBX1
     XBX6=XBX4/XBX2
C
     XBX7=(1.+Z1*XBX1+Z2*XBX1**2+Z3*XBX1**3+Z4*XBX1**4
+*    Z5*XBX1**5+Z6*XBX1**6)**16
     XBX8=(1.+Z1*XBX2+Z2*XBX2**2+Z3*XBX2**3+Z4*XBX2**4
+*    Z5*XBX2**5+Z6*XBX2**6)**16
C
     EFXBX1=1.-(1./XBX7(
     EFXBX2=1.-(1./XBX8(
C
     GG1(I)=GG1(I-1)-(2.*Q11*(((RI1(IN)*.28209479*(
)*    XBX5-XBX6))+(RI1(IN)*0.5*(EFXBX1-EFXBX2((((
C
```

```
      EXAMA=XCOR(IN(
      IF(EXAMA.LT.0.0) GO TO 63
      SGNN=1.0
      GO TO 633
 63   SGNN=-1.0
 633  GG2(I)=GG2(I-1)+2.*POT1(I)*(.5*SGNN*(EFXBX2-
*     EFXBX1((
 8115 CONTINUE
 1815 CONTINUE
C**************************************************************
      DO 1715 I=N1+1,KI2
      IF(I.EQ.ki2) GO TO 52
      GO TO 53
 52   DO 8007 IN=1,2
C
      AW1(IN,I)=(XCOR(IN)-XI(II((
      GW1(IN,I)=(T(KI2)-T(I-1))*SI(II(
      DW1(IN,I)=0.0
      FW1(IN,I)=0.0
      QW(IN,I)=(4.*COND/SI(II((
```

```
C*******************************
     TAG1=DABS(GW1(IN,I)-AW1(IN,I))/(DSQRT(DABS(QW(IN,I(
**     GW1(IN,I((((
     TAG2=DABS(GW1(IN,I)+AW1(IN,I))/(DSQRT(DABS(QW(IN,I(
**     GW1(IN,I((((
     TAG10=(1.+Z1*TAG1+Z2*TAG1**2+Z3*TAG1**3+Z4*TAG1**4
+*     Z5*TAG1**5+Z6*TAG1**6)**16
     TAG20=(1.+Z1*TAG2+Z2*TAG2**2+Z3*TAG2**3+Z4*TAG2**4
+*     Z5*TAG2**5+Z6*TAG2**6)**16
     EFTAG1=1.-(1./TAG10(
     EFTAG2=1.-(1./TAG20(
     TAG101=(GW1(IN,I)-AW1(IN,I((
     IF(TAG101.LT.0.0) GO TO 8652
     GO TO 8653
 8652 EFTAG1=-EFTAG1
C
 8653 if(aw1(in,i).eq.0.0) go to 6007
     go to 6008
 6007 tagtl1=(eftag1)+(eftag2(
     colyy2(in,i)=(.5*2.*dfi(ii)/si(ii))*tagtl1
```

```
      colvv1(in,i)=.5*2.*(-eftag2(
      go to 8007
C
 6008 IF(AW1(IN,I).LT.0.0) GO TO 9652
      GO TO 9653
 9652 EFTAG2=-EFTAG2
      TFXF=DEXP(4.*AW1(IN,I)/QW(IN,I((
      TAGTOL=(EFTAG1-1.)+(TFXF*(EFTAG2+1.0((
      COLYY2(IN,I)=(.5*2.*DFI(II)/SI(II))*TAGTOL
      COLVV1(IN,I)=(.5*2.*TFXF*(-EFTAG2-1.0((
      GO TO 8007
 9653 TFXF=DEXP(4.*AW1(IN,I)/QW(IN,I((
      TAGTOL=(EFTAG1+1.)+(TFXF*(EFTAG2-1.0((
      COLYY2(IN,I)=(.5*2.*DFI(II)/SI(II))*TAGTOL
      COLVV1(IN,I)=(.5*2.*TFXF*(-EFTAG2+1.0((
 8007 CONTINUE
      GO TO 1715
C*************************************************************
CGG5 :sum of pot. int. with source  at  x=x(t (
C    and field x=x(t) for times t(m)to t(n-1(
```

```
CGG8 :sum of pot. int. with source at  x=0.0&
C    field at x=x(t) for times t(m) to t(n-1(
CGG6 :sum of flux int. with source at x=x(t& (
C    field at x=x(t) for times t(m) to t(n-1(
CGG9 :sum of flux int. with source at x=0.0&
C    field at x=x(t) for times t(m) to t(n-1(
C ********************************************
C
 53 GG5(N1)=0.0
    GG6(N1)=0.0
    GG8(N1)=0.0
    GG9(N1)=0.0
C
    DO 1808 IN=1,2
    IF(IN.EQ.1) GO TO 1951
C
    AW1(IN,I)=((T(KI2)-T(I-1))*SI(I))+(XCOR(IN)-XI(I-1((
    GW1(IN,I)=((T(KI2)-T(I-1))*SI(I((
    DW1(IN,I)=((T(KI2)-T(I-1))*SI(I))+(XI(I)-XI(I-1((
    FW1(IN,I)=(T(KI2)-T(I))*SI(I(
```

```
      QW(IN,I)=(4.*COND/SI(I((
C
      FW10=(FW1(IN,I)+AW1(IN,I))/(DSQRT(QW(IN,I*(
*      FW1(IN,I(((
      FW20=(FW1(IN,I)-AW1(IN,I))/(DSQRT(QW(IN,I*(
*      FW1(IN,I(((
      FW11=DABS(FW10(
      FW22=DABS(FW20(
      FW111=(1.+Z1*FW11+Z2*FW11**2+Z3*FW11**3+Z4*FW11**4
+*      Z5*FW11**5+Z6*FW11**6)**16
      FW222=(1.+Z1*FW22+Z2*FW22**2+Z3*FW22**3+Z4*FW22**4
+*      Z5*FW22**5+Z6*FW22**6)**16
      ERFFW1=1.-(1./FW111(
      ERFFW2=1.-(1./FW222(
      IF(FW20.LT.0.0)GO TO 1718
      GO TO 1720
 1718 ERFFW2=-ERFFW2
C
 1720 if(fw10.lt.0.0) go to 2718
      go to 2819
```

```
 2718 erffw1=-erffw1
 2819 TX30=(GW1(IN,I)+AW1(IN,I))/(DSQRT(QW(IN,I*(
*     GW1(IN,I(((
      WL40=(GW1(IN,I)-AW1(IN,I))/(DSQRT(QW(IN,I*(
*     GW1(IN,I(((
C
      TX3=DABS(TX30(
      WL4=DABS(WL40(
      TX33=(1.+Z1*TX3+Z2*TX3**2+Z3*TX3**3+Z4*TX3**4
+*    Z5*TX3**5+Z6*TX3**6)**16
      WL44=(1.+Z1*WL4+Z2*WL4**2+Z3*WL4**3+Z4*WL4**4
+*    Z5*WL4**5+Z6*WL4**6)**16
C-------------------------------------
      TXX1=DEXP(4.*AW1(IN,I)/QW(IN,I((
C-------------------------------------
      ERFTX3=1.-(1./TX33(
      ERFWL4=1.-(1./WL44(
C-------------------------------------
      IF(WL40.LT.0.0) GO TO 4011
      GO TO 4012
```

```
 4011 ERFWL4=-ERFWL4
C------------------------------------
 4012 if(tx30.lt.0.0) go to 3718
    go to 3818
 3718 erftx3=-erftx3
 3818 if(fw10.eq.0.0) go to 6012
    go to 6013
 6012 GG5(I)=GG5(I-1)+(2.*.5*DFI(I)/si(i))
))*1    TXX1*(ERFTX3))+(ERFWL4-ERFFW2((
C
    GG6(I)=GG6(I-1)-hmlt*(TXX1*(-ERFTX3((
    go to 1808
 6013 GG5(I)=GG5(I-1)+(2.*.5*DFI(I)/si(i))
))*1    TXX1*(ERFTX3-ERFFW1))+(ERFWL4-ERFFW2((
C
    GG6(I)=GG6(I-1)-hmlt*(TXX1*(-ERFTX3+ERFFW1((
    GO TO 1808
C
 1951 AW1(IN,I)=((T(KI2)-T(I-1))*SI(I))+(XCOR(IN)-XI(I-1((
    GW1(IN,I)=((T(KI2)-T(I-1))*SI(I((
```

```
      DW1(IN,I)=((T(KI2)-T(I-1))*SI(I))+(XI(I)-XI(I-1((
      FW1(IN,I)=(T(KI2)-T(I))*SI(I(
      QW(IN,I)=(4.*COND/SI(I((
C---------------------------------------------------
      FW10=(FW1(IN,I)+AW1(IN,I))/(DSQRT(QW(IN,I*(
*      FW1(IN,I(((
      FW20=(FW1(IN,I)-AW1(IN,I))/(DSQRT(QW(IN,I*(
*      FW1(IN,I(((
      FW11=DABS(FW10(
      FW22=DABS(FW20(
      FW111=(1.+Z1*FW11+Z2*FW11**2+Z3*FW11**3+Z4*FW11**4
+*     Z5*FW11**5+Z6*FW11**6)**16
      FW222=(1.+Z1*FW22+Z2*FW22**2+Z3*FW22**3+Z4*FW22**4
+*     Z5*FW22**5+Z6*FW22**6)**16
      ERFFW1=1.-(1./FW111(
      ERFFW2=1.-(1./FW222(
      IF(FW20.LT.0.0)GO TO 1798
      GO TO 6720
 1798 ERFFW2=-ERFFW2
C---------------------------------------------------
```

```
 6720 if(fw10.lt.0.0) go to 6730
      go to 6731
 6730 erffw1=-erffw1
 6731 TX30=(GW1(IN,I)+AW1(IN,I))/(DSQRT(QW(IN,I*(
*     GW1(IN,I(((
      WL40=(GW1(IN,I)-AW1(IN,I))/(DSQRT(QW(IN,I*(
*     GW1(IN,I(((
      TX3=DABS(TX30(
      WL4=DABS(WL40(
      TX33=(1.+Z1*TX3+Z2*TX3**2+Z3*TX3**3+Z4*TX3**4
+*     Z5*TX3**5+Z6*TX3**6)**16
      WL44=(1.+Z1*WL4+Z2*WL4**2+Z3*WL4**3+Z4*WL4**4
+*     Z5*WL4**5+Z6*WL4**6)**16
C-------------------------------------
      TXX1=DEXP(4.*AW1(IN,I)/QW(IN,I((
C-------------------------------------
      ERFTX3=1.-(1./TX33(
      ERFWL4=1.-(1./WL44(
      IF(WL40.LT.0.0) GO TO 4811
      GO TO 412
```

```
 4811 ERFWL4=-ERFWL4
C------------------------------------------------------
  412 if(tx30.lt.0.0) go to 6732
      go to 6733
 6732 erftx3=-erftx3
 6733 GG8(I)=GG8(I-1)+(2.*.5*DFI(I)/si(i))
))*1     TXX1*(ERFTX3-ERFFW1))+(ERFWL4-ERFFW2((
C
      GG9(I)=GG9(I-1)-hmlt*(TXX1*(-ERFTX3+ERFFW1((
 1808 CONTINUE
 1715 CONTINUE
C
CTOT1=a11
CTOT2=a12
CTOT3=a21
CTOT4=a22
CTOT5=b1
CTOT6=b2
C********
C
```

```
      TOT1=(1.-COLYY3(1,KI2((
      TOT2=COLVV1(1,KI2)-COLVV2(1,KI2(
      TOT3=-COLYY3(2,KI2(
      TOT4=1.+COLVV1(2,KI2)-COLVV2(2,KI2(
C
      TOT5=DINT(1)+SOLX1(1,N1)+COLYY1(1,KI2)+COLYY2(1,KI2(
+*     GG7(1,KI2-1)+GG3(KI2-1)+GG4(KI2-1)+GG8(ki2-1)+GG9
)*    ki2-1(
      TOT6=DINT(2)+SOLX1(2,N1)+COLYY1(2,KI2)+COLYY2(2,KI2(
+*     GG7(2,KI2-1)+GG1(KI2-1)+GG2(KI2-1)+GG5(ki2-1)+GG6
)*    ki2-1(
C
      TOT7=(TOT1*TOT4)-(TOT3*TOT2(
C
      TOTB2=(TOT5*TOT4)-(TOT6*TOT2(
C
      TOTB1=(TOT1*TOT6)-(TOT3*TOT5(
      UMOV(KI2)=TOTB1/TOT7
      POT1(KI2)=TOTB2/TOT7
C----------------------------------------------------------------
```

```
      ERROR=DABS((UMOV(KI2)-HMLT)/HMLT(
      IF(ERROR.LT.EXX) GO TO 2026
C
      DO 981 IN=1,2
      DO 54 I=N1+1,KI2
      AW1(IN,I)=0.0
      GW1(IN,I)=0.0
      DW1(IN,I)=0.0
      FW1(IN,I)=0.0
      COLYY1(IN,I)=0.0
      COLYY2(IN,I)=0.0
      COLYY3(IN,I)=0.0
      COLVV1(IN,I)=0.0
      COLVV2(IN,I)=0.0
      GG7(IN,I)=0.0
   54 CONTINUE
  981 CONTINUE
      DO 1510 IN=1,2
      DO 2256 I=1,N1
      SOLX1(IN,I)=0.0
```

```
 2256 CONTINUE
 1510 CONTINUE
      DO 1245 I1=N1,KI2
      GG1(I1)=0.0
      GG2(I1)=0.0
      GG3(I1)=0.0
      GG4(I1)=0.0
      GG5(I1)=0.0
      GG6(I1)=0.0
      GG8(I1)=0.0
      GG9(I1)=0.0
 1245 CONTINUE
      II=II+1
      DFI(II)=(HAMB-UMOV(KI2))-(SI(II-1((
      SI(II)=dabs(Q22-DFI(II((
      JJ=JJ+1
      XI(II)=XI(II-1)-(SI(II)*DT(
      GO TO 2037
c^^^^^^^^^^^^^^^^^^^^^^^^^^^^^^^^^^^^^^^^^^^^^^^^^^^^^^^^^^^^^^^^^^^^^^^^
 2026 SI(KI2)=SI(II(
```

```
    DFI(KI2)=DFI(II(
    XI(KI2)=XI(II(
    DFI(KI2+1)=DFI(KI2(
    SI(KI2+1)=SI(KI2(
    xi(ki2+1)=xi(ki2(
    DO 921 IN=1,2
    DO 94 I=N1+1,KI2
    AW1(IN,I)=0.0
    GW1(IN,I)=0.0
    DW1(IN,I)=0.0
    FW1(IN,I)=0.0
    COLYY1(IN,I)=0.0
    COLYY2(IN,I)=0.0
    COLYY3(IN,I)=0.0
    COLVV1(IN,I)=0.0
    COLVV2(IN,I)=0.0
    GG7(IN,I)=0.0
 94 CONTINUE
 921 CONTINUE
    DO 6510 IN=1,2
```

```
      DO 56 I=1,N1
      SOLX1(IN,I)=0.0
 56   CONTINUE
6510  CONTINUE
      DO 9245 I1=N1,KI2
      GG1(I1)=0.0
      GG2(I1)=0.0
      GG3(I1)=0.0
      GG4(I1)=0.0
      GG5(I1)=0.0
      GG6(I1)=0.0
      GG8(I1)=0.0
      GG9(I1)=0.0
9245  CONTINUE
      WRITE(IPR,5028) T(KI2),Xi(KI2),SI(KI2),DFI(KI2),POT1(KI2)
      if(POT1(KI2).GT.HMLT) GO TO 91
5988  CONTINUE
 91   CLOSE (INP)
      CLOSE (IPR)
      STOP
      END
```

SUBJECT INDEX

A

B

S. G. Ahmed

C

F

G

K

L

M

N

Q

R

S

T

V

W

Y

Z